Motives, Quantum Field Theory, and Pseudodifferential Operators

Clay Mathematics Proceedings
Volume 12

Motives, Quantum Field Theory, and Pseudodifferential Operators

Conference on Motives, Quantum Field Theory, and Pseudodifferential Operators
June 2–13, 2008
Boston University, Boston, Massachusetts

Alan Carey
David Ellwood
Sylvie Paycha
Steven Rosenberg
Editors

American Mathematical Society
Clay Mathematics Institute

Cover design courtesy of Selene Rosenberg.

2000 *Mathematics Subject Classification.* Primary 14–02, 14F42, 16T05, 33F10, 58–02, 58J42, 81–02, 81T15, 81T18, 81Q30.

Library of Congress Cataloging-in-Publication Data

Motives, quantum field theory, and pseudodifferential operators : Conference on Motives, Quantum Field Theory, and Pseudodifferential Operators, June 2–13, 2008, Boston University, Boston, Massachusetts / Alan Carey ... [et al.], editors.
p. cm. — (Clay mathematics proceedings ; v. 12)
Includes bibliographical references.
ISBN 978-0-8218-5199-9 (alk. paper)
1. Geometry, Algebraic. 2. Motives (Mathematics) 3. Quantum theory—Mathematics. 4. Pseudodifferential operators. I. Carey, Alan L.

QA564.M6836 2010
530.14′30151—dc22

2010040806

Published by the American Mathematical Society, Providence, RI,
for the Clay Mathematics Institute, Cambridge, MA.
Printed in the United States of America.

∞ The paper used in this book is acid-free and falls within the guidelines
established to ensure permanence and durability.
Visit the AMS home page at http://www.ams.org/
Visit the Clay Mathematics Institute home page at http://www.claymath.org/

10 9 8 7 6 5 4 3 2 1   15 14 13 12 11 10

Contents

Preface

This volume grew out of lectures given at the conference "Motives, Quantum Field Theory, and Pseudodifferential Operators," held at Boston University on June 2 – 12, 2008. Over 90 participants listened to both introductory and advanced lectures in these subjects. It is becoming increasingly clear that the three conference topics are deeply linked, but since few mathematicians and physicists are conversant in any two of the conference areas, it seemed worthwhile to collect the lectures into a volume of proceedings.

In general, quantum field theory (QFT) is a puzzling subject for mathematicians. It is a broad field with established rules of remarkable predicative power in the laboratory. Moreover, over the past 30 years QFT has developed into a "conjecture machine" in algebraic geometry, symplectic geometry and number theory [**9**]. However, to date there is no rigorous mathematical foundation for the standard path integral techniques in QFT in physically relevant dimensions, either in the exact or topological approaches predominant in string theory or in the perturbative approach so successful in particle physics.

Thinking like physicists, we can (and did) avoid the foundational questions and concentrate instead on the business of perturbative QFT, namely the calculation of Feynman integrals associated to the infinite number of Feynman graphs of a particular QFT. Since around 1990, by stepping back a bit from the extensive calculations in the literature, mathematicians and mathematical physicists have uncovered surprising algebraic and analytic structures within perturbative QFT, and have produced still incompletely explained overlaps with analytic number theory. These discoveries include the Hopf algebra structure of perturbative QFT due to Connes-Kreimer [**5, 6**] and the appearance of multiple zeta values as specific Feynman integrals [**3**]. There results indicate a deep connection between perturbative QFT and the theory of motives. Investigations of this connection include e.g. [**2, 7**], but the theory is certainly incomplete at present. The connection between perturbative QFT and pseudodifferential operators (ΨDOs) is similarly incomplete; while elementary Feynman integrals, resp. multiple zeta functions, can be interpreted as integrals, resp. sums, of symbols of ΨDOs and obey algebraic rules seen in the combinatorics of the Connes-Kreimer Hopf algebra, the extent and depth of these relationships are currently unknown.

From the beginning, we felt the need to present an overview of the three conference fields, and the writeups of these lectures appear in the first part of the volume. André's article outlines those aspects of the theory of motives most relevant to Feynman integral computations. The surprising appearance of multiple zeta values as coefficients in the Taylor expansion of suitably regularized Feynman integrals suggests that these integrals are in fact periods for some motive underlying the theory. Since the theory of motives is greatly enhanced by artihmetic methods such as assembling information from counting p points on varieties, we see hints of a deep connection between long established computational techniques in physics and arithmetic geometry.

The introductory article by Kreimer makes the general considerations in André's article more concrete for specific QFTs. After motivating the general setup of perturbative QFT, Kreimer discusses how a particular QFT produces Feynman integrals which are periods for a specific sequence of blowups of projective spaces. This article outlines the Hopf algebra structures which both encode time honored Feynman integral manipulations and greatly simplify calculations. The companion article by Manchon strips away much of the physical and analytic aspects to focus more directly on the relevant algebra.

Finally, the survey article by Lesch both outlines the basic theory of ΨDOs and focuses on the theory of regularized traces of ΨDOs. There is a simple minded overlap between QFT and ΨDOs, namely that both fields must develop a systematic method to regularize, i.e. extract a finite part from, divergent integrals. For example, in ΨDO theory one often wants to compute a meaningful trace of a non-trace class operator by some sort of integral expression. However, it is impossible to extend the ordinary integral for L^1 functions to a functional on pseudodifferential symbols with the desired integration by parts and translation invariance. The standard replacements for the usual integral are the Wodzicki residue or the canonical integral of Kontsevich-Vishik, which applies only to non-integer order symbols. (Physicists choose this second option for the corresponding Feynman integrals, avoiding integer order symbols via dimensional regularization, which perturbs the integer dimension of spacetime to a complex dimension.) In particular, it is impossible to extend the ordinary operator trace to all ΨDOs, and one is forced to either use the Wodzicki residue or the Kontsevich-Vishik trace. The Wodzicki residue is locally computable and so is in the spirit of QFT's locality requirement, but does not extend the operator trace; in contrast, the Kontsevich-Vishik trace extends the operator trace, but is not local and is not defined for integer order ΨDOs. Again, it is difficult to draw deep conclusions at present about the exact interface between ΨDOs and QFT.

The second part of this volume contains more technical articles on current research related to the conference themes. The article of Bergström-Brown focuses on the Poincaré series for the moduli space $\mathcal{M}_{0,n}$ and the Tate motive underlying an associated moduli space. This moduli space is a fundamental object in string theory, which leads us to the unresolved question of the relationship between QFT and string theory: it has been claimed that QFT is some sort of limit of string theory, but the physics arguments are not formulated in terms that can be checked with current mathematics. In any case, the appearance of arithmetic methods in string theory [**10**] mirrors the corresponding appearance in QFT, and the article of Roth-Yui concretely uses motivic techniques from arithmetic geometry to calculate string theory generating functions. The article of Bouwknegt-Hannabuss-Mathai is also motivated by aspects of string theory, in this case by T-duality, and relies heavily on noncommutative geometry techniques. In fact, noncommutative geometry has been proposed in [**4, 7**] as a framework to understand the relationships between QFT and motives, so we once again note the similarities between rigorous mathematical approaches to the two presently ill-defined physical theories.

Another group of articles covers topics directly related to QFT. In contrast to the many papers in mathematics which talk about QFT, the papers of Bluemlein and Schneider actually do the QFT computations needed to compare the theory with laboratory results. These state of the art results go far beyond "doing really hard integrals." Indeed, these papers involve the nontrivial combinatorics of nested harmonic sums, and Schneider produces a novel human-computer generated recurrence relation for these sums. The papers of Foissy and van Suijlekom provide a mathematical counterpart. Foissy examines the fine algebraic structure of the Hopf algebra of rooted trees, and van Suijlekom extends the Connes-Kreimer Hopf algebra to incorporate gauge theory aspects which are necessary for fermionic theories. Finally, the paper of Mickelsson treats gauge theoretic QFT geometrically, relating the modern theory of gerbes to classical quantization questions. It is striking to see the range of mathematical issues in analysis, algebra and geometry that are motivated by the mathematically non-existent theory of QFT!

The final group of articles involves ΨDO techniques. Scott's paper places his and others' work on exotic determinants and other spectral invariants associated to ΨDOs into the more abstract framework of logarithmic structures associated to TQFTs. The article of Albin-Melrose extends their earlier work on ΨDOs on manifolds with boundary, treating the technical issues of Fredholmness of operators, and prove an index theorem in their context. The adiabatic techniques in this paper are refinements of quantum mechanical techniques, and have become an important tool in global analysis since Witten's supersymmetric derivation of the Morse inequalities [**15**]. Ponge's paper produces invariants in conformal, pseudohermitian and CR geometry from the logarithmic singularity of the kernel of appropriate ΨDOs. This analysis of singularities is an essential step in understanding any regularization scheme for divergent integrals, and there is a surprising amount of local and global geometry contained in the singularities.

Participants in this conference could not be expected to be experts in all three areas, and neither can the reader. For background material on motives, one can consult André's book [**1**] and the more technical book by Levine [**12**]. Different mathematicians have their favorite QFT text, including Ryder [**13**] for a mathematics-friendly physics approach and Deligne *et al.* [**8**] for a physics-friendly mathematical approach. For ΨDOs, one can consult Shubin [**14**] for the basics (and beyond) or Gilkey [**11**] for geometric applications.

In summary, the conference and these proceedings fail to present an overarching theory uniting motives, QFT and ΨDOs, simply because no such theory exists at present. Nevertheless, we hope that this volume gives a thorough overview of the clues that point to the existence of a comprehensive theory. We hope some readers will add evidence to this detective story or even solve the case completely!

We were very fortunate to have had financial support from the Boston University Department of Mathematics and Statistics, the Boston University Graduate School of Arts and Sciences, the Clay Mathematics Institute, and the National Science Foundation. The Scientific Advisory Committee of Spencer Bloch, Louis Boutet de Monvel, Alain Connes, Dirk Kreimer, and Noriko Yui greatly helped to produce a coherent conference over widely separated fields. We would also like to express our deep thanks to Kathleen Heavey for her administrative work for the conference, Amanda Battese for her administrative work at CMI, Vida Salahi for her production work on these proceedings, and Arthur Greenspoon for his lightning quick copyediting.

Alan Carey, David Ellwood, Sylvie Paycha, Steve Rosenberg
August 2010

References

[1] Y. André. *Une introduction aux motifs (motifs purs, motifs mixtes, priodes)*, volume 17 of *Panoramas et Synthéses*. Société Mathématique de France, Paris, 2002.
[2] S. Bloch, H. Esnault, and D. Kreimber. On motives associated to graph polynomials. *Comm. Math. Phys.*, 267:181–225, 2006.
[3] D. Broadhurst and D. Kreimber. Association of multiple zeta values with positive knots via feynman diagrams up to 9 loops. *Phys. Lett. B*, 393:403–412, 1997.
[4] A. Connes, C. Consani, and M. Marcolli. Noncommutative geometry and motives: the thermodynamics of endomotives. *Adv. in Math.*, 214:761–831, 2007.
[5] A. Connes and D. Kreimer. Renormalization in Quantum Field Theory and the Riemann-Hilbert Problem I: The Hopf algebra structure of Graphs and the Main Theorem. *Comm. Math. Phys.*, 210:249–273, 2000.
[6] A. Connes and D. Kreimer. Renormalization in quantum field theory and the Riemann-Hilbert problem. II. The β-function, diffeomorphisms and the renormalization group. *Comm. Math. Phys.*, 216(1):215–241, 2001.
[7] A. Connes and M. Marcolli. Quantum field theory and motives. *J. Geom. Phys.*, 56:55–85, 2005.
[8] P. Deligne *et al. Quantum Fields and Stings: A Course for Mathematicians.* American Mathematical Society, Providence, RI, 1999.
[9] D. Freed, D. Morrison, and I. M. Singer. *Quantum Field Theory, Supersymmetry, and Enumerative Geometry*, volume 11 of *IAS/Park City Mathematics Series.* American Mathematical Society, Providence, 2006.
[10] E. Frenkel. Gauge theory and Langlands duality. *Seminaire Bourbaki*, page arXiv:0906.2747.
[11] P. B. Gilkey. *Invariance Theory, the Heat Equation, and the Atiyah-Singer Index Theorem.* Studies in Advanced Mathematics. CRC Press, Boca Raton, 1994.
[12] M. Levine. *Mixed Motives*, volume 57 of *Mathematical Surveys and Monographs.* American Mathematical Society, Providence, 1998.
[13] L. H. Ryder. *Quantum Field Theory.* Cambridge U. Press, Cambridge, 1996.
[14] M. A. Shubin. *Pseudodifferential Operators and Spectral Theory.* Springer-Verlag, Berlin, 1987.
[15] E. Witten. Supersymmetry and Morse theory. *J. Diff. Geom.*, 17:661–692, 1982.

Introductory Articles

Clay Mathematics Proceedings
Volume 12, 2010

An Introduction to Motivic Zeta Functions of Motives

Yves André

ABSTRACT. It oftens occurs that Taylor coefficients of (dimensionally regularized) Feynman amplitudes I with rational parameters, expanded at an integral dimension $D = D_0$, are not only periods (Belkale-Brosnan, Bogner-Weinzierl) but actually multiple zeta values (Broadhurst, Kreimer, ..., Bierenbaum-Weinzierl, Brown).

In order to determine, at least heuristically, whether this is the case in concrete instances, the philosophy of motives - more specifically, the theory of mixed Tate motives - suggests an arithmetic approach (Kontsevich): counting points of algebraic varieties related to I modulo sufficiently many primes p and checking that the number of points varies polynomially in p.

On the other hand, Kapranov has introduced a new "zeta function", the role of which is precisely to "interpolate" between zeta functions of reductions modulo different primes p.

In this survey, we outline this circle of motivic ideas and some of their recent developments.

This article is divided in two parts.

In the second and main part, we survey motivic zeta functions of motives, which "interpolate" between Hasse-Weil zeta functions of reductions modulo different primes p of varieties defined by polynomial equations with rational coefficients.

In the first and introductory part, we give some hints about the relevance of the concepts of motives and motivic zeta functions in questions related to computations of Feynman integrals in connection with multiple zeta values.

Acknowledgments. I am grateful to P. Brosnan for some enlightening remarks about Feynman amplitudes, and to the organizers of the Boston Conference for their invitation and advice.

1. Periods and motives

1.1. Introduction. Relations between Feynman integrals and (Grothendieck) motives are manifold and mysterious. The most direct conceptual bridge relies on the notion of *period*, in the sense of arithmetic geometry; that is, integrals where

2010 *Mathematics Subject Classification.* 81Q, 32G, 19F, 19E, 14F.

Key words and phrases. Feynman integral, period, multiple zeta value, motive, zeta function, L-function, Tamagawa number, moduli stack of bundles.

both the integrand and the domain are defined in terms of polynomials with rational (or algebraic) coefficients. The ubiquitous multiple zeta values encountered in the computation of Feynman amplitudes are emblematic examples of periods.

Periods are just complex numbers, but they carry a rich hidden structure reflecting their geometric origin. They occur as entries of a canonical isomorphism between the complexification of two rational cohomologies attached to algebraic varieties defined over $\mathbb{Q}$: *algebraic De Rham cohomology*, defined in terms of algebraic differential forms[1], and *Betti*[2] *cohomology*, defined in terms of topological cochains.

Periods are thus best understood in the framework of *motives*, which are supposed to play the role of pieces of universal cohomology of algebraic varieties. For instance, the motive of the projective space $\mathbf{P}^n$ splits into $n+1$ pieces (so-called Tate motives) whose periods are $1, 2\pi i, \ldots, (2\pi i)^n$.

To each motive over $\mathbb{Q}$ is associated a square matrix of periods (well defined up to left or right multiplication by matrices with rational coefficients), and a deep conjecture of Grothendieck predicts that the period matrix actually determines the motive.

For instance, multiple zeta values (*cf.* (1.12) below) are periods of so-called mixed Tate motives over $\mathbb{Z}$, which are iterated extensions of Tate motives. Grothendieck's conjecture implies that these are the only motives (over $\mathbb{Q}$) with (rational linear combinations of) multiple zeta values as periods.

In the philosophy of motives, cohomologies are thought of interchangeable realizations (functors with vector values), and one should take advantage of switching from one cohomology to another. Aside from de Rham or Betti cohomology, one may also consider etale cohomology, together with the action of the absolute Galois group $Gal(\bar{\mathbb{Q}}/\mathbb{Q})$; this amounts, more or less, to considering the number N_p of points of the reduction modulo p for almost all prime numbers p.

A deep conjecture of Tate, in the same vein as Grothendieck's conjecture, predicts that the numbers N_p determine the motive, up to semi-simplification. For mixed Tate motives[3], the N_p are polynomials in p, and Tate's conjecture implies the converse.

To decide whether periods of a specific algebraic variety over $\mathbb{Q}$, say a hypersurface, are (rational linear combinations of) multiple zeta values may be a difficult problem about concrete integrals. The philosophy of motives suggests, as a test, to look at the number of points N_p of this hypersurface modulo p, and check whether they grow polynomially in p (if not, there is no chance for the periods to be multiple zeta values: this would contradict either Grothendieck's or Tate's conjecture). Recently, various efficient algorithms have been devised for computing N_p, at least for curves or for some hypersurfaces, *cf. e.g.* [**27**][**33**].

1.2. Periods. A *period* is a complex number whose real and imaginary parts are absolutely convergent multiple integrals

$$\alpha = \frac{1}{\pi^m} \int_\Sigma \Omega$$

[1] first defined by Grothendieck

[2] or singular

[3] not necessarily over $\mathbb{Z}$

where Σ is a domain in $\mathbb{R}^n$ defined by polynomial inequations with rational coefficients, Ω is a rational differential n-form with rational coefficients, and m is a natural integer[4]. The set of periods is a countable subring of $\mathbb{C}$ which contains $\bar{\mathbb{Q}}$.

This is the definition proposed in [**32**][5]. There are some variants, which turn out to be equivalent. For instance, one could replace everywhere "rational" by "algebraic". Also one could consider a (not necessarily closed) rational k-form in n variables with rational coefficients, integrated over a domain in $\mathbb{R}^n$ defined by polynomial equations and inequations with rational coefficients (interpreting the integral of a function as the volume under the graph, one can also reduce to the case when $k = n$ and Ω is a volume form).

More geometrically, the ring of periods is generated by $\frac{1}{\pi}$ and the numbers of the form $\int_\gamma \omega$ where $\omega \in \Omega^n(X)$ is a top degree differential form on a smooth algebraic variety X defined over $\mathbb{Q}$, and $\gamma \in H_n(X(\mathbb{C}), Y(\mathbb{C}); \mathbb{Q})$ for some divisor $Y \subset X$ with normal crossings *cf.* [**32**, p. 3, 31]).

In many examples, *e.g.* multiple zeta values (see below), the integrals have singularities along the boundary, hence are not immediately periods in the above sense. However, it turns out that any convergent integral $\int_\gamma \omega$ where $\omega \in \Omega^n(X \setminus Y)$ is a top degree differential form on the complement of a closed (possibly reducible) subvariety Y of a smooth algebraic variety X defined over $\mathbb{Q}$, and γ is a semialgebraic subset of $X(\mathbb{R})$ defined over $\mathbb{Q}$ (with non-empty interior), is a period (*cf.* [**9**, th. 2.6]).

1.3. Periods and motives. Periods arise as entries of a matrix of the comparison isomorphism, given by integration of algebraic diffential forms over chains, between algebraic De Rham and ordinary Betti relative cohomology

$$H_{DR}(X,Y) \otimes \mathbb{C} \overset{\varpi_{X,Y}}{\cong} H_B(X,Y) \otimes \mathbb{C}. \tag{1.1}$$

X being a smooth algebraic variety over $\mathbb{Q}$, and Y being a closed (possibly reducible) subvariety[6].

This is where motives enter the stage. They are intermediate between algebraic varieties and their linear invariants (cohomology). One expects the existence of an abelian category $\mathrm{MM}(\mathbb{Q})$ of *mixed motives* (over $\mathbb{Q}$, with rational coefficients), and of a functor

$$h : Var(\mathbb{Q}) \to \mathrm{MM}(\mathbb{Q})$$

(from the category of algebraic varieties over $\mathbb{Q}$) which plays the role of universal cohomology (more generally, to any pair (X, Y) consisting of a smooth algebraic variety and a closed subvariety, one can attach a motive $h(X, Y)$ which plays the role of the universal relative cohomology of the pair).

The morphisms in $\mathrm{MM}(\mathbb{Q})$ should be related to algebraic correspondences. In addition, the cartesian product on $Var(\mathbb{Q})$ corresponds via h to a certain tensor product $\otimes$ on $\mathrm{MM}(\mathbb{Q})$, which makes $\mathrm{MM}(\mathbb{Q})$ into a *tannakian category*, *i.e.* it

[4]the very name "period" comes from the case of elliptic periods (in the case of an elliptic curve defined over $\bar{\mathbb{Q}}$, the periods of elliptic functions in the classical sense are indeed periods in the above sense)

[5]except for the factor $\frac{1}{\pi^m}$. We prefer to call *effective period* an integral α in which $m = 0$, to parallel the distinction motive versus effective motive

[6]by the same trick as above, or using the Lefschetz's hyperplane theorem, one can express a period of a closed form of any degree as a period of a top degree differential form

has the same formal properties as the category of representations of a group. The (positive or negative) $\otimes$-powers of $h^2(\mathbf{P}^1)$ (and their direct sums) are called the pure *Tate motives.*

The cohomologies H_{DR} and H_B factor through h, giving rise to two $\otimes$-functors

$$H_{DR},\, H_B : \mathrm{MM}(\mathbb{Q}) \to Vec_{\mathbb{Q}}$$

with values in the category of finite-dimensional $\mathbb{Q}$-vector spaces. Moreover, corresponding to (1.1), there is an isomorphism in $Vec_{\mathbb{C}}$

$$\varpi_M : H_{DR}(M) \otimes \mathbb{C} \cong H_B(M) \otimes \mathbb{C} \tag{1.2}$$

which is $\otimes$-functorial in the motive M. The entries of a matrix of ϖ_M with respect to some basis of the $\mathbb{Q}$-vector space $H_{DR}(M)$ (*resp.* $H_B(M)$) are the *periods* of M.

One can also consider, for each prime number ℓ, the ℓ-adic etale realization

$$H_\ell : \mathrm{MM}(\mathbb{Q}) \to Rep_{\mathbb{Q}_\ell}(Gal(\bar{\mathbb{Q}}/\mathbb{Q}))$$

with values in the category of finite-dimensional $\mathbb{Q}_\ell$-vector spaces endowed with continuous action of the absolute Galois group of $\mathbb{Q}$. As for varieties over $\mathbb{Q}$, it makes sense to reduce any motive $M \in MM(\mathbb{Q})$ modulo sufficiently large primes p, and to "count the number of points of M modulo p". This number can be evaluated using the trace of Frobenius element at p acting on $H_\ell(M)$.

1.4. Motivic Galois groups, period torsors, and Grothendieck's period conjecture. Let $\langle M\rangle$ be the tannakian subcategory of $\mathrm{MM}(\mathbb{Q})$ generated by a motive M: its objects are given by algebraic constructions on M (sums, subquotients, duals, tensor products).

One defines the *motivic Galois group* of M to be the group scheme

$$G_{mot}(M) := Aut^{\otimes}\, H_{B|\langle M\rangle} \tag{1.3}$$

of automorphisms of the restriction of the $\otimes$-functor H_B to $\langle M\rangle$.

This is a linear algebraic group over $\mathbb{Q}$: in heuristic terms, $G_{mot}(M)$ is just *the Zariski-closed subgroup of $GL(H_B(M))$ consisting of matrices which preserve motivic relations in the algebraic constructions on $H_B(M)$.*

Similarly, one can consider both H_{DR} and H_B, and define the *period torsor* of M to be

$$P_{mot}(M) := Isom^{\otimes}\left(H_{DR|\langle M\rangle}, H_{B|\langle M\rangle}\right) \in Var(\mathbb{Q}) \tag{1.4}$$

of isomorphisms of the restrictions of the $\otimes$-functors H_{DR} and H_B to $\langle M\rangle$. This is a torsor under $G_{mot}(M)$, and it has a canonical complex point:

$$\varpi_M \in P_{mot}(M)(\mathbb{C}). \tag{1.5}$$

Grothendieck's period conjecture asserts that *the smallest algebraic subvariety of $P_{mot}(M)$ defined over $\mathbb{Q}$ and containing ϖ is $P_{mot}(M)$ itself.*

In more heuristic terms, this means that any polynomial relations with rational coefficients between periods should be of motivic origin (the relations of motivic origin being precisely those which define $P_{mot}(M)$). This implies that a motive $M \in MM(\mathbb{Q})$ can be recovered from its periods.

The conjecture is also equivalent to: *$P_{mot}(M)$ is connected (over $\mathbb{Q}$) and*

$$\mathrm{tr.deg}_{\mathbb{Q}}\, \mathbb{Q}[\mathrm{periods}(M)] = \dim G_{mot}(M). \tag{1.6}$$

For further discussion, see [**2**, ch. 23].

1.4.1. EXAMPLE. The motive of $\mathbb{P}^n$ decomposes as

$$h(\mathbb{P}^n) = \mathbb{Q}(0) \oplus \ldots \oplus \mathbb{Q}(-n), \tag{1.7}$$

with periods $1, 2\pi i, \ldots, (2\pi i)^n$. Its motivic Galois group is the mutiplicative group $\mathbb{G}_m$. In this case, Grothendieck's conjecture amounts to the transcendence of π.

1.4.2. REMARK. By definition, periods are convergent integrals of a certain type. They can be transformed by algebraic changes of variable, or using additivity of the integral, or using Stokes formula.

M. Kontsevich conjectured that *any polynomial relation with rational coefficients between periods can be obtained by way of these elementary operations from calculus* (*cf.* [**32**]). Using ideas of M. Nori [7], it can be shown that this conjecture is actually equivalent to Grothendieck's conjecture (*cf.* [**2**, ch. 23]).

Grothendieck's conjecture can be developed further into a Galois theory for periods, *cf.* [**3**][**4**].

1.5. Periods and Feynman amplitudes. Let Γ be a finite graph (without self-loop), with set of vertices V and set of edges E. Let Ψ_Γ be its classical Kirchhoff (determinant) polynomial, *i.e.* the homogeneous polynomial of degree $b_1(\Gamma)$ defined by

$$\Psi_\Gamma = \sum_T \prod_{e \notin T} x_e, \tag{1.8}$$

where T runs through the spanning trees of a given graph Γ ((x_e) is a set of indeterminates indexed by the edges of Γ).

Let D_0 be an even integer (in practice, $D_0 = 4$). The graph Γ can be considered as a (scalar) Feynman graph without external momenta. According to the Feynman rules, when all masses are equal to 1, the corresponding D_0-dimensional Feynman amplitude is written as

$$I_\Gamma(D_0) = \int_{\mathbb{R}^{D_0|E|}} \prod_{e\in E} (1 + |p_e|^2)^{-1} \prod_{v\in V} \delta(\sum_{e\to v} p_e - \sum_{v\to e} p_e) \prod_{e\in E} d^{D_0} p_e. \tag{1.9}$$

Its dimensional-regularization, for D close to D_0, can be evaluated, using the technique of Feynman parameters, to be

$$I_\Gamma(D) = \frac{\pi^{b_1(\Gamma)\cdot D/2} \cdot \Gamma(|E| - b_1(\Gamma)D/2)}{\Gamma(|E|)} \cdot J_\Gamma(D) \tag{1.10}$$

where

$$J_\Gamma(D) = \int_{\Delta_{|E|}} \Psi_\Gamma^{-D/2} \prod_{e\in E} dx_e, \tag{1.11}$$

(an integral over the standard simplex $\Delta_{|E|}$ in $|E|$ variables).

This integral converges for $D << 0$, but may diverge for $D = D_0$. The convergence issue was already studied in detail fifty years ago by S. Weinberg [**36**][8]; see also [**13**][**14**][**30**] (one has to cope with the singularities of $\Psi_\Gamma^{-D_0/2} \prod dx_e$ on $\Delta_{|E|}$, by means of a sequence of blow-ups).

[7]and granting the expected equivalence of various motivic settings

[8]we are indebted to P. Brosnan for this reference

On the other hand, P. Belkale and P. Brosnan gave a meaning to $J_\Gamma(D)$ in general, using analytic continuation [**9**]. Moreover, they showed that the Taylor coefficients of $J_\Gamma(D)$ at D_0 are always periods (of varieties closely related to he hypersurface $X_\Gamma : \Psi_\Gamma = 0$: for higher Taylor coefficients, one has to add an indeterminate, *cf.* [**9**, p. 2660]).

In [**15**], this result was extended (by another method) to the case when Γ is a semi-graph (*i.e.* in the presence of external momenta) and when the masses are not necessarily equal to 1 but are commensurable to each other.

1.5.1. REMARK. Taking into account these results, polynomial relations between Feynman amplitudes attached to different graphs Γ (like the relations which lead to Kreimer's Hopf algebra) can be intepreted as period relations. According to Grothendieck's conjecture, they should be of motivic origin, *i.e.* come from relations between the motives attached to the hypersurfaces X_Γ (and related varieties). The work [**13**] provides some indirect evidence for this.

1.6. Multiple zeta values, Feynman amplitudes and Hasse-Weil zeta functions. Multiple zeta values

$$\zeta(s_1, \dots, s_k) = \sum_{n_1 > \cdots > n_k \geq 1} \frac{1}{n_1^{s_1} \dots n_k^{s_k}} \tag{1.12}$$

(where s_i are integers ≥ 1, with $s_1 \geq 2$) can be written in integral form: setting

$$\omega_0 = \frac{dt}{t}\ \omega_1 = \frac{dt}{1-t},\ \omega_r = \omega_0^{\wedge(r-1)} \wedge \omega_1 \text{ pour } r \geq 2,$$

one has

$$\zeta(s_1, \dots, s_k) = \int_{1 > t_1 > \cdots > t_s > 0} \omega_{s_1} \dots \omega_{s_k}, \tag{1.13}$$

which is thus a period. This is actually the period of a mixed Tate motive over $\mathbb{Z}$, *i.e.* an iterated extension in $MM(\mathbb{Q})$ of pure Tate motives, which is unramified with respect to the Galois action on etale cohomology (*cf.* [**2**, ch. 25] for more detail).

These numbers have long been known to occur in a pervasive manner in the computation of Feynman amplitudes (*cf. e.g.* [**17**], [**37**]). In particular, it has been shown in [**10**] that all Taylor coefficients of $J(D)$ at $D = 4$ are (rational linear combinations of) multiple zeta values when the semi-graph Γ is a wheel with one spoke and two external edges, and this was extended to some higher-loop semi-graphs in [**18**].

Kontsevich once speculated that the periods of the hypersurface $X_\Gamma : \Psi_\Gamma = 0$ were (rational linear combinations of) multiple zeta values.

According to Grothendieck's period conjecture, this would imply that the motive of X_Γ is a mixed Tate motive over $\mathbb{Z}$. If this is the case, the number $\sharp X_\Gamma(\mathbb{F}_p)$ of points of the reduction of X_Γ modulo p should be polynomial in p; equivalently, the poles of the Hasse-Weil zeta function of $X_\Gamma \otimes \mathbb{F}_p$ should be integral powers of p.

This has been checked for graphs with less than 12 edges by J. Stembridge [**33**], but disproved in general by Belkale and Brosnan [**8**].

However, this leaves open the interesting general question, for any $X \in Var(\mathbb{Q})$, of *controlling* $\sharp X(\mathbb{F}_p)$ *uniformly in* p *- or equivalently, of the variation of* $Z(X \otimes \mathbb{F}_p, t)$ *with* p.

As we shall see, there are mathematical tools well-suited to tackle this question: the *Kapranov zeta function*, and its variant the *motivic zeta function*.

1.6.1. REMARK. The relationship between Feynman diagrams and motives has been investigated much further. For instance, in [**1**], P. Aluffi and M. Marcolli propose an algebro-geometric version of the Feynman rules, which takes place in a certain K_0-ring built from immersed conical varieties.

2. Motivic zeta functions

2.1. The ring of varieties. The idea of building a ring out of varieties by viewing pasting as the addition is very old. In the case of algebraic varieties over a field k, this leads to the ring $K_0(Var(k))$[9]: the generators are denoted by $[X]$, one for each isomorphism class of k-variety; the relations are generated by

$$[X - Y] = [X] - [Y] \tag{2.1}$$

when Y is a closed subvariety of X. With the product given by

$$[X \times Y] = [X] \cdot [Y], \tag{2.2}$$

$K_0(Var(k))$ becomes a ring.

It is standard to denote by $\mathbb{L}$ the class $[\mathbf{A}^1]$ of the affine line.

2.1.1. EXAMPLES. 1) One has

$$[GL_n] = (\mathbb{L}^n - 1) \cdots (\mathbb{L}^n - \mathbb{L}^{n-1}) = (\mathbb{L} - 1) \cdot [SL_n]. \tag{2.3}$$

2) In the case of a Zariski locally trivial fibration $X \to S$ with fiber Y, one has $[X] = [S] \cdot [Y]$. This applies to GL_n-fibrations (which are locally trivial); in order to recover $[S]$ from $[X]$, taking into account the formula for $[GL_n]$, it will often be convenient to localize $K_0(Var(k))$ by $\mathbb{L}$ and $\mathbb{L}^n - 1$, $n > 0$.

3) (McWilliams, Belkale-Brosnan [**8**]): the class of the space of forms of rank r in n variables is

$$[Sym_r^n] = \prod_1^s \frac{\mathbb{L}^{2i}}{\mathbb{L}^{2i} - 1} \prod_0^{2s-1} (\mathbb{L}^{n-i} - 1) \tag{2.4}$$

if $0 \leq r = 2s \leq n$,

$$[Sym_r^n] = \prod_1^s \frac{\mathbb{L}^{2i}}{\mathbb{L}^{2i} - 1} \prod_0^{2s} (\mathbb{L}^{n-i} - 1) \tag{2.5}$$

if $0 \leq r = 2s + 1 \leq n$.

4) In [**8**], it is shown that the classes of the hypersurfaces X_Γ attached to all graphs Γ (*cf.* §1.5) generate the localization of $K_0(Var(\mathbb{Q}))$ by $\mathbb{L}$ and $\mathbb{L}^n - 1$, $n > 0$.

[9] which occurs in some early letters from Grothendieck to Serre about motives

5) (Bloch [**12**]): in contrast to this result, for any fixed positive integer n, the weighted sum $\sum \frac{n!}{\sharp \mathrm{Aut}\,\Gamma} \cdot [X_\Gamma]$, running over all connected graphs Γ with n vertices, no multiple edge and no tadpole (edge with just one vertex), belongs to $\mathbb{Z}[\mathbb{L}] \subset K_0(Var(\mathbb{Q}))$.

The structure of the ring $K_0(Var(k))$ is rather mysterious. It is slightly better understood when k is of characteristic zero, using strong versions of the resolution of singularities.

2.1.2. PROPOSITION. [**11**] *If* char $k = 0$, $K_0(Var(k))$ *admits the following presentation: generators are classes of smooth projective varieties X, with the blow-up relations:*

$$[Bl_Y X] - [E] = [X] - [Y] \tag{2.6}$$

where E denotes the exceptional divisor in the blow-up $Bl_Y X$ of X along the smooth subvariety Y.

2.2. Relation to motives. In the category $MM(k)$ of mixed motives over k with rational coefficients[10], relations (2.1), (2.2), (2.3), (2.6) have more sophisticated counterparts, which actually reduce to analogous relations if one passes to $K_0(MM(k)$, the Grothendieck group constructed in terms of extensions of mixed motives.

In fact, one expects that the functor[11] $h : Var(k) \to MM(k)$ gives rise to a ring homomorphism

$$K_0(Var(k)) \to K_0(MM(k)). \tag{2.7}$$

This can be made rigorous, if char $k = 0$, using the previous proposition and a category $M_\sim(k)$ of *pure motives* (*i.e.* of motives of smooth projective k-varieties, with morphisms given by algebraic correspondences modulo some fixed equivalence relation $\sim$). One gets a canonical ring homomorphism

$$\mu_c : K_0(Var(k)) \to K_0(M_\sim(k)) \tag{2.8}$$

(where $K_0(M_\sim(k))$ denotes the Grothendieck group constructed in terms of direct sums of pure motives; which is actually a ring with respect to the multiplication induced by tensoring motives). It sends $\mathbb{L}$ to $[\mathbb{Q}(-1)]$.

Recent work by H. Gillet and C. Soulé [**22**] allows one to drop the assumption on char k.

2.2.1. REMARK. Conjecturally, $K_0(M_\sim(k)) = K_0(MM(k))$ and does not depend on the chosen equivalence $\sim$ used in the definition of the $M_\sim(k)$. In fact, this independence would follow from a conjecture due to S. Kimura and P. O'Sullivan, which predicts that any pure motive $M \in M_\sim(k)$ decomposes (non-canonically) as $M_+ \oplus M_-$, where $S^n M_- = \bigwedge^n M_+ = 0$ for $n >> 0$ (*cf. e.g.* [**2**, ch. 12][12]; here S^n and $\bigwedge^n$ denote the n-th symmetric and antisymmetric powers, respectively). This is for instance the case for motives of products of curves.

[10] there are actually several candidates for this category, some conditional, some not

[11] more accurately, its variant with compact supports

[12] for the coarsest equivalence $\sim$ (the so-called numerical equivalence), this conjecture amounts to the following: the even Künneth projector $H(X) \to H^{even}(X) \to H(X)$ is algebraic

In the sequel, we shall deal with sub-$\otimes$-categories of $M_\sim(k)$ which satisfy this conjecture, and we will drop $\sim$ from the notation $K_0(M_\sim(k))$.

2.3. Kapranov zeta functions. When k is a finite field, counting k-points of varieties factors through a ring homomorphism

$$\nu : K_0(Var(k)) \to \mathbb{Z}, \quad [X] \mapsto \sharp X(k), \tag{2.9}$$

which factors through $K_0(M(k))$. One of the expressions of the *Hasse-Weil zeta function*, which encodes the number of points of X in all finite extensions of k, is

$$Z(X,t) = \sum_0^\infty \sharp((S^n X)(k))\, t^n \in \mathbb{Z}[[t]], \tag{2.10}$$

and it belongs to $\mathbb{Q}(t)$ (Dwork).

M. Kapranov had the idea [**26**] to replace, in this expression, $\sharp((S^n X)(k))$ by the class of $S^n X$ itself in $K_0(Var(k))$[13]). More precisely, he attached to any ring homomorphism

$$\mu : K_0(Var(k) \to R$$

the series

$$Z_\mu(X,t) := \sum_0^\infty \mu[S^n X]\, t^n \in R[[t]], \tag{2.11}$$

which satisfies the equation

$$Z_\mu(X \coprod X', t) = Z_\mu(X,t).Z_\mu(X',t).$$

When $\mu = \nu$ (counting k-points), one recovers the Hasse-Weil zeta function.

When $k = \mathbb{C}$ and $\mu = \chi_c$ (Euler characteristic), $Z_\mu(X,t) = (1-t)^{-\chi_c(X)}$ (MacDonald).

The universal case (the *Kapranov zeta function*) corresponds to $\mu = id$. When $k = \mathbb{Q}$, one can reduce X modulo $p >> 0$ and count $\mathbb{F}_p$-points of the reduction. The Kapranov zeta function then specializes to the Hasse-Weil zeta function of the reduction, and thus may be seen as some kind of interpolation of these Hasse-Weil zeta functions when p varies.

See [**19**] for further discussion in the perspective of motivic integration.

2.4. The Kapranov zeta function of a curve. Let us assume that X is a smooth projective curve of genus g, defined over the field k. The Kapranov zeta function

$$Z_\mu(X,t) := \sum_0^\infty \mu[S^n X]\, t^n \tag{2.12}$$

has the same features as the usual Hasse-Weil zeta function:

[13]for X quasiprojective, say, in order to avoid difficulties with symmetric powers

2.4.1. PROPOSITION. [**26**]

$$Z_\mu(X,t) = \frac{P_\mu(X,t)}{(1-t)(1-\mathbb{L}t)} \tag{2.13}$$

where P_μ is a polynomial of degree $2g$, and one has the functional equation

$$Z_\mu(X,t) = \mathbb{L}^{g-1}t^{2g-2}Z_\mu(X,\mathbb{L}^{-1}t^{-1}). \tag{2.14}$$

Sketch of proof of (2.13) (assuming for simplicity that X has a k-rational point x_0): the mapping

$$X^n \to J(X),\ (x_1,\dots,x_n) \mapsto [x_1]+\dots+[x_n]-n[x_0]$$

factors through $S^nX \to J(X)$, which is a projective bundle if $n \geq 2g-1$.

Moreover, one has an injection

$$S^n(X) \hookrightarrow S^{n+1}(X),\quad x_1+\dots+x_n \mapsto x_0+x_1+\dots+x_n$$

and the complement of its image is a vector budle of rank $n+1-g$ over $J(X)$. This implies $[S^{n+1}(X)]-[S^n(X)] = [J(X)]\mathbb{L}^{n+1-g}$ hence, by telescoping, that $Z_\mu(X,T)(1-T)(1-\mathbb{L}T)$ is a polynomial of degree $\leq 2g$. □

When k is a finite field, the Hasse-Weil zeta function of X can also be written in the form

$$Z(X,t) = \sum_{D\geq 0} t^{\deg D} \text{ (sum over effective divisors)} \tag{2.15}$$

$$= \sum_{\mathcal{L}} h^0(\mathcal{L})_q.\, t^{\deg \mathcal{L}} \text{ (sum over line bundles)}, \tag{2.16}$$

where one uses the standard notation $n_q = 1+q+\dots+q^{n-1}$.

R. Pellikan [**31**] had the idea to substitute, in this expression, q by an indeterminate u. He proved that

$$Z(X,t,u) := \sum_{\mathcal{L}} h^0(\mathcal{L})_u.\, t^{\deg \mathcal{L}} \tag{2.17}$$

is a rational function of the form $\frac{P(X,t,u)}{(1-t)(1-ut)}$.

Finally, F. Baldassarri, C. Deninger and N. Naumann [**6**] unified the two generalizations (2.12) (Kapranov) and (2.17) (Pellikaan) of the Hasse-Weil zeta function (2.15) by setting:

$$Z_\mu(X,t,u) := \sum_{n,d} [Pic_n^d]n_u.\, t^d \in R[[t,u]] \tag{2.18}$$

(where Pic_n^d classifies line bundles of degree d with $h^0(\mathcal{L}) \geq n$, and $n_u = 1+u+\dots+u^{n-1}$), and they proved that this is again a rational function of the form $\frac{P_\mu(X,t,u)}{(1-t)(1-ut)}$.

One thus has a commutative diagram of specializations

$$\begin{array}{ccccc} & & Z_\mu(X,t) & & \\ & \overset{u\mapsto q}{\nearrow} & & \overset{\mu\mapsto\nu}{\searrow} & \\ Z_\mu(X,t,u) & & & & Z(X,t). \\ & \underset{}{\overset{\mu\mapsto\nu}{\searrow}} & & \overset{u\mapsto q}{\nearrow} & \\ & & Z(X,t,u) & & \end{array}$$

On the other hand, M. Larsen and V. Lunts investigated the Kapranov zeta function of products of curves.

2.4.2. PROPOSITION. [**28**][**29**] *If X is a product of two curves of genus > 1, $Z_\mu(X,t)$ is* not *rational for $\mu = id$.*

In the sequel, following [**2**, 13.3], we remedy this by working with a μ which is "sufficiently universal", but for which one can nevertheless hope that $Z_\mu(X,t)$ is always rational. Namely, we work with μ_c (*cf.* (2.8)): in other words, we replace the ring of varieties by the K-ring of pure motives.

2.5. Motivic zeta functions of motives.

Thus, let us define, for any pure motive M over k (with rational coefficients), its *motivic zeta function* to be the series

$$Z_{mot}(M,t) := \sum_0^\infty [S^n M].t^n \in K_0(\mathcal{M}(k))[[t]]. \tag{2.19}$$

One has

$$Z_{mot}(M \oplus M',t) = Z_{mot}(M,t)\cdot Z_{mot}(M',t).$$

2.5.1. PROPOSITION. [**2**, 13.3] *If M is finite-dimensional in the sense of Kimura-O'Sullivan (i.e. $M = M_+ \oplus M_-$, $S^n M_- = \bigwedge^n M_+ = 0$ for $n >> 0$), then $Z_{mot}(M,t)$ is rational.*

(This applies for instance to motives of products of curves - and conjecturally to any motive).

Moreover, B. Kahn [**25**] (*cf.* also [**24**]) has established a functional equation of the form

$$Z_{mot}(M^\vee,t^{-1}) = (-1)^{\chi_+(M)}\cdot \det M \cdot t^{\chi(M)}\cdot Z_{mot}(M,t) \tag{2.20}$$

(where $\det M = \bigwedge^{\chi_+} M_+ \otimes (S^{-\chi_-}M_-)^{-1}$).

2.6. Motivic Artin L-functions.

One can play this game further. Hasse-Weil zeta functions of curves can be decomposed into (Artin) L-functions. A. Dhillon and J. Mináč upgraded this formalism at the level of motivic zeta functions [**21**].

Starting in slightly greater generality, let V be a $\mathbb{Q}$-vector space of finite dimension. To any pure motive M, one attaches another one $V\otimes M$, defined by

$$Hom(V\otimes M, M') = Hom_F(V, Hom(M,M')).$$

Let G be a finite group acting on M, and let $\rho: G \to GL(V)$ be a homomorphism.

The *motivic L-function* attached to M and ρ is

$$L_{mot}(M,\rho,t) := Z_{mot}((V\otimes M)^G,t). \tag{2.21}$$

This definition extends to characters χ of G (that is, $\mathbb{Z}$-linear combinations of ρ's), and gives rise to a formalism analogous to the usual formalism of Artin L-functions. Namely, one has the following identities in $K_0(\mathcal{M}(k))(t)$:

$$L_{mot}(M,\chi+\chi',t) = L_{mot}(M,\chi,t)\cdot L_{mot}(M,\chi',t) \tag{2.22}$$

$$L_{mot}(M,\chi',t) = L_{mot}(M, Ind_{G'}^G\chi',t) \tag{2.23}$$

(for G' a subgroup of G),

$$L_{mot}(M,\chi'',t) = L_{mot}(M,\chi,t) \tag{2.24}$$

($G' \triangleleft G$, χ coming from a character χ'' of G/G'),

$$Z_{mot}(M,t) = \prod_{\chi \text{ irr.}} L_{mot}(M,\chi,t)^{\chi(1)}. \tag{2.25}$$

2.6.1. Example. Let X be again a smooth projective curve, and let G act on X. Then G acts on the motive $h(X)$ of X via $(g^*)^{-1}$. By definition $L_{mot}(X,\chi,t)$ is the motivic L-function of $h(X)$.

If k is finite, $\nu(L_{mot}(X,\chi,t))$ is nothing but the Artin non-abelian L-function $L(X,\chi,t)$, defined by the formula (where F denotes the Frobenius)

$$\log L(X,\chi,t) = \sum \nu_n(X)\frac{t^n}{n},\ \ \nu_n(X) = \frac{1}{\sharp G}\sum \chi(g^{-1})\sharp Fix(gF^n).$$

This leads to the definition of *motivic Artin symbols* and to a motivic avatar of Chebotarev's density theorem [**21**].

2.7. The class of the motive of a semisimple group G. Let G be a connected split semisimple algebraic group over k. Let $T \subset G$ be a maximal torus (of dimension r), with character group $X(T)$. The Weyl group W acts on the symmetric algebra $S(X(T)_{\mathbb{Q}}))$, and its invariants are generated in degree $d_1,\dots,d_r$ (for $G = SL_n$, one has $d_1 = 2,\dots,d_r = n$).

One has the classical formulas

$$(t-1)^r \sum_{w\in W} t^{\ell(w)} = \prod_1^r (t^{d_i}-1),\ \ \sum d_i = \frac{1}{2}(\dim G + r). \tag{2.26}$$

K. Behrend and A. Dhillon gave the following generalization of formula (2.3) for $[SL_n]$.

2.7.1. Proposition. *In $K_0(Var(k))[\mathbb{L}^{-1}]$ or $K_0(M(k))$, one has*

$$[G] = \mathbb{L}^{\dim G}\prod_1^r (1-\mathbb{L}^{-d_i}). \tag{2.27}$$

(In $K_0(M(k))$, it is preferable to write $[\mathbb{Q}(m)]$ instead of $\mathbb{L}^{-m}$).

Sketch of proof: let B be a Borel subgroup, U its unipotent radical. Then $[G] = [G/B]\cdot[T]\cdot[U]$ (in $K_0(Var(k))$ or $K_0(M(k))$). On the other hand, one computes easily $[U] = \mathbb{L}^{\frac{1}{2}(\dim G - r)}$, $[T] = (\mathbb{L}-1)^r$, and using the Bruhat decomposition, $[G/B] = \sum_{w\in W}\mathbb{L}^{\ell(w)}$. Combining these formulas with (2.26), one gets (2.27). □

2.7.2. Remark. With proper interpretation, (2.27) can be reformulated as a formula for the class of the classifying stack of G-torsors over k, in a suitable localization of $K_0(M(k))$:

$$[BG] = [G]^{-1} = [\mathbb{Q}(\dim G)]\prod_i (1-[\mathbb{Q}(d_i)])^{-1}. \tag{2.28}$$

Here BG is understood as the Artin quotient stack $[\operatorname{Spec} k/G]$. The classes in $K_0(M(k))$ of such quotient stacks with linear stabilizers are well-defined, according to [**7**]; the key point is that for any connected linear algebraic group G and any

G-torsor $P \to X$, one has the relation $[P] = [X] \cdot [G]$ in the K-group of Voevodsky motives (*cf.* [**7**, app. A]), which coincides with $K_0(M(k))$ according to [**16**].

2.8. G-torsors over a curve X, and special values of $Z_{mot}(X)$. Let us now look at G-torsors not over the point, but over a smooth projective curve X of genus g.

More precisely, let G be a simply connected split semisimple algebraic group over k (*e.g.* SL_n), and let $Bun_{G,X}$ be the moduli stack of G-torsors on X (which is smooth of dimension $(g-1).\dim G$).

This Artin stack admits an infinite stratification by pieces of the form $[X_i/GL_{n_i}]$, whose dimensions tend to $-\infty$. According to Behrend and Dhillon [**7**], this allows to define unambiguously the class

$$[Bun_{G,X}] := \sum [X_i][GL_{n_i}]^{-1} \tag{2.29}$$

in a suitable completion of $K_0[M(k)]$ with respect to $[\mathbb{Q}(1)]$, taking into account the fact that $[GL_n]^{-1} = [BGL_n] = \mathbb{Q}(n^2) \cdot (1 + \cdots) \in \mathbb{Z}[[\mathbb{Q}(1)]]$.

2.8.1. CONJECTURE. *(Behrend-Dhillon)*

$$[Bun_{G,X}] = [\mathbb{Q}((1-g).\dim G)] \prod_i Z_{mot}(X, [\mathbb{Q}(d_i)]). \tag{2.30}$$

This has to be compared with (2.28), where the d_i have the same meaning;. Note that the special values $Z_{mot}(X, [\mathbb{Q}(d_i)])$ are well-defined since $Z_{mot}(X,t)$ is rational with poles at 1 and $[\mathbb{Q}(1)]$ only.

2.8.2. PROPOSITION. [**7**] *The conjecture holds for $X = \mathbb{P}^1$ and any G, and for $G = SL_n$ and any X.*

Let us consider the case of SL_n to fix ideas (*cf.* [**20**]), and comment briefly on some specializations of the motivic formula (2.30).

1) For $k = \mathbb{C}$, $\mu = \chi_c$, the formula specializes to a formula for the Euler characteristic of $Bun_{G,X}$, which can be established via gauge theory à la Yang-Mills (Atiyah-Bott [**5**], see also Teleman [**34**]).

More precisely, $H^*(Bun_{G,X}) \cong H^*(G)^{\otimes 2g} \otimes H^*(BG) \otimes H^*(\Omega G)$.

2) For $k = \mathbb{F}_q$, $\mu = \nu$ (counting points), the formula specializes to a formula for the number of k-points of $Bun_{G,X}$ (Harder, *cf.* [**23**]).

More precisely, $Bun_{G,X}$ can be viewed as the transformation groupoid of $G(K)$ on $G(\mathbb{A}_K)/\mathcal{K}$, for $K = k(X)$, $\mathcal{K} = \prod_x G(\hat{\mathcal{O}}_{X,x})$; so that $\sharp Bun_{G,X}(k) = \frac{vol(G(K)\backslash G(\mathbb{A}_K))}{vol(\mathcal{K})}$. One has $vol(\mathcal{K}) = q^{(1-g).(n^2-1)} \prod_2^n \zeta_K(i)^{-1}$, and the Tamagawa number $vol(G(K)\backslash G(\mathbb{A}_K))$ is 1, whence

$$\sharp Bun_{G,X}(k) = q^{(g-1).(n^2-1)} \prod_2^n \zeta_K(i). \tag{2.31}$$

In the last section of [**5**], one finds a precise comparison between the Morse-Yang-Mills approach to the cohomology of $Bun_{G,X}$ and the (Harder-Narasimhan) arithmetic approach via the computation of the number of points modulo primes p, followed by some interesting speculation. The (Behrend-Dhillon) motivic approach goes one step forward in this direction.

References

[1] Aluffi, Paolo; Marcolli, Matilde. Algebro-geometric Feynman rules, arXiv:08112514.

[2] André, Yves. *Une introduction aux motifs (motifs purs, motifs mixtes, périodes)*. Panoramas et Synthèses 17. Société Mathématique de France, Paris, 2004.

[3] –. Ambiguity theory, old and new. Bollettino U.M.I. (8) I (2008)

[4] –. Galois theory, motives, and transcendental number theory, in *Renormalization and Galois theories*, Connes, Fauvet, Ramis (Eds), IRMA Lectures in Mathematics and Theoretical Physics XV, (2009).

[5] Atiyah, Michael; Bott, Raoul. Yang-Mills and bundles over algebraic curves. *Geometry and analysis*, pp. 11–20, Indian Acad. Sci., Bangalore, 1980.

[6] Baldassarri, Francesco; Deninger, Christopher; Naumann, Niko. A motivic version of Pellikaan's two variable zeta function. *Diophantine geometry*, 35–43, CRM Series, 4, Ed. Norm., Pisa, 2007.

[7] Behrend, Kai; Dhillon, Ajneet. On the motivic class of the stack of bundles. Adv. Math. 212 (2007), no. 2, 617–644.

[8] Belkale, Prakash; Brosnan, Patrick. Matroids, motives and a conjecture of Kontsevich, Duke Math. J. **116** (2003), no. 1, 147–188.

[9] – – . Periods and Igusa local zeta functions, International Mathematics Research Notices. Vol. 2003, no. 49, pp. 2655-2670. 2003.

[10] Bierenbaum, Isabella; Weinzierl, Stefan. The massless two-loop two-point function, Eur. Phys. J. Mod. Phys. C **6** (1995) 519.

[11] Bittner, Franziska, The universal Euler characteristic for varieties of characteristic zero, Compos. Math. **140** (2004), no. 4, 1011–1032.

[12] Bloch, Spencer. Motives associated to sums of graphs, ArXiv 08101313.

[13] – ; Esnault, Hélène; Kreimer, Dirk. On motives associated to graph polynomials. Comm. Math. Phys. 267 (2006), no. 1, 181–225.

[14] Bogner, Christian; Weinzierl, Stefan. Resolution of singularities for multi-loop integrals. arXiv:0709.4092

[15] – – . Periods and Feynman integrals, arXiv:0711.4863.

[16] Bondarko, Mikhail. Differential graded motives: weight complex, weight filrtations and spectral sequences for realizations: Voevodsky vs. Hanamura. J. Inst. Math. Jussieu, DOI: 10.1017/S147474800800011X.

[17] Broadhurst, David; Kreimer, Dirk. Association of multiple zeta values with positive knots via Feynman diagrams up to 9 loops. Phys. Lett. B, 393 (1997), 403-412.

[18] Brown, Francis. The massless higher-loop two-point function, ArXiv 08041660.

[19] Denef, Jan; Loeser, François. On some rational generating series occuring in arithmetic geometry. Geometric aspects of Dwork theory. Vol. I, II, 509–526, de Gruyter, Berlin, 2004.

[20] Dhillon, Ajneet. On the cohomology of moduli of vector bundles and the Tamagawa number of SL_n. Canad. J. Math. 58 (2006), no. 5, 1000–1025.

[21] – ; Minác, Ján. A motivic Chebotarev density theorem. New York J. Math. 12 (2006), 123–141

[22] Gillet, Henri; Soulé, Christophe. Motivic weight complexes for arithmetic varieties, ArXiv: math/0804.4853.

[23] Harder, Günther; Narasimhan, Mudumbai. On the cohomology groups of moduli spaces of vector bundles on curves. Math. Ann. 212 (1974/75), 215–248.

[24] Heinloth, Franziska. A note on functional equations for zeta functions with values in Chow motives. Ann. Inst. Fourier 57 (2007), no. 6, 1927–1945.

[25] Kahn, Bruno. Zeta functions and motives, III. Pure and Applied Math. Quat. 5, 1 (2009), 507-570.

[26] Kapranov, Mikhail. The elliptic curve in the S-duality theory and Eisenstein series for Kac-Moody groups. ArXiv AG/0001005.

[27] Kedlaya, Kiran. Counting points on hyperelliptic curves using Monsky-Washnitzer cohomology, Journal of the Ramanujan Mathematical Society 16 (2001), 323338.

[28] Larsen, Michael; Lunts, Valery. Motivic measures and stable birational geometry. Mosc. Math. J. 3 (2003), no. 1, 85–95, 259.

[29] — — . Rationality criteria for motivic zeta functions. Compos. Math. 140 (2004), no. 6, 1537–1560.

[30] Marcolli, Matilde. Motivic renormalization and singularities. Arxiv:math-ph0804824.

[31] Pellikaan, Ruud. On special divisors and the two variable zeta function of algebraic curves over finite fields. *Arithmetic, geometry and coding theory* (Luminy, 1993), 175–184, de Gruyter, Berlin, 1996.

[32] Kontsevich, Maxim; Zagier, Don. Periods. *Mathematics unlimited—2001 and beyond*, 771–808, Springer, Berlin, 2001.

[33] Stembridge, John. Counting points on varieties over finite fields related to a conjecture of Kontsevich. Ann. Comb. 2 (1998), no. 4, 365–385.

[34] Teleman, Constantin. Borel-Weil-Bott theory on the moduli stack of G-bundles over a curve. Invent. Math. 138 (1998), 1-57.

[35] Todorov, Ivan. *Analytic properties of Feynman diagrams in quantum field theory.* International series of monogrpahs in natural philosophy 38, Pergamon press, Oxford 1971.

[36] Weinberg, Steven. High-energy behavior in quantum field theory, Phys. Rev. **118** (1960), 838-849.

[37] Weinzierl, Stefan. Feynman integrals and multiple polylogarithms. arXiv:0705.0900.

DÉPARTEMENT DE MATHÉMATIQUES ET APPLICATIONS, ÉCOLE NORMALE SUPÉRIEURE, 45 RUE D'ULM, 75230 PARIS CEDEX 05, FRANCE.

E-mail address: `andre@dma.ens.fr`

Clay Mathematics Proceedings
Volume 12, 2010

Algebra for Quantum Fields

Dirk Kreimer

ABSTRACT. We give an account of the current state of the approach to quantum field theory via Hopf algebras and Hochschild cohomology. We emphasize the versatility and mathematical foundations of this algebraic structure, and collect algebraic structures here in one place which are either scattered throughout the literature, or only implicit in previous writings. In particular we point out mathematical structures which can be helpful to further develop our mathematical understanding of quantum fields.

Introduction

Acknowledgments. It is a pleasure to thank my former students Christoph Bergbauer, Kurusch Ebrahimi-Fard and Karen Yeats for discussions and advice.

The reason for QFT. What is quantum field theory (QFT) about? For that matter, what is quantum physics about? The answer, given with the necessary grain of pragmatism, is simple: sum over all histories connecting a chosen initial state with a particular final state. Square that complex-valued sum.

Various attempts had been made to make this paradigm precise: the desired "sum over histories" has not yet reached its final form though, and mathematicians are still, and rightfully so, baffled by QFT.

Physicists have created myriads of examples meanwhile where one can stretch one's mind and flex one's muscles on what is usually called the path integral. Many results point to rather fascinating structures, carefully formulated in a self-consistent way. A mathematical definition of said path integral beyond perturbaton theory is lacking though, leaving the author with considerable unease.

Early on, it was recognized that the desired Green functions in field theory are constrained by quantum equations of motion, the Dyson–Schwinger equations. The latter suffer from short-distance singularities, leading to the need for renormalization.

It took us physicists a while to learn how to handle this routinely, and in a mathematically consistent way. Progress was made by elaborate attempts at low orders of perturbation theory, and the above equations of motion took a backseat in contemporary physics, while formal approaches starting from the functional integral

---

2010 *Mathematics Subject Classification.* Primary 81Q30, Secondary 16T05, 81T15, 81T18.
Work supported in parts by grant NSF-DMS/0603781. Author supported by CNRS.

(constructive methods) typically failed to come to conclusions for renormalizable theories, which remain theories of prime interest.

Meanwhile, perturbative renormalization was embarrassingly effective in describing reality, and kept QFT, understood as an expansion in Feynman graphs, in its role as the best-tested and most-used workhorse in the stables of theoretical physics. It is commonly denied the status of being a theory these days though, as at the moment of writing it is not yet defined in mathematically satisfying terms.

It is a personal belief of the author that this is not testimony for bigger (read extended) things hiding behind local quantum fields, but rather testimony to the subtlety by which nature hides its concepts.

On an optimistic note, I indeed believe that the clear mathematical understanding we have now of the practice of perturbative renormalization paves the way for a mathematically consistent approach to QFT which bridges the gap between what practitioners of QFT have learned, and what is respectable mathematics.

The approach exhibited in the following is based solely on the representation theory of the Poincaré group and the requirement that interactions are local.

Before we start, let me emphasize that here is not the place to comment in detail on progress with analytic aspects.

Still, let me mention two encouraging developments: non-perturbative aspects can now be studied using Dyson–Schwinger equations in a much more effective manner [**29, 31, 32, 3, 4**], and the relation to periods and motivic theory became a (little!) bit clearer in collaboration with Esnault and Bloch [**10, 11**].

In particular, Feynman integrals are periods [**2, 17, 10, 11**] (considered as a function of masses and external momenta, they are periods when those parameters take rational values [**13**], though a much better argument should be made for the Taylor coefficients in the expansion in such variables).

Better still, these periods are interesting: in a suitable parametric representation based on edge variables A_e for edges e [**10, 9, 11**], they appear as periods of the mixed Hodge structure on the middle-dimensional cohomology

$$H^{2m-1}(P\backslash Y_\Gamma, B\backslash B \cap Y_\Gamma)$$

constructed from blow-ups $P \to \cdots \to \mathbb{P}^{2m-1}$ which separate the boundary of the chain of integration (contained in $\Delta = \cup_{e\in E(\Gamma)} A_e = 0$) from the singularities of the graph hypersurface X_Γ, with Y_Γ the strict transform of X_Γ and B the inverse image of Δ.

For example, the complete graph on four vertices (a contribution to the vertex function of ϕ^4 theory) has a period $6\zeta(3)$ contained in $\zeta(3)\mathbb{Q}$, [**10, 9**]. Such results have been recently extended by Dzimitry Doryn [**21**].

Now back to the underlying algebraic structures. Let's first get edges and vertices for our graphs.

1. Free QFT, interacting QFT

Classical geometry does not rule the day when it comes to quantum fields. Much to the contrary, the often beautiful classical geometry of fields, gauge fields in particular, must emerge as a classical limit of quantum field theory. Hence we speak about QFT without us taking recourse to classical fields. We ignore the geometry of the classical spacetime manifold over which we want to construct QFT, and just remember that it has a four-dimensional tangent and cotangent space locally, isomorphic to flat Minkowski space. It is over such local fibres that one formulates QFT.

Our first concern is to understand the elementary amplitudes which we use to describe the observable physics which results from quantum field theory.

They come in two garden varieties: amplitudes for propagation, and amplitudes for scattering. The former are provided by free quantum field theory: free propagators, in momentum space, are obtained as the inverse of the free covariant wave equations. Hence, for Minkowski space, it is Wigner's representation theory of the Lorentz and Poincaré groups which rules the day, providing us with free propagators for massless and massive bosons and fermions.

The latter, amplitudes for scattering, are again provided by the representation theory of the Poincaré group, augmented by the requirement for locality.

Let us look at a simple example to see how this comes about. Assume we take from free quantum fields the covariances for a free propagating electron, positron and photon. Assume we want to couple these in a local interaction. Representation theory tells us that this interaction will have to transform as a Lorentz vector v_μ, coupling the spin-one photon to a spin-1/2 electron and positron. Also, knowing the scaling weights of free photons, electrons and positrons as determined from the accompanying free field monomials, such a vertex itself must have zero scaling weight, as the scaling weights of those monomials add up to the dimension of spacetime. Indeed,

$$[\bar{\psi}\partial\!\!\!/\psi] = 4 \quad \Rightarrow \quad [\bar{\psi}] = [\psi] = 3/2, \tag{1}$$

$$\frac{1}{4}[F^2] = 4 \quad \Rightarrow \quad [A] = 1, \tag{2}$$

$$[v_\mu \bar{\psi} A_\mu \psi] = 4 \Rightarrow [v_\mu] = 0. \tag{3}$$

So what would be the Feynman rule, in momentum space say, for such an amplitude? If the electron has momentum p_1, the positron momentum p_2, and the photon momentum $q = -p_1 - p_2$, the vertex can be a linear combination of twelve invariants

$$v_\mu = c_1\gamma_\mu + c_2\frac{q_\mu q\!\!\!/}{q^2} + \cdots . \tag{4}$$

But if we have to have a local theory, any graph for a quantum correction for the unknown vertex built from that unknown vertex and the known propagators will be, by a simple powercounting exercise – we know the scaling weight of our unknown vertex at least – logarithmically divergent.

If we are to absorb this logarithmic divergence by a local counterterm, this gives us information on the desired Feynman rule. Let us work it out. To keep the example simple, let us assume we suspect that the vertex is of the form

$$v_\mu = v_\mu(q) = c_1\gamma_\mu + c_2\frac{q_\mu q\!\!\!/}{q^2}. \tag{5}$$

Let us consider the one-loop 1PI graph –the lowest order quantum correction– to find the sought after Feynman rule.

With three vertices in the graph we have $2^3 = 8$ integrals to do which appear in the limit

$$\lim_{\Lambda\to+\infty}\frac{1}{\ln\Lambda}\Phi\left(\quad\right) \sim \lim_{\Lambda\to+\infty}\frac{1}{\ln\Lambda}\int_{-\Lambda}^{\Lambda} v_\alpha(k)\frac{1}{\not k}v_\mu(k)\frac{1}{\not k}v_\beta(k)D_{\alpha\beta}((k+q)^2)d^4k$$

$$= f(c_1,c_2)\gamma_\mu.$$

In a local theory, the coefficient of $\ln\Lambda$ from the integral must be proportional to the desired vertex. Hence, dividing and taking the limit, we confirm that the term $\sim q_\mu \not q$ vanishes like $1/\ln\Lambda$ in all eight terms. We hence conclude $c_2 = 0$, and this gives a good idea how locality is needed for quantum field theory to stabilize at dintinguished Feynman rules in a self-similar manner. The same result would have been obtained if we had done the example with the full 12^3 terms of the full vertex, as it must be for a renormalizable theory.

We also conclude that the price for Feynman rules determined by locality is that we indeed pick up local short-distance singularities. That leaves us the freedom to set a scale, which is no big surprise: looking only at quantum fields for a typical fibre, the cotangent space, we hence miss the only parameter around to set a scale: the curvature of the underlying manifold. The extension of such local notions to the whole manifold awaits future understanding of quantum gravity. This might well start from understanding how gravity with its peculiar powercounting behaves as a Hopf algebra [**26**].

Let us now proceed to see what comes with those edges and vertices as prescribed above: graphs, obviously.

2. 1PI graphs, Hopf algebras

Having thus elementary scattering and propagation amplitudes available, we can set up a quantum theory: we define incoming and outgoing asymptotic states, and sum over all unobserved intermediate states. This is standard material for a physicist, and we leave it to the reader to acquaint himself with the necessary details on the LSZ formalism and other such aspects [**45, 14, 12, 25**].

While many textbooks on contemporary physics proceed using the path integral to define Green functions for amplitudes, for connected amplitudes and for 1PI amplitudes, we emphasize that these Green functions can be given a mathematically precise meaning through the study of the Hopf algebra structure underlying the graphs constructed from the representation theory mentioned above.[1]

So, having Feynman rules for edges and vertices, the above gives us Feynman rules for n-PI graphs, graphs which do not disconnect upon removal of any n internal edges. Amplitudes for connected graphs are obtained from 1-PI Green functions by connecting them via free covariances, and disconnected graphs finally by exponentiation. It is for 1-PI graphs that the underlying algebraic structure of field theory becomes fully visible.

The basic such algebraic structures then at our disposal are:
i) the Hopf and Lie algebras coming with such graphs,[**40, 18, 19, 20**]
ii) the corresponding Hochschild cohomology and the sub-Hopf algebras generated

[1]This might implicitly define the path integral, which has to be seen in future work. Too often, in the authors opinion, the path integral is in the context of quantum field theory only a reparametrization of our lack of understanding, giving undue prominence to the classical Lagrangian.

by the grading,[**8, 23**]
iii) the co-ideals corresponding to symmetries in the Lagrangian,[**33, 43**]
iv) the coradical filtration and Dynkin operators governing the renormalization group and leading log expansion,[**16, 30, 42**]
v) the semi-direct product structure between superficially convergent and divergent graphs, [**19**]
vi) and finally the core Hopf algebra [**11, 27**], suggesting co-ideals leading to recursions à la Britto–Cachazo–Feng–Witten (BCFW) [**15**], showing that loops and legs speak to each other in many ways: it is indeed the hope of the author that the rather disparate structures we observe in experience with multi-loop vs multi-leg expansions will combine finally in a common mathematical framework [**28**].

We omitted in this list Rota–Baxter algebras [**22**], which are useful for minimal subtracton (MS) schemes but less so in renormalization schemes based on on-shell or momentum space subtractions. The reader can find a detailed study of Rota–Baxter algebras in the above-cited work of Ebrahimi-Fard and Manchon, while the use of momentum space subtractions was exhibited recently beyond perturbation theory in [**31**]. We also omit the algebraic structure of field theory in coordinate space, see [**5, 7, 6**] for a clarification on how to connect it with the approach described here.

In this contribution, we will mainly review combinatorial and algebraic aspects developed in recent years. We include a few results only implicit in published work so far. A summary of analytic and algebro-geometric achievements will have to be given elsewhere.

Let us now illustrate these algebraic structures. For that we develop our muscles on quantum electrodynamics (QED) graphs for the vertex, fermion and photon selfenergy, up to two loops each. Here they are:

(6) $c_1^{\bar\psi A\!\!\!/\psi} =$  ,

(7) $c_2^{\bar\psi A\!\!\!/\psi} =$  ,

(8) $c_1^{\bar\psi\psi} =$  ,

(9) $c_2^{\bar\psi\psi} =$  ,

(10) $c_1^{\frac{1}{4}F^2} =$  ,

(11) $c_2^{\frac{1}{4}F^2} =$ 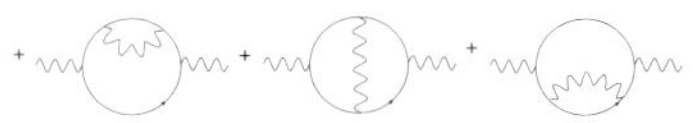 .

2.1. The Hopf algebra. We define a family of Hopf algebras $\mathcal{H}$. Each Hopf algebra $H \in \mathcal{H}$ is generated by generators given by 1-PI graphs and its algebra structure is given as the free commutative $\mathbb{Q}$-algebra over those generators, with the empty graph furnishing the unit $\mathbb{I}$.

For a graph Γ, we let $\Gamma^{[0]}$ be the set of its vertices, $\Gamma^{[1]}_{\text{int}}$ be the set of its internal edges, and $\Gamma^{[1]}_{\text{ext}}$ be the set of its external edges. Each edge is assigned an arbitrary orientation (all physics is independent of that choice), so that we can speak of a source $s(e)$ and target $t(e)$ for an edge e. For each internal edge $e \in \Gamma^{[1]}_{\text{int}}$, $s(e) \in \Gamma^{[0]}$ and $t(e) \in \Gamma^{[0]}$. We do not require that $s(e) \neq t(e)$. For each $e \in \Gamma^{[1]}_{\text{ext}}$, $t(e) \in \Gamma^{[0]}$ but $s(e) \notin \Gamma^{[0]}$.

To each internal edge e we assign a weight $w(e)$, to each vertex v we assign similarly a weight $w(v)$. We write $\sum_{w\in\Gamma} w$ for the sum over all these edge and vertex weights. Then, we define

$$\omega_{2n}(\Gamma) := -2n|\Gamma| + \sum_{w\in\Gamma} w. \tag{12}$$

Here, a grading $|\Gamma|$ is used which is provided by the number of independent cycles in a graph Γ, its lowest Betti number, and we hence write

$$H = \underbrace{H^0}_{\mathbb{Q}\mathbb{I}} \oplus \underbrace{\left(\bigoplus_{j=1}^{\infty} H^j\right)}_{\mathrm{Aug}(H)}. \tag{13}$$

So H is reduced to scalars off the augmentation ideal $\mathrm{Aug}(H)$. We let $\langle\Gamma\rangle$ be the linear span of the generators.

We distinguish these Hopf algebras $H = H_{2n}$ by an even integer $0 \leq 2n$, $n \in \mathbb{N}$. They are all based on the same set of generators, hence have an identical algebra structure. There are slight differences in their coalgebra structure though, as we give them a coproduct depending on $2n$:

$$\Delta_{2n}(\Gamma) := \Gamma\otimes\mathbb{I} + \mathbb{I}\otimes\Gamma + \sum_{\gamma=\prod_i \gamma_i\subset\Gamma,\omega_{2n}(\gamma_i)\leq 0} \gamma\otimes\Gamma/\gamma. \tag{14}$$

The sum is over all disjoint unions of 1-PI subgraphs γ_i such that for each γ_i, $\omega_{2n}(\gamma_i) \leq 0$. In the limit $n \to \infty$, we hence obtain the core Hopf algebra H_{core} with coproduct

$$\Delta_{\mathrm{core}}(\Gamma) = \Gamma\otimes\mathbb{I} + \mathbb{I}\otimes\Gamma + \sum_{\gamma=\prod_i \gamma_i\subset\Gamma} \gamma\otimes\Gamma/\gamma. \tag{15}$$

We also use the reduced coproducts

$$\Delta'_{2n}(\Gamma) := \sum_{\gamma=\prod_i \gamma_i\subset\Gamma,\omega_{2n}(\gamma_i)\leq 0} \gamma\otimes\Gamma/\gamma. \tag{16}$$

This gives us a tower of quotient Hopf algebras [**11**]

$$H_0 \subset H_2 \subset H_4 \subset \cdots \subset H_{2n} \cdots \subset H_{\mathrm{core}}. \tag{17}$$

In the following, we often omit the subscript $2n$ as it is either clear which integer we speak about, or the statement holds for arbitrary $2n$ in an obvious manner.

Note that H_0 is the trivial Hopf algebra in which every graph is primitive. It is the free commutative and cocommutative bialgebra of polynomials in all its generators $\in \langle\Gamma\rangle$. Fittingly, its use in zero-dimensional field theory is an excellent tool to count graphs [**1**].

For any other such $H_{2n} \in \mathcal{H}$, $n < \infty$, we find that the Hopf algebra decomposes into a semi-direct product

$$H_{2n} = H_{2n}^{\mathrm{ren}} \times H_{2n}^{\mathrm{ab}}, \tag{18}$$

where H_{2n}^{ren} is generated by graphs Γ such that $\omega_{2n}(\Gamma) \leq 0$ and H_{2n}^{ab} is the abelian factor generated by graphs such that $\omega_{2n}(\Gamma) > 0$. See [**19**].

Let us explain the above tower a bit more. The core Hopf algebra allows us to shrink any 1-PI subgraph γ_i to a point, and hence is built on graphs with internal vertices of arbitrary valence, coupling an arbitrary numbers of edges and all types of edges for which we had free covariances. Again, locality and representation theory provide for such vertices Feynman rules as before, which are in general a sum over all local operators which are in accordance with the quantum numbers of those covariances. We can distinguish those operators by labeled vertices, which does not hinder us from setting up the Hopf algebra as before. For the core Hopf algebra, all primitives which we find in the linear span $\langle\Gamma\rangle$ have degree one,

$$\Delta'_{\text{core}}(\Gamma) \neq 0 \Rightarrow |\Gamma| > 1. \tag{19}$$

Note that for any chosen finite $2n$, the results can be very different. A renormalizable theory is distinguished by the fact that for some finite n_0,

$$\omega_{2n_0}(\Gamma) = \omega_{2n_0}(\Gamma'),\ \forall\,\Gamma,\Gamma' \text{ with } \mathrm{res}(\Gamma) = \mathrm{res}(\Gamma'). \tag{20}$$

Here, $\mathrm{res}(\Gamma)$ is the map which assigns to a graph Γ the vertex obtained by shrinking all internal edges to zero length. What remains is the external edges connected to the same point. If the number of external edges is greater than two, this gives us a vertex. If it is two, we identify those two connected edges to a single edge.

If such an n_0 exists, we call $2n_0$ the critical dimension of the theory. Particle physics so far is concerned with theories critical at $n_0 = 2$, i.e. in four dimensions of spacetime.

In such a case, all graphs with the same type of external edges evaluate to the same result under evaluation by ω_{2n_0}. $\omega_{2n_0}(\Gamma)$ then takes values $\in \{-r_0, \ldots, +\infty\}$, where $-r_0$ is the value achieved for vacuum graphs, and we obtain arbitrary positive values on considering graphs with a sufficient number of external edges.

For $n > n_0$, for any configuration of external edges we find, at sufficiently high degree $|\Gamma|$, graphs such that $\omega_{2n}(\Gamma) \leq 0$. The theory becomes non-renormalizable.

If $n < n_0$, only a finite number of graphs fulfill $\omega_{2n}(\Gamma) \leq 0$ and the theory is super-renormalizable.

In any case, for a Hopf algebra H_{2n}, continuing our appeal to self-similarity, we consider graphs made out of vertices such that $\omega_{2n}(\Gamma) \leq 0$. This defines a Hopf algebra H_{2n}^{ren}. Graphs made out of such vertices but with sufficiently many external edges such that $\omega_{2n}(\Gamma) > 0$ then provide a semi-direct product $H_{2n} = H_{2n}^{\text{ren}} \times H_{2n}^{\text{ab}}$. This Hopf algebra is a quotient of the core Hopf algebra, eliminating any graph with undesired vertices.

So, already at this elementary level, there is a nice interplay between the above-mentioned representation theory of the Lorentz and Poincaré groups and such towers of Hopf algebras, as it is this representation theory which determines the covariances and their possible local vertices, and hence the quotient algebras we get.

Let us now continue to list the other structural maps of these Hopf algebras. An antipode:

$$S(\Gamma) = -\Gamma - \sum_{\gamma\subset\Gamma} S(\gamma)\Gamma/\gamma. \tag{21}$$

A counit $\bar{e} : H \to \mathbb{Q}$ and unit $E : \mathbb{Q} \to H^0 \subset H$:

$$\bar{e}(q\mathbb{I}) = q, \bar{e}(X) = 0, X \in \mathrm{Aug}(H),\ E(q) = q\mathbb{I}. \tag{22}$$

Finally, an example:

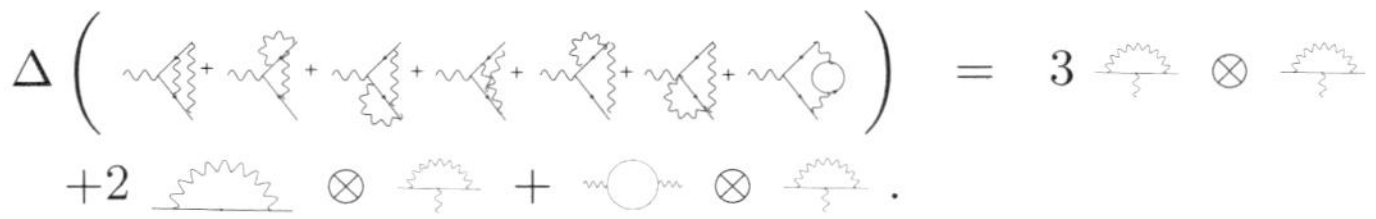

As a final remark, note that there are many more quotient Hopf algebras, by restricting generators to planar, or parquet, or whatever graphs.

Also, we will find all the Hopf algebras needed for an operator product expansion as quotient Hopf algebras, using that, for monomials (in operator-valued fields and their derivatives) $\mathcal{O}_a, \mathcal{O}_b$, the expansion of vacuum expectation values (vev's) of products of two monomials at different spacetime points x, y into localized field monomials

$$\langle \mathcal{O}_a(x)\mathcal{O}_b(0)\prod_i \mathcal{O}_{d_i}(y_i)\rangle = \sum_c \mathcal{C}^c_{ab}(x)\,\langle \mathcal{O}_c(0)\prod_i \mathcal{O}_{d_i}(y_i)\rangle, \tag{23}$$

(for $|x| < |y_i|\ \forall i$ and suitable (generalized) functions on spacetime $\mathcal{C}^c_{ab}$ determined only by the operators labelled a, b, c) proceeds on a set of graphs having as local vertices the tree-level vev's of the operators $\mathcal{O}_c$, again in accordance with Wigner's representation theory. Note that all such vertices appear naturally in the core Hopf algebra (as we have quotients Γ/γ), and hence the core Hopf algebra is the endpoint in this tower of Hopf algebras, which allows us to formulate a full field algebra in the sense of operator product expansions. In passing, we mention that the operator product expansion has a connection to vertex algebras as recently established by Hollands and Olbermann [**24**].

Such expansions in the core Hopf algebra also then underlie any study of the renormalization group flow in the sense of Wilson from the Hopf algebraic viewpoint. Let us finish this section with a cautionary remark: the difference between Minkowski or Euclidean signature is rather irrelevant for most combinatorial considerations below. It is crucial though in the operator product expansions, where the set of operators $\mathcal{O}_c$ above needs much more careful consideration in the Minkowskian case for expansions on the lightcone.

2.2. The Lie algebra L such that $U^*(L) = H$. As a graded commutative Hopf algebra ([**35**]), any $H \in \mathcal{H}$ can be regarded as the dual $U^*(L)$ of the universal enveloping algebra $U(L)$ of a Lie algebra L. The tower $\mathcal{H}$ of quotient Hopf algebras H_{2n} corresponds to a tower $\mathcal{L}$ of sub-Lie algebras L_{2n}. We write for each $L \in \mathcal{L}$,

$$U(L) = Q\mathbb{I} \oplus L \oplus \left(\bigoplus_{k=2}^{\infty} L^{\hat{\otimes}^k}\right), \tag{24}$$

where $\hat{\otimes}^k$ indicates the symmetrized k-fold tensor product of L as usual for an universal enveloping algebra, obtained by dividing the tensor algebra $L^{\otimes^k}$ by the ideal $l_1 \otimes l_2 - l_2 \otimes l_1 - [l_1, l_2] = 0$.

We manifest the duality by a pairing between generators of L and generators of H,

$$(Z_\gamma, \Gamma) = \delta_{\gamma,\Gamma}, \tag{25}$$

the Kronecker pairing. It extends to $U(L)$ thanks to the coproduct.

There is an underlying pre-Lie algebra structure:

$$[Z_{\Gamma_1}, Z_{\Gamma_2}] = Z_{\Gamma_1} \otimes Z_{\Gamma_2} - Z_{\Gamma_2} \otimes Z_{\Gamma_1} \tag{26}$$

with

$$[Z_{\Gamma_1}, Z_{\Gamma_2}] = Z_{\Gamma_2 \star \Gamma_1 - \Gamma_1 \star \Gamma_2}. \tag{27}$$

Here, $\Gamma_i \star \Gamma_j$ sums over all ways of gluing Γ_j into Γ_i, which can be written as

$$\Gamma_i \star \Gamma_j = \sum_{\Gamma} n(\Gamma_i, \Gamma_j, \Gamma)\Gamma. \tag{28}$$

For any $\Gamma \in H$, we have

$$([Z_{\Gamma_1}, Z_{\Gamma_2}], \Gamma) = (Z_{\Gamma_1} \otimes Z_{\Gamma_2} - Z_{\Gamma_2} \otimes Z_{\Gamma_1}, \Delta(\Gamma)), \tag{29}$$

for consistency.

With such section coefficients $n(\Gamma_i, \Gamma_j, \Gamma)$ we then have

$$\Delta(\Gamma) = \sum_{h,g} n(h, g, \Gamma) g \otimes h. \tag{30}$$

The (necessarily finite, as Δ respects the grading) sum is over all graphs h including the empty graph and all monomials in graphs g.

Note that we can regard a graph Γ obtained by inserting Γ_j into Γ_i as an extension

$$0 \to \Gamma_j \to \Gamma \to \Gamma_i \to 0. \tag{31}$$

A proper mathematical discussion of this idea has been given recently by Kobi Kremnizer and Matt Szczesny [**34**].

2.3. Hochschild cohomology. The Hochschild cohomology is encaptured by non-trivial one-cocycles $B_+^\gamma : H \to \mathrm{Aug}(H)$. The one-cocycle condition (see [**8**]) means

$$bB_+^\gamma = 0 \Leftrightarrow \Delta B_+^\gamma(X) = B_+^\gamma(X) \otimes \mathbb{I} + (\mathrm{id} \otimes B_+^\gamma)\Delta(X). \tag{32}$$

We define $\forall \gamma \in \langle \Gamma \rangle$, such that $\Delta'(\gamma) = 0$, linear maps

$$B_+^\gamma(X) := \sum_{\Gamma \in \langle \Gamma \rangle} \frac{\mathbf{bij}(\gamma, X, \Gamma)}{|X|_\vee} \frac{1}{\mathrm{maxf}(\Gamma)} \frac{1}{(\gamma|X)} \Gamma. \tag{33}$$

Here, the sum is over the linear span $\langle \Gamma \rangle$ of generators of H. Furthermore,
i) $\mathrm{maxf}(\Gamma)$ is the number of maximal forests of Γ defined as the integer

$$\mathrm{maxf}(\Gamma) = \sum_{p, \gamma \in \langle \Gamma \rangle, \Delta'(p) = 0} (Z_\gamma, \Gamma')(Z_p, \Gamma''), \tag{34}$$

(we used Sweedler's notation $\Delta(\Gamma) = \Gamma' \otimes \Gamma''$)
ii) $|X|_\vee$ is the number of distinct graphs obtained by permuting external edges of a graph,
iii) $\mathbf{bij}(\gamma, X, \Gamma)$ is the number of bijections between the external edges of X and half-edges of γ such that Γ results,
iv) and $(\gamma|X)$ is the number of insertion places for X in γ.
Finally, for any r which can appear as a residue $\mathbf{res}(\Gamma)$, we define

$$B_+^{r;k} = \sum_{\mathbf{res}(\gamma) = r, |\gamma| = k} \frac{1}{\mathrm{Aut}(\gamma)} B_+^\gamma, \tag{35}$$

which sums over all B_+^γ with a specified external leg structure and loop number, weighted by the rank $\mathrm{Aut}(\gamma)$ of their automorphism groups.

We want to understand these notions. We will do so by going through an example (see [**33**] for a more thorough exploration):

(36) $\Gamma =$  .

We will investigate

(37) B_+  $=?$

and

(38) B_+ 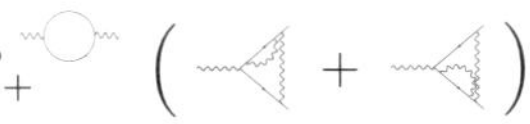 $=?$

Let us start with (37). We have

(39)  $= 1.$

As fermion lines are oriented and hence all external edges distinguished, we cannot permute external edges and obtain a different graph contributing to the same amplitude. Now let us count the bijections.

(40) **bij**  $, X) = 1,$

for all

$X \in$  .

Indeed, to glue the argument X of $B_+^\gamma(X)$ into γ, we identify the factors $X = \prod_i \gamma_i$. The multiset $\text{res}(\gamma_i)$ identifies a number of edges and vertices. From the internal edges and vertices of γ we choose a corresponding set m which contains the same type and number of internal edges and vertices.

We then consider the external edges of elements γ_i of X and count bijections between this set and the similar set defined from m. Summing over all choices of m and counting all bijections at a given m such that Γ is obtained gives **bij** by definition. In the example, there is a unique such bijection for each of the four different graphs X.

(41)  $= 4.$

This counts the number of insertion places.  has two internal vertices and two internal edges, hence four possible choices of an insertion place.

Next, the maximal forests: we count the number of different subsets γ of 1PI subgraphs such that Γ/γ is a primitive element, $\Delta'(\Gamma/\gamma) = 0$.

(42) $\text{maxf}\,(X) = \text{maxf}$ 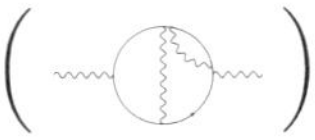 $= 2,$

for any of the four graphs X as above. For each of the four graphs there are two such possibilities. We indicate them in a way which makes the underlying tree structure ([**40, 18**]) obvious:

(43) 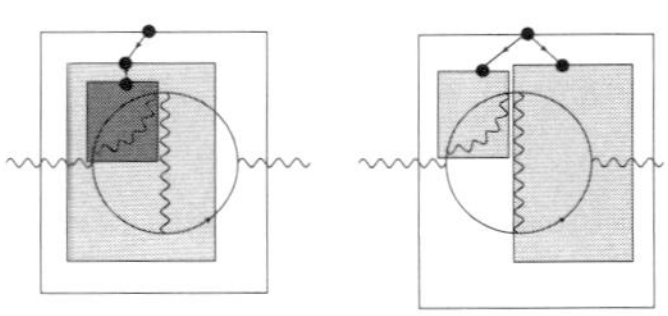 .

This is one major asset of systematically building graphs from images of Hochschild closed one-cocycles: it resolves for us overlapping divergences into rooted trees.

Let us now collect:

(44)

(45)

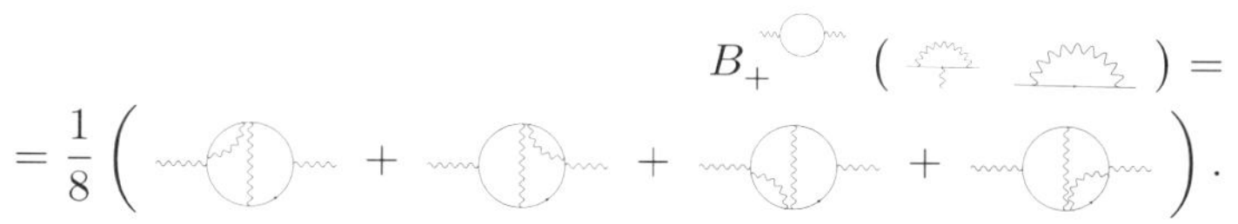

The reader will notice that this fails to satisfy the desired cocycle property. To understand the reason for this failure and the solution to this problem, we turn to (38). We have

(46)

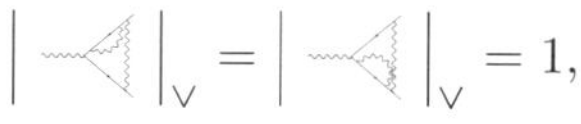

as before.

(47)

(48)

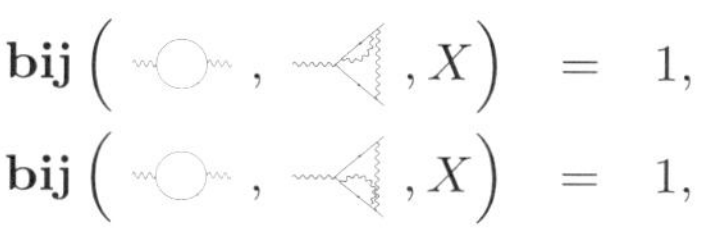

where X can still be any of the four graphs defined above.

Next,

(49)

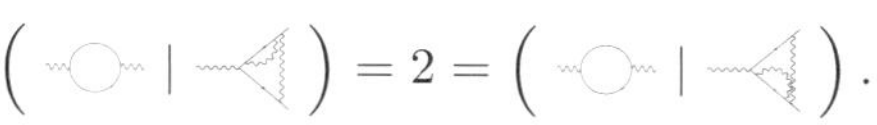

There are now two insertion places for the vertex graph to be inserted into the one-loop photon selfenergy graph.

The maximal forests remain unchanged as we are generating the same graphs X in the two examples. Hence

(50)

(51)

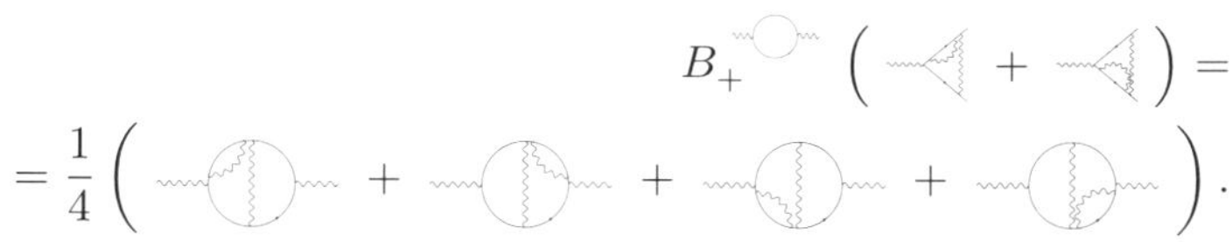

Again, this fails to satisfy the cocycle property. But let us now consider

(52)

(53)

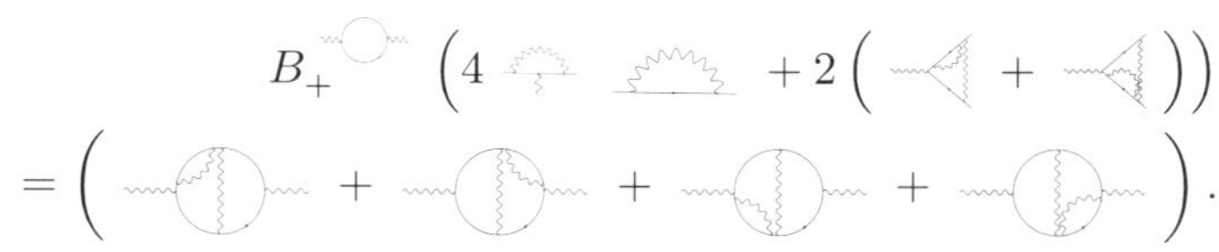

We see that with these weights we do fulfill the cocycle condition. For this, it is actually sufficient that the ratio of the weights is two-to-one. Taking those weights to be four and two gives the result with the proper weights needed in the perturbative expansion of the photon propagator. It was a major result of [**33**] that these weights always work out in field theory such that we do have the desired perturbative expansion and cocycle properties. So while the maps B_+^γ are one-cocycles for Hopf algebras generated by suitable subsets of graphs, one finds that the maps $B_+^{r;k}$ are proper cocycles for a Hopf algebra generated by sums of graphs with given external leg structure and loop number.

So, working out

(54) B_+  () $= \frac{1}{2}$ 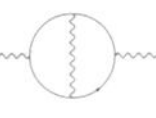 ,

and

(55) B_+  () $= \frac{1}{2}$  ,

we indeed confirm

(56) ΔB_+  $(X) = B_+$  $(X) \otimes \mathbb{I} + \Big(\mathrm{id} \otimes B_+$  $\Big)\Delta(X),$

for

(57) $X = 4$   $+\, 2\Big($  $+$ 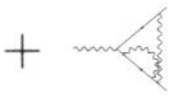 $\Big).$

We will understand soon how the weights 4 and 2 in (53) come about.

As a final exercise the reader might finally wish to confirm

(58) B_+  (2  + 2 ) = 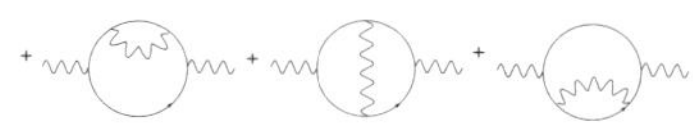 ,

ΔB_+ 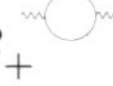 (2  + 2 ) = B_+  (2  + 2 ) $\otimes \mathbb{I}$

$+\ \Big(\mathrm{id} \otimes B_+$ 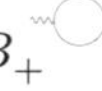 $\Big)\, \Delta$ (2  + 2 ).

To put it shortly:

(59) $$B_+^{\frac{1}{4}F^2;1}(2c_1^{\bar\psi \not A \psi} + 2c_1\bar\psi\psi) = c_2^{\bar\psi \not A \psi},$$

where we indicated the residue **res**() of the one-loop primitive graph  by its corresponding monomial $\frac{1}{4}F^2$ in the Lagrangian of QED, and there is indeed only one primitive at first loop order,

(60) $B_+^{\frac{1}{4}F^2;1} = B_+$  .

2.4. Sub-Hopf algebras. In the example above, we looked at the sum of all 1-PI graphs contributing to a chosen amplitude r at a given loop order k . This gives us Hopf algebra elements $c_k^r \in H^k$ as particular linear combinations of degree-homogeneous elements. Such Hopf algebra elements generate a sub-Hopf algebra.

For example in QED we have

$$(61)\quad \Delta'(c_k^{\bar\psi \not A \psi}) = \sum_{j=1}^{k-1}\left[(2(k-j)+1)c_j^{\bar\psi \not A \psi} + 2(k-j)c_j^{\bar\psi\psi} + (k-j)c_j^{\frac{1}{4}F^2}\right] \otimes c_{k-j}^{\bar\psi \not A \psi} + \text{terms nonlinear on the lhs}$$

$$(62)\quad \Delta'(c_k^{\bar\psi\psi}) = \sum_{j=1}^{k-1}\left[(2(k-j))c_j^{\bar\psi \not A \psi} + (2(k-j)-1)c_j^{\bar\psi\psi} + (k-j)c_j^{\frac{1}{4}F^2}\right] \otimes c_{k-j}^{\bar\psi\psi} + \text{terms nonlinear on the lhs}$$

$$(63)\quad \Delta'(c_k^{\frac{1}{4}F^2}) = \sum_{j=1}^{k-1}\left[(2(k-j))c_j^{\bar\psi \not A \psi} + (2(k-j)-1)c_j^{\bar\psi\psi} + (k-j)c_j^{\frac{1}{4}F^2}\right] \otimes c_{k-j}^{\frac{1}{4}F^2} + \text{terms nonlinear on the lhs.}$$

We omit to give explicit expressions for the terms nonlinear on the lhs of the co-product. They are not really needed, as we will soon see when we study the Dynkin operator. Similar to these sub-Hopf algebras, one can determine the corresponding quotient Lie algebras.

2.5. Co-ideals. Often, sub-Hopf algebras like the above only emerge when divided by suitable co-ideals. An immediate application is a derivation of Ward–Takahashi and Slavnov–Taylor identities in this context [**33, 43**]. Lifting the idea of capturing relations between Green functions to the core Hopf algebra leads to the celebrated BCFW recursion relations [**28**]. All this needs much further work. The upshot is that dividing by a suitable co-ideal I, Feynman rules $\Phi : H \to \mathbb{C}$ can be well-formulated as maps

$$(64)\qquad \Phi : H/I \to \mathbb{C}.$$

Let us consider as an example (following van Suijlekom [**44**]) the ideal and co-ideal I in QED given by

$$(65)\qquad i_k := c_k^{\bar\psi \not A \psi} + c_k^{\bar\psi\psi} = 0,\ \forall k > 0.$$

So, for example,

(66) $i_1 =$  $+$  .

For I to be a co-ideal we need

$$(67)\qquad \Delta(I) \subset (H \otimes I) \oplus (I \otimes H).$$

Let us look at $\Delta(i_2)$ for an example:

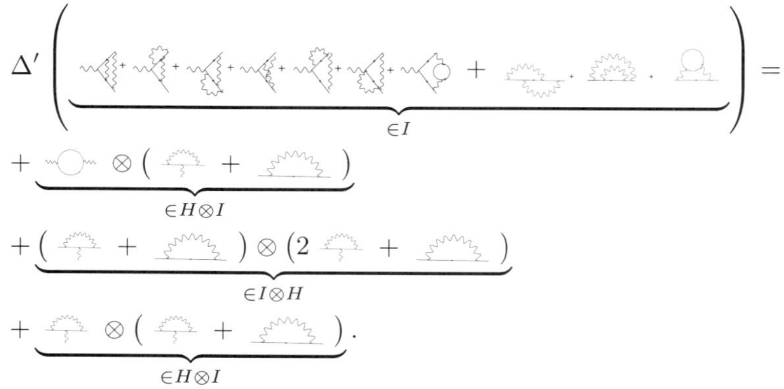

For a thorough discussion of the role of co-ideals and their interplay with Hochschild cohomology in renormalization and core Hopf algebras, see [**28**] and references there.

2.6. Co-radical filtration and the Dynkin operator. For our graded commutative Hopf algebras H there is a co-radical filtration. We consider iterations $[\Delta']^k : H \to \mathrm{Aug}(H)^{\otimes(k+1)}$ of the map $\Delta' : H \to \mathrm{Aug}(H) \otimes \mathrm{Aug}(H)$, and filter Hopf algebra elements by the smallest integer k such that they lie in the kernel of such a map. We can write the Hopf algebra as a direct sum over the corresponding graded spaces $H^{[j]}$,

$$H = \bigoplus_{j=0}^{\infty} H^{[j]}. \tag{68}$$

Elements $q\mathbb{I}$ are in $H^{[0]}$, primitive elements are in $H^{[1]}$, and so on.

A Hochschild one-cocycle is now a map

$$B_+^\gamma : H^{[j]} \to H^{[j+1]}. \tag{69}$$

Note that for example in $H^{[2]}$,

$$B_+^x(\mathbb{I})B_+^y(\mathbb{I}) = B_+^x \circ B_+^y(\mathbb{I}) + B_+^y \circ B_+^x(\mathbb{I}), \tag{70}$$

with the difference between the lhs and the rhs being an element in $H^{[1]}$.

In [**11**] this was used to reduce the study of renormalization theory to the study of flags of subdivergent sectors. This is closely connected to the Dynkin operator [**16, 30, 42**]

$$D : H \to \langle\Gamma\rangle,\ D := S \star Y = m(S \otimes Y)\Delta. \tag{71}$$

Here, $Y(\Gamma) = |\Gamma|\Gamma$ for all homogeneous elements, extended by linearity.

Indeed, the above difference can be calculated as

$$\begin{aligned} D(B_+^x \circ B_+^y(\mathbb{I}) + B_+^y \circ B_+^x(\mathbb{I})) &= (|x| + |y|)\left(B_+^x \circ B_+^y(\mathbb{I}) + B_+^y \circ B_+^x(\mathbb{I})\right. \\ &\quad \left. - B_+^x(\mathbb{I})B_+^y(\mathbb{I})\right). \end{aligned} \tag{72}$$

In physics, this leads to the next-to-leading log expansion, see [**16**], upon recognising that the Feynman rules send elements in H to polynomials in suitable variables $L = \ln q^2/\mu^2$, say, such that elements in $H^{[k]}$ are mapped to the terms $\sim L^k$.

There is an interesting remark to be made concerning the fact that the Dynkin operator vanishes on products. This allows for all things concerning renormalization (including for example the derivation of the renormalization group [**20**]) to rely on a linearized coproduct

$$\Delta_{\text{lin}} := (P_{\text{lin}} \otimes \text{id})\Delta : H \to H \otimes H, \tag{73}$$

with $P_{\text{lin}} : H \to \langle \Gamma \rangle$ the projector into the linear span of generators.

Obviously, this is not a coassociative map.

$$(\Delta_{\text{lin}} \otimes \text{id})\Delta_{\text{lin}} \neq (\text{id} \otimes \Delta_{\text{lin}})\Delta_{\text{lin}}. \tag{74}$$

To control this loss of associativity is a fascinating task on which we hope to report in the future.

2.7. Unitarity of the S-matrix. A fact which will need much more attention in the future from the viewpoint of mixed Hodge structures is the fact that Feynman amplitudes are boundary values of analytic functions. We hence have dispersion relations available, and can relate, in the spirit of the Cutkosky rules, branchcut ambiguities to cuts on diagrams.

In particular, following the guidance of the core Hopf algebra whose primitives are the one-loop cycles in the graph, the structure of the following matrix should reveal the desired relation between Feynman amplitudes and (variations of) mixed Hodge structures.

Actually, let us study a simple example where the renormalization Hopf algebra suffices (as the extra co-graphs in the core Hopf algebra would all be tadpoles [**27**]):

(75) 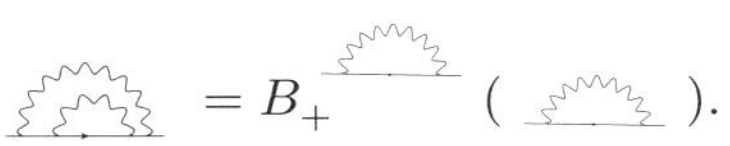

Then, the two-particle cuts on $\Gamma :=$ 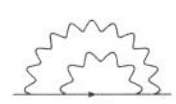 are given by the two-particle cuts on the primtives appearing in the one-cocycles:

(76) 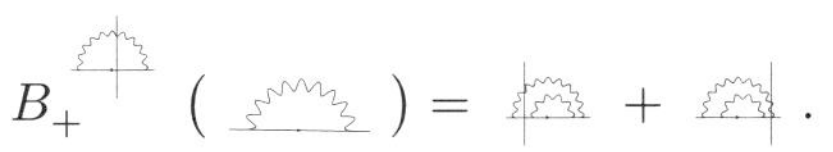

The whole imaginary part can be obtained from this plus the three-particle cut  .

This can be combined into a nice matrix M^Γ which indeed suggests to study the connection to mixed Hodge structures more deeply.

(77) 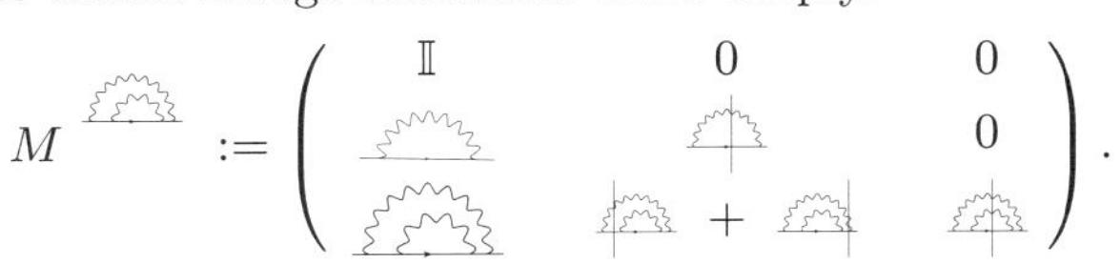

In each column we cut one loop at a time, such that suitable linear combinations of columns will express the branchcut ambiguities of the first column.

We hope that such matrices will come in handy in an attempt to deepen the connection between Hodge theory and quantum fields, which started with the study of limiting mixed Hodge structures and renormalization in a recent collaboration between Spencer Bloch and the author [**11**]. While there it was the nilpotent orbit theorem which was at work in the background, we hope that the reader gets an idea from the above how we hope to further the connection to Hodge structures. This hopefully succeeds in giving a precise mathematical backing to renormalizability and unitarity simultaneously, a treat notoriously missing in all attempts at quantum field theory (and gauge theories in particular) at present.

2.8. Fix-point equations. Let us finish this paper by listing the final fix-point equations (we give them for QED, and refer the reader to [**33, 43, 28**] for the general case) which generate the whole Feynman graph expansion of QED. We distinguish between the two formfactors of the massive fermion, $m\bar{\psi}\psi$ for its mass and $\bar{\psi}\partial\!\!\!/\psi$ for its wave function renormalization. Let

$$\mathcal{R}_{\mathrm{QED}} := \{\bar{\psi}\partial\!\!\!/\psi, m\bar{\psi}\psi, \bar{\psi}A\!\!\!/\psi, \frac{1}{4}F^2\}. \tag{78}$$

Then

$$X^r(\alpha) = \mathbb{I} \pm \sum_{k=1}^{\infty} \alpha^k B_+^{r;k}(X^r(\alpha)Q^{2k}(\alpha)), \tag{79}$$

where we take the plus sign for $r = \bar{\psi}A\!\!\!/\psi$ and the minus sign otherwise, if r corresponds to an edge. We let

$$Q = \frac{X^{\bar{\psi}A\!\!\!/\psi}}{X^{\bar{\psi}\partial\!\!\!/\psi}\sqrt{X^{\frac{1}{4}F^2}}}. \tag{80}$$

Upon evaluation by renormalized Feynman rules this delivers the invariant charge of QED. The resulting maps $B_+^{r;k}$ are Hochschild closed:

$$bB_+^{i,K} = 0. \tag{81}$$

Dividing by the (co-)ideal I simplifies Q;

$$Q = \frac{1}{\sqrt{X^{\frac{1}{4}F^2}}}. \tag{82}$$

See for example [**31**] for a far-reaching application of these techniques in QED.

Let us finally mention that upon adding suitable exact terms, $B_+^{r;k} \to B_+^{r;k} + L_0^{r;k}$ with $L_0^{r;k} = b\phi^{r,k}$, b being the Hochschild differential $b^2 = 0$, $\phi^{r,k} : H \to \mathbb{C}$, we can capture the change of parameters in the Feynman rules by suitable such coboundaries [**41**].

References

[1] E. N. Argyres, A. F. W. van Hameren, R. H. P. Kleiss and C. G. Papadopoulos, *Zero-dimensional field theory,* Eur. Phys. J. C **19** (2001) 567 [arXiv:hep-ph/0101346].

[2] Belkale, P., and Brosnan, P.: *Matroids, motives, and a conjecture of Kontsevich,* Duke Math. J. 116 (2003), no. 1, 147188.

[3] M. P. Bellon and F. A. Schaposnik, *Renormalization group functions for the Wess-Zumino model: up to 200 loops through Hopf algebras,* Nucl. Phys. B **800** (2008) 517.

[4] M. Bellon, *Approximate differential equations for renormalization group functions,* arXiv:0907.2296 [hep-th].

[5] C. Bergbauer, Diploma Thesis: *Epstein-Glaser renormalization, the Hopf algebra of rooted trees and the Fulton-MacPherson compactification of configuration spaces,* FU Berlin, available upon request.

[6] C. Bergbauer, Dissertation: *Combinatorial and geometric aspects of Feynman graphs and Feynman integrals,* FU Berlin 2009, (`http://www.diss.fu-berlin.de/diss/receive/FUDISS_thesis_000000010972`), C. Bergbauer, R. Brunetti, D. Kreimer, in preparation.

[7] C. Bergbauer and D. Kreimer, *The Hopf algebra of rooted trees in Epstein-Glaser renormalization,* Annales Henri Poincaré **6** (2005) 343 [arXiv:hep-th/0403207].

[8] C. Bergbauer and D. Kreimer, *Hopf algebras in renormalization theory: Locality and Dyson-Schwinger equations from Hochschild cohomology,* IRMA Lect. Math. Theor. Phys. **10** (2006) 133 [arXiv:hep-th/0506190].

[9] S. Bloch, *Motives associated to graphs*, Jpn. J. Math. 2 (2007), no. 1, 165196
[10] S. Bloch, H. Esnault and D. Kreimer, *On Motives associated to graph polynomials,* Commun. Math. Phys. **267** (2006) 181 [arXiv:math/0510011].
[11] S. Bloch and D. Kreimer, *Mixed Hodge structures and renormalization in phyiscs*, Comm. in Number Theory and Physics, Vol. 2.4, p.637-718; arXiv:0804.4399 [hep-th].
[12] J.D. Bjorken, S.D. Drell, *Relativistische Quantenfeldtheorie*, B.I. Hochschultaschenbúcher, Band **101** 1967.
[13] C. Bogner and S. Weinzierl, *Periods and Feynman integrals,* J. Math. Phys. **50** (2009) 042302 [arXiv:0711.4863 [hep-th]].
[14] N.N. Bogoliubov, D.V. Shirkov, *Introduction to the theory of quantized fields*, J. Wiley, New York, 1980.
[15] R. Britto, F. Cachazo, B. Feng and E. Witten, *Direct proof Of tree-level recursion relation In Yang-Mills theory,* Phys. Rev. Lett. **94** (2005) 181602 [arXiv:hep-th/0501052]
[16] D. J. Broadhurst and D. Kreimer, *Combinatoric explosion of renormalization tamed by Hopf algebra: 30-loop Padé-Borel resummation,* Phys. Lett. B **475** (2000) 63 [arXiv:hep-th/9912093].
[17] F. Brown, *The massless higher-loop two-point function,* Commun. Math. Phys. **287** (2009) 925 [arXiv:0804.1660 [math.AG]].
[18] A. Connes and D. Kreimer, *Hopf algebras, renormalization and noncommutative geometry,* Commun. Math. Phys. **199** (1998) 203 [arXiv:hep-th/9808042].
[19] A. Connes and D. Kreimer, *Renormalization in quantum field theory and the Riemann-Hilbert problem. I: The Hopf algebra structure of graphs and the main theorem,* Commun. Math. Phys. **210** (2000) 249 [arXiv:hep-th/9912092].
[20] A. Connes and D. Kreimer, *Renormalization in quantum field theory and the Riemann-Hilbert problem. II: The beta-function, diffeomorphisms and the renormalization group,* Commun. Math. Phys. **216** (2001) 215 [arXiv:hep-th/0003188].
[21] Dzmitry Doryn, *Cohomology of graph hypersurfaces associated to certain Feynman graphs,* Ph.D. thesis at the University of Duisburg-Essen 2008, arXiv:0811.0402v1 [math.AG].
[22] K. Ebrahimi-Fard and D. Manchon, *The combinatorics of Bogoliubov's recursion in renormalization,* arXiv:0710.3675 [math-ph].
[23] L. Foissy, *Faà di Bruno subalgebras of the Hopf algebra of planar trees from combinatorial Dyson-Schwinger equations,* arXiv:0707.1204.
[24] S. Hollands and H. Olbermann, *Perturbative Quantum Field Theory via Vertex Algebras,* arXiv:0906.5313 [math-ph].
[25] C. Itzykson, J.-B. Zuber, Quantum Field Theory, McGraw-Hill, 1980
[26] D. Kreimer, *A remark on quantum gravity,* Annals Phys. **323** (2008) 49 [arXiv:0705.3897 [hep-th]].
[27] D. Kreimer, *The core Hopf algebra,* To appear in the proceedings of Noncommutative Geometry Conference in Honor of Alain Connes (NCG 2007), Paris, France, 29 Mar - 6 Apr 2007, Clay Math. Inst., in print. arXiv:0902.1223 [hep-th]
[28] D. Kreimer, W.D. van Suijlekom, *Recursive relations in the core Hopf algebra,* arXiv:0903.2849 [hep-th], Nucl. Phys. **B820** (2009), 682-693, DOI information: 10.1016/j.nuclphysb.2009.04.025.
[29] D. Kreimer and K. Yeats, *Recursion and growth estimates in renormalizable quantum field theory,* Commun. Math. Phys. **279** (2008) 401; [arXiv:hep-th/0612179].
[30] D. Kreimer and K. Yeats, *An étude in non-linear Dyson-Schwinger equations,* Nucl. Phys. Proc. Suppl. **160** (2006) 116 [arXiv:hep-th/0605096].
[31] D. Kreimer, G. Van Baalen, D. Uminsky, K. Yeats, *The QED beta-function from global solutions to Dyson-Schwinger equations,* Annals of Physics (2008) accepted ms, arXiv 0805.0826 [hep-th].
[32] D. Kreimer, G. Van Baalen, D. Uminsky, K. Yeats, *The QCD beta-function from global solutions to Dyson-Schwinger equations,* submitted, arXiv:0906.1754 [hep-th].
[33] D. Kreimer, *Anatomy of a gauge theory,* Annals Phys. **321** (2006) 2757 [arXiv:hep-th/0509135].
[34] K. Kremnizer, M. Szczesny, *Feynman graphs, rooted trees, and Ringel-Hall algebras,* to appear in Comm. Math. Phys., arXiv:0806.1179.
[35] D. Manchon, *Hopf algebras, from basics to applications to renormalization,* arXiv:math/0408405.

[36] Karen Yeats, *Growth estimates for Dyson–Schwinger equations,* Thesis, Boston U. 2008, arXiv:0810.2249.
[37] W. D. van Suijlekom, *Renormalization of gauge fields: A Hopf algebra approach,* Commun. Math. Phys. **276** (2007) 773 [arXiv:hep-th/0610137].
[38] W.D. van Suijlekom, *Representing Feynman graphs on BV-algebras,* arXiv:0807:0999, to appear.
[39] E. W. Nigel Glover and C. Williams, *One-loop gluonic amplitudes from single unitarity cuts,* JHEP **0812** (2008) 067 [arXiv:0810.2964 [hep-th]]; and references there.
[40] D. Kreimer, *On the Hopf algebra structure of perturbative quantum field theories,* Adv. Theor. Math. Phys. **2** (1998) 303 [arXiv:q-alg/9707029].
[41] D. Kreimer, in preparation.
[42] F. Patras, *Dynkin operators and renormalization group actions in pQFT,* http://hal.archives-ouvertes.fr/hal-00341511/fr/.
[43] W. D. van Suijlekom, *Renormalization of gauge fields: A Hopf algebra approach,* Commun. Math. Phys. **276** (2007) 773 [arXiv:hep-th/0610137].
[44] W. van Suijlekom, *The Hopf algebra of Feynman graphs in QED,* Lett. Math. Phys. **77** (2006) 265 [arXiv:hep-th/0602126].
[45] A. S. Wightman, *Quantum field theory in terms of vacuum expectation values,* Phys. Rev. **101** (1956) 860.

IHES, 35 RTE. DE CHARTRES, 91440 BURES-SUR-YVETTE, FRANCE, AND BOSTON UNIVERSITY
E-mail address: `kreimer@ihes.fr`

Clay Mathematics Proceedings
Volume 12, 2010

Pseudodifferential Operators and Regularized Traces

Matthias Lesch

Abstract. This is a survey on trace constructions on various operator algebras with an emphasis on regularized traces on algebras of pseudodifferential operators. For motivation our point of departure is the classical Hilbert space trace which is the unique semifinite *normal* trace on the algebra of bounded operators on a separable Hilbert space. Dropping the normality assumption leads to the celebrated Dixmier traces.

Then we give a leisurely introduction to pseudodifferential operators. The parameter dependent calculus is emphasized and it is shown how this calculus leads naturally to the asymptotic expansion of the resolvent trace of an elliptic differential operator.

The Hadamard partie finie regularization of an integral is explained and used to extend the Hilbert space trace to the Kontsevich-Vishik canonical trace on pseudodifferential operators of non–integral order.

Then the stage is well prepared for the residue trace of Wodzicki-Guillemin and its purely functional analytic interpretation as a Dixmier trace by Alain Connes.

We also discuss existence and uniqueness of traces for the algebra of parameter dependent pseudodifferential operators; the results are surprisingly different.

Finally, we will discuss the analogue of the regularized traces on the symbolic level and study the de Rham cohomology of $\mathbb{R}^n$ with coefficients being symbol functions. This generalizes a recent result of S. Paycha concerning the characterization of the Hadamard partie finie integral and the residue integral in light of the Stokes property.

Contents

2010 *Mathematics Subject Classification.* Primary 58J42; Secondary 58J40, 58J35, 58B34.
Key words and phrases. pseudodifferential operator, noncommutative residue, Dixmier trace.
The author gratefully acknowledges the hospitality of the Department of Mathematics at The University of Colorado at Boulder where this paper was written.
The author was partially supported by the Hausdorff Center for Mathematics (Bonn).

List of Figures

1. Introduction

Traces on an algebra are important linear functionals which come up in various incarnations in various branches of mathematics, e.g. group characters, norm and trace in field extensions, many trace formulas, to mention just a few.

On a separable Hilbert space $\mathscr{H}$ there is a canonical trace (tracial weight, see Section 2) Tr defined on non–negative operators by

$$\operatorname{Tr}(T) := \sum_{j=0}^{\infty} \langle Te_j, e_j \rangle, \tag{1.1}$$

where $(e_j)_{j\geq 0}$ is an orthonormal basis. This is the unique semifinite normal trace on the algebra $\mathscr{B}(\mathscr{H})$ of bounded operators on $\mathscr{H}$. In the 1930's Murray and von Neumann [**MuvN36**], [**MuvN37**], [**vN40**], [**MuvN43**] studied traces on weakly closed $*$–subalgebras (now known as von Neumann algebras) of $\mathscr{B}(\mathscr{H})$. They showed that on a von Neumann *factor* there is up to a normalization a unique semifinite normal trace.

Guillemin [**Gui85**] and Wodzicki [**Wod84**], [**Wod87**] discovered independently that a similar uniqueness statement holds for the algebra of pseudodifferential operators on a compact manifold. The *residue trace*, however, has nothing to do with the Hilbert space trace: it vanishes on trace class operators.

In the 60s Dixmier [**Dix66**] had already proved that the uniqueness statement for the Hilbert space trace fails if one gives up the assumption that the trace is normal.

In the late 80's and early 90's then the Dixmier trace had a celebrated comeback when Alain Connes [**Con88**] proved that in important cases the residue trace coincides with a Dixmier trace.

The aim of this note is to survey some of these results. We will not touch von Neumann algebras, however, any further.

The paper is organized as follows:

In Section 2 our point of departure is the classical Hilbert space trace. We give a short proof that it is up to a factor the unique normal tracial weight on the algebra $\mathscr{B}(\mathscr{H})$ of bounded operators on a separable Hilbert space $\mathscr{H}$.

Then we reproduce Dixmier's very elegant construction which shows that non–normal tracial weights are abundant. We do confine ourselves however to those Dixmier traces which will later turn out to be related to the residue trace.

Section 3 presents the basic calculus of pseudodifferential operators with parameter on a closed manifold.

In Section 4 we pause the discussion of pseudodifferential operators and look at the problem of extending the Hilbert space trace to pseudodifferential operators of higher order. A pseudodifferential operator A of order $< -\dim M$ on a closed manifold M is of trace class and its trace is given by integrating is Schwartz kernel $k_A(x,y)$ over the diagonal

$$\mathrm{Tr}(A) = \int_M k_A(x,x)dx. \tag{1.2}$$

We will show that the classical Hadamard partie finie regularization of integrals allows to extend Eq. (1.2) to all pseudodifferential operators of non–integral order. This is the celebrated Kontsevich-Vishik canonical trace.

Section 5 on asymptotic analysis then shows how the parameter dependent pseudodifferential calculus leads naturally to the asymptotic expansion of the resolvent trace of an elliptic differential operator. For the resolvent of elliptic pseudodifferential operators a refinement, due to Grubb and Seeley, of the parametric calculus is necessary. Without going into the details of this refined calculus we will explain why additional $\log\lambda$ terms appear in the asymptotic expansion of $\mathrm{Tr}(B(P-\lambda)^{-N})$ if B or P are pseudodifferential rather than differential operators. These $\log\lambda$ terms are at the heart of the noncommutative residue trace. The straightforward relations between the resolvent expansion, the heat trace expansion and the meromorphic continuation of the ζ–function, which are based on the Mellin transform respectively a contour integral method, are also briefly discussed.

In Section 6 we state the main result about the existence and uniqueness of the residue trace. We present it in a slightly generalized form due to the author for log–polyhomogeneous pseudodifferential operators. A formula for the relation between the residue trace of a power of the Laplacian and the Einstein–Hilbert action due to KALAU–WALZE [**KaWa95**] and KASTLER [**Kas95**] is proved in an example.

Then we give a proof of Connes' Trace Theorem which states that on pseudodifferential operators of order minus $\dim M$ on a closed manifold M the residue trace is proportional to the Dixmier trace.

Having seen the significance of the parameter dependent calculus it is natural to ask whether the algebras of parameter dependent pseudodifferential operators have an analogue of the residue trace. Somewhat surprisingly the results for these algebras are quite different: there are many traces on this algebra, however, there is a unique symbol–valued trace from which many other traces can be derived. This result resembles very much the center valued trace in von Neumann algebra theory. Furthermore, in contrast to the non–parametric case the L^2–Hilbert space trace extends to a trace on the whole algebra. This part of the paper surveys results from a joint paper with MARKUS J. PFLAUM [**LePf00**].

Finally, in the short Section 7 we will discuss the analogue of the regularized traces on the symbolic level and announce a generalization of a recent result of S. Paycha concerning the characterization of the Hadamard partie finie integral and the residue integral in light of the Stokes property. The result presented here allows one to calculate de Rham cohomology groups of forms on $\mathbb{R}^n$ whose coefficients lie in a certain symbol space. We will show that both the Hadamard partie finie integral and the residue integral provide an integration along the fiber on the cone $\mathbb{R}_+^* \times M$ and as a consequence there is an analogue of the Thom isomorphism.

ACKNOWLEDGMENTS. I would like to thank the organizers of the conference on Motives, Quantum Field Theory and Pseudodifferential Operators for inviting me to contribute these notes. Also I would like to thank the anonymous referee for taking his job very seriously and for making very detailed remarks on how to improve the paper. I think the paper has benefited considerably from those remarks.

2. The Hilbert space trace (tracial weight)

2.1. Basic definitions. Let $\mathscr{H}$ be a separable Hilbert space. Denote by $\mathscr{B}(\mathscr{H})$ the algebra of bounded operators on $\mathscr{H}$. Let $\mathscr{A}$ be a C^*–subalgebra, that is, a norm closed self–adjoint ($a \in \mathscr{A} \Rightarrow a^* \in \mathscr{A}$) subalgebra. It follows that $\mathscr{A}$ is invariant under continuous functional calculus, e.g. if $a \in \mathscr{A}$ is non–negative then $\sqrt{a} \in \mathscr{A}$.

Denote by $\mathscr{A}_+ \subset \mathscr{A}$ the set of non–negative elements. $\mathscr{A}_+$ is a cone in the following sense:

(1) $T \in \mathscr{A}_+, \lambda \in \mathbb{R}_+ \Rightarrow \lambda T \in \mathscr{A}_+$,
(2) $S, T \in \mathscr{A}_+, \lambda, \mu \in \mathbb{R}_+ \Rightarrow \lambda S + \mu T \in \mathscr{A}_+$.

A *weight* on $\mathscr{A}$ is a map

$$\tau : \mathscr{A}_+ \longrightarrow \mathbb{R}_+ \cup \{\infty\}, \quad \mathbb{R}_+ := [0, \infty), \tag{2.1}$$

such that

$$\tau(\lambda S + \mu T) = \lambda\tau(S) + \mu\tau(T), \quad \lambda, \mu \geq 0, \ S, T \in \mathscr{A}_+. \tag{2.2}$$

A weight is called *tracial* if

$$\tau(TT^*) = \tau(T^*T), \quad T \in \mathscr{A}_+. \tag{2.3}$$

It follows from (2.3) that for a unitary $U \in \mathscr{A}$ and $T \in \mathscr{A}_+$

$$\tau(UTU^*) = \tau((UT^{1/2})(UT^{1/2})^*) = \tau((UT^{1/2})^*(UT^{1/2})) = \tau(T). \tag{2.4}$$

(2.2) implies that τ is monotone in the sense that if $0 \leq S \leq T$ then

$$\tau(T) = \tau(S) + \tau(T - S) \geq \tau(S). \tag{2.5}$$

REMARK 2.1. In the literature tracial weights are often just called traces. We adopt here the convention of KADISON and RINGROSE [**KaRi97**, Chap. 8].

We reserve the word trace for a linear functional $\tau : \mathscr{R} \longrightarrow \mathbb{C}$ on a $\mathbb{C}$–algebra $\mathscr{R}$ which satisfies $\tau(AB) = \tau(BA)$ for $A, B \in \mathscr{R}$. A priori a tracial weight τ is only defined on the positive cone of $\mathscr{A}$ and it may take the value ∞. Below we will see that there is a natural ideal in $\mathscr{A}$ on which τ is a trace.

2.1.1. *The canonical tracial weight on bounded operators on a Hilbert space.* Let $(e_j)_{j \in \mathbb{Z}_+}$ be an orthonormal basis of the Hilbert space $\mathscr{H}$; $\mathbb{Z}_+ := \{0, 1, 2, \ldots\}$. For $T \in \mathscr{B}_+(\mathscr{H})$ put

$$\operatorname{Tr}(T) := \sum_{j=0}^{\infty} \langle Te_j, e_j \rangle. \tag{2.6}$$

$\operatorname{Tr}(T)$ is indeed independent of the choice of the orthonormal basis and it is a tracial weight on $\mathscr{B}(\mathscr{H})$ (PEDERSEN [**Ped89**, Sec. 3.4]).

2.1.2. *Trace ideals.* We return to the general set–up of a tracial weight on a C^*–subalgebra $\mathscr{A} \subset \mathscr{B}(\mathscr{H})$. Put

$$\mathscr{L}^1_+(\mathscr{A},\tau) := \big\{T \in \mathscr{A}_+ \;\big|\; \tau(T) < \infty\big\} \tag{2.7}$$

and denote by $\mathscr{L}^1(\mathscr{A},\tau)$ the linear span of $\mathscr{L}^1_+(\mathscr{A},\tau)$. Furthermore, let

$$\mathscr{L}^2(\mathscr{A},\tau) := \big\{T \in \mathscr{A} \;\big|\; \tau(T^*T) < \infty\big\}. \tag{2.8}$$

Using the inequality

$$\begin{aligned}(S+T)^*(S+T) &\leq (S+T)^*(S+T) + (S-T)^*(S-T)\\ &= 2(S^*S + T^*T)\end{aligned} \tag{2.9}$$

and the polarization identity

$$4T^*S = \sum_{k=0}^{3} i^k (S + i^k T)^*(S + i^k T) \tag{2.10}$$

one proves exactly as for the tracial weight Tr in [**Ped89**, Sec. 3.4]:

PROPOSITION 2.2. *$\mathscr{L}^1(\mathscr{A},\tau)$ and $\mathscr{L}^2(\mathscr{A},\tau)$ are two–sided self–adjoint ideals in $\mathscr{A}$.*

Moreover for $T, S \in \mathscr{L}^2(\mathscr{A},\tau)$ one has $TS, ST \in \mathscr{L}^1(\mathscr{A},\tau)$ and

$$\tau(ST) = \tau(TS).$$

The same formula holds for $T \in \mathscr{L}^1(\mathscr{A},\tau)$ and $S \in \mathscr{B}(\mathscr{H})$.

In particular $\tau \restriction \mathscr{L}^p(\mathscr{A},\tau), p = 1, 2,$ is a trace.

2.2. Uniqueness of Tr **on** $\mathscr{B}(\mathscr{H})$**.** As for finite–dimensional matrix algebras one now shows that up to a normalization there is a unique trace on the ideal of finite rank operators.

LEMMA 2.3. *Let $\mathcal{FR}(\mathscr{H})$ be the ideal of finite rank operators on $\mathscr{H}$. Any trace $\tau : \mathcal{FR}(\mathscr{H}) \longrightarrow \mathbb{C}$ is proportional to* Tr $\restriction \mathcal{FR}(\mathscr{H})$.

PROOF. Let $P, Q \in \mathscr{B}(\mathscr{H})$ be rank one orthogonal projections. Choose $v \in \operatorname{im} P, w \in \operatorname{im} Q$ with $\|v\| = \|w\| = 1$ and put

$$T := \langle v, \cdot\rangle\, w. \tag{2.11}$$

Then $T \in \mathcal{FR}(\mathscr{H})$ and $T^*T = P, TT^* = Q$. Consequently τ takes the same value $\lambda_\tau \geq 0$ on all orthogonal projections of rank one.

If $T \in \mathcal{FR}(\mathscr{H})$ is self–adjoint then $T = \sum_{j=1}^N \mu_j P_j$ with rank one orthogonal projections P_j. Thus

$$\tau(T) = \lambda_\tau \sum_{j=1}^{N} \mu_j = \lambda_\tau \operatorname{Tr}(T). \tag{2.12}$$

Since each $T \in \mathcal{FR}(\mathscr{H})$ is a linear combination of self–adjoint elements of $\mathcal{FR}(\mathscr{H})$ we reach the conclusion. □

The properties of Tr we have mentioned so far are not sufficient to show that a tracial weight on $\mathscr{B}(\mathscr{H})$ is proportional to Tr. The property which implies this is *normality*:

PROPOSITION 2.4. 1. Tr *is* normal, *that is, if* $(T_n)_{n\in\mathbb{Z}_+} \subset \mathscr{B}_+(\mathscr{H})$ *is an increasing sequence with* $T_n \to T \in \mathscr{B}_+(\mathscr{H})$ *strongly then* $\operatorname{Tr}(T) = \sup_{n\in\mathbb{Z}_+} \operatorname{Tr}(T_n)$.

2. *Let* τ *be a normal tracial weight on* $\mathscr{B}(\mathscr{H})$. *Then there is a constant* $\lambda_\tau \in \mathbb{R}_+ \cup \{\infty\}$ *such that for* $T \in \mathscr{B}_+(\mathscr{H})$ *we have* $\tau(T) = \lambda_\tau \operatorname{Tr}(T)$.

REMARK 2.5. In the somewhat pathological case $\lambda = \infty$ the tracial weight τ_∞ is given by

$$\tau_\infty(T) = \begin{cases} \infty, & T \in \mathscr{B}_+(\mathscr{H}) \setminus \{0\}, \\ 0, & T = 0. \end{cases} \tag{2.13}$$

In all other cases τ is *semifinite*, that means for $T \in \mathscr{B}_+(\mathscr{H})$ there is an increasing sequence $(T_n)_{n\in\mathbb{Z}_+}$ with $\tau(T_n) < \infty$ and $T_n \nearrow T$ strongly. Here, T_n may be chosen of finite rank.

PROOF. 1. Let $(e_k)_{k\in\mathbb{Z}_+}$ be an orthonormal basis of $\mathscr{H}$. Since $T_n \to T$ strongly we have $\langle T_n e_k, e_k\rangle \nearrow \langle T e_k, e_k\rangle$. The Monotone Convergence Theorem for the counting measure on $\mathbb{Z}_+$ then implies

$$\operatorname{Tr}(T) = \sum_{k=0}^{\infty} \langle T e_k, e_k\rangle = \sup_{n\in\mathbb{Z}_+} \sum_{k=0}^{\infty} \langle T_n e_k, e_k\rangle = \sup_{n\in\mathbb{Z}_+} \operatorname{Tr}(T_n). \tag{2.14}$$

2. Let $\tau : \mathscr{B}_+(\mathscr{H}) \longrightarrow \mathbb{R}_+ \cup \{\infty\}$ be a normal tracial weight. As in the proof of Lemma 2.3 one shows that $\tau \restriction \mathcal{FR}(\mathscr{H}) = \lambda_\tau \operatorname{Tr} \restriction \mathcal{FR}(\mathscr{H})$ for some $\lambda_\tau \in \mathbb{R}_+ \cup \{\infty\}$.

Choose an increasing sequence of orthogonal projections $(P_n)_{n\in\mathbb{Z}_+}$, $\operatorname{rank} P_n = n$. Given $T \in \mathscr{B}_+(\mathscr{H})$ the sequence of finite rank operators $(T^{1/2} P_n T^{1/2})_{n\in\mathbb{Z}_+}$ is increasing and it converges strongly to T. Since τ is assumed to be normal we thus find

$$\begin{aligned} \tau(T) &= \sup_{n\in\mathbb{Z}_+} \tau(T^{1/2} P_n T^{1/2}) \\ &= \sup_{n\in\mathbb{Z}_+} \lambda_\tau \operatorname{Tr}(T^{1/2} P_n T^{1/2}) = \lambda_\tau \operatorname{Tr}(T). \end{aligned}$$ □

REMARK 2.6. The uniqueness of the trace Tr we presented here is in fact a special case of a rich theory of traces for weakly closed self–adjoint subalgebras of $\mathscr{B}(\mathscr{H})$ (von Neumann algebras) due to MURRAY and VON NEUMANN [**MuvN36**], [**MuvN37**], [**vN40**], [**MuvN43**].

2.3. The Dixmier Trace. In view of Proposition 2.4 it is natural to ask whether there exist non–normal tracial weights on $\mathscr{B}(\mathscr{H})$. A cheap answer to this question would be to define for $T \in \mathscr{B}_+(\mathscr{H})$

$$\tau(T) := \begin{cases} \operatorname{Tr}(T), & T \in \mathcal{FR}(\mathscr{H}), \\ \infty, & T \notin \mathcal{FR}(\mathscr{H}). \end{cases} \tag{2.15}$$

Then τ is certainly a non–trivial non–normal tracial weight on $\mathscr{B}(\mathscr{H})$.

To make the problem non–trivial, one should ask whether there exists a non–trivial non–normal tracial weight on $\mathscr{B}(\mathscr{H})$ which vanishes on trace class operators. This was answered affirmatively by J. DIXMIER in the short note [**Dix66**]. We briefly describe Dixmier's very elegant argument.

Denote by $\mathscr{K}(\mathscr{H})$ the ideal of compact operators. We abbreviate

$$\mathscr{L}^p(\mathscr{H}) := \mathscr{L}^p(\mathscr{B}(\mathscr{H}), \operatorname{Tr}), \tag{2.16}$$

see Section 2.1.2. A compact operator T is in $\mathscr{L}^1(\mathscr{H})$ if and only if $\sum_{j=1}^{\infty} \mu_j(T) < \infty$. Here $\mu_j(T), j \geq 1$, denotes the sequence of eigenvalues of $|T|$ counted with multiplicity.

By $\mathscr{L}^{(1,\infty)}(\mathscr{H}) \supset \mathscr{L}^1(\mathscr{H})$ one denotes the space of $T \in \mathscr{K}(\mathscr{H})$ for which

$$\sum_{j=1}^{N} \mu_j(T) = O(\log N), \quad N \to \infty.$$

For an operator $T \in \mathscr{L}^{(1,\infty)}(\mathscr{H})$ the sequence

$$\alpha_N(T) := \frac{1}{\log(N+1)} \sum_{j=1}^{N} \mu_j(T), \quad N \geq 1,$$

is thus bounded.

PROPOSITION 2.7 (J. DIXMIER [**Dix66**]). *Let $\omega \in l^\infty(\mathbb{Z}_+ \setminus \{0\})^*$ be a linear functional satisfying*

(1) *ω is a* state, *that is, a positive linear functional with* $\omega(1, 1, \dots) = 1$.
(2) $\omega((\alpha_N)_{N\geq 1}) = 0$ *if* $\lim_{N\to\infty} \alpha_N = 0$.
(3)

$$\omega(\alpha_1, \alpha_2, \alpha_3, \dots) = \omega(\alpha_1, \alpha_1, \alpha_2, \alpha_2, \dots). \tag{2.17}$$

Put for non-negative $T \in \mathscr{L}^{(1,\infty)}(\mathscr{H})$

$$\begin{aligned} \operatorname{Tr}_\omega(T) &:= \omega\Big(\big(\frac{1}{\log(N+1)} \sum_{j=1}^{N} \mu_j(T)\big)_{N\geq 1}\Big) \\ &=: \lim_\omega \frac{1}{\log(N+1)} \sum_{j=1}^{N} \mu_j(T). \end{aligned} \tag{2.18}$$

Then Tr_ω *extends by linearity to a trace on $\mathscr{L}^{(1,\infty)}(\mathscr{H})$. If $T \in \mathscr{L}^1(\mathscr{H})$ is of trace class then* $\operatorname{Tr}_\omega(T) = 0$. *Furthermore,*

$$\operatorname{Tr}_\omega(T) = \lim_{N\to\infty} \frac{1}{\log(N+1)} \sum_{j=1}^{N} \mu_j(T), \tag{2.19}$$

if the limit on the right hand side exists.

Finally, by putting $\operatorname{Tr}_\omega(T) = \infty$ *if $T \in \mathscr{B}_+(\mathscr{H}) \setminus \mathscr{L}^{(1,\infty)}(\mathscr{H})$ one extends* Tr_ω *to $\mathscr{B}_+(\mathscr{H})$ and hence one obtains a non-normal tracial weight on $\mathscr{B}(\mathscr{H})$.*

PROOF. Let us make a few comments on how this result is proved: First the existence of a state ω with the properties (1), (2), and (3) can be shown by a fixed point argument; in this simple case even Schauder's Fixed Point Theorem would suffice. Alternatively, the theory of Cesàro means leads to a more constructive proof of the existence of ω, CONNES [**Con94**, Sec. 4.2.γ].

Next we note that (1) and (2) imply that if $(\alpha_N)_{N\geq 1}$ is convergent then $\omega((\alpha_N)_{N\geq 1}) = \lim_{N\to\infty} \alpha_N$. Thus changing finitely many terms of $(\alpha_N)_{N\geq 1}$ (i.e.

adding a sequence of limit 0) does not change its ω–limit. Together with the positivity of ω this implies

$$\text{if } \alpha_N \leq \beta_N \text{ for } N \geq N_0 \text{ then } \omega((\alpha_N)_{N\geq 1}) \leq \omega((\beta_N)_{N\geq 1}). \tag{2.20}$$

The previously mentioned facts imply furthermore

$$\liminf_{N\to\infty} \alpha_N \leq \omega((\alpha_N)_{N\geq 1}) \leq \limsup_{N\to\infty} \alpha_N. \tag{2.21}$$

Now let $T_1, T_2 \in \mathscr{L}^{(1,\infty)}$ be non–negative operators and put

$$\begin{aligned} \alpha_N &:= \frac{1}{\log(N+1)} \sum_{j=1}^{N} \mu_j(T_1), \quad \beta_N := \frac{1}{\log(N+1)} \sum_{j=1}^{N} \mu_j(T_2), \\ \gamma_N &:= \frac{1}{\log(N+1)} \sum_{j=1}^{N} \mu_j(T_1+T_2). \end{aligned} \tag{2.22}$$

Using the min-max principle one shows the inequalities

$$\sum_{j=1}^{N} \mu_j(T_1+T_2) \leq \sum_{j=1}^{N} \mu_j(T_1) + \mu_j(T_2) \leq \sum_{j=1}^{2N} \mu_j(T_1+T_2), \tag{2.23}$$

cf. HERSCH [**Her61a, Her61b**], thus

$$\gamma_N \leq \alpha_N + \beta_N, \tag{2.24}$$

$$\alpha_N + \beta_N \leq \frac{\log(2N+1)}{\log(N+1)} \gamma_{2N}. \tag{2.25}$$

(2.24) gives $\omega((\gamma_N)_{N\geq 1}) \leq \omega((\alpha_N)_{N\geq 1}) + \omega((\beta_N)_{N\geq 1})$.

The proof of the converse inequality makes essential use of the crucial assumption (2.17). Together with (2.25) and (2.20) we find

$$\begin{aligned} \omega((\alpha_N)_{N\geq 1}) + \omega((\beta_N)_{N\geq 1}) &\leq \omega(\gamma_2, \gamma_4, \gamma_6, \dots) \\ &= \omega(\gamma_2, \gamma_2, \gamma_4, \gamma_4, \dots), \end{aligned} \tag{2.26}$$

so, in view of 2.7 (2), it only remains to remark that

$$\lim_{N\to\infty} (\gamma_{2N} - \gamma_{2N-1}) = 0.$$

Thus Tr_ω is additive on the cone of positive operators. Since $\operatorname{Tr}_\omega(T)$ depends only on the spectrum, it is certainly invariant under conjugation by unitary operators. Now it is easy to see that Tr_ω extends by linearity to a trace on $\mathscr{L}^{(1,\infty)}(\mathscr{H})$. The other properties follow easily. □

3. Pseudodifferential operators with parameter

3.1. From differential operators to pseudodifferential operators. Historically, pseudodifferential operators were invented to understand differential operators. Suppose given a differential operator

$$P = \sum_{|\alpha|\leq d} p_\alpha(x)\, i^{-|\alpha|} \frac{\partial^\alpha}{\partial x^\alpha} \tag{3.1}$$

in an open set $U \subset \mathbb{R}^n$. Representing a function $u \in C_0^\infty(U)$ in terms of its Fourier transform

$$u(x) = \int_{\mathbb{R}^n} e^{i\langle x,\xi\rangle} \hat{u}(\xi) đ\xi, \quad đ\xi = (2\pi)^{-n} d\xi, \tag{3.2}$$

where $\hat{u}(\xi) = \int_{\mathbb{R}^n} e^{-i\langle x,\xi\rangle} u(x)dx$, we find

$$\begin{aligned} Pu(x) &= \int_{\mathbb{R}^n} e^{-i\langle x,\xi\rangle} p(x,\xi)\hat{u}(\xi)đ\xi \\ &= \int_{\mathbb{R}^n} \Big(\int_U e^{i\langle x-y,\xi\rangle} p(x,\xi) u(y) dy \Big) đ\xi \\ &=: \big(\mathrm{Op}(p)u\big)(x). \end{aligned} \tag{3.3}$$

Here

$$p(x,\xi) = \sum_{|\alpha|\le d} p_\alpha(x)\xi^\alpha \tag{3.4}$$

denotes the *complete symbol* of P. The right hand side of (3.3) shows that P is a pseudodifferential operator with complete symbol function $p(x,\xi)$.

Note that $p(x,\xi)$ is a polynomial in ξ. One now considers pseudodifferential operators with more general symbol functions such that inverses of differential operators are included into the calculus. E.g. a first approximation to the resolvent $(P - \lambda^d)^{-1}$ is given by $\mathrm{Op}((p(\cdot,\cdot) - \lambda^d)^{-1})$. For constant coefficient differential operators this is indeed the exact resolvent.

Let us now describe the most commonly used symbol spaces. In view of the resolvent example above we are going to consider symbols with an auxiliary parameter.

3.2. Basic calculus with parameter. We first recall the notion of conic manifolds and conic sets from DUISTERMAAT [**Dui96**, Sec. 2]. A conic manifold is a smooth principal fiber bundle $\Gamma \to B$ with structure group $\mathbb{R}_+^* := (0,\infty)$. It is always trivializable. A subset $\Gamma \subset \mathbb{R}^N \setminus \{0\}$ which is a conic manifold by the natural $\mathbb{R}_+^*$-action on $\mathbb{R}^N \setminus \{0\}$ is called a conic set. The base manifold of a conic set $\Gamma \subset \mathbb{R}^N \setminus \{0\}$ is diffeomorphic to $S\Gamma := \Gamma \cap S^{N-1}$. By a cone $\Gamma \subset \mathbb{R}^N$ we will always mean a conic set or the closure of a conic set in $\mathbb{R}^N$ such that Γ has nonempty interior. Thus $\mathbb{R}^N$ and $\mathbb{R}^N \setminus \{0\}$ are cones, but only the latter is a conic set. $\{0\}$ is a zero–dimensional cone.

3.2.1. *Symbols.* Let $U \subset \mathbb{R}^n$ be an open subset and $\Gamma \subset \mathbb{R}^N$ a cone. A typical example we have in mind is $\Gamma = \mathbb{R}^n \times \Lambda$, where $\Lambda \subset \mathbb{C}$ is an open cone.

We denote by $S^m(U;\Gamma)$, $m \in \mathbb{R}$, the space of symbols of Hörmander type $(1,0)$ (HÖRMANDER [**Hör71**], GRIGIS–SJØSTRAND [**GrSj94**]). More precisely, $S^m(U;\Gamma)$ consists of those $a \in C^\infty(U \times \Gamma)$ such that for multi–indices $\alpha \in \mathbb{Z}_+^n, \gamma \in \mathbb{Z}_+^N$ and compact subsets $K \subset U, L \subset \Gamma$ we have an estimate

$$\big|\partial_x^\alpha \partial_\xi^\gamma a(x,\xi)\big| \le C_{\alpha,\gamma,K,L}(1+|\xi|)^{m-|\gamma|}, \quad x \in K, \xi \in L^c. \tag{3.5}$$

Here $L^c = \{t\xi \mid \xi \in L, t \ge 1\}$. The best constants in (3.5) provide a set of semi-norms which endow $S^\infty(U;\Gamma) := \bigcup_{m\in\mathbb{C}} S^m(U;\Gamma)$ with the structure of a Fréchet algebra. We mention the following variants of the space $S^\bullet$:

3.2.2. *Classical symbols* $\mathrm{CS}^m(U;\Gamma)$. A symbol $a \in \mathrm{S}^m(U;\Gamma)$ is called *classical* if there are $a_{m-j} \in \mathrm{C}^\infty(U \times \Gamma)$ with

$$a_{m-j}(x, r\xi) = r^{m-j} a_{m-j}(x,\xi), \quad r \geq 1, |\xi| \geq 1, \tag{3.6}$$

such that for $N \in \mathbb{Z}_+$

$$a - \sum_{j=0}^{N-1} a_{m-j} \in \mathrm{S}^{m-N}(U;\Gamma). \tag{3.7}$$

The latter property is usually abbreviated $a \sim \sum_{j=0}^{\infty} a_{m-j}$.

Many authors require the functions in (3.6) to be homogeneous everywhere on $\Gamma \setminus \{0\}$. Note however that if $\Gamma = \mathbb{R}^p$ and $f : \Gamma \to \mathbb{C}$ is a function which is homogeneous of degree α then f cannot be smooth at 0 unless $\alpha \in \mathbb{Z}_+$. So such a function is not a symbol in the strict sense. We prefer the functions in the expansion (3.7) to be smooth everywhere and homogeneous only for $r \geq 1$ and $|\xi| \geq 1$.

The space of classical symbols of order m is denoted by $\mathrm{CS}^m(U;\Gamma)$. In view of the asymptotic expansion (3.7) we have $\mathrm{CS}^{m'}(U;\Gamma) \subset \mathrm{CS}^m(U;\Gamma)$ only if $m - m' \in \mathbb{Z}_+$ is a non–negative integer.

3.2.3. log–*polyhomogeneous symbols* $\mathrm{CS}^{m,k}(U;\Gamma)$. $a \in \mathrm{S}^m(U;\Gamma)$ is called log–*polyhomogeneous* (cf. LESCH [**Les99**]) of order (m,k) if it has an asymptotic expansion in $\mathrm{S}^\infty(U;\Gamma)$ of the form

$$a \sim \sum_{j=0}^{\infty} a_{m-j} \quad \text{with } a_{m-j} = \sum_{l=0}^{k} b_{m-j,l}, \tag{3.8}$$

where $a_{m-j} \in \mathrm{C}^\infty(U \times \Gamma)$ and $b_{m-j,l}(x,\xi) = \tilde{b}_{m-j,l}(x,\xi/|\xi|)|\xi|^{m-j}\log^l|\xi|$ for $|\xi| \geq 1$.

By $\mathrm{CS}^{m,k}(U;\Gamma)$ we denote the space of log–polyhomogeneous symbols of order (m,k). Classical symbols are those of log degree 0, i.e. $\mathrm{CS}^m(U;\Gamma) = \mathrm{CS}^{m,k}(U;\Gamma)$.

3.2.4. *Symbols which are holomorphic in the parameter.* If $\Gamma = \mathbb{R}^n \times \Lambda$, where $\Lambda \subset \mathbb{C}$ is a cone one may additionally require symbols to be holomorphic in the Λ variable. This aspect is important if one deals with the resolvent of an elliptic differential operator since the latter depends analytically on the resolvent parameter. This class of symbols is not emphasized in this paper.

3.2.5. *Pseudodifferential operators with parameter.* Fix $a \in \mathrm{S}^m(U;\mathbb{R}^n \times \Gamma)$ (respectively $\in \mathrm{CS}^m(U;\mathbb{R}^n \times \Gamma)$). For each fixed $\mu_0 \in \Gamma$ we have $a(\cdot,\cdot,\mu_0) \in \mathrm{S}^m(U;\mathbb{R}^n)$ (respectively $\in \mathrm{CS}^m(U;\mathbb{R}^n)$) and hence we obtain a family of pseudodifferential operators parametrized over Γ by putting

$$\begin{aligned} \big[\mathrm{Op}(a(\mu_0))\,u\big](x) &:= \big[A(\mu_0)\,u\big](x) \\ &:= \int_{\mathbb{R}^n} e^{i\langle x,\xi\rangle}\, a(x,\xi,\mu_0)\,\hat{u}(\xi)\,đ\xi \\ &= \int_{\mathbb{R}^n}\int_U e^{i\langle x-y,\xi\rangle}\, a(x,\xi,\mu_0)\,u(y)dyđ\xi. \end{aligned} \tag{3.9}$$

Note that the Schwartz kernel $K_{A(\mu_0)}$ of $A(\mu_0) = \mathrm{Op}(a(\mu_0))$ is given by

$$K_{A(\mu_0)}(x,y,\mu_0) = \int_{\mathbb{R}^n} e^{i\langle x-y,\xi\rangle}\, a(x,\xi,\mu_0)\,đ\xi. \tag{3.10}$$

In general the integral is to be understood as an oscillatory integral, for which we refer the reader to [**Shu01**], [**GrSj94**]. The integral exists in the usual sense if $m + n < 0$.

The extension to manifolds and vector bundles is now straightforward, although historically it took quite a while until the theory of singular integral operators had evolved into a theory of pseudodifferential operators on vector bundles over smooth manifolds (CALDERÓN-ZYGMUND [**CaZy57**], SEELEY [**See59, See65**], KOHN-NIRENBERG [**KoNi65**]). For a smooth manifold M and a vector bundle E over M we define the space $\mathrm{CL}^m(M, E; \Gamma)$ of classical parameter dependent pseudodifferential operators between sections of E in the usual way by patching together local data:

DEFINITION 3.1. Let E be a complex vector bundle of finite fiber dimension N over a smooth closed manifold M and let $\Gamma \subset \mathbb{R}^p$ be a cone. A *classical pseudodifferential operator of order m with parameter $\mu \in \Gamma$* is a family of operators $B(\mu) : \Gamma^\infty(M; E) \longrightarrow \Gamma^\infty(M; E)$, $\mu \in \Gamma$, such that locally $B(\mu)$ is given by

$$\big[B(\mu)\, u\big](x) = (2\pi)^{-n} \int_{\mathbb{R}^n} \int_U e^{i\langle x-y,\xi\rangle} b(x, \xi, \mu) u(y) dy d\xi$$

with b an $N \times N$ matrix of functions belonging to $\mathrm{CS}^m(U, \mathbb{R}^n \times \Gamma)$.

$\mathrm{CL}^{m,k}(M, E; \Gamma)$ is defined similarly, although we will discuss $\mathrm{CL}^{m,k}$ only in the non–parametric case. Of course, operators may act between sections of different vector bundles E, F. In that case we write $\mathrm{CL}^{m,k}(M, E, F; \Gamma)$.

REMARK 3.2. 1. In case $\Gamma = \{0\}$ we obtain the usual (classical) pseudodifferential operators of order m on U. Here we write $\mathrm{CL}^m(M, E)$ instead of $\mathrm{CL}^m(M, E; \{0\})$ respectively $\mathrm{CL}^m(M, E, F)$ instead of $\mathrm{CL}^m(M, E, F; \{0\})$.

2. Parameter dependent pseudodifferential operators play a crucial role e.g., in the construction of the resolvent expansion of an elliptic operator (GILKEY [**Gil95**]).

A *pseudodifferential operator with parameter* is more than just a map from Γ to the space of pseudodifferential operators, cf. Corollary 3.8 and Remark 3.9.

To illustrate this let us consider a single elliptic operator $A \in \mathrm{CL}^m(U)$. For simplicity let the symbol $a(x, \xi)$ of A be positive definite. Then we can consider the "parametric symbol" $b(x, \xi, \lambda) = a(x, \xi) - \lambda^m$ for $\lambda \in \Lambda := \mathbb{C} \setminus \mathbb{R}_+$.

However, in general b lies in $\mathrm{CS}^m(U; \Lambda)$ only if A is a differential operator. The reason is that b will satisfy the estimates (3.5) only if $a(x, \xi)$ is polynomial in ξ, because then $\partial_\xi^\beta a(x, \xi) = 0$ if $|\beta| > m$. If $a(x, \xi)$ is not polynomial in ξ, however, (3.5) will in general not hold if $\beta > m$.

This problem led GRUBB and SEELEY [**GrSe95**] to invent their calculus of *weakly parametric* pseudodifferential operators. $b(x, \xi, \lambda) = a(x, \xi) - \lambda^m$ is weakly parametric for any elliptic A with positive definite leading symbol (or more generally if A satisfies Agmon's angle condition). The class of weakly parametric operators is beyond the scope of this survey, however.

3. The definition of the parameter dependent calculus is not uniform in the literature. It will be crucial in the sequel that differentiating by the parameter reduces the order of the operator. This is the convention, e.g. of GILKEY [**Gil95**] but differs from the one in SHUBIN [**Shu01**]. In LESCH–PFLAUM [**LePf00**, Sec. 3] it is shown that parameter dependent pseudodifferential operators can be viewed as translation invariant pseudodifferential operators on $U \times \Gamma$ and therefore our

convention of the parameter dependent calculus contains MELROSE's suspended algebra from [**Mel95**].

PROPOSITION 3.3. $\mathrm{CL}^{\bullet,\bullet}(M,E;\Gamma)$ *is a bi–filtered algebra, that is,*

$$AB \in \mathrm{CL}^{m+m',k+k'}(M,E;\Gamma)$$

for $A\in \mathrm{CL}^{m,k}(M,E;\Gamma)$ *and* $B\in\mathrm{CL}^{m',k'}(M,E;\Gamma)$.

The following result about the L^2–continuity of a parameter dependent pseudodifferential operator is crucial. We denote by $L^2_s(M,E)$ the Hilbert space of sections of E of Sobolev class s.

THEOREM 3.4. *Let* $A\in\mathrm{CL}^m(M,E;\Gamma)$. *Then for fixed* $\mu\in\Gamma$ *the operator* $A(\mu)$ *extends by continuity to a bounded linear operator* $L^2_s(M,E)\longrightarrow L^2_{s-m}(M,E)$, $s\in\mathbb{R}$.

Furthermore, for $m\le 0$ *one has the following uniform estimate in* μ*: for* $0\le\vartheta\le 1, \mu_0\in\Gamma$, *there is a constant* $C(s,\vartheta)$ *such that*

$$\|A(\mu)\|_{s,s+\vartheta|m|}\le C(s,\vartheta,\mu_0)(1+|\mu|)^{-(1-\vartheta)|m|},\quad |\mu|\ge|\mu_0|,\ \mu\in\Gamma.$$

Here $\|A(\mu)\|_{s,s+\vartheta|m|}$ *denotes the norm of the operator* $A(\mu)$ *as a map from the Sobolev space* $L^2_s(M,E)$ *into* $L^2_{s+\vartheta|m|}(M,E)$.

If $\Gamma=\mathbb{R}^n$ then we can omit the μ_0 in the formulation of the Theorem (i.e. $\mu_0=0$). For a proof of Theorem 3.4 see e.g. [**Shu01**, Theorem 9.3].

3.2.6. *The parametric leading symbol.* The leading symbol of a classical pseudodifferential operator A of order m with parameter is now defined as follows: if A has complete symbol $a(x,\xi,\mu)$ with expansion $a\sim\sum\limits_{j=0}^{\infty}a_{m-j}$ then

$$\begin{aligned}\sigma^m_A(x,\xi,\mu)&:=\lim_{r\to\infty}r^{-m}a(x,r\xi,r\mu)\\ &=(|\xi|^2+|\mu|^2)^{m/2}a_m(x,\frac{(\xi,\mu)}{\sqrt{|\xi|^2+|\mu|^2}}).\end{aligned}\tag{3.11}$$

σ^m_A has an invariant meaning as a smooth function on

$$T^*M\times\Gamma\setminus\big\{(x,0,0)\ \big|\ x\in M\big\}$$

which is homogeneous in the following sense:

$$\sigma^m_A(x,r\xi,r\mu)=r^m\sigma^m_A(x,\xi,\mu)\text{ for }(\xi,\mu)\ne(0,0),\ r>0.$$

This symbol is determined by its restriction to the sphere in

$$S(T^*M\times\Gamma)=\big\{(\xi,\mu)\in T^*M\times\Gamma\ \big|\ |\xi|^2+|\mu|^2=1\big\}$$

and there is an exact sequence

$$0\longrightarrow\mathrm{CL}^{m-1}(M;\Gamma)\hookrightarrow\mathrm{CL}^m(M;\Gamma)\xrightarrow{\sigma}C^\infty(S(T^*M\times\Gamma))\longrightarrow 0;\tag{3.12}$$

the vector bundle E being omitted from the notation just to save horizontal space.

EXAMPLE 3.5. Let us look at an example to illustrate the difference between the parametric leading symbol and the leading symbol for a single pseudodifferential operator. Let

$$a(x,\xi)=\sum_{|\alpha|\le m}a_\alpha(x)\xi^\alpha\tag{3.13}$$

be the complete symbol of an elliptic *differential* operator. Then (cf. Remark (3.2) 2.)

$$b(x,\xi,\lambda) = a(x,\xi) - \lambda^m \tag{3.14}$$

is a symbol of a parameter dependent (pseudo)differential operator $B(\lambda)$ with parameter λ in a suitable cone $\Lambda \subset \mathbb{C}$. The parameter dependent leading symbol of B is $\sigma_B^m(x,\xi,\lambda) = a_m(x,\xi) - \lambda^m$ while for fixed λ the leading symbol of the single operator $B(\lambda)$ is $\sigma_{B(\lambda)}^m(x,\xi) = a_m(x,\xi) = \sigma_B^m(x,\xi,\lambda = 0)$.

In fact we have in general:

LEMMA 3.6. *Let $A \in \mathrm{CL}^m(M,E;\Gamma)$ with parameter dependent leading symbol $\sigma_A^m(x,\xi,\mu)$. For fixed $\mu_0 \in \Gamma$ the operator $A(\mu_0) \in \mathrm{CL}^m(M,E)$ has leading symbol $\sigma_{A(\mu_0)}^m(x,\xi) = \sigma_A^m(x,\xi,0)$.*

PROOF. It suffices to prove this locally in a chart U for a scalar operator A. Since the leading symbols are homogeneous it suffices to consider ξ with $|\xi| = 1$.

So suppose that A has complete symbol $a(x,\xi,\mu)$ in U. Write $a(x,\xi,\mu) = a_m(x,\xi,\mu)+\tilde a(x,\xi,\mu)$ with $\tilde a \in \mathrm{CS}^{m-1}(U;\mathbb{R}^n\times\Gamma)$ and $a_m(x,r\xi,r\mu) = r^m a_m(x,\xi,\mu)$ for $r \geq 1$, $|\xi|^2 + |\mu|^2 \geq 1$. Then for fixed $\mu_0 \in \Gamma$ we have $\tilde a(\cdot,\cdot,\mu_0) \in \mathrm{CS}^{m-1}(U;\mathbb{R}^n)$ and hence $\lim_{r\to\infty} r^{-m}\tilde a(x,r\xi,\mu_0) = 0$. Consequently

$$\begin{aligned}\sigma_{A(\mu_0)}^m(x,\xi) &= \lim_{r\to\infty} r^{-m} a_m(x,r\xi,\mu_0)\\ &= \lim_{r\to\infty} a_m(x,\xi,\mu_0/r) = a_m(x,\xi,0).\end{aligned}$$ □

3.2.7. *Parameter dependent ellipticity.* This is now defined as the invertibility of the parametric leading symbol. The basic example of a pseudodifferential operator with parameter is the resolvent of an elliptic differential operator (cf. Remark 3.2 and Example 3.5). The following two results can also be found in [**Shu01**, Section II.9].

THEOREM 3.7. *Let M be a closed manifold and E,F complex vector bundles over M. Let $A \in \mathrm{CL}^m(M,E,F;\Gamma)$ be* elliptic. *Then there exists a $B \in \mathrm{CL}^{-m}(M,F,E;\Gamma)$ such that $AB - I \in \mathrm{CL}^{-\infty}(M,F;\Gamma)$, $BA - I \in \mathrm{CL}^{-\infty}(M,E;\Gamma)$.*

Note that in view of Theorem 3.4 this implies the estimates

$$\|B(\mu)A(\mu) - I\|_{s,t} + \|A(\mu)B(\mu) - I\|_{s,t} \leq C(s,t,N)(1+|\mu|)^{-N} \tag{3.15}$$

for all $s,t \in \mathbb{R}$, $N > 0$. This result has an important implication:

COROLLARY 3.8. *Under the assumptions of Theorem 3.7, for each $s \in \mathbb{R}$ there is a $\mu_0 \in \Gamma$ such that for $|\mu| \geq |\mu_0|$ the operator*

$$A(\mu) : L_s^2(M,E) \longrightarrow L_{s-m}^2(M,F)$$

is invertible.

PROOF. In view of (3.15) there is a $\mu_0 = \mu_0(s)$ such that

$$\|(BA - I)(\mu)\|_s < 1 \text{ and } \|(AB - I)(\mu)\|_{s-m} < 1,$$

for $|\mu| \geq |\mu_0|$ and hence $AB : L_s^2 \longrightarrow L_s^2$ and $BA : L_{s-m}^2 \longrightarrow L_{s-m}^2$ are invertible. □

REMARK 3.9. This result causes an interesting constraint on those pseudodifferential operators which may appear as special values of an elliptic parametric family. Namely, if $A \in \mathrm{CL}^m(M, E, F; \Gamma)$ is parametric elliptic then for each μ the operator $A(\mu) \in \mathrm{CL}^m(M, E, F)$ is elliptic. Furthermore, by the previous Corollary and the stability of the Fredholm index we have $\operatorname{ind} A(\mu) = 0$ for all μ.

4. Extending the Hilbert space trace to pseudodifferential operators

We pause the discussion of pseudodifferential operators and look at the Hilbert space trace Tr on pseudodifferential operators.

4.1. Tr **on operators of order** $< -\dim M$**.** Consider the local situation, i.e. a compactly supported operator $A = \mathrm{Op}(a) \in \mathrm{CL}^{m,k}(U, E)$ in a local chart.

If $m < -\dim M$ then A is trace class and the trace is given by integrating the kernel of A over the diagonal:

$$
\begin{aligned}
\operatorname{Tr}(A) &= \int_U \operatorname{tr}_{E_x}\big(k_A(x,x)\big)dx \\
&= \int_U \int_{\mathbb{R}^n} \operatorname{tr}_{E_x}\big(a(x,\xi)\big) đ\xi dx,
\end{aligned} \tag{4.1}
$$

where we have used (3.10).

The right hand side is indeed coordinate invariant. To explain this consider a coordinate transformation $\kappa : U \to V$. Denote variables in U by x, y and variables in V by $\tilde{x}, \tilde{y}$. It is not so easy to write down the symbol of $\kappa_* A$. However, an amplitude function (these are "symbols" which depend on x and y, otherwise the basic formula (3.9) still holds) for $\kappa_* A$ is given by

$$
(\tilde{x}, \tilde{y}, \xi) \mapsto a(\kappa^{-1}\tilde{x}, \phi(\tilde{x}, \tilde{y})^{-1}\xi) \frac{|\det D\kappa^{-1}(\tilde{x}, \tilde{y})|}{|\det \phi(\tilde{x}, \tilde{y})|}, \tag{4.2}
$$

cf. [**Shu01**, Sec. 4.1, 4.2], where $\phi(\tilde{x}, \tilde{y})$ is smooth with $\phi(\tilde{x}, \tilde{x}) = D\kappa^{-1}(\tilde{x})^t$. Comparing the trace densities in the two coordinate systems requires a *linear* coordinate change in the ξ–variable. Indeed,

$$
\begin{aligned}
\operatorname{Tr}(\kappa_* A) &= \int_V \int_{\mathbb{R}^n} \operatorname{tr}_{E_{\tilde{x}}}\big(a(\kappa^{-1}\tilde{x}, \phi(\tilde{x}, \tilde{x})^{-1}\xi)\big) đ\xi d\tilde{x} \\
&= \int_V \int_{\mathbb{R}^n} \operatorname{tr}_{E_{\tilde{x}}}\big(a(\kappa^{-1}\tilde{x}, \xi)\big) đ\xi \, |\det D\kappa^{-1}(\tilde{x})| d\tilde{x}, \\
&= \int_U \int_{\mathbb{R}^n} \operatorname{tr}_{E_x}\big(a(x, x, \xi)\big) đ\xi \, dx = \operatorname{Tr}(A).
\end{aligned} \tag{4.3}
$$

Therefore, the trace of a pseudodifferential operator $A \in \mathrm{CL}^{m,k}(M, E)$ of order $m < -\dim M =: -n$ on the closed manifold M may be calculated from the complete symbol of A in coordinates as follows. Choose a finite open cover by coordinate neighborhoods $U_j, j = 1, \ldots, r$, and a subordinated partition of unity $\varphi_j, j = 1, \ldots, r$. Furthermore, let $\psi_j \in \mathrm{C}_0^\infty(U_j)$ with $\psi_j \varphi_j = \varphi_j$. Denoting by $a_j(x, \xi)$ the complete symbol in the coordinate system on U_j we obtain

$$
\operatorname{Tr}(A) = \sum_{j=1}^r \operatorname{Tr}(\varphi_j A \psi_j) = \sum_{j=1}^r \int_{U_j} \int_{\mathbb{R}^n} \varphi_j(x) \operatorname{tr}_{E_x}\big(a_j(x, \xi)\big) đ\xi \, dx. \tag{4.4}
$$

A priori the previous argument is valid only for operators of order $m < -n$. However, the symbol function $a_j(x, \xi)$ is rather well–behaved in ξ. If for a class of pseudodifferential operators we can regularize $\int_{\mathbb{R}^n} a_j(x,\xi)đ\xi$ in such a way that the change of variables (4.3) works then indeed (4.4) extends the trace to this class of operators. Such a regularization is provided by:

4.2. The Hadamard partie finie regularized integral. The problem of regularizing divergent integrals is in fact quite old. The method we are going to present here goes back to HADAMARD who used his method to regularize integrals which arose when solving the wave equation [**Had32**].

Given a function $f \in \mathrm{CS}^{m,k}(\mathbb{R}^p)$, e.g. $a(x,\cdot)$ above for fixed x. Then f has an asymptotic expansion

$$f(x) \sim_{|x|\to\infty} \sum_{j=0}^{\infty}\sum_{l=0}^{k} f_{jl}(x/|x|)|x|^{m-j}\log^l |x|. \tag{4.5}$$

Integrating over balls of radius R gives the asymptotic expansion

$$\int_{|x|\le R} f(x)dx \sim_{R\to\infty} \sum_{j=0}^{\infty}\sum_{l=0}^{k+1} \tilde{f}_{jl} R^{m+n-j}\log^l R. \tag{4.6}$$

The *regularized integral* $⨍_{\mathbb{R}^p} f(x)dx$ is, by definition, the constant term in this asymptotic expansion. Some authors call the regularized integral *partie finie integral* or *cut–off integral.*

It has a couple of peculiar properties, cf. [**Mel95**], which were further investigated in [**Les99**, Sec. 5] and [**LePf00**]. The most notable features are a modified change of variables rule for linear coordinate changes and, as a consequence, the fact that Stokes' theorem does not hold in general:

PROPOSITION 4.1. [**Les99**, Prop. 5.2] *Let $A \in \mathrm{GL}(p,\mathbb{R})$ be a regular matrix. Furthermore, let $f \in \mathrm{CS}^{m,k}(\mathbb{R}^p)$ with expansion* (4.5). *Then we have the change of variables formula*

$$\begin{aligned}(4.7)\quad & ⨍_{\mathbb{R}^p} f(A\xi)d\xi \\ &= |\det A|^{-1}\left(⨍_{\mathbb{R}^p} f(\xi)d\xi + \sum_{l=0}^{k}\frac{(-1)^{l+1}}{l+1}\int_{S^{p-1}} f_{-p,l}(\xi)\log^{l+1}|A^{-1}\xi|d\xi\right).\end{aligned}$$

The following proposition, which substantiates the mentioned fact that Stokes' Theorem does not hold for $⨍$, was stated as a Lemma in [**LePf00**]. A couple of years later it was rediscovered by MANCHON, MAEDA, and PAYCHA [**MMP05**], [**Pay05**].

PROPOSITION 4.2. [**LePf00**, Lemma 5.5] *Let $f \in \mathrm{CS}^{m,k}(\mathbb{R}^p)$ with asymptotic expansion* (4.5). *Then*

$$⨍_{\mathbb{R}^p} \frac{\partial f}{\partial \xi_j} d\xi = \int_{S^{p-1}} f_{1-p,k}(\xi)\xi_j d\mathrm{vol}_S(\xi).$$

We will come back to this below when we discuss the residue trace.

4.3. The Kontsevich–Vishik canonical trace. Using the Hadamard partie finie integral we can now follow the scheme outlined in Subsection 4.1. Let $A \in \mathrm{CL}^{a,k}(M,E)$ be a log–polyhomogeneous pseudodifferential operator on a closed manifold M. If $a \notin \mathbb{Z}$ we put, using the notation of (4.4) and (4.3),

$$\mathrm{TR}(A) := \sum_{j=1} \int_{U_j} ⨍_{\mathbb{R}^n} \varphi_j(x) \operatorname{tr}_{E_x}\big(a_j(x,\xi)\big) đ\xi \, dx. \tag{4.8}$$

By Proposition 4.1 one shows exactly as in (4.3) that $\mathrm{TR}(A)$ is well–defined.

In fact we have (essentially) proved the following:

THEOREM 4.3 (KONTSEVICH–VISHIK [**KoVi95**], [**KoVi94**], LESCH [**Les99**, Sec. 5]). *There is a linear functional* TR *on*

$$\bigcup_{a\in\mathbb{C}\setminus\{-n,-n+1,-n+2,\dots\},k\geq 0} \mathrm{CL}^{a,k}(M,E)$$

such that

(i) *In a local chart* TR *is given by* (4.1), *with* $\int_{\mathbb{R}^n}$ *to be replaced by the cut–off integral* $⨍_{\mathbb{R}^n}$.
(ii) $\mathrm{TR} \restriction \mathrm{CL}^{a,k}(M,E) = \mathrm{Tr} \restriction \mathrm{CL}^{a,k}(M,E)$ *if* $a < -\dim M$.
(iii) $\mathrm{TR}([A,B]) = 0$ *if* $A \in \mathrm{CL}^{a,k}(M,E), B \in CL^{b,l}(M,E)$, $a+b \notin \mathbb{Z}$.

We mention a stunning application of this result [**KoVi95**, Cor. 4.1]. Let G be a domain in the complex plane and let $A(z), B(z)$ be holomorphic families of operators in $\mathrm{CL}^{\bullet,k}(M,E)$ with $\operatorname{ord} A(z) = \operatorname{ord} B(z) = z$. We do not formalize the notion of a holomorphic family here. What we have in mind are e.g. families of complex powers $A(z) = A^z$. Assume that G contains points z with $\operatorname{Re} z < -\dim M$. Then $\mathrm{TR}(A(z))$ is the analytic continuation of $\mathrm{Tr}(A(\cdot)) \restriction G \cap \{z \in \mathbb{C} \mid \operatorname{Re} z < -\dim M\}$; a similar statement holds for $B(z)$.

If for a point $z_0 \in G \setminus \{-n, -n+1, \dots\}$ we have $A(z_0) = B(z_0)$ we can conclude that the value of the analytic continuation of $\mathrm{Tr}(A(\cdot)) \restriction G \cap \{z \in \mathbb{C} \mid \operatorname{Re} z < -\dim M\}$ to z_0 coincides with the value of the corresponding analytic continuation of $\mathrm{Tr}(B(\cdot)) \restriction G \cap \{z \in \mathbb{C} \mid \operatorname{Re} z < -\dim M\}$. Namely, we obviously have $\mathrm{TR}(A(z_0)) = \mathrm{TR}(B(z_0))$. The author does not know of a direct proof of this fact.

Proposition 4.1 shows that if A is of integral order additional terms show up when making the linear change of coordinates (4.3), indicating that TR cannot be extended to a trace on the algebra of pseudodifferential operators. The following no go result shows that the order constraints in Theorem 4.3 are indeed sharp:

PROPOSITION 4.4. *There is no trace* τ *on the algebra* $\mathrm{CL}^0(M)$ *of classical pseudodifferential operators of order* 0 *such that* $\tau(A) = \mathrm{Tr}(A)$ *if* $A \in \mathrm{CL}^{-\infty}(M)$.

PROOF. We reproduce here the very easy proof: from Index Theory we use the fact that on M there exists an elliptic system $T \in \mathrm{CL}^0(M,\mathbb{C}^r)$ of non–vanishing Fredholm index; in general we cannot find a scalar elliptic operator with non–trivial index. Let $S \in \mathrm{CL}^0(M,\mathbb{C}^r)$ be a pseudodifferential parametrix (cf. Theorem 3.7) such that $I - ST, I - TS \in \mathrm{CL}^{-\infty}(M,\mathbb{C}^r)$. τ and Tr extend to traces on $\mathrm{CL}^0(M,\mathbb{C}^r) = \mathrm{CL}^0(M) \otimes \mathrm{M}(r,\mathbb{C})$ via $\tau(A \otimes X) = \tau(A)\,\mathrm{Tr}(X)$, $A \in \mathrm{CL}^a(M), X \in \mathrm{M}(r,\mathbb{C})$ and $\mathrm{Tr}(X)$ is the usual trace on matrices. Since smoothing operators are of trace class one has

$$\operatorname{ind} T = \operatorname{Tr}(I - ST) - \operatorname{Tr}(I - TS) \tag{4.9}$$

and we arrive at the contradiction

$$\begin{aligned} 0 \neq \operatorname{ind} T &= \operatorname{Tr}(I - ST) - \operatorname{Tr}(I - TS) \\ &= \tau(I - ST) - \tau(I - TS) = \tau([T, S]) = 0. \end{aligned}$$

□

5. Pseudodifferential operators with parameter: Asymptotic expansions

We take up Section 3 and continue the discussion of pseudodifferential operators with parameter.

5.1. The Resolvent Expansion. The following result is the main technical result needed for the residue trace. It goes back to MINAKSHISUNDARAM and PLEIJEL [**MiPl49**] who follow carefully HADAMARD's method of the construction of a fundamental solution for the wave equation [**Had32**]. It is at the heart of the Local Index Theorem and therefore has received much attention. In the form stated below it is essentially due to SEELEY [**See67**], see also [**GrSe95**]. The (straightforward) generalization to log–polyhomogeneous symbols was done by the author [**Les99**]. Of the latter the published version contains annoying typos, the arxiv version is correct.

THEOREM 5.1. 1. *Let $U \subset \mathbb{R}^n$ open, $\Gamma \subset \mathbb{R}^p$ a cone, and $a \in \mathrm{CS}^{m,k}(U;\Gamma)$, $m + n < 0$, $A = \mathrm{Op}(a)$. Let $k_A(x;\mu) := \int_{\mathbb{R}^n} a(x,\xi,\mu) d\xi$ be the Schwartz kernel (cf. Eq. (3.10)) of A on the diagonal. Then $k_A \in \mathrm{CS}^{m+n,k}(U;\Gamma)$. In particular there is an asymptotic expansion*

$$k_A(x, x; \mu) \sim_{|\mu|\to\infty} \sum_{j=0}^{\infty} \sum_{l=0}^{k} e_{m-j,l}(x, \mu/|\mu|) |\mu|^{m+n-j} \log^k |\mu|. \tag{5.1}$$

2. *Let M be a compact manifold, $\dim M =: n$, and $A \in \mathrm{CL}^{m,k}(M, E; \Gamma)$. If $m + n < 0$ then $A(\mu)$ is trace class for all $\mu \in \Gamma$ and $\operatorname{Tr} A(\cdot) \in \mathrm{CS}^{m+n,k}(\Gamma)$. In particular,*

$$\operatorname{Tr} A(\mu) \sim_{|\mu|\to\infty} \sum_{j=0}^{\infty} \sum_{l=0}^{k} e_{m-j,l}(\mu/|\mu|) |\mu|^{m+n-j} \log^k |\mu|.$$

3. *Let $P \in \mathrm{CL}^m(M, E)$ be an elliptic classical pseudodifferential operator and assume for simplicity that with respect to some Riemannian structure on M and some Hermitian structure on E the operator P is self–adjoint and non–negative. Furthermore, let $B \in \mathrm{CL}^{b,k}(M, E)$ be a pseudodifferential operator. Let $\Lambda = \{\lambda \in \mathbb{C} \mid |\arg \lambda| \geq \varepsilon\}$ be a sector in $\mathbb{C} \setminus \mathbb{R}_+$. Then for $N > (b + n)/m, n := \dim M$, the operator $B(P - \lambda)^{-N}$ is of trace class and there is an asymptotic expansion*

$$\begin{aligned} \operatorname{Tr}(B(P - \lambda)^{-N}) \sim_{\lambda\to\infty} & \sum_{j=0}^{\infty} \sum_{l=0}^{k+1} c_{jl} \lambda^{\frac{n+b-j}{m} - N} \log^l \lambda + \\ & + \sum_{j=0}^{\infty} d_j \, \lambda^{-j-N} \end{aligned}, \quad \lambda \in \Lambda. \tag{5.2}$$

Furthermore, $c_{j,k+1} = 0$ if $(j - b - n)/m \notin \mathbb{Z}_+$.

PROOF. We present a proof of 1. and 2. and sketch the proof of 3. in a special case.

Since $a \in \mathrm{CS}^{m,k}(U;\Gamma)$ we have Eq. (3.8). Thus we write

$$a = \sum_{j=0}^{N} a_{m-j} + R_N, \tag{5.3}$$

with $R_N \in \mathrm{S}^{m-N}(U;\Gamma)$. In fact, $R_N \in \mathrm{S}^{m-N-1+\varepsilon}(U;\Gamma)$ for every $\varepsilon > 0$, but we don't need this below. Now pick $L \subset \Gamma, K \subset U$, compact and a multi–index α. Then for $x \in K$ the kernel $k_{A,N}$ of R_N satisfies

$$\begin{aligned}
&\Big|\partial_\mu^\alpha k_{A,N}(x,x;\mu)\Big| \\
&= \Big|\int_{\mathbb{R}^n} \partial_\mu^\alpha R_N(x,\xi,\mu) đ\xi\Big| \\
&\le C_{\alpha,K,L}\int_{\mathbb{R}^n} (1+(|\xi|^2+|\mu|^2)^{1/2})^{m-|\alpha|-N}\, đ\xi \\
&\le C_{\alpha,K,L}(1+|\mu|)^{m+n-|\alpha|-N}.
\end{aligned} \tag{5.4}$$

Now consider one of the summands of (3.8). We write it in the form

$$b_{m-j,l}(x,\xi,\mu) = \tilde b_{m-j,l}(x,\xi,\mu)\log^l(|\xi|^2+|\mu|^2), \tag{5.5}$$

with

$$\tilde b_{m-j,l}(x,r\xi,r\mu) = r^{m-j}\tilde b_{m-j,l}(x,\xi,\mu), \qquad \text{for } r \ge 1, |\xi|^2+|\mu|^2 \ge 1. \tag{5.6}$$

Then the contribution $k_{m-j,l}$ of $b_{m-j,l}$ to the kernel of A satisfies

$$\begin{aligned}
&k_{m-j,l}(x,x;r\mu) \\
&= \int_{\mathbb{R}^n} \tilde b_{m-j,l}(x,\xi,r\mu)\,\log^l(|\xi|^2+r^2|\mu|^2)\,đ\xi \\
&= r^{m-j}\int_{\mathbb{R}^n} \tilde b_{m-j,l}(x,r^{-1}\xi,\mu)\big(\log r^2+\log(|r^{-1}\xi|^2+|\mu|^2)\big)^l\,đ\xi \\
&= r^{m+n-j}\int_{\mathbb{R}^n} \tilde b_{m-j,l}(x,\xi,\mu)\big(\log r^2+\log(|\xi|^2+|\mu|^2)\big)^l\,đ\xi,
\end{aligned} \tag{5.7}$$

proving the expansion (5.1).

2. follows simply by integrating (5.1). In view of (5.4) the expansion (5.1) is uniform on compact subsets of U and hence may be integrated over compact subsets. Covering the compact manifold M by finitely many charts then gives the claim.

3. We cannot give a full proof of 3. here; but we at least want to explain where the additional log terms in (5.2) come from. Note that even if $B \in \mathrm{CL}^b(M,E)$ is classical there are log terms in (5.2). In general the highest log power occurring on the rhs of (5.2) is one higher than the log degree of B.

For simplicity let us assume that P is a differential operator. This ensures that $(P-\lambda^m)^{-N}$ (note the λ^m instead of λ) is in the parametric calculus (cf. Remarks 3.2 2., 3.5). We first describe the local expansion of the symbol of $B(P-\lambda^m)^{-N}$. To obtain the claim as stated one then has to replace λ^m by λ and integrate over M: choose a chart and denote the complete symbol of B by $b(x,\xi)$ and the complete parametric symbol of $(P-\lambda^m)^{-N}$ by $q(x,\xi,\lambda)$. Then the symbol of the product is given by

$$(b*q)(x,\xi,\lambda)\sim\sum_{\alpha\in\mathbb{Z}_+^n}\frac{i^{-\alpha}}{\alpha!}\big(\partial_\xi^\alpha b(x,\xi)\big)\big(\partial_x^\alpha q(x,\xi,\lambda)\big). \tag{5.8}$$

Expanding the rhs into its homogeneous components gives

$$\begin{aligned}&(b*q)(x,\xi,\lambda)\\ &\sim\sum_{j=0}^\infty\sum_{|\alpha|+l+l'=j}\frac{i^{-\alpha}}{\alpha!}\underbrace{\underbrace{\big(\partial_\xi^\alpha b_{b-l}(x,\xi)\big)}_{(b-l-|\alpha|)-(\log)\text{homogeneous}}\underbrace{\big(\partial_x^\alpha q_{-mN-l'}(x,\xi,\lambda)\big)}_{(-mN-l')-\text{homogeneous}}}_{(b-mN-j)-(\log)\text{homogeneous}}.\end{aligned} \tag{5.9}$$

The contribution to the Schwartz kernel of $B(P-\lambda^m)^{-N}$ of a summand is given by

$$\frac{i^{-\alpha}}{\alpha!}\int_{\mathbb{R}^n}\big(\partial_\xi^\alpha b_{b-l}(x,\xi)\big)\big(\partial_x^\alpha q_{-mN-l'}(x,\xi,\lambda)\big)\,đ\xi. \tag{5.10}$$

We will see that the asymptotic expansion of each of these integrals a priori contributes to the term λ^{-N} in the expansion (5.2). So additional considerations, which we will not present here, are necessary to show that by expanding the individual integrals (5.10) one indeed obtains the asymptotic expansion (5.2).

The asymptotic expansion of (5.10) will be singled out as Lemma 5.2 below. The proof of it will in particular explain why the highest possible log-power in (5.2) is one higher than the log-degree of B. □

The following expansion Lemma is maybe of interest in its own right. Its proof will explain the occurrence of higher log powers in the resolvent respectively heat expansions. The homogeneous version of the Lemma can again be found in [**GrSe95**]. We generalize it here slightly to the log–polyhomogeneous setting (cf. [**Les99**]).

LEMMA 5.2. *Let $B\in C^\infty(\mathbb{R}^n), Q\in C^\infty(\mathbb{R}^n\times[1,\infty))$ and assume that B,Q have the following properties*

$$\begin{aligned}B(\xi)&=\tilde B(\xi/|\xi|)|\xi|^b\log^k|\xi|,\quad |\xi|\ge1,\\ Q(r\xi,r\lambda)&=r^qQ(\xi,\lambda),\quad r\ge1,\lambda\ge1,\\ |Q(\xi,1)|&\le C(|\xi|+1)^{-q},\end{aligned} \tag{5.11}$$

where $b,q\in\mathbb{R}$ and $b+q+n<0$. Then the following asymptotic expansion holds:

$$\begin{aligned}F(\lambda)&=\int_{\mathbb{R}^n}B(\xi)Q(\xi,\lambda)d\xi\\ &\sim_{\lambda\to\infty}\sum_{j=0}^{k+1}c_j\lambda^{q+b+n}\log^j\lambda+\sum_{j=0}^\infty d_j\lambda^{q-j}.\end{aligned} \tag{5.12}$$

$c_{k+1}=0$ if b is not an integer $\le -n$.

The coefficients c_j,d_j will be explained in the proof.

PROOF. The integral on the lhs of (5.12) exists since $b+q+n<0$.

We split the domain of integration into the three regions:
$1\le\lambda\le|\xi|, |\xi|\le1$, and $1\le|\xi|\le\lambda$.

$1 \le \lambda \le |\xi|$: Here we are in the domain of homogeneity and a change of variables yields

$$\begin{aligned}
&\int_{\lambda\le|\xi|} B(\xi)Q(\xi,\lambda)d\xi\\
&= \lambda^q \int_{\lambda\le|\xi|} \tilde{B}(\xi/|\xi|)|\xi|^b\big(\log^k|\xi|\big)Q(\xi/\lambda,1)d\xi\\
&= \lambda^{q+b+n} \int_{1\le|\xi|} \tilde{B}(\xi/|\xi|)|\xi|^b\big(\log\lambda+\log|\xi|\big)^k Q(\xi,1)d\xi,\\
&= \sum_{j=0}^{k} \alpha_j \lambda^{q+b+n}\log^j\lambda,
\end{aligned} \tag{5.13}$$

giving a contribution to the coefficient c_j for $0 \le j \le k$.

$|\xi| \le 1$: For the remaining two cases we employ the Taylor expansion of the smooth function $\eta \mapsto Q(\eta,1)$ about $\eta = 0$:

$$Q(\eta,1) = \sum_{j=0}^{N} Q_j(\eta) + R_N(\eta), \tag{5.14}$$

where $Q_j(\eta) \in \mathbb{C}[\eta_1,,\ldots,\eta_n]$ are homogeneous polynomials of degree j and R_N is a smooth function satisfying $R_N(\eta) = O(|\eta|^{N+1}),\ \eta \to 0$. Respectively, for $\xi \in \mathbb{R}^n, \lambda \ge 1$,

$$Q(\xi,\lambda) = Q(\xi/\lambda,1)\,\lambda^q = \sum_{j=0}^{N} Q_j(\xi)\,\lambda^{q-j} + R_N(\xi/\lambda)\,\lambda^q. \tag{5.15}$$

Plugging (5.15) into the integral for $|\xi| \le 1$ we find

$$\begin{aligned}
&\int_{|\xi|\le1} B(\xi)Q(\xi,\lambda)d\xi =\\
&= \sum_{j=0}^{N} \int_{|\xi|\le1} B(\xi)Q_j(\xi)d\xi\,\lambda^{q-j} + O(\lambda^{q-N-1}), \quad \lambda\to\infty,
\end{aligned} \tag{5.16}$$

giving a contribution to the coefficient d_j.

$1 \le |\xi| \le \lambda$: We again use the Taylor expansion (5.15) with N large enough such that $b+N+1 > -n$ to ensure $\int_{|\xi|\le1} |\xi|^b \log^j|\xi|\,|R_N(\xi)|d\xi < \infty$ for all j. Let $B^h(\xi) := \tilde{B}(\xi/|\xi|)|\xi|^b\log^k|\xi|$ be the homogeneous extension of $B(\xi)$ to all $\xi \ne 0$. Then

$$\int_{|\xi|\le1} \big(|B(\xi)| + |B^h(\xi)|\big)\lambda^q|R_N(\xi/\lambda)|d\xi = O(\lambda^{q-N-1}), \quad \lambda\to\infty, \tag{5.17}$$

and thus

$$
\begin{aligned}
&\int_{1\le|\xi|\le\lambda} B(\xi)\lambda^q R_N(\xi/\lambda)d\xi\\
&= \int_{0\le|\xi|\le\lambda} B^h(\xi)\lambda^q R_N(\xi/\lambda)d\xi + O(\lambda^{q-N-1})\\
&= \int_{|\xi|\le 1} \tilde{B}(\xi/|\xi|)|\xi|^b\big(\log\lambda+\log|\xi|\big)^k R_N(\xi)d\xi\ \lambda^{q+b+n}+\\
&+O(\lambda^{q-N-1}),\quad \lambda\to\infty.
\end{aligned}
\tag{5.18}
$$

So the contribution of the "remainder" R_N to the expansion is not small, rather it contributes to the coefficient c_j of the $\lambda^{q+b+n}\log^j\lambda$ term for $0\le j\le k$. Note that so far we have not obtained any contribution to the coefficient c_{k+1}.

Such a contribution will show up only now when we finally deal with the summands in the Taylor expansion. Using polar coordinates we find

$$
\begin{aligned}
&\int_{1\le|\xi|\le\lambda} B(\xi)Q_j(\xi)d\xi\ \lambda^{q-j}\\
&= \lambda^{q-j}\int_1^\lambda\int_{S^{n-1}} \tilde{B}(\omega)r^b\big(\log^k r\big)Q_j(r\omega)r^{n-1}d\,\mathrm{vol}_{S^{n-1}}(\omega)dr\\
&= C_j\lambda^{q-j}\int_1^\lambda r^{b+n-1+j}\log^k r dr\\
&= C_j\lambda^{q-j}\begin{cases}\sum\limits_{\sigma=0}^{k}\alpha'_\sigma\lambda^{b+n+j}\log^\sigma\lambda+\beta_j, & b+n+j\ne 0,\\ \frac{1}{k+1}\log^{k+1}\lambda, & b+n+j=0.\end{cases}
\end{aligned}
\tag{5.19}
$$

As a side remark note the explicit formula

$$
\begin{aligned}
&\int_1^\lambda r^\alpha\log^k r dr\\
&=\begin{cases}\sum\limits_{j=0}^{k}\frac{(-1)^j k!}{(k-j)!(\alpha+1)^{j+1}}\lambda^{\alpha+1}\log^{k-j}\lambda+\frac{(-1)^{k+1}k!}{(\alpha+1)^{k+1}}, & \alpha\ne-1,\\ \frac{1}{k+1}\log^{k+1}\lambda, & \alpha=-1.\end{cases}
\end{aligned}
\tag{5.20}
$$

The constant term in (5.20) respectively β_j on the rhs of (5.19) was omitted in [**Les99**, Eq. 3.16]. Fortunately the error was inconsequential for the formulation of the expansion result because β_j is just another contribution to the coefficient d_j. □

5.2. Resolvent expansion vs. heat expansion. From the resolvent expansion one can easily derive the heat expansion and the meromorphic continuation of the ζ–function. In fact under a mild additional assumption the resolvent expansion can be derived from the heat expansion of the meromorphic continuation of the ζ–function (cf. e.g. LESCH [**Les97**, Theorem 5.1.4 and 5.1.5], BRÜNING–LESCH [**BrLe99**, Lemma 2.1 and 2.2]).

Let B, P be as above. Next let γ be a contour in the complex plane as sketched in Figure 1. Then Be^{-tP} has the following contour integral representation:

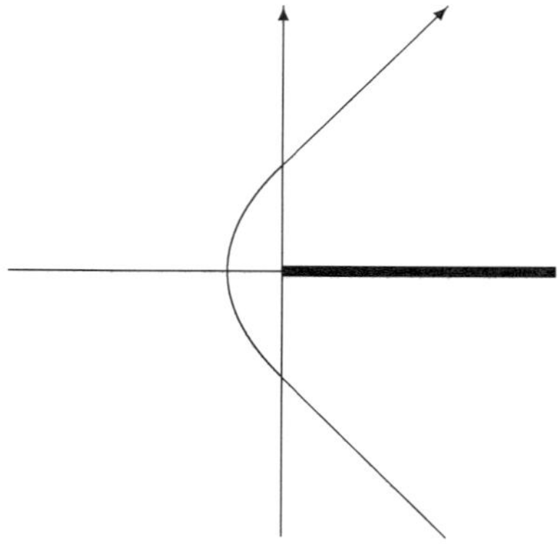

FIGURE 1. Contour of integration for calculating Be^{-tP} from the resolvent.

$$(5.21)\qquad \begin{aligned} Be^{-tP} &= \frac{-1}{2\pi i}\int_\gamma e^{-t\lambda}B(P-\lambda)^{-1}d\lambda \\ &= -(-t)^{-N+1}\frac{(N-1)!}{2\pi i}\int_\gamma e^{-t\lambda}B(P-\lambda)^{-N}d\lambda. \end{aligned}$$

Taking the trace on both sides and plugging in the asymptotic expansion of $\operatorname{Tr}(B(P-\lambda)^{-N})$ one easily finds

$$(5.22)\qquad \operatorname{Tr}(Be^{-tP}) \sim_{t\to 0+} \sum_{j=0}^{\infty}\sum_{l=0}^{k+1} a_{jl}(B,P)t^{\frac{j-b-n}{m}}\log^l t + \sum_{j=0}^{\infty}\tilde{d}_j(B,P)\, t^j.$$

$a_{j,k+1}=0$ if $(j-b-n)/m \notin \mathbb{Z}_+$.

5.3. Heat expansion vs. ζ–function. Finally we briefly explain how the meromorphic continuation of the ζ–function can be obtained from the heat expansion. As before let $B\in \mathrm{CL}^{b,k}(M,E)$ and let $P\in\mathrm{CL}^m(M,E)$ be an elliptic operator which is self–adjoint with respect to some Riemannian structure on M and some Hermitian structure on E. Furthermore, assume that $P\geq 0$ is non–negative. Let $\Pi_{\ker P}$ be the orthogonal projection onto $\ker P$ and put for $\operatorname{Re} s>0$

$$(5.23)\qquad P^{-s} := \big(I-\Pi_{\ker P}\big)\big(P+\Pi_{\ker P}\big)^{-s}.$$

I.e. $P^{-s}\restriction \ker P = 0$ and for $\xi\in\operatorname{im}P$ we let $P^{-s}\xi$ be the unique $\eta\in\ker P^{\perp}$ with $P^s\eta=\xi$. The ζ–function of (B,P) is defined (up to a Γ–factor) as the *Mellin transform* of the heat trace $\operatorname{Tr}(B(I-\Pi_{\ker P})e^{-tP})$:

$$(5.24)\qquad \begin{aligned} \zeta(B,P;s) &= \operatorname{Tr}\big(BP^{-s}\big) \\ &= \frac{1}{\Gamma(s)}\int_0^\infty t^{s-1}\operatorname{Tr}\big(B(I-\Pi_{\ker P})e^{-tP}\big)dt, \qquad \operatorname{Re} s\gg 0. \end{aligned}$$

$\operatorname{Tr}\big(B(I-\Pi_{\ker P})e^{-tP}\big)$ decays exponentially as $t\to\infty$. The meromorphic continuation is thus obtained by plugging the short time asymptotic expansion (5.22) into the rhs of (5.24) (cf. e.g. [**Les97**, Sec. II.1]):

$$\begin{aligned}\Gamma(s)\zeta(B,P;s) &= \int_0^1 t^{s-1}\operatorname{Tr}(Be^{-tP})dt \\ &\quad - \frac{1}{s}\operatorname{Tr}\big(B\Pi_{\ker P}\big) + \text{ Entire function}(s), \\ &\sim \sum_{j=0}^{\infty}\sum_{j=0}^{k+1}\frac{a'_{jl}(B,P)}{(s-\frac{n+b-j}{m})^{j+1}} + \sum_{j=0}^{\infty}\frac{\tilde{d}'_j(B,P)}{s+j},\end{aligned} \tag{5.25}$$

where the formal sum on the right is meant to display the principal parts of the Laurent series at the poles of $\Gamma(s)\zeta(B,P;s)$.

The Γ–function has simple poles in $\mathbb{Z}_- = \{0,-1,-2,\dots\}$, hence the $\tilde{d}'_j$ do not contribute to the poles of $\zeta(B,P;s)$. The a'_{jl} depend linearly on the a_{jl} and consequently $a'_{j,k+1} = 0$ if $(n+b-j)/m$ is *not* a pole of the Γ–function. Let us summarize:

THEOREM 5.3. *Let M be a compact closed manifold of dimension n. Let $B \in \mathrm{CL}^{b,k}(M,E)$ and let $P \in \mathrm{CL}^m(M,E)$ be an elliptic operator which is self-adjoint with respect to some Riemannian structure on M and some Hermitian structure on E. Then the ζ-function $\zeta(B,P;s)$ is meromorphic for $s \in \mathbb{C}$ with poles of order at most $k+1$ in $(n+b-j)/m$.*

6. Regularized traces

6.1. The Residue Trace (Noncommutative Residue). We have seen in Proposition 4.4 that the Hilbert space trace Tr cannot be extended to all classical pseudodifferential operators.

However, in his seminal papers [**Wod84**], [**Wod87**] M. WODZICKI was able to show that, up to a constant, the algebra $\mathrm{CL}^{\bullet}(M)$ has a unique trace which he called the noncommutative residue; we prefer to call it residue trace. The residue trace was independently discovered by V. GUILLEMIN [**Gui85**] as a byproduct of his axiomatic approach to the Weyl asymptotics. In [**Les99**] the author generalized the residue trace to the algebra $\mathrm{CL}^{\bullet,\bullet}(M,E)$. Strictly speaking there is no residue trace on the full algebra $\mathrm{CL}^{\bullet,\bullet}(M,E)$. Rather one has to restrict to operators with a given bound on the log degree.

In detail: let $A \in \mathrm{CL}^{a,k}(M,E)$ and let $P \in \mathrm{CL}^m(M,E)$ elliptic, non–negative and invertible, cf. Subsection 5.3. Put

$$\begin{aligned}&\operatorname{Res}_k(A,P) \\ &\quad := m^{k+1}\operatorname{Res}_{k+1}\operatorname{Tr}(AP^{-s})|_{s=0} \\ &\quad = m^{k+1}(-1)^{k+1}(k+1)! \times \text{ coefficient of } \log^{k+1} t \text{ in the} \\ &\qquad \text{asymptotic expansion of } \operatorname{Tr}(Ae^{-tP}) \text{ as } t \to 0.\end{aligned} \tag{6.1}$$

In [**Les99**] it was assumed in addition that the leading symbol of P is scalar. This assumption allows one to use Duhamel's principle and to systematically exploit the fact that the order of a commutator $[A,P]$ is at most $\operatorname{ord} A + \operatorname{ord} P - 1$. Using the resolvent approach it was shown in GRUBB [**Gru05**] that for defining Res_k and to derive its properties one does not need to assume that P has scalar leading symbol.

The main properties of Res_k can now be summarized as follows:

THEOREM 6.1 (Wodzicki–Guillemin; *log*–polyhomogeneous case [**Les99**]). *Let* $A \in \mathrm{CL}^{a,k}(M,E)$ *and let* $P \in \mathrm{CL}^m(M,E)$ *be elliptic, non–negative and invertible.*

1. $\operatorname{Res}_k(A,P) =: \operatorname{Res}_k(A)$ *is independent of* P, *i.e.*

$$\operatorname{Res}_k : \mathrm{CL}^{\bullet,k}(M,E) \longrightarrow \mathbb{C}$$

is a linear functional.

2. *If* $A \in \mathrm{CL}^{a,k}(M,E), B \in \mathrm{CL}^{b,l}(M,E)$ *then* $\operatorname{Res}_k([A,B]) = 0$. *In particular,* $\operatorname{Res} := \operatorname{Res}_0$ *is a trace on* $\mathrm{CL}^\bullet(M,E)$.

3. *For* $A \in \mathrm{CL}^{a,k}(M,E)$ *the k-th residue* $\operatorname{Res}_k(A)$ *vanishes if*

$$a \notin -\dim M + \mathbb{Z}_+.$$

4. *In a local chart one puts*

$$\omega_k(A)(x) = \frac{(k+1)!}{(2\pi)^n}\Big(\int_{|\xi|=1} \operatorname{tr}_{E_x}(a_{-n,k}(x,\xi))|d\xi|\Big)|dx|. \tag{6.2}$$

Then $\omega_k(A) \in \Gamma^\infty(M,|\Omega|)$ *is a density (in particular independent of the choice of coordinates), which depends functorially on* A. *Moreover*

$$\operatorname{Res}_k(A) = \int_M \omega_k(A). \tag{6.3}$$

5. *If* M *is connected and* $n = \dim M > 1$ *then* Res_k *induces an isomorphism* $\mathrm{CL}^{a,k}(M)/[\mathrm{CL}^{a,k}(M), \mathrm{CL}^{1,0}(M)] \longrightarrow \mathbb{C}$. *In particular,* Res *is up to scalar multiples the only trace on* $\mathrm{CL}^\bullet(M)$.

EXAMPLE 6.2. 1. Let A be a classical pseudodifferential operator of order $-n = -\dim M$ which is assumed to be elliptic, non–negative and invertible. To calculate the residue trace of A we may use $P := A^{-1}$. Thus

$$\operatorname{Res}(A) = n \operatorname{Res} \operatorname{Tr}(A^{1+s})|_{s=0} = n \operatorname{Res} \zeta(A^{-1}; s)|_{s=1} > 0, \tag{6.4}$$

where $\zeta(A^{-1}; s) = \zeta(I, A^{-1}; s)$ is the ζ–function of the elliptic operator A^{-1}. The positivity follows from Eq. (6.2).

2. Let Δ be the Laplacian on a closed Riemannian manifold (M,g). Then the heat expansion (5.22) (with $B = I$ and $P = \Delta$) simplifies: since Δ is a differential operator there are no log terms and by a parity argument every other heat coefficient vanishes [**Gil95**]. Thus we have an asymptotic expansion

$$\operatorname{Tr}(e^{-t\Delta}) \sim_{t\to 0} \sum_{j=0}^\infty a_j(\Delta) t^{(j-n)/2}, \quad a_{2j+1}(\Delta) = 0. \tag{6.5}$$

The $a_j(\Delta)$ are enumerated such that (6.5) is consistent with (5.22). The first few $a_j(\Delta)$ have been calculated although the computational complexity increases drastically with j (cf. e.g. [**Gil95**]). One has

$$\begin{aligned} a_0(\Delta) &= c_n \operatorname{vol}(M)\\ a_2(\Delta) &= c_n' \int_M \operatorname{scal}(M,g) d\operatorname{vol}. \end{aligned} \tag{6.6}$$

The latter is known as the *Einstein-Hilbert action* in the physics literature. Therefore the following relation between the heat coefficients (and in particular the EH action) and the residue trace has received some attention from the physics community, e.g. KALAU–WALZE [**KaWa95**], KASTLER [**Kas95**]. We find for real α

$$\begin{aligned}
\operatorname{Res}(\Delta^\alpha) &= 2\lim_{s\to 0} s\operatorname{Tr}(\Delta^{\alpha-s}) \\
&= 2\lim_{s\to 0} s\zeta(I,\Delta;s-\alpha) \\
&= 2\lim_{s\to 0}\frac{s}{\Gamma(s-\alpha)}\int_0^1 t^{s-\alpha-1}\big(\operatorname{Tr}(e^{-t\Delta})-\dim\ker\Delta\big)dt \qquad (6.7)\\
&= 2\sum_{j=0}^{\infty}\lim_{s\to 0}\frac{a_j(\Delta)s}{\Gamma(s-\alpha)(s-\alpha+\frac{j-n}{2})} \qquad (6.8)\\
&= \begin{cases}\frac{2a_j(\Delta)}{\Gamma(\frac{n-j}{2})}, & \alpha=\frac{j-n}{2}<0,\\ 0, & \text{otherwise.}\end{cases} \qquad (6.9)
\end{aligned}$$

Here we have used that the ζ–function of Δ has only simple poles (cf. Theorem 5.3). Furthermore, in (6.7) we use that due to the exponential decay of $(\operatorname{Tr}(\epsilon^{-t\Delta})-\dim\ker\Delta)$ the function $s\mapsto\int_1^\infty t^{s-\alpha-1}(\operatorname{Tr}(e^{-t\Delta})-\dim\ker\Delta)dt$ is entire and hence does not contribute to the residue at $s=0$. Furthermore, note that the sum in (6.8) is finite.

In view of (6.6) we have the following special cases of (6.9):

$$\operatorname{Res}(\Delta^{-n/2})=\frac{2a_0(\Delta)}{\Gamma(\frac{n}{2})}=c_n\operatorname{vol}(M), \tag{6.10}$$

$$\operatorname{Res}(\Delta^{1-n/2})=c_n'\operatorname{EH}(M,g), \tag{6.11}$$

where EH denotes the above mentioned Einstein-Hilbert action. It is formula (6.11) which caused physicists to become enthusiastic about this business. Needless to say, the calculation we present here goes through for any Dirac Laplacian. One only has to replace the scalar curvature in (6.6) by the second local heat coefficient, which can be calculated for any Dirac Laplacian.

We wanted to show that the relation between the heat asymptotic and the poles of the ζ–function, which is an easy consequence of the Mellin transform, leads to a straightforward proof of (6.11). There also exist "hard" proofs of this fact which check that the *local* Einstein-Hilbert action coincides with the residue density of the operator $\Delta^{1-n/2}$ [**KaWa95**],[**Kas95**].

6.2. Connes' Trace Theorem. The famous trace Theorem of Connes gives a relation between the Dixmier trace and the Wodzicki–Guillemin residue trace for pseudodifferential operators of order minus $\dim M$. It was extended by CAREY et. al. [**CPS03**], [**CRSS07**] to the von Neumann algebra setting.

THEOREM 6.3 (Connes' Trace Theorem [**Con88**]). *Let M be a closed manifold of dimension n and let E be a smooth vector bundle over M. Furthermore let $P\in\operatorname{CL}^{-n}(M,E)$ be a pseudodifferential operator of order $-n$. Then $P\in\mathscr{L}^{(1,\infty)}(L^2(M,E))$ and for any ω satisfying the assumptions of the previous Proposition one has*

$$\operatorname{Tr}_\omega(P)=\frac{1}{n}\operatorname{Res}P. \tag{6.12}$$

We give a sketch of the proof of Connes' Theorem using a Tauberian argument. This was mentioned without proof in [**Con94**, Prop. 4.2.β.4] and has been elaborated in various ways by many authors. The argument we present here is an adaption of an argument in [**CPS03**] to the type I case.

Let us mention the following simple version of Ikehara's Tauberian Theorem:

THEOREM 6.4 ([**Shu01**, Sec. II.14]). *Let $F : [1, \infty) \to \mathbb{R}$ be an increasing function such that*

(1) $\zeta_F(s) = \int_1^\infty \lambda^{-s} dF(\lambda)$ *is analytic for* $\operatorname{Re} s > 1$,
(2) $\lim_{s \to 1+} (s-1)\zeta_F(s) = L$.

Then

$$\lim_{\lambda \to \infty} \frac{F(\lambda)}{\lambda} = L. \tag{6.13}$$

COROLLARY 6.5. *Let $F : [1, \infty) \to \mathbb{R}$ be an increasing function such that $\int_1^\infty e^{-t\lambda} dF(\lambda) = \frac{L}{t} + O(t^{\varepsilon - 1}), t \to 0+$, for some $\varepsilon > 0$. Then Ikehara's Theorem applies to F and (6.13) holds.*

PROOF. The ζ–function of F satisfies

$$\begin{aligned}
\zeta_F(s) &= \int_1^\infty \lambda^{-s} dF(\lambda) \\
&= \int_1^\infty \frac{1}{\Gamma(s)} \int_0^\infty t^{s-1} e^{-t\lambda} dt\, dF(\lambda) \\
&= \int_0^1 \frac{t^{s-1}}{\Gamma(s)} \int_1^\infty e^{-t\lambda} dF(\lambda)\, dt + \text{ holomorphic near } s = 1 \\
&\sim \frac{1}{\Gamma(s)} \frac{L}{s-1} \text{ near } s = 1.
\end{aligned}$$ □

PROOF OF CONNES' TRACE THEOREM. Each $P \in \mathrm{CL}^{-n}(M, E)$ is a linear combination of at most 4 non–negative operators: to see this we first write $P = \frac{1}{2}(P + P^*) + \frac{1}{2i}(P - P^*)$ as a linear combination of two self–adjoint operators. So consider a self–adjoint $P = P^*$. We choose an elliptic operator $Q \in \mathrm{CL}^{-n}(M, E)$ with $Q > 0$ and positive definite leading symbol. Since we are on a compact manifold it then follows that $c \cdot Q - P \geq 0$ for c large enough. Hence $P = c \cdot Q - (c \cdot Q - P)$ is the desired decomposition of P as a difference of non–negative operators.

So it suffices to prove the claim for a non–negative operator P. Then $P + \varepsilon Q$ is elliptic and invertible for each $\varepsilon > 0$. By an approximation argument we are ultimately left with the problem of proving the claim for an *elliptic* positive operator $P \in \mathrm{CL}^{-n}(M, E)$.

Let $\mu_1 \geq \mu_2 \geq \mu_3 \geq \cdots > 0$ be the eigenvalues of P counted with multiplicity. We consider the counting function

$$F(\lambda) = \#\{j \in \mathbb{N} \mid \mu_j^{-1} \leq \lambda\}. \tag{6.14}$$

The associated ζ–function

$$\zeta_F(s) = \int_1^\infty \lambda^{-s} dF(\lambda) = \operatorname{Tr}(P^s) - \sum_{\mu_j > 1} \mu_j^s \tag{6.15}$$

is, up to the entire function $\sum_{\mu_j > 1} \mu_j^s$, the ζ–function of the elliptic operator P^{-1}. Thus by Theorem 5.3 the function ζ_F is holomorphic for $\operatorname{Re} s > 1$ and it has a meromorphic extension to the complex plane, and 1 is a simple pole with

$$\lim_{s\to 1}(s-1)\zeta_F(s) = \frac{1}{n}\operatorname{Res}(P) \neq 0, \tag{6.16}$$

cf. Example 6.2 1. Thus Ikehara's Theorem 6.4 applies to F and hence

$$\lim_{\lambda\to\infty}\frac{F(\lambda)}{\lambda} = \frac{1}{n}\operatorname{Res}(P). \tag{6.17}$$

Claim:

$$\lim_{j\to\infty} j\mu_j = \frac{1}{n}\operatorname{Res}(P) =: L. \tag{6.18}$$

To see this let $\varepsilon > 0$ be given. Then there exists a λ_0 such that for $\lambda \geq \lambda_0$

$$1-\varepsilon \leq \frac{F(\lambda)}{\lambda L} \leq 1+\varepsilon. \tag{6.19}$$

Thus

$$\exists_{\lambda_0}\forall_{\lambda\geq\lambda_0}\quad (1-\varepsilon)\lambda L \leq \#\{j\in\mathbb{N} \mid \mu_j^{-1}\leq\lambda\} \leq (1+\varepsilon)\lambda L. \tag{6.20}$$

Hence for $j \geq (1+\varepsilon)\lambda L$ we have $\mu_j^{-1} \geq \lambda$ and for $j \leq (1-\varepsilon)\lambda L$ we have $\mu_j^{-1} \leq \lambda$. For a given fixed j_0 large enough we therefore infer

$$(1-\varepsilon)L \leq j\mu_j \leq (1+\varepsilon)L, \quad j \geq j_0, \tag{6.21}$$

proving the Claim.

Now consider

$$\beta(u) = \int_1^{e^u} \lambda^{-1} dF(\lambda) = \sum_{\mu_j \geq e^{-u}} \mu_j. \tag{6.22}$$

We check that Ikehara's Tauberian Theorem applies to β:

$$\begin{aligned}\int_1^\infty e^{-s\lambda} d\beta(\lambda) &= \int_1^\infty e^{-(s+1)\lambda} dF(e^\lambda)\\ &= \int_e^\infty x^{-s-1} dF(x) = \zeta_F(1+s)\\ &= \frac{\operatorname{Res}(P)}{ns} + O(1), \quad s\to 0.\end{aligned} \tag{6.23}$$

Thus Corollary 6.5 implies

$$\frac{1}{u}\sum_{\mu_j\geq e^{-u}} \mu_j = \frac{\beta(u)}{u} \xrightarrow{u\to\infty} \frac{1}{n}\operatorname{Res}(P). \tag{6.24}$$

To infer Connes' Trace Theorem from (6.24) we choose j_0 such that (6.21) holds for $\varepsilon = 1/2$ and $j \geq j_0$. Then put for N large enough $u_N := \log\frac{N}{(1-\varepsilon)L}$. Hence we have $\mu_j \geq \mu_N \geq e^{-u_N}$ for $1 \leq j \leq N$ and thus

$$\begin{aligned}\frac{1}{\log(N+1)}\sum_{j=1}^N \mu_j &\leq \frac{1}{\log(N+1)}\sum_{\mu_j\geq\exp(-u_N)} \mu_j\\ &= \frac{u_N}{\log N+1}\frac{1}{u_N}\sum_{\mu_j\geq\exp(-u_N)}\mu_j\\ &\longrightarrow L, \quad \text{for } N\to\infty,\end{aligned} \tag{6.25}$$

by (6.24) and since $u_N/\log(N+1) \to 1$. This proves

$$\limsup_{N\to\infty} \frac{1}{\log(N+1)} \sum_{j=1}^{N} \mu_j \leq L = \frac{1}{n}\operatorname{Res}(P). \tag{6.26}$$

Arguing with $u_N = \log \frac{N}{(1+\varepsilon)L}$ instead of $u_N = \log \frac{N}{(1-\varepsilon)L}$ one shows

$$\liminf_{N\to\infty} \frac{1}{\log(N+1)} \sum_{j=1}^{N} \mu_j \geq L = \frac{1}{n}\operatorname{Res}(P), \tag{6.27}$$

and Connes' Trace Theorem is proved. □

The attentive reader might have noticed that we did not use the full strength of the Claim (6.18). We only used that there exist positive constants c_1, c_2 such that $c_1 \leq j\mu_j \leq c_2$ for $j \geq j_0$.

6.3. Parametric case: The symbol valued trace. In contrast to Proposition 4.4 the situation is entirely different for the algebra of parametric pseudodifferential operators.

Fix a compact smooth manifold M without boundary of dimension n. Denote the coordinates in $\mathbb{R}^p$ by $\mu_1, \ldots, \mu_p$ and let $\mathbb{C}[\mu_1, \ldots, \mu_p]$ be the algebra of polynomials in $\mu_1, \ldots, \mu_p$. By a slight abuse of notation we denote by μ_j also the operator of multiplication by the j-th coordinate function. Then we have maps

$$\begin{aligned} &\partial_j : \mathrm{CL}^m(M, E; \mathbb{R}^p) \to \mathrm{CL}^{m-1}(M, E; \mathbb{R}^p), \\ &\mu_j : \mathrm{CL}^m(M, E; \mathbb{R}^p) \to \mathrm{CL}^{m+1}(M, E; \mathbb{R}^p). \end{aligned} \tag{6.28}$$

Also ∂_j and μ_j act naturally on the parametric symbols over the one–point space $\mathrm{CS}^{\bullet,\bullet}(\mathbb{R}^p) := \mathrm{CS}^{\bullet,\bullet}(\{\mathrm{pt}\}; \mathbb{R}^p)$ and on polynomials $\mathbb{C}[\mu_1, \ldots, \mu_p]$. Thus they act on the quotient $\mathrm{CS}^{\bullet,\bullet}(\mathbb{R}^p)/\mathbb{C}[\mu_1, \ldots, \mu_p]$. After these preparations we can summarize one of the main results of [**LePf00**].

Let E be a smooth vector bundle on M and consider $A \in \mathrm{CL}^m(M, E; \mathbb{R}^p)$ with $m + n < 0$. Then for $\mu \in \mathbb{R}^p$ the operator $A(\mu)$ is trace class; hence we may define the function $\mathrm{TR}(A) : \mu \mapsto \mathrm{Tr}(A(\mu))$. The map TR is obviously tracial, i.e. $\mathrm{TR}(AB) = \mathrm{TR}(BA)$, and commutes with ∂_j and μ_j. In fact, the following theorem holds.

THEOREM 6.6. [**LePf00**, Theorems 2.2, 4.6 and Lemma 5.1] *There is a unique linear extension*

$$\mathrm{TR} : \mathrm{CL}^{\bullet}(M, E; \mathbb{R}^p) \to \mathrm{CS}^{\bullet,\bullet}(\mathbb{R}^p)/\mathbb{C}[\mu_1, \ldots, \mu_p]$$

of TR *to operators of all orders such that*

(1) $\mathrm{TR}(AB) = \mathrm{TR}(BA)$, *i.e.* TR *is tracial.*
(2) $\mathrm{TR}(\partial_j A) = \partial_j \mathrm{TR}(A)$ *for* $j = 1, \ldots, p$.

This unique extension TR *satisfies furthermore:*

(3) $\mathrm{TR}(\mu_j A) = \mu_j \mathrm{TR}(A)$ *for* $j = 1, \ldots, p$.
(4) $\mathrm{TR}(\mathrm{CL}^m(M, E; \mathbb{R}^p)) \subset \mathrm{CS}^{m+p,1}(\mathbb{R}^p)/\mathbb{C}[\mu_1, \ldots, \mu_p]$.

This Theorem is an example where functions with log–polyhomogeneous expansions occur naturally. Note that although an operator $A \in \mathrm{CL}^m(M, E; \mathbb{R}^p)$ has a homogeneous symbol expansion without log terms the trace function $\mathrm{TR}(A)$ is log–polyhomogeneous.

SKETCH OF PROOF. The main observation for the proof is that differentiating by the parameter (6.28) lowers the degree and hence differentiating often enough we obtain a parametric family of trace class operators:

Given $A \in \mathrm{CL}^m(M, E; \mathbb{R}^p)$, then $\partial^\alpha A \in \mathrm{CL}^{m-|\alpha|}(M, E, \mathbb{R}^p)$ is of trace class if $m - |\alpha| + \dim M < 0$. Now integrate the function $\mathrm{TR}(\partial^\alpha A)(\mu)$ back. Since we mod out polynomials this procedure is independent of α and the choice of anti–derivatives. This integration procedure also explains the possible occurrence of log terms in the asymptotic expansion and hence why TR ultimately takes values in $\mathrm{CS}^{\bullet,\bullet}(\mathbb{R}^p)$. For details, see [**LePf00**, Sec. 4]. □

TR is not a trace in the usual sense since it maps into a quotient space of the space of parametric symbols over a point. However, composing any linear functional on $\mathrm{CS}^{\bullet,\bullet}(\mathbb{R}^p)/\mathbb{C}[\mu_1, \dots, \mu_p]$ with TR yields a trace on $\mathrm{CL}^\bullet(M, E; \mathbb{R}^p)$. A very natural choice for such a trace is the Hadamard partie finie integral $⨍$ introduced in Subsection 4.2. Let us first note that for a polynomial $P(\mu) \in \mathbb{C}[\mu_1, \dots \mu_p]$ of degree r the function

$$\int_{|\mu|\le R} P(\mu)d\mu = \sum_{j=p}^{p+r} a_j R^j \tag{6.29}$$

is a polynomial of degree $p + r$ without constant term. In particular

$$⨍_{\mathbb{R}^p} P(\mu)d\mu = 0 \tag{6.30}$$

and hence $⨍_{\mathbb{R}^p}$ induces a linear functional on the quotient space $\mathrm{CS}^{\bullet,\bullet}(\mathbb{R}^p)/\mathbb{C}[\mu_1, \dots, \mu_p]$.

Thus putting for $A \in \mathrm{CL}^\bullet(M, E; \mathbb{R}^p)$

$$\overline{\mathrm{TR}}(A) := ⨍_{\mathbb{R}^p} \mathrm{TR}(A)(\mu)d\mu \tag{6.31}$$

we obtain a trace $\overline{\mathrm{TR}}$ on $\mathrm{CL}^\bullet(M, E; \mathbb{R}^p)$ which extends the natural trace on operators of order $< -\dim M - p$

$$(\int \mathrm{Tr})(A) := \int_{\mathbb{R}^p} \mathrm{Tr}(A(\mu))d\mu. \tag{6.32}$$

However, since $⨍$ is not closed on $\mathrm{CS}^{\bullet,\bullet}(\mathbb{R}^p)$ (Prop. 4.2), $\overline{\mathrm{TR}}$ is not closed on $\mathrm{CL}^\bullet(M, E; \mathbb{R})$. Therefore we obtain derived traces

$$\partial_j\overline{\mathrm{TR}}(A) := \widetilde{\mathrm{TR}}_j(A) := ⨍_{\mathbb{R}^p} \mathrm{TR}(\partial_j A)(\mu)d\mu. \tag{6.33}$$

The relation between $\overline{\mathrm{TR}}$ and $\widetilde{\mathrm{TR}}_j$ can be explained more elegantly in terms of differential forms on $\mathbb{R}^p$ with coefficients in $\mathrm{CL}^\infty(M, E; \mathbb{R}^p)$ (see LESCH, MOSCOVICI and PFLAUM [**LMJ09**]). Let $\Lambda^\bullet := \Lambda^\bullet(\mathbb{R}^p)^* = \mathbb{C}[d\mu_1, \dots, d\mu_p]$ be the exterior algebra of the vector space $(\mathbb{R}^p)^*$ and put

$$\Omega_p := \mathrm{CL}^\infty(M, E; \mathbb{R}^p) \otimes \Lambda^\bullet. \tag{6.34}$$

Then, Ω_p consists of pseudodifferential operator-valued differential forms, the coefficients of $d\mu_I$ being elements of $\mathrm{CL}^\infty(M, E; \mathbb{R}^p)$.

For a p-form $A(\mu)d\mu_1 \wedge \dots \wedge d\mu_p$ we define the *regularized trace* by

$$\overline{\mathrm{TR}}(A(\mu)d\mu_1 \wedge \dots \wedge d\mu_p) := ⨍_{\mathbb{R}^p} \mathrm{TR}(A)(\mu)d\mu_1 \wedge \dots \wedge d\mu_p. \tag{6.35}$$

On forms of degree less than p the regularized trace is defined to be 0. $\overline{\mathrm{TR}}$ is a *graded trace* on the differential algebra (Ω_p, d). In general, $\overline{\mathrm{TR}}$ is not closed. However, its boundary,

$$\widetilde{\mathrm{TR}} := d\overline{\mathrm{TR}} := \overline{\mathrm{TR}} \circ d\,,$$

called the *formal trace*, is a closed graded trace of degree $p-1$. It is shown in [**LePf00**, Prop. 5.8], [**Mel95**, Prop. 6] that $\widetilde{\mathrm{TR}}$ is *symbolic*, i.e. it descends to a well-defined closed graded trace of degree $p-1$ on

$$\partial\Omega_p := \mathrm{CL}^{\infty}(M, E; \mathbb{R}^p)/\,\mathrm{CL}^{-\infty}(M, E; \mathbb{R}^p) \otimes \Lambda^{\bullet}. \tag{6.36}$$

The properties of the formal trace $\widetilde{\mathrm{TR}}$ resemble those of the residue trace.

Denoting by r the quotient map $\Omega_p \to \partial\Omega_p$ we see that Stokes' formula with 'boundary'

$$\overline{\mathrm{TR}}(d\omega) = \widetilde{\mathrm{TR}}(r\omega) \tag{6.37}$$

now holds by construction for any $\omega \in \Omega$.

Finally we mention an interesting linear form on $\mathrm{CS}^{\bullet,\bullet}(\mathbb{R}^p)/\mathbb{C}[\mu_1, \dots, \mu_p]$ in the spirit of the residue trace. Let

$$\Omega^r\,\mathrm{CS}^{\bullet,\bullet}(\mathbb{R}^p) = \mathrm{CS}^{\bullet,\bullet}(\mathbb{R}^p) \otimes \Lambda^{\bullet} \tag{6.38}$$

be the r–forms on $\mathbb{R}^p$ with coefficients in $\mathrm{CS}^{\bullet,\bullet}(\mathbb{R}^p)$. We extend the notion of homogeneous functions to differential forms in the obvious way. If $\omega = f d\mu_{i_1} \wedge \dots \wedge d\mu_{i_r}$ is a form of degree r and $f \in \mathrm{CS}^{a,k}(\mathbb{R}^p)$ then we define the *total degree* of ω to be $r + a$. The exterior derivative preserves the total degree and each $\omega \in \Omega^{\bullet}\,\mathrm{CS}^{\bullet,\bullet}(\mathbb{R}^p)$ of total degree a has an asymptotic expansion

$$\omega \sim \sum_{j=0}^{\infty} \omega_{a-j} \tag{6.39}$$

where ω_{a-j} are forms of total degree $a - j$ which are log–polyhomogeneous in the sense of (3.8), see (3.6). More concretely, if $f \in \mathrm{CS}^{a,k}(\mathbb{R}^p)$ then for $\omega = f\,d\mu_1 \wedge \dots d\mu_r$ we have

$$\omega_{a+r-j} = f_{a-j}. \tag{6.40}$$

Accordingly we define $\omega_{a+r-j,l} := f_{a-j,l}$.

Finally let $X = \sum_{j=1}^{p} \mu_j \frac{\partial}{\partial\mu_j}$ be the Liouville vector field on $\mathbb{R}^p$.

After these preparations we put for $\omega = f d\mu_1 \wedge \dots \wedge d\mu_p \in \Omega^p\,\mathrm{CS}^{\bullet,\bullet}(\mathbb{R}^p)$

$$\mathrm{res}(\omega) := \frac{1}{(2\pi)^p} \int_{S^{p-1}} i_X(\omega_0) = \frac{1}{(2\pi)^p} \int_{S^{p-1}} f_{-p,0} d\,\mathrm{vol}_S\,. \tag{6.41}$$

On forms of degree $< p$ we put $\mathrm{res}(\omega) = 0$.

Proposition 6.7. *If $f \in \mathbb{C}[\mu_1, \dots, \mu_p]$ is a polynomial then*

$$\mathrm{res}(f d\mu_1 \wedge \dots \wedge d\mu_p) = 0.$$

If $\omega \in \Omega^{\bullet}\,\mathrm{CS}^{a,0}(\mathbb{R}^p)$ then $\mathrm{res}(d\omega) = 0$.

The second statement is due to Manchon, Maeda and Paycha [**MMP05**].

PROOF. For $f \in \mathbb{C}[\mu_1, \dots, \mu_p]$ the component of homogeneity degree 0 of $f d\mu_1 \wedge \dots \wedge d\mu_p$ is obviously 0.

Using Cartan's identity we have

$$\begin{aligned} \operatorname{res}(d\omega) &= \int_{S^{p-1}} i_X(d\omega_0) = \int_{S^{p-1}} (i_X d + d i_X)(\omega_0) \\ &= \int_{S^{p-1}} \mathcal{L}_X \omega_0 = 0, \end{aligned} \tag{6.42}$$

since the Lie derivative of a form of homogeneity degree 0 with respect to the Liouville vector field X is 0. □

Composing the res functional with TR we obtain another trace on the algebra $\mathrm{CL}^\bullet(M, E; \mathbb{R}^p)$ which despite of the previous Proposition is not closed. The point here is that the range of TR is not contained in $\mathrm{CS}^\bullet(\mathbb{R}^p)$ but rather in $\mathrm{CS}^{\bullet,1}(\mathbb{R}^p)$.

The significance of this functional and its relation to the noncommutative residue is still to be clarified.

7. Differential forms whose coefficients are symbol functions

Proposition 6.7 says that the res functional on $\Omega^\bullet \mathrm{CS}^\bullet(\mathbb{R}^n)$ descends to a linear functional on the n–th de Rham cohomology of differential forms with coefficients in $\mathrm{CS}^\bullet(\mathbb{R}^n)$. In PAYCHA [**Pay05**] it is shown that the space of linear functionals on $\mathrm{CS}^\bullet(\mathbb{R}^n)$ having the Stokes property is one–dimensional. From this statement in fact the uniqueness of the residue trace can be derived. Translated into our terminology this means that the dual of the n–th de Rham cohomology group of $\mathbb{R}^n$ with coefficients in $\mathrm{CS}^\bullet(\mathbb{R}^n)$ is spanned by res. In particular the n-th de Rham cohomology group of $\mathbb{R}^n$ with coefficients in $\mathrm{CS}^\bullet(\mathbb{R}^n)$ is one–dimensional. In [**Pay05**] it is shown furthermore that the uniqueness statement for linear functionals having the Stokes property is basically equivalent to the uniqueness statement for the residue trace.

We take up this theme and study in a rather general setting the de Rham cohomology of differential forms whose coefficients are symbol functions. The results announced here are inspired by [**Pay05**] but are more general. We pursue here an axiomatic approach. Details will appear elsewhere.

7.1. Differential forms with prescribed asymptotics.

DEFINITION 7.1. Let $\mathscr{A} \subset C^\infty[0,\infty)$ be a Fréchet space with the following properties.

(1) $\mathrm{C}_0^\infty([0,\infty)) \subset \mathscr{A} \subset \mathrm{C}^\infty([0,\infty))$ are continuous embeddings. $\mathrm{C}^\infty([0,\infty))$ carries the usual Fréchet topology of uniform convergence of all derivatives on compact sets and $\mathrm{C}_0^\infty(\mathbb{R})$ has the standard LF-space topology as inductive limit of the Fréchet spaces $\{f \in \mathrm{C}^\infty([0,\infty)) \mid \operatorname{supp} f \subset [0,N]\}$, $N \in \mathbb{N}$.

   We denote by $\mathscr{A}_0 = \{f \in \mathscr{A} \mid \operatorname{supp} f \subset (0,\infty)\}$.

(2) The derivative $\partial := \frac{d}{dx}$ maps $\mathscr{A}$ into $\mathscr{A}$.

(3) There is a non–trivial linear functional $⨍ : \mathscr{A} \to \mathbb{C}$ with the following properties:

   (a) The restriction of $⨍$ to $\mathrm{C}_0^\infty([0,\infty))$ is a multiple of the integral $\int_0^\infty$. That is, there is a $\lambda \in \mathbb{C}$ such that for $f \in \mathrm{C}_0^\infty([0,\infty))$ we have $⨍ f = \lambda \int_0^\infty f(x)dx$.

(b) $⨍$ is *closed* on $\mathscr{A}_0$. That is, for $f \in \mathscr{A}_0$ we have $⨍ f' = 0$.
(c) If $f \in \mathscr{A}_0$ and $⨍ f = 0$ then the function $F := \int_0^\bullet f \in \mathscr{A}$.

REMARK 7.2. It follows from (1) that if $\chi \in \mathrm{C}^\infty([0,\infty))$ with $\chi(x) = 1, x \geq x_0$ and $f \in \mathscr{A}$ then $\chi f \in \mathscr{A}$ because $(1-\chi)f \in \mathrm{C}_0^\infty([0,\infty)) \subset \mathscr{A}$.

2. Since $\mathscr{A}$ is Fréchet it follows from (1) and (2) and the Closed Graph Theorem that $\frac{d}{dx} : \mathscr{A} \to \mathscr{A}$ is continuous.

3. If λ in (3a) is nonzero we can renormalize $⨍$ such that $\lambda = 1$. Thus we are left with two major cases: $\lambda = 1$ and $\lambda = 0$. In the first case $⨍$ is a regularization of the ordinary integral while in the second case $⨍$ is an analogue of the residue trace. This will be explained below in the examples.

EXAMPLE 7.3. 1. The Schwartz space $\mathscr{S}(\mathbb{R})$, $⨍ = \int$.

2. Let $\mathrm{CS}^a([0,\infty))$, $a \in [0,\infty)$ be the classical symbols of order a. This space carries a natural Fréchet topology. If $a \notin \{-1,0,1,\dots\}$ then let $⨍$ be the regularized integral in the partie finie sense described in Subsection 4.2. This integral is continuous with respect to the Fréchet topology on $\mathrm{CS}^a([0,\infty))$.

If $a \in \{-1,0,1,\dots\}$ then let $⨍$ be the residue integral (cf. (6.2)), i.e. if

$$f(x) \sim_{x\to\infty} \sum_{j=0}^\infty f_{a-j} x^{a-j} \tag{7.1}$$

then

$$⨍ f := f_{-1}. \tag{7.2}$$

One can vary this example. With some care one can also deal with log–polyhomogeneous symbols. Moreover, there are classes of symbols of integral order where the regularized integral has the Stokes property [**Pay05**]. These "odd class symbols" also fit into the present framework.

From now on $\mathscr{A}$ will always denote a Fréchet space as in Def. 7.1.

Starting from $\mathscr{A}$ we can construct associated spaces of functions on $\mathbb{R}^n$ respectively on cones over a manifold.

Let M be an oriented compact manifold. By $\mathscr{A}_0([0,\infty)\times M)$ we denote the space of functions $f \in \mathrm{C}^\infty([0,\infty)\times M)$ such that

- There is an $\varepsilon > 0$ such that $f(r,p) = 0$ for $r < \varepsilon, p \in M$.
- For fixed $p \in M$ we have $f(\cdot,p) \in \mathscr{A}$.

Note that for $f \in \mathscr{A}_0([0,\infty)\times M)$ the map $M \to \mathscr{A}, p \mapsto f(\cdot,p)$ is smooth. This follows from the Closed Graph Theorem.

As a consequence we have a continuous integration along the fiber

$$⨍_{([0,\infty)\times M)/M} : \mathscr{A}_0([0,\infty)\times M) \longrightarrow \mathrm{C}^\infty(M), \quad f \mapsto ⨍ f(\cdot,p). \tag{7.3}$$

We put

$$\mathscr{A}_0(\mathbb{R}^n) = \{\pi^* f \mid f \in \mathscr{A}_0([0,\infty)\times S^{n-1}\}, \tag{7.4}$$

where $\pi : \mathbb{R}^n \setminus \{0\} \longrightarrow [0,\infty)\times S^{n-1}, x \mapsto (\|x\|, x/\|x\|)$ is the polar coordinate diffeomorphism.

Furthermore we put $\mathscr{A}(\mathbb{R}^n) := \mathrm{C}_0^\infty(\mathbb{R}^n) + \mathscr{A}_0(\mathbb{R}^n)$. $\mathscr{A}_0(\mathbb{R}^n)$ carries a natural LF-topology while $\mathscr{A}(\mathbb{R}^n)$ carries a natural Fréchet topology.

REMARK 7.4. Composing the integral (7.3) with an integral over M yields a natural integral on $\mathscr{A}_0([0,\infty)) \times M)$. In the case of $M = S^{n-1}$ and the standard integral on S^{n-1} this integral even extends to an integral on $\mathscr{A}(\mathbb{R}^n)$ which has the Stokes property. If $\mathscr{A} = \mathrm{CS}^a([0,\infty))$ the so constructed integral on $\mathscr{A}(\mathbb{R}^n)$ is the Hadamard regularized integral if $a \notin \{-1,0,1,\dots\}$ and the residue integral if $a \in \{-1,0,1,\dots\}$. Thus our approach allows us to discuss these two, a priori rather different, regularized integrals within one common framework.

Finally we denote by $\Omega^k\mathscr{A}_0([0,\infty)\times M)$ the space of differential forms whose coefficients are locally in $\mathscr{A}_0([0,\infty)\times U)$ for any chart $U \subset M$. A more global description in terms of projective tensor products is also possible:

$$\mathscr{A}_0([0,\infty)\times M) = \mathscr{A}_0 \otimes_\pi \mathrm{C}^\infty(M), \tag{7.5}$$

respectively

$$\Omega^\bullet\mathscr{A}_0([0,\infty)\times M) = (\mathscr{A}_0 \oplus \mathscr{A}_0 dr) \otimes_\pi \Omega^\bullet(M). \tag{7.6}$$

By Def. 7.1, (2) the exterior derivative maps $\Omega^k\mathscr{A}_{(0)}(X)$ to $\Omega^{k+1}\mathscr{A}_{(0)}(X)$ for $X = [0,\infty)\times M$, respectively $X = \mathbb{R}^n$. The corresponding cohomology groups are denoted by $H^k\Omega^\bullet\mathscr{A}_{(0)}(X)$. Our goal is to calculate these cohomology groups.

DEFINITION 7.5. We call the $\mathscr{A}$ of *type I* if λ in Def. 7.1 (3a) is 1 and of *type II* if λ is 0.

LEMMA 7.6. *$\mathscr{A}$ is of type II if and only if the constant function 1 is in $\mathscr{A}$. Moreover we have for $k = 0, 1$*

$$H^k\mathscr{A}([0,\infty)) \simeq \begin{cases} 0 & , \text{ if } \mathscr{A} \text{ is of type I,} \\ \mathbb{C} & , \text{ if } \mathscr{A} \text{ is of type II.} \end{cases} \tag{7.7}$$

$H^k\mathscr{A}([0,\infty))$ (obviously) vanishes for $k \geq 2$. Furthermore $⨍$ induces an isomorphism $H^1\mathscr{A}_0([0,\infty)) \simeq \mathbb{C}$.

7.2. Integration along the fiber and statement of the main result.

7.2.1. *Integration along the fiber.* The integration (7.3) extends to an integration along the fiber of differential forms as follows (cf. [**BoTu82**]):

A k–form $\omega \in \Omega^k\mathscr{A}_0([0,\infty)\times M)$ is, locally on M, a sum of differential forms of the form

$$\omega = f_1(r,p)\pi^*\eta_1 + f_2(r,p)\pi^*\eta_2 \wedge dr \tag{7.8}$$

with $f_j \in \mathscr{A}_0([0,\infty)\times M), \eta_1 \in \Omega^k(M), \eta_2 \in \Omega^{k-1}(M)$. For such forms we put

$$\pi_*\omega := \Big(⨍_{([0,\infty)\times M)/M} f_2\Big)\pi^*\eta_2. \tag{7.9}$$

LEMMA 7.7. *π_* extends to a well–defined homomorphism*

$$\Omega^k\mathscr{A}_0([0,\infty)\times M) \longrightarrow \Omega^{k-1}\mathscr{A}_0([0,\infty)\times M).$$

Furthermore, π_* commutes with exterior differentiation, i.e.

$$d_M \circ \pi_* = \pi_* \circ d_{\mathbb{R}_+\times M}.$$

For the proof of this Lemma the closedness of $⨍$ is crucial.

7.2.2. *Statement of the main result.* We are now able to state our main result:

THEOREM 7.8. Type I: *If $\mathscr{A}$ is of type I then the natural inclusion $\Omega_c^\bullet(\mathbb{R}^n) \hookrightarrow \Omega^\bullet\mathscr{A}(\mathbb{R}^n)$ of compactly supported forms induces an isomorphism in cohomology.*

Type II: *If $\mathscr{A}$ is of type II then*

$$H^k\mathscr{A}(\mathbb{R}^n) \simeq \begin{cases} \mathbb{C}, & k = 0, 1, n, \\ 0, & \textit{otherwise}. \end{cases} \tag{7.10}$$

In both cases $⨍$ induces an isomorphism $H^n\mathscr{A}(\mathbb{R}^n) \longrightarrow \mathbb{C}$.

REMARK 7.9. 1. The groups $H^k\mathscr{A}(\mathbb{R}^n)$ can be described more explicitly. Namely, the natural inclusion $\Omega^\bullet\mathscr{A}_0(\mathbb{R}^n) \hookrightarrow \Omega^\bullet\mathscr{A}(\mathbb{R}^n)$ induces isomorphisms

$$H^k\mathscr{A}_0(\mathbb{R}^n) \longrightarrow H^k\mathscr{A}(\mathbb{R}^n)$$

for $k \geq 1$. Furthermore, integration along the fiber induces isomorphisms

$$\pi_* : H^k\mathscr{A}_0(\mathbb{R}^n) \longrightarrow H^{k-1}(S^{n-1}), \quad \text{for } k \geq 1.$$

Thus there is a natural extension of integration along the fiber to closed forms $\pi_* : \Omega^k_{\text{cl}}\mathscr{A}(\mathbb{R}^n) \to \Omega^{k-1}(S^{n-1})$. The isomorphisms $H^k\mathscr{A}_0(\mathbb{R}^n) \longrightarrow \mathbb{C}, \quad k = 1, n$ are given by integration along the fiber.

2. This Theorem generalizes the results of [**Pay05**, Sec. 1] on the characterization of the residue integral and the regularized integral in terms of the Stokes property.

3. The proof of the Theorem is based on the Thom isomorphism below.

7.2.3. *The Thom isomorphism.* We consider again a Fréchet space $\mathscr{A}$ as in Def. 7.1. Having established integration along the fiber the Thom isomorphism is proved along the lines of the classical case of smooth compactly supported forms. The result is as follows:

THEOREM 7.10. *Let $\mathscr{A}$ be a Fréchet algebra as in Def. 7.1. Let M be a compact oriented manifold of dimension n. Furthermore let*

$$\pi_* : \Omega^k\mathscr{A}_0([0, \infty) \times M) \longrightarrow \Omega^{k-1}([0, \infty) \times M)$$

be integration along the fiber as defined in Section 7.2.1.

Then π_* induces an isomorphism

$$H^k\mathscr{A}_0([0, \infty) \times M) \longrightarrow H^{k-1}_{\text{dR}}(M) \tag{7.11}$$

for all $k \geq 0$ (meaning $H^0\mathscr{A}_0([0, \infty) \times M) \simeq \{0\}$.)

References

[BoTu82] R. Bott and L. W. Tu, *Differential forms in algebraic topology*, Graduate Texts in Mathematics, vol. 82, Springer-Verlag, New York, 1982. MR 658304 (83i:57016)

[BrLe99] J. Brüning and M. Lesch, *On the η-invariant of certain nonlocal boundary value problems*, Duke Math. J. **96** (1999), no. 2, 425–468. `arxiv:9609001 [dg-ga,math.DG]`, MR 1666570 (99m:58180)

[CaZy57] A.-P. Calderón and A. Zygmund, *Singular integral operators and differential equations*, Amer. J. Math. **79** (1957), 901–921. MR 0100768 (20 #7196)

[Con88] A. Connes, *The action functional in noncommutative geometry*, Comm. Math. Phys. **117** (1988), no. 4, 673–683. MR 953826 (91b:58246)

[Con94] A. Connes, *Noncommutative geometry*, Academic Press Inc., San Diego, CA, 1994. MR 1303779 (95j:46063)

[CPS03] A. CAREY, J. PHILLIPS, and F. SUKOCHEV, *Spectral flow and Dixmier traces*, Adv. Math. **173** (2003), no. 1, 68–113. arxiv:0205076 [math.OA], MR 1954456 (2004e:58049)

[CRSS07] A. L. CAREY, A. RENNIE, A. SEDAEV, and F. SUKOCHEV, *The Dixmier trace and asymptotics of zeta functions*, J. Funct. Anal. **249** (2007), no. 2, 253–283. arxiv:0611629 [math.OA], MR 2345333

[DIX66] J. DIXMIER, *Existence de traces non normales*, C. R. Acad. Sci. Paris Sér. A-B **262** (1966), A1107–A1108. MR 0196508 (33 #4695)

[DUI96] J. J. DUISTERMAAT, *Fourier integral operators*, Progress in Mathematics, vol. 130, Birkhäuser Boston Inc., Boston, MA, 1996. MR 1362544 (96m:58245)

[GIL95] P. B. GILKEY, *Invariance theory, the heat equation, and the Atiyah-Singer index theorem*, second ed., Studies in Advanced Mathematics, CRC Press, Boca Raton, FL, 1995. MR 1396308 (98b:58156)

[GRSE95] G. GRUBB and R. T. SEELEY, *Weakly parametric pseudodifferential operators and Atiyah-Patodi-Singer boundary problems*, Invent. Math. **121** (1995), no. 3, 481–529. MR 1353307 (96k:58216)

[GRSJ94] A. GRIGIS and J. SJÖSTRAND, *Microlocal analysis for differential operators*, London Mathematical Society Lecture Note Series, vol. 196, Cambridge University Press, Cambridge, 1994, An introduction. MR 1269107 (95d:35009)

[GRU05] G. GRUBB, *A resolvent approach to traces and zeta Laurent expansions*, Spectral geometry of manifolds with boundary and decomposition of manifolds, Contemp. Math., vol. 366, Amer. Math. Soc., Providence, RI, 2005, pp. 67–93. arxiv:0311081 [math.AP], MR 2114484 (2006f:58042)

[GUI85] V. GUILLEMIN, *A new proof of Weyl's formula on the asymptotic distribution of eigenvalues*, Adv. in Math. **55** (1985), no. 2, 131–160. MR 772612 (86i:58135)

[HAD32] J. HADAMARD, *Le problème de Cauchy et les équations aux dérivées partielles linéaires hyperboliques*, Hermann, Paris, 1932.

[HER61A] J. HERSCH, *Caractérisation variationnelle d'une somme de valeurs propres consécutives; généralisation d'inégalités de Pólya-Schiffer et de Weyl*, C. R. Acad. Sci. Paris **252** (1961), 1714–1716. MR 0126065 (23 #A3362)

[HER61B] ———, *Inégalités pour des valeurs propres consécutives de systèmes vibrants inhomogènes allant "en sens inverse" de celles de Pólya-Schiffer et de Weyl*, C. R. Acad. Sci. Paris **252** (1961), 2496–2498. MR 0123908 (23 #A1229)

[HÖR71] L. HÖRMANDER, *Fourier integral operators. I*, Acta Math. **127** (1971), no. 1-2, 79–183. MR 0388463 (52 #9299)

[KARI97] R. V. KADISON and J. R. RINGROSE, *Fundamentals of the theory of operator algebras. Vol. II*, Graduate Studies in Mathematics, vol. 16, American Mathematical Society, Providence, RI, 1997, Advanced theory, Corrected reprint of the 1986 original. MR 1468230 (98f:46001b)

[KAS95] D. KASTLER, *The Dirac operator and gravitation*, Comm. Math. Phys. **166** (1995), no. 3, 633–643. MR 1312438 (95j:58181)

[KAWA95] W. KALAU and M. WALZE, *Gravity, non-commutative geometry and the Wodzicki residue*, J. Geom. Phys. **16** (1995), no. 4, 327–344. arxiv:9312031 [gr-qc], MR 1336738 (96c:58016)

[KONI65] J. J. KOHN and L. NIRENBERG, *An algebra of pseudo-differential operators*, Comm. Pure Appl. Math. **18** (1965), 269–305. MR 0176362 (31 #636)

[KOVI94] M. KONTSEVICH and S. VISHIK, *Determinants of elliptic pseudo-differential operators*, arxiv:9404046 [hep-th].

[KOVI95] ———, *Geometry of determinants of elliptic operators*, Functional analysis on the eve of the 21st century, Vol. 1 (New Brunswick, NJ, 1993), Progr. Math., vol. 131, Birkhäuser Boston, Boston, MA, 1995, pp. 173–197. arxiv:9406140 [hep-th], MR 1373003 (96m:58264)

[LEPF00] M. LESCH and M. J. PFLAUM, *Traces on algebras of parameter dependent pseudodifferential operators and the eta-invariant*, Trans. Amer. Math. Soc. **352** (2000), no. 11, 4911–4936. MR 1661258 (2001b:58042)

[LES97] M. LESCH, *Operators of Fuchs type, conical singularities, and asymptotic methods*, Teubner-Texte zur Mathematik [Teubner Texts in Mathematics], vol. 136, B. G.

Teubner Verlagsgesellschaft mbH, Stuttgart, 1997. `arxiv:9607005 [dg-ga, math.DG]`, MR 1449639 (98d:58174)

[Les99] ———, *On the noncommutative residue for pseudodifferential operators with log-polyhomogeneous symbols*, Ann. Global Anal. Geom. **17** (1999), no. 2, 151–187. `arXiv:9708010 [dg-ga,math.DG]`, MR 1675408 (2000b:58050)

[LMJ09] M. Lesch, H. Moscovici, and P. M. J., *Relative pairing in cyclic cohomology and divisor flows*, J. K-Theory **3** (2009), 359–407. `arXiv:0603500 [math.KT]`

[Mel95] R. B. Melrose, *The eta invariant and families of pseudodifferential operators*, Math. Res. Lett. **2** (1995), no. 5, 541–561. MR 1359962 (96h:58169)

[MiPl49] S. Minakshisundaram and Å. Pleijel, *Some properties of the eigenfunctions of the Laplace-operator on Riemannian manifolds*, Canadian J. Math. **1** (1949), 242–256. MR 0031145 (11,108b)

[MMP05] D. Manchon, Y. Maeda, and S. Paycha, *Stokes' formulae on classical symbol valued forms and applications*, `arxiv:0510454 [math.DG]`.

[MuvN36] F. J. Murray and J. von Neumann, *On rings of operators*, Ann. of Math. (2) **37** (1936), no. 1, 116–229. MR MR1503275

[MuvN37] ———, *On rings of operators. II*, Trans. Amer. Math. Soc. **41** (1937), no. 2, 208–248. MR 1501899

[MuvN43] ———, *On rings of operators. IV*, Ann. of Math. (2) **44** (1943), 716–808. MR 0009096 (5,101a)

[Pay05] S. Paycha, *The noncommutative residue and canonical trace in the light of Stokes' and continuity properties*, `arXiv:0706.2552 [math.OA]`.

[Ped89] G. K. Pedersen, *Analysis now*, Graduate Texts in Mathematics, vol. 118, Springer-Verlag, New York, 1989. MR 971256 (90f:46001)

[See59] R. T. Seeley, *Singular integrals on compact manifolds*, Amer. J. Math. **81** (1959), 658–690. MR 0110022 (22 #905)

[See65] ———, *Integro-differential operators on vector bundles*, Trans. Amer. Math. Soc. **117** (1965), 167–204. MR 0173174 (30 #3387)

[See67] ———, *Complex powers of an elliptic operator*, Singular Integrals (Proc. Sympos. Pure Math., Chicago, Ill., 1966), Amer. Math. Soc., Providence, R.I., 1967, pp. 288–307. MR 0237943 (38 #6220)

[Shu01] M. A. Shubin, *Pseudodifferential operators and spectral theory*, second ed., Springer-Verlag, Berlin, 2001, Translated from the 1978 Russian original by Stig I. Andersson. MR 1852334 (2002d:47073)

[vN40] J. v. Neumann, *On rings of operators. III*, Ann. of Math. (2) **41** (1940), 94–161. MR 0000898 (1,146g)

[Wod84] M. Wodzicki, *Local invariants of spectral asymmetry*, Invent. Math. **75** (1984), no. 1, 143–177. MR 728144 (85g:58089)

[Wod87] M. Wodzicki, *Noncommutative residue. I. Fundamentals*, K-theory, arithmetic and geometry (Moscow, 1984–1986), Lecture Notes in Math., vol. 1289, Springer, Berlin, 1987, pp. 320–399. MR 923140 (90a:58175)

Mathematisches Institut, Universität Bonn, Beringstr. 6, D-53115 Bonn, Germany
E-mail address: `ml@matthiaslesch.de, lesch@math.uni-bonn.de`
URL: `www.matthiaslesch.de, www.math.uni-bonn.de/people/lesch`

Clay Mathematics Proceedings
Volume 12, 2010

Renormalization in Connected Graded Hopf Algebras: An Introduction

Dominique Manchon

ABSTRACT. We give an account of the Connes-Kreimer renormalization in the context of connected graded Hopf algebras. We first explain the Birkhoff decomposition of characters in the more general context of connected filtered Hopf algebras, then specialize down to the graded case in order to introduce the notions of locality, renormalization group and Connes-Kreimer's Beta function. The connection with Rota-Baxter and dendriform algebras will also be outlined. This introductory/survey article is based on joint work with Kurusch Ebrahimi-Fard, Li Guo and Frédéric Patras ([**19**], [**16**], [**21**], [**22**], [**23**]).

CONTENTS

1. Introduction

In any physical system in interaction, it is crucial to make a distinction between actually measured parameters and *bare* parameters, i.e. the value these parameters would take in the absence of any interaction with the environment. *Renormalization* can be briefly defined as any device enabling us to pass from the bare parameters to the actually observed parameters, which will be called *renormalized*. We can have an idea of it by considering a spherical balloon moving in a fluid (water, the air, any gas...), as considered by G. Green as early as 1836 ([**26**], see also [**9**] and [**10**]): at very low speed (such that the friction is negligible), everything happens as if an extra mass $\frac{M}{2}$ had been added to the balloon mass m_0, where M is the mass of the fluid volume replaced by the balloon. The total force $F = mg$ acting on the balloon (with $m = m_0 + \frac{M}{2}$) splits into gravity $F_0 = m_0 g_0$ and Archimedes' force $-Mg_0$, where $g_0 \simeq 9.81\ ms^{-2}$ is the gravity on the Earth's surface. Bare parameters are the mass m_0, the gravity force F_0 and acceleration g_0, whereas the renormalized parameters are:

$$m = m_0 + \frac{M}{2}, \qquad F = (1 - \frac{M}{m_0})F_0, \qquad g = \frac{m_0 - M}{m_0 + \frac{M}{2}} g_0. \tag{1}$$

Let us remark that the initial aceleration g decreases from g_0 to $-2g_0$ when the interaction, represented by the fluid mass M, increases from 0 to $+\infty$. An extra difficulty arises in quantum field theory, even in its perturbative approach: bare parameters are usually infinite! They are typically given by divergent integrals[1] like, for example,

$$\int_{\mathbb{R}^4} \frac{1}{1 + \|p\|^2} \, dp. \tag{2}$$

These infinite quantities illustrate the fact that switching off the interactions in quantum field theory is impossible except as a mental exercise[2]. One must then subtract another infinite quantity from the bare parameter in order to recover the (observed, hence finite) renormalized parameter. This process very often splits into two steps:

(1) *regularization*, which replaces the infinite bare parameter by a function of an auxiliary variable z, which tends to infinity when z tends to some z_0.

(2) Renormalization itself, of purely combinatorial nature. For *renormalizable* theories, it extracts a finite part from the function above when z tends to z_0.

There are a lot of ways to regularize: let us mention the cut-off regularization, which consists in considering integrals like (2) on a ball of radius z (with $z_0 = +\infty$), and dimensional regularization ([**28**], [**4**]), which "integrates on a space of complex dimension z", where z_0 is the spatial dimension d (for example $d = 4$ for the Minkowski space-time)[3]. In this case the function which appears is meromorphic in z with a pole in z_0.

Renormalization is given by the BPHZ algorithm (BPHZ for N. Bogoliubov, O. Parasiuk, K. Hepp and W. Zimmermann, [**3**], [**27**], [**51**]). The combinatorical objects here are *Feynman graphs*, classified according to their loop number L. The *Feynman rules* associate to each graph[4] some quantity to be regularized and renormalized. One has first to choose a *renormalization scheme*, i.e. the finite part for the "simplest" quantities, corresponding to one-loop graphs ($L = 1$). One can then renormalize the other quantities by induction on L. When regularized Feynman rules give meromorphic functions of one complex variable z (which is the

---

[1]More precisely, the physical parameters are given by a *series* in the coupling constants (representing the interaction), each term of which is a divergent integral. We focus here on the renormalization of each of those terms, leaving aside the question of renormalizing the whole series.

[2]In contrast to the balloon considered above, for which the interaction can be brought very close to zero by letting it evolve in a quasi-perfect vacuum.

[3]This "space of dimension z" has been recently given a rigorous meaning in terms of type II factors and spectral triples ([**10**] § 19.2).

[4]together with an extra datum: its *external momenta*.

case, for instance, for dimensional regularization), a most popular regularization scheme is the *minimal subtraction scheme*, which consists in taking the value at z_0 after removing the polar part. D. Kreimer first observed [**30**] that Feynman graphs are organized in a connected graded Hopf algebra. The BPHZ algorithm is then reinterpreted in terms of a Birkhoff decomposition for the regularized Feynman rules understood as an $\mathcal{A}$-valued character of the Hopf algebra, where $\mathcal{A}$ is some algebra of functions of the variable z (e.g. meromorphic functions of the complex variable z for dimensional regularization) [**8**].

2. A summary of Birkhoff–Connes–Kreimer factorization

We introduce the crucial property of connectedness for bialgebras. The main interest resides in the possibility to implement recursive procedures in connected bialgebras, the induction taking place with respect to a filtration (e.g. the coradical filtration) or a grading. An important example of these techniques is the recursive construction of the antipode, which then "comes for free", showing that any connected bialgebra is in fact a connected Hopf algebra. The recursive nature of Bogoliubov's formula in the BPHZ [**3, 27, 51**] approach to perturbative renormalization ultimately comes from the connectedness of the underlying Hopf algebra, respectively the corresponding pro-nilpotency of the Lie algebra of infinitesimal characters.

For details on bialgebras and Hopf algebras we refer the reader to the standard references, e.g. [**48**]. The use of bialgebras and Hopf algebras in combinatorics can be traced back at least to the seminal work of Joni and Rota [**29**].

2.1. Connected graded bialgebras. Let k be a field with characteristic zero. A *graded Hopf algebra* on k is a graded k-vector space

$$\mathcal{H} = \bigoplus_{n\geq 0} \mathcal{H}_n$$

endowed with a product $m : \mathcal{H}\otimes\mathcal{H} \to \mathcal{H}$, a coproduct $\Delta : \mathcal{H} \to \mathcal{H}\otimes\mathcal{H}$, a unit $u : k \to \mathcal{H}$, a co-unit $\varepsilon : \mathcal{H} \to k$ and an antipode $S : \mathcal{H} \to \mathcal{H}$ fulfilling the usual axioms of a Hopf algebra [**48**], and such that:

$$\begin{aligned} m(\mathcal{H}_p\otimes\mathcal{H}_q) &\subset \mathcal{H}_{p+q}, \\ \Delta(\mathcal{H}_n) &\subset \bigoplus_{p+q=n} \mathcal{H}_p\otimes\mathcal{H}_q, \\ S(\mathcal{H}_n) &\subset \mathcal{H}_n. \end{aligned}$$

If we do not ask for the existence of an antipode S on $\mathcal{H}$ we get the definition of a *graded bialgebra*. In a graded bialgebra $\mathcal{H}$ we shall consider the increasing filtration:

$$\mathcal{H}^n = \bigoplus_{p=0}^{n} \mathcal{H}_p.$$

Suppose moreover that $\mathcal{H}$ is *connected*, i.e. $\mathcal{H}_0$ is one-dimensional. Then we have:

$$\operatorname{Ker}\varepsilon = \bigoplus_{n\geq 1} \mathcal{H}_n.$$

PROPOSITION 1. *For any $x \in \mathcal{H}^n, n \geq 1$ we can write:*

$$\Delta x = x \otimes \mathbf{1} + \mathbf{1} \otimes x + \widetilde{\Delta} x, \qquad \widetilde{\Delta} x \in \bigoplus_{\substack{p+q=n,\\ p\neq 0, q\neq 0}} \mathcal{H}_p \otimes \mathcal{H}_q.$$

The map $\widetilde{\Delta}$ is coassociative on $\operatorname{Ker}\varepsilon$ and $\widetilde{\Delta}_k := (I^{\otimes k-1} \otimes \widetilde{\Delta})(I^{\otimes k-2} \otimes \widetilde{\Delta}) \cdots \widetilde{\Delta}$ sends $\mathcal{H}^n$ into $(\mathcal{H}^{n-k})^{\otimes k+1}$.

PROOF. Thanks to connectedness we clearly can write:

$$\Delta x = a(x \otimes \mathbf{1}) + b(\mathbf{1} \otimes x) + \widetilde{\Delta} x$$

with $a, b \in k$ and $\widetilde{\Delta} x \in \operatorname{Ker}\varepsilon \otimes \operatorname{Ker}\varepsilon$. The co-unit property then tells us that, with $k \otimes \mathcal{H}$ and $\mathcal{H} \otimes k$ canonically identified with $\mathcal{H}$:

$$x = (\varepsilon \otimes I)(\Delta x) = bx, \qquad x = (I \otimes \varepsilon)(\Delta x) = ax,$$

hence $a = b = 1$. We shall use the following two variants of Sweedler's notation:

$$\Delta x = \sum_{(x)} x_1 \otimes x_2, \qquad \widetilde{\Delta} x = \sum_{(x)} x' \otimes x'',$$

the second being relevant only for $x \in \operatorname{Ker}\varepsilon$. If x is homogeneous of degree n we can suppose that the components x_1, x_2, x', and x'' in the expressions above are homogeneous as well, and we have then $|x_1| + |x_2| = n$ and $|x'| + |x''| = n$. We easily compute:

$$\begin{aligned}(\Delta \otimes I)\Delta(x) &= x \otimes \mathbf{1} \otimes \mathbf{1} + \mathbf{1} \otimes x \otimes \mathbf{1} + \mathbf{1} \otimes \mathbf{1} \otimes x \\ &\quad + \sum_{(x)} x' \otimes x'' \otimes \mathbf{1} + x' \otimes \mathbf{1} \otimes x'' + \mathbf{1} \otimes x' \otimes x'' \\ &\quad + (\widetilde{\Delta} \otimes I)\widetilde{\Delta}(x)\end{aligned}$$

and

$$\begin{aligned}(I \otimes \Delta)\Delta(x) &= x \otimes \mathbf{1} \otimes \mathbf{1} + \mathbf{1} \otimes x \otimes \mathbf{1} + \mathbf{1} \otimes 1 \otimes x \\ &\quad + \sum_{(x)} x' \otimes x'' \otimes \mathbf{1} + x' \otimes \mathbf{1} \otimes x'' + \mathbf{1} \otimes x' \otimes x'' \\ &\quad + (I \otimes \widetilde{\Delta})\widetilde{\Delta}(x),\end{aligned}$$

hence the co-associativity of $\widetilde{\Delta}$ comes from that of Δ. Finally, it is easily seen by induction on k that for any $x \in \mathcal{H}^n$ we can write:

$$\widetilde{\Delta}_k(x) = \sum_x x^{(1)} \otimes \cdots \otimes x^{(k+1)},$$

with $|x^{(j)}| \geq 1$. The grading imposes

$$\sum_{j=1}^{k+1} |x^{(j)}| = n,$$

so the maximum possible for any degree $|x^{(j)}|$ is $n - k$. □

2.2. Connected filtered bialgebras. A *filtered Hopf algebra* on k is a k-vector space together with an increasing $\mathbb{Z}_+$-indexed filtration:

$$\mathcal{H}^0 \subset \mathcal{H}^1 \subset \cdots \subset \mathcal{H}^n \subset \cdots, \quad \bigcup_n \mathcal{H}^n = \mathcal{H}$$

endowed with a product $m : \mathcal{H} \otimes \mathcal{H} \to \mathcal{H}$, a coproduct $\Delta : \mathcal{H} \to \mathcal{H} \otimes \mathcal{H}$, a unit $u : k \to \mathcal{H}$, a co-unit $\varepsilon : \mathcal{H} \to k$ and an antipode $S : \mathcal{H} \to \mathcal{H}$ fulfilling the usual axioms of a Hopf algebra, and such that

$$m(\mathcal{H}^p \otimes \mathcal{H}^q) \subset \mathcal{H}^{p+q}, \qquad \Delta(\mathcal{H}^n) \subset \sum_{p+q=n} \mathcal{H}^p \otimes \mathcal{H}^q, \qquad \text{and } S(\mathcal{H}^n) \subset \mathcal{H}^n.$$

If we do not ask for the existence of an antipode S on $\mathcal{H}$ we get the definition of a *filtered bialgebra*. For any $x \in \mathcal{H}$ we set:

$$|x| := \min\{n \in \mathbb{N},\ x \in \mathcal{H}^n\}.$$

Any graded bialgebra or Hopf algebra is obviously filtered by the canonical filtration associated to the grading:

$$\mathcal{H}^n := \bigoplus_{i=0}^{n} \mathcal{H}_i,$$

and in that case, if x is a homogeneous element, x is of degree n if and only if $|x| = n$. We say that the filtered bialgebra $\mathcal{H}$ is connected if $\mathcal{H}^0$ is one-dimensional. There is an analogue of Proposition 1 in the connected filtered case, the proof of which is very similar:

PROPOSITION 2. *For any $x \in \mathcal{H}^n, n \geq 1$ we can write:*

$$\Delta x = x \otimes \mathbf{1} + \mathbf{1} \otimes x + \widetilde{\Delta} x, \qquad \widetilde{\Delta} x \in \sum_{\substack{p+q=n,\\ p\neq 0, q\neq 0}} \mathcal{H}^p \otimes \mathcal{H}^q.$$

The map $\widetilde{\Delta}$ is coassociative on $\operatorname{Ker}\varepsilon$ *and $\widetilde{\Delta}_k = (I^{\otimes k-1} \otimes \widetilde{\Delta})(I^{\otimes k-2} \otimes \widetilde{\Delta}) \cdots \widetilde{\Delta}$ sends $\mathcal{H}^n$ into $(\mathcal{H}^{n-k})^{\otimes k+1}$.*

The coradical filtration endows any pointed Hopf algebra $\mathcal{H}$ with a structure of filtered Hopf algebra (S. Montgomery, [**39**] Lemma 1.1). If $\mathcal{H}$ is moreover irreducible (i.e. if the image of k under the unit map u is the unique one-dimensional simple subcoalgebra of $\mathcal{H}$) this filtered Hopf algebra is moreover connected.

2.3. The convolution product. An important result is that any connected filtered bialgebra is indeed a filtered Hopf algebra, in the sense that the antipode comes for free. We give a proof of this fact as well as a recursive formula for the antipode with the help of the *convolution product*: let $\mathcal{H}$ be a (connected filtered) bialgebra, and let $\mathcal{A}$ be any k-algebra (which will be called the *target algebra*). The convolution product on the space $\mathcal{L}(\mathcal{H}, \mathcal{A})$ of linear maps from $\mathcal{H}$ to $\mathcal{A}$ is given by:

$$\begin{aligned} \varphi * \psi(x) &= m_{\mathcal{A}}(\varphi \otimes \psi)\Delta(x) \\ &= \sum_{(x)} \varphi(x_1)\psi(x_2). \end{aligned}$$

PROPOSITION 3. *The map $e = u_{\mathcal{A}} \circ \varepsilon$, given by $e(\mathbf{1}) = \mathbf{1}_{\mathcal{A}}$ and $e(x) = 0$ for any $x \in \operatorname{Ker}\varepsilon$, is a unit for the convolution product. Moreover the set $G(\mathcal{A}) := \{\varphi \in \mathcal{L}(\mathcal{H}, \mathcal{A}),\ \varphi(\mathbf{1}) = \mathbf{1}_{\mathcal{A}}\}$ endowed with the convolution product is a group.*

PROOF. The first statement is straightforward. To prove the second let us consider the formal series:

$$\begin{aligned}\varphi^{*-1}(x) &= \big(e-(e-\varphi)\big)^{*-1}(x)\\ &= \sum_{m\geq 0}(e-\varphi)^{*m}(x).\end{aligned}$$

Using $(e-\varphi)(\mathbf{1})=0$ we have immediately $(e-\varphi)^{*m}(\mathbf{1})=0$, and for any $x\in \operatorname{Ker}\varepsilon$:

$$(e-\varphi)^{*n}(x)=m_{\mathcal{A},n-1}(\underbrace{\varphi\otimes\cdots\otimes\varphi}_{n\text{ times}})\widetilde{\Delta}_{n-1}(x).$$

When $x\in\mathcal{H}^p$ this expression then vanishes for $n\geq p+1$. The formal series ends up then with a finite number of terms for any x, which proves the result. □

COROLLARY 1. *Any connected filtered bialgebra $\mathcal{H}$ is a filtered Hopf algebra. The antipode is defined by*

$$S(x)=\sum_{m\geq 0}(u\circ\varepsilon - I)^{*m}(x). \tag{3}$$

It is given by $S(\mathbf{1})=\mathbf{1}$ and recursively by either of the two formulas for $x\in\operatorname{Ker}\varepsilon$:

$$S(x)=-x-\sum_{(x)}S(x')x'' \quad\text{and}\quad S(x)=-x-\sum_{(x)}x'S(x'').$$

PROOF. The antipode, when it exists, is the inverse of the identity for the convolution product on $\mathcal{L}(\mathcal{H},\mathcal{H})$. One just needs then to apply Proposition 3 with $\mathcal{A}=\mathcal{H}$. The two recursive formulas follow directly from the two equalities

$$m(S\otimes I)\Delta(x)=\ 0\ =m(I\otimes S)\Delta(x)$$

fulfilled by any $x\in\operatorname{Ker}\varepsilon$. □

Let $\mathfrak{g}(\mathcal{A})$ be the subspace of $\mathcal{L}(\mathcal{H},\mathcal{A})$ formed by the elements α such that $\alpha(\mathbf{1})=0$. It is clearly a subalgebra of $\mathcal{L}(\mathcal{H},\mathcal{A})$ for the convolution product. We have:

$$G(\mathcal{A})=e+\mathfrak{g}(\mathcal{A}). \tag{4}$$

From now on we shall suppose that the ground field k is of characteristic zero. For any $x\in\mathcal{H}^n$ the exponential:

$$\exp^*(\alpha)(x)=\sum_{k\geq 0}\frac{\alpha^{*k}(x)}{k!}$$

is a finite sum (ending up at $k=n$). It is a bijection from $\mathfrak{g}(\mathcal{A})$ onto $G(\mathcal{A})$. Its inverse is given by

$$\log^*(e+\alpha)(x)=\sum_{k\geq 1}\frac{(-1)^{k-1}}{k}\alpha^{*k}(x).$$

This sum again ends up at $k=n$ for any $x\in\mathcal{H}^n$. Let us introduce a decreasing filtration on $\mathcal{L}=\mathcal{L}(\mathcal{H},\mathcal{A})$:

$$\mathcal{L}^n:=\{\alpha\in\mathcal{L},\ \alpha_{|_{\mathcal{H}^{n-1}}}=0\}.$$

Clearly $\mathcal{L}^0 = \mathcal{L}$ and $\mathcal{L}^1 = \mathfrak{g}(\mathcal{A})$. We define the valuation val φ of an element φ of $\mathcal{L}$ as the largest integer k such that φ is in $\mathcal{L}^k$. We shall consider in the sequel the ultrametric distance on $\mathcal{L}$ induced by the filtration:

$$d(\varphi, \psi) = 2^{-\operatorname{val}(\varphi-\psi)}. \tag{5}$$

For any $\alpha, \beta \in \mathfrak{g}(\mathcal{A})$ let $[\alpha, \beta] = \alpha * \beta - \beta * \alpha$.

PROPOSITION 4. *We have the inclusion*

$$\mathcal{L}^p * \mathcal{L}^q \subset \mathcal{L}^{p+q},$$

and moreover the metric space $\mathcal{L}$ endowed with the distance defined by (5) is complete.

PROOF. Take any $x \in \mathcal{H}^{p+q-1}$, and any $\alpha \in \mathcal{L}^p$ and $\beta \in \mathcal{L}^q$. We have

$$(\alpha * \beta)(x) = \sum_{(x)} \alpha(x_1)\beta(x_2).$$

Recall that we denote by $|x|$ the minimal n such that $x \in \mathcal{H}^n$. Since $|x_1| + |x_2| = |x| \leq p+q-1$, either $|x_1| \leq p-1$ or $|x_2| \leq q-1$, so the expression vanishes. Now if (ψ_n) is a Cauchy sequence in $\mathcal{L}$ it is immediate to see that this sequence is *locally stationary*, i.e. for any $x \in \mathcal{H}$ there exists $N(x) \in \mathbb{N}$ such that $\psi_n(x) = \psi_{N(x)}(x)$ for any $n \geq N(x)$. Then the limit of (ψ_n) exists and is clearly defined by

$$\psi(x) = \psi_{N(x)}(x).$$

□

As a corollary the Lie algebra $\mathcal{L}^1 = \mathfrak{g}(\mathcal{A})$ is *pro-nilpotent*, in a sense that it is the projective limit of the Lie algebras $\mathfrak{g}(\mathcal{A})/\mathcal{L}^n$, which are nilpotent.

2.4. Characters and infinitesimal characters. Let $\mathcal{H}$ be a connected filtered Hopf algebra over k, and let $\mathcal{A}$ be a *commutative* k-algebra. We shall consider unital algebra morphisms from $\mathcal{H}$ to the target algebra $\mathcal{A}$, which we shall call, slightly abusively, *characters*. We recover of course the usual notion of character when the algebra $\mathcal{A}$ is the ground field k. The notion of character involves only the algebra structure of $\mathcal{H}$. On the other hand the convolution product on $\mathcal{L}(\mathcal{H}, \mathcal{A})$ involves only the *coalgebra* structure on $\mathcal{H}$. Let us consider now the full Hopf algebra structure on $\mathcal{H}$ and see what happens to characters with the convolution product:

PROPOSITION 5. *Let $\mathcal{H}$ be a connected filtered Hopf algebra over k, and let $\mathcal{A}$ be a commutative k-algebra. Then the characters from $\mathcal{H}$ to $\mathcal{A}$ form a group $G_1(\mathcal{A})$ under the convolution product, and for any $\varphi \in G_1(\mathcal{A})$ the inverse is given by*

$$\varphi^{*-1} = \varphi \circ S.$$

We call *infinitesimal characters with values in the algebra $\mathcal{A}$* those elements α of $\mathcal{L}(\mathcal{H}, \mathcal{A})$ such that

$$\alpha(xy) = e(x)\alpha(y) + \alpha(x)e(y).$$

PROPOSITION 6. *Let $G_1(\mathcal{A})$ (resp. $\mathfrak{g}_1(\mathcal{A})$) be the set of characters of $\mathcal{H}$ with values in $\mathcal{A}$ (resp. the set of infinitesimal characters of $\mathcal{H}$ with values in $\mathcal{A}$). Then $G_1(\mathcal{A})$ is a subgroup of $G(\mathcal{A})$, the exponential restricts to a bijection from $\mathfrak{g}_1(\mathcal{A})$ onto $G_1(\mathcal{A})$, and $\mathfrak{g}_1(\mathcal{A})$ is a Lie subalgebra of $\mathfrak{g}(\mathcal{A})$.*

PROOF. Part of these results are a reformulation of Proposition 5 and some points are straightforward. The only non-trivial point concerns $\mathfrak{g}_1(\mathcal{A})$ and $G_1(\mathcal{A})$. Take two infinitesimal characters α and β with values in $\mathcal{A}$ and compute

$$\begin{aligned}(\alpha*\beta)(xy) &= \sum_{(x)(y)} \alpha(x_1y_1)\beta(x_2y_2)\\ &= \sum_{(x)(y)} \big(\alpha(x_1)e(y_1)+e(x_1)\alpha(y_1)\big).\big(\beta(x_2)e(y_2)+e(x_2)\alpha(y_2)\big)\\ &= (\alpha*\beta)(x)e(y)+\alpha(x)\beta(y)+\beta(x)\alpha(y)+e(x)(\alpha*\beta)(y).\end{aligned}$$

Using the commutativity of $\mathcal{A}$ we immediately get

$$[\alpha,\beta](xy)=[\alpha,\beta](x)e(y)+e(x)[\alpha,\beta](y),$$

which shows that $\mathfrak{g}_1(\mathcal{A})$ is a Lie algebra. Now for $\alpha\in\mathfrak{g}_1(\mathcal{A})$ we have

$$\alpha^{*n}(xy)=\sum_{k=0}^{n}\binom{n}{k}\alpha^{*k}(x)\alpha^{*(n-k)}(y),$$

as easily seen by induction on n. A straightforward computation then yields

$$\exp^*(\alpha)(xy)=\exp^*(\alpha)(x)\exp^*(\alpha)(y).$$

□

2.5. Renormalization in connected filtered Hopf algebras. We describe in this section the renormalization à la Connes–Kreimer ([**30**], [**7**], [**8**]) in the abstract context of connected filtered Hopf algebras: the objects to be renormalised are characters with values in a commutative unital target algebra $\mathcal{A}$ endowed with a *renormalization scheme*, i.e. a splitting $\mathcal{A}=\mathcal{A}_-\oplus\mathcal{A}_+$ into two subalgebras which play a symmetrical role, except that one has to decide in wich one to put the unit $\mathbf{1}$. An important example is given by the *minimal subtraction* (MS) *scheme* on the algebra $\mathcal{A}$ of meromorphic functions of one variable z, where $\mathcal{A}_+$ is the algebra of meromorphic functions which are holomorphic at $z=0$, and where $\mathcal{A}_-=z^{-1}\mathbb{C}[z^{-1}]$ stands for the "polar parts". Any $\mathcal{A}$-valued character φ admits a unique *Birkhoff decomposition*

$$\varphi=\varphi_-^{*-1}*\varphi_+,$$

where φ_+ is an $\mathcal{A}_+$-valued character, and where $\varphi_-(\operatorname{Ker}\varepsilon)\subset\mathcal{A}_-$. In the MS scheme case described just above, the renormalized character is the scalar-valued character given by the evaluation of φ_+ at $z=0$ (whereas the evaluation of φ at $z=0$ does not necessarily make sense).

THEOREM 1. *Factorization of the group $G(\mathcal{A})$*

(1) *Let $\mathcal{H}$ be a connected filtered Hopf algebra. Let $\mathcal{A}$ be a commutative unital algebra with a renormalization scheme such that $\mathbf{1}\in\mathcal{A}_+$, and let $\pi:\mathcal{A}\to\mathcal{A}$ be the projection onto $\mathcal{A}_-$ parallel to $\mathcal{A}_+$. Let $G(\mathcal{A})$ be as earlier the group of those $\varphi\in\mathcal{L}(\mathcal{H},\mathcal{A})$ such that $\varphi(\mathbf{1})=\mathbf{1}_{\mathcal{A}}$ endowed with the convolution product. Any $\varphi\in G(\mathcal{A})$ admits a unique Birkhoff decomposition*

$$\varphi=\varphi_-^{*-1}*\varphi_+, \tag{6}$$

where φ_- sends $\mathbf{1}$ to $\mathbf{1}_\mathcal{A}$ and $\operatorname{Ker}\varepsilon$ *into $\mathcal{A}_-$, and where φ_+ sends $\mathcal{H}$ into $\mathcal{A}_+$. The maps φ_- and φ_+ are given on* $\operatorname{Ker}\varepsilon$ *by the following recursive formulas:*

$$\varphi_-(x) = -\pi\Big(\varphi(x) + \sum_{(x)} \varphi_-(x')\varphi(x'')\Big)$$
$$\varphi_+(x) = (I-\pi)\Big(\varphi(x) + \sum_{(x)} \varphi_-(x')\varphi(x'')\Big).$$

where I is the identity map.

(2) *If φ is a character, the components φ_- and φ_+ occurring in the Birkhoff decomposition of φ are characters as well.*

PROOF. The proof goes along the same lines as the proof of Theorem 4 of [**8**]: for the first assertion it is immediate from the definition of π that φ_- sends $\operatorname{Ker}\varepsilon$ into $\mathcal{A}_-$, and that φ_+ sends $\operatorname{Ker}\varepsilon$ into $\mathcal{A}_+$. It only remains to check equality $\varphi_+ = \varphi_- * \varphi$, which is an easy computation:

$$\begin{aligned}\varphi_+(x) &= (I-\pi)\Big(\varphi(x) + \sum_{(x)} \varphi_-(x')\varphi(x'')\Big).\\ &= \varphi(x) + \varphi_-(x) + \sum_{(x)} \varphi_-(x')\varphi(x'')\\ &= (\varphi_- * \varphi)(x).\end{aligned}$$

The proof of assertion (2) can be carried out exactly as in [**8**] and relies on the following *Rota–Baxter relation* in $\mathcal{A}$:

$$\pi(a)\pi(b) = \pi\big(\pi(a)b + a\pi(b)\big) - \pi(ab), \tag{7}$$

which is easily verified by decomposing a and b into their $\mathcal{A}_\pm$-parts. We will derive a more conceptual proof in Paragraph 2.7 below. □

REMARK 1. *Define the* Bogoliubov preparation map *as the map $B : G(\mathcal{A}) \to \mathcal{L}(\mathcal{H},\mathcal{A})$ given by*

$$B(\varphi) = \varphi_- * (\varphi - e), \tag{8}$$

such that for any $x \in$ $\operatorname{Ker}\varepsilon$ *we have*

$$B(\varphi)(x) = \varphi(x) + \sum_{(x)} \varphi_-(x')\varphi(x'').$$

The components of φ in the Birkhoff decomposition read

$$\varphi_- = e - \pi \circ B(\varphi), \qquad \varphi_+ = e + (I-\pi)\circ B(\varphi). \tag{9}$$

On $\operatorname{Ker}\varepsilon$ *they reduce to $-\pi \circ B(\varphi)$, $(I-\pi)\circ B(\varphi)$, respectively. Plugging equation (8) inside (9) and setting $\alpha := e - \varphi$ we get the following expression for φ_-:*

$$\begin{aligned}\varphi_- &= e + P(\varphi_- * \alpha) \\ &= e + P(\alpha) + P\big(P(\alpha)*\alpha\big) + \cdots + \underbrace{P\Big(P\big(\ldots P(}_{n \text{ times}}\alpha)*\alpha\big)\cdots *\alpha\Big) + \cdots\end{aligned} \tag{10}$$

and for φ_+ we find

$$\varphi_+ = e - \tilde{P}(\varphi_- * \alpha) \tag{11}$$

$$
\begin{aligned}
(12) \qquad &= e + \tilde{P}(\varphi_+ * \beta) \\
&= e + \tilde{P}(\beta) + \tilde{P}\big(\tilde{P}(\beta) * \beta\big) - \cdots + \underbrace{\tilde{P}\Big(\tilde{P}\big(\ldots\tilde{P}(}_{n \text{ times}}\beta) * \beta\big)\cdots * \beta\Big) + \cdots
\end{aligned}
$$

with $\beta := -\varphi^{-1} * \alpha = e - \varphi^{-1}$, *and where* $\tilde{P}$ *and* P *are projections on* $\mathcal{L}(\mathcal{H}, \mathcal{A})$ *defined by* $\tilde{P}(\alpha) = (I - \pi) \circ \alpha$ *and* $P(\alpha) = \pi \circ \alpha$, *respectively.*

REMARK 2. *Although subalgebras* $\mathcal{A}_+$ *and* $\mathcal{A}_-$ *can obviously be interchanged so that* $\mathbf{1}$ *always belongs to* $\mathcal{A}_+$, *it is better to keep the notation* $\mathcal{A}_-$ *for the counterterms and* $\mathcal{A}_+$ *for the renormalized quantities. Hence the unit* $\mathbf{1}$ *of the target algebra* $\mathcal{A}$ *can belong to* $\mathcal{A}_-$ *in some renormalization schemes: the most common example of this situation in physics is the* zero-momentum subtraction scheme *which can be briefly recast as follows (see* [**4**] *Paragraph 3.4.2): the target algebra* $\mathcal{A}$ *is the algebra of functions which are rational with respect to external* and *internal momenta (denoted by letters* p *and* k *respectively), well-defined at the origin, and polynomial with respect to an extra indeterminate* λ. *The product is given by:*

$$
(13) \qquad f.g(\lambda, k_1, k_2, p_1, p_2) := f(\lambda, k_1, p_1)g(\lambda, k_2, p_2).
$$

The subalgebra $\mathcal{A}_+$ *is the subalgebra of functions* $f = \sum_{d=0}^{r} \lambda^d f_d$ *such that* f_d, *as a function of* external *momenta, vanishes at the origin at order* $\geq d+1$. *The subalgebra* $\mathcal{A}_-$ *is the subalgebra of functions* $f = \sum_{d=0}^{r} \lambda^d f_d$ *such that* f_d *is a polynomial in external momenta of degree* $\leq d$. *The character* φ *from the Hopf algebra of Feynman graphs (with external momenta) to* $\mathcal{A}$ *is given by* $\Gamma \mapsto \lambda^{d(\Gamma)} I_\Gamma$, *where* I_Γ *is the integrand for the Feynman rules, and* $d(\Gamma)$ *is the superficial degree of divergence of the graph. Hence the renormalization is performed before integrating with respect to internal momenta, and is based on subtracting the terms of degree* $\leq d(\Gamma)$ *w.r.t. external momenta in the taylor expansion of the integrand. This is the original setting in which the BPHZ algorithm has been first developed* [**3**], [**27**], [**51**]. *The on-shell scheme can be understood in a similar way, by considering Taylor expansions around the physical mass of the theory instead of around the origin (see* [**4**], *Paragraphs 3.1.3 and 3.3.1).*

2.6. The Baker–Campbell–Hausdorff recursion. Let $\mathcal{L}$ be any complete filtered Lie algebra. Thus $\mathcal{L}$ has a decreasing filtration $(\mathcal{L}_n)$ of Lie subalgebras such that $[\mathcal{L}_m, \mathcal{L}_n] \subseteq \mathcal{L}_{m+n}$ and $\mathcal{L} \cong \varprojlim \mathcal{L}/\mathcal{L}_n$ (i.e., $\mathcal{L}$ is complete with respect to the topology induced by the filtration). Let A be the completion of the enveloping algebra $\mathcal{U}(\mathcal{L})$ for the decreasing filtration naturally coming from that of $\mathcal{L}$. The functions

$$
\exp : A_1 \to 1 + A_1, \qquad \exp(a) = \sum_{n=0}^{\infty} \frac{a^n}{n!},
$$

$$
\log : 1 + A_1 \to A_1, \qquad \log(1+a) = -\sum_{n=1}^{\infty} \frac{(-a)^n}{n}
$$

are well-defined and are inverses of each other. The Baker–Campbell–Hausdorff (*BCH*) formula writes for any $x, y \in \mathcal{L}_1$ [**44, 50**],

$$
\exp(x)\exp(y) = \exp\big(C(x,y)\big) = \exp\big(x + y + \mathrm{BCH}(x,y)\big),
$$

where $\mathrm{BCH}(x,y)$ is an element of $\mathcal{L}_2$ given by a Lie series the first few terms of which are:

$$\mathrm{BCH}(x,y)=\frac{1}{2}[x,y]+\frac{1}{12}[x,[x,y]]+\frac{1}{12}[y,[y,x]]-\frac{1}{24}[x,[y,[x,y]]]+\cdots$$

Now let $P:\mathcal{L}\to\mathcal{L}$ be any linear map preserving the filtration of $\mathcal{L}$. We define $\tilde{P}$ to be $\mathrm{Id}_{\mathcal{L}}-P$. For $a\in\mathcal{L}_1$, define $\chi(a)=\lim_{n\to\infty}\chi_{(n)}(a)$ where $\chi_{(n)}(a)$ is given by the BCH recursion

$$\begin{aligned}&\chi_{(0)}(a):=a,\\ (14)\qquad &\chi_{(n+1)}(a)=a-\mathrm{BCH}\big(P(\chi_{(n)}(a)),\,(\mathrm{Id}_{\mathcal{L}}-P)(\chi_{(n)}(a))\big),\end{aligned}$$

and where the limit is taken with respect to the topology given by the filtration. Then the map $\chi:\mathcal{L}_1\to\mathcal{L}_1$ satisfies

$$\chi(a)=a-\mathrm{BCH}\big(P(\chi(a)),\,\tilde{P}(\chi(a))\big). \tag{15}$$

This map appeared in [**13**], [**12**], where more details can be found, see also [**37, 38**]. The following proposition ([**16**], [**37**]) gives further properties of the map χ.

PROPOSITION 7. *For any linear map $P:\mathcal{L}\to\mathcal{L}$ preserving the filtration of $\mathcal{L}$ there exists a (usually non-linear) unique map $\chi:\mathcal{L}_1\to\mathcal{L}_1$ such that $(\chi-\mathrm{Id}_{\mathcal{L}})(\mathcal{L}_i)\subset\mathcal{L}_{2i}$ for any $i\geq 1$, and such that, with $\tilde{P}:=\mathrm{Id}_{\mathcal{L}}-P$ we have*

$$\forall a\in\mathcal{L}_1,\quad a=C\Big(P\big(\chi(a)\big),\,\tilde{P}\big(\chi(a)\big)\Big). \tag{16}$$

This map is bijective, and its inverse is given by

$$\chi^{-1}(a)=C\big(P(a),\,\tilde{P}(a)\big)=a+\mathrm{BCH}\big(P(a),\,\tilde{P}(a)\big). \tag{17}$$

PROOF. Equation (16) can be rewritten as

$$\chi(a)=F_a\big(\chi(a)\big),$$

with $F_a:\mathcal{L}_1\to\mathcal{L}_1$ defined by

$$F_a(b)=a-\mathrm{BCH}\big(P(b),\tilde{P}(b)\big).$$

This map F_a is a contraction with respect to the metric associated with the filtration: indeed if $b,\varepsilon\in\mathcal{L}_1$ with $\varepsilon\in\mathcal{L}_n$, we have

$$F_a(b+\varepsilon)-F_a(b)=\mathrm{BCH}\big(P(b),\,\tilde{P}(b)\big)-\mathrm{BCH}\big(P(b+\varepsilon),\,\tilde{P}(b+\varepsilon)\big).$$

The right-hand side is a sum of iterated commutators in each of which ε does appear at least once. So it belongs to $\mathcal{L}_{n+1}$. So the sequence $F_a^n(b)$ converges in $\mathcal{L}_1$ to a unique fixed point $\chi(a)$ for F_a.

Let us remark that for any $a\in\mathcal{L}_i$, then, by a straightforward induction argument, $\chi_{(n)}(a)\in\mathcal{L}_i$ for any n, so $\chi(a)\in\mathcal{L}_i$ by taking the limit. Then the difference $\chi(a)-a=\mathrm{BCH}\big(P(\chi(a)),\,\tilde{P}(\chi(a))\big)$ clearly belongs to $\mathcal{L}_{2i}$. Now consider the map $\psi:\mathcal{L}_1\to\mathcal{L}_1$ defined by $\psi(a)=C\big(P(a),\,\tilde{P}(a)\big)$. It is clear from the definition of χ that $\psi\circ\chi=\mathrm{Id}_{\mathcal{L}_1}$. Then χ is injective and ψ is surjective. The injectivity of ψ will be an immediate consequence of the following lemma.

LEMMA 1. *The map ψ increases the ultrametric distance given by the filtration.*

PROOF. For any $x, y \in \mathcal{L}_1$ the distance $d(x,y)$ is given by 2^{-n} where $n = \sup\{k \in \mathbb{N},\, x-y \in \mathcal{L}_k\}$. We have then to prove that $\psi(x)-\psi(y) \notin \mathcal{L}_{n+1}$. But

$$\begin{aligned}\psi(x)-\psi(y) &= x-y+\mathrm{BCH}\big(P(x),\, \tilde{P}(x)\big) - \mathrm{BCH}\big(P(y),\, \tilde{P}(y)\big)\\ &= x-y+\Big(\mathrm{BCH}\big(P(x),\, \tilde{P}(x)\big) - \mathrm{BCH}\big(P(x)-P(x-y),\, \tilde{P}(x)-\tilde{P}(x-y)\big)\Big).\end{aligned}$$

The rightmost term inside the large brackets clearly belongs to $\mathcal{L}_{n+1}$. As $x-y \notin \mathcal{L}_{n+1}$ by hypothesis, this proves the claim. □

The map ψ is then a bijection, so χ is also bijective, which proves Proposition 7. □

COROLLARY 2. *For any $a \in \mathcal{L}_1$ we have the following equality taking place in $1 + A_1 \subset A$:*

$$\exp(a) = \exp\big(P(\chi(a))\big)\exp\big(\tilde{P}(\chi(a))\big). \tag{18}$$

Putting (10) and (18) together we get for any $\alpha \in \mathcal{L}_1$ the following *non-commutative Spitzer identity*:

$$e+P(\alpha)+\cdots+\underbrace{P\Big(P\big(\ldots P(\alpha)*\alpha\big)\cdots*\alpha\Big)}_{n \text{ times}}+\cdots = \exp\Big[-P\Big(\chi\big(\log(e-\alpha)\big)\Big)\Big]. \tag{19}$$

This identity is valid for any filtration-preserving Rota–Baxter operator P in a complete filtered Lie algebra (see section 4). For a detailed treatment of these aspects, see [**13**], [**12**], [**16**], [**22**].

2.7. Application to perturbative renormalization. Suppose now that $\mathcal{L} = \mathcal{L}(\mathcal{H}, \mathcal{A})$ (with the setup and notations of paragraph 2.5), and that the operator P is now the projection defined by $P(a) = \pi \circ a$. It is clear that Corollary 2 applies in this setting and that the first factor on the right-hand side of (18) is an element of $G_1(\mathcal{A})$, the group of $\mathcal{A}$-valued characters of $\mathcal{H}$, which sends $\operatorname{Ker}\varepsilon$ into $\mathcal{A}_-$, and that the second factor is an element of G_1 which sends $\mathcal{H}$ into $\mathcal{A}_+$. Going back to Theorem 1 and using uniqueness of the decomposition (6) we see then that (18) in fact is the Birkhoff–Connes–Kreimer decomposition of the element $\exp^*(a)$ in G_1. Indeed, starting with the infinitesimal character a in the Lie algebra $\mathfrak{g}_1(\mathcal{A})$ equation (18) gives the Birkhoff–Connes–Kreimer decomposition of $\varphi = \exp^*(a)$ in the group $G_1(\mathcal{A})$ of $\mathcal{A}$-valued characters of $\mathcal{H}$, i.e.,

$$\varphi_- = \exp^*\big(-P(\chi(a))\big) \quad\text{and}\quad \varphi_+ = \exp^*\big(\tilde{P}(\chi(a))\big) \quad\text{such that}\quad \varphi = \varphi_-^{-1} * \varphi_+,$$

thus proving the second assertion in Theorem 1. Comparing Corollary 2 and Theorem 1 the reader may wonder about the role played by the Rota–Baxter relation (7) for the projector P. In the following section we will show that it is this identity that allows us to write the exponential $\varphi_- = \exp^*\big(-P(\chi(a))\big)$ as a recursion, that is, $\varphi_- = e + P(\varphi_- * \alpha)$, with $\alpha = e - \varphi$. Equivalently, this amounts to the fact that the group $G_1(\mathcal{A})$ factorizes into two subgroups $G_1^-(\mathcal{A})$ and $G_1^+(\mathcal{A})$, such that $\varphi_\pm \in G_1^\pm(\mathcal{A})$.

3. Locality, the renormalization group and the Beta function

3.1. The Dynkin operator. Any connected graded Hopf algebra $\mathcal{H}$ admits a natural biderivation Y defined by $Y(x) = nx$ for $x \in \mathcal{H}_n$. The map $\varphi \mapsto \varphi \circ Y$ is a derivation of $\big(\mathcal{L}(\mathcal{H}, \mathcal{A}), *\big)$. When the ground field is $k = \mathbb{R}$ or $\mathbb{C}$ the biderivation Y gives rise to the one-parameter subgroup of automorphisms of $\mathcal{H}$ given by

$\theta_t(x) = e^{nt}x$ for $x \in \mathcal{H}_n$, and $\varphi \mapsto \varphi \circ \theta_t$ is an automorphism of $\big(\mathcal{L}(\mathcal{H}, \mathcal{A}), *\big)$ for any $t \in k$.

The *Dynkin operator* is defined as the endomorphism $D = S * Y$ of $\mathcal{H}$ (where S is the antipode). One can show that for any commutative unital algebra $\mathcal{A}$ the correspondence $\varphi \mapsto \varphi \circ D$ gives rise to a bijection Ξ from the group of characters $G_{\mathcal{A}}$ onto the Lie algebra of infinitesimal characters $\mathfrak{g}_{\mathcal{A}}$. When $k = \mathbb{R}$ ou $\mathbb{C}$ the inverse $\Xi^{-1} = \Gamma : \mathfrak{g}_{\mathcal{A}} \to G_{\mathcal{A}}$ is given by the following formula ([**38**] § 8.2):

$$\Gamma(\alpha) = e + \sum_{n\geq 1} \int_{0\leq v_n \leq \cdots \leq v_1 \leq +\infty} (\alpha \circ \theta_{-v_1}) * \cdots * (\alpha \circ \theta_{-v_n})\, dv_1 \cdots dv_n. \tag{20}$$

Applying this to an element x in $\mathcal{H}$ decomposed into its homogeneous components x_k (one can suppose $x_0 = 0$), and using the equality

$$\int_{0\leq v_l \leq \cdots \leq v_1 \leq +\infty} e^{-k_1 v_1} \cdots e^{-k_l v_l}\, dv_1 \cdots dv_l = \frac{1}{k_1(k_1+k_2)\cdots(k_1+\cdots+k_l)}, \tag{21}$$

one easily infers the explicit formula [**17**] :

$$\Gamma(\alpha) = e + \sum_{n\geq 1} \sum_{k_1,\ldots,k_l \in \mathbb{N}^*,\ k_1+\cdots+k_l = n} \frac{\alpha_{k_1} * \cdots * \alpha_{k_l}}{k_1(k_1+k_2)\cdots(k_1+\cdots+k_l)}, \tag{22}$$

with $\alpha_k = \alpha \circ \pi_k$, and where for any $k \geq 0$ one denotes by π_k the projection of $\mathcal{H}$ onto the homogeneous component $\mathcal{H}_k$ of degree k. One can then easily verify the same formula on the field of rational numbers, and then on any field k of characteristic zero. The Dynkin operator was introduced in the general setting of commutative or cocommutative Hopf algebras by F. Patras and Chr. Reutenauer ([**42**], see also [**17**]). Several properties, such as the explicit formula (22) above, still make sense for any connected graded Hopf algebra.

3.2. The renormalization group and the Beta function. We suppose $k = \mathbb{R}$ or $k = \mathbb{C}$ here. We will consider the one-parameter group $\varphi \mapsto \varphi \circ \theta_{tz}$ of automorphisms of the algebra $\big(\mathcal{L}(\mathcal{H}, \mathcal{A}), *\big)$ i.e.,

$$\varphi^t(x)(z) := e^{tz|x|}\varphi(x)(z). \tag{23}$$

Differentiating at $t = 0$ we get

$$\frac{d}{dt}\Big|_{t=0} \varphi^t = z(\varphi \circ Y). \tag{24}$$

Let $G_{\mathcal{A}}$ be any of the two groups $G(\mathcal{A})$ or $G_1(\mathcal{A})$ (see Paragraph 2.4). We denote by $G_{\mathcal{A}}^{\mathrm{loc}}$ the set of *local* elements of $G_{\mathcal{A}}$, i.e. those $\varphi \in G_{\mathcal{A}}$ such that the negative part of the Birkhoff decomposition of φ^t does not depend on t, namely

$$G_{\mathcal{A}}^{\mathrm{loc}} = \Big\{\varphi \in G_{\mathcal{A}} \Big|\ \frac{d}{dt}(\varphi^t)_- = 0\Big\}.$$

In particular the dimensional-regularized Feynman rules verify this property: in physical terms, the counterterms do not depend on the choice of the arbitrary mass parameter μ ('t Hooft's mass) one must introduce in dimensional regularization in order to get dimensionless expressions, which is indeed a manifestation of locality (see [**9**]). We also denote by $G_{\mathcal{A}-}^{\mathrm{loc}}$ the elements φ of $G_{\mathcal{A}}^{\mathrm{loc}}$ such that $\varphi = \varphi_-^{*-1}$. Since composition on the right with Y is a derivation for the convolution product, the map Ξ of the preceding paragraph verifies a cocycle property:

$$\Xi(\varphi * \psi) = \Xi(\psi) + \psi^{*-1} * \Xi(\varphi) * \psi. \tag{25}$$

We summarise some key results of [9] in the following proposition:

PROPOSITION 8. *(1) For any $\varphi \in G_{\mathcal{A}}$ there is a one-parameter family h_t in $G_{\mathcal{A}}$ such that $\varphi^t = \varphi * h_t$, and we have*

$$\dot{h}_t := \frac{d}{dt} h_t = h_t * z\Xi(h_t) + z\Xi(\varphi) * h_t. \tag{26}$$

(2) $z\Xi$ restricts to a bijection from $G^{\mathrm{loc}}_{\mathcal{A}}$ onto $\mathfrak{g}_{\mathcal{A}} \cap \mathcal{L}(\mathcal{H}, \mathcal{A}_+)$. Moreover it is a bijection from $G^{\mathrm{loc}}_{\mathcal{A}\,-}$ onto those elements of $\mathfrak{g}_{\mathcal{A}}$ with values in the constants, i.e.

$$\mathfrak{g}^c_{\mathcal{A}} = \mathfrak{g}_{\mathcal{A}} \cap \mathcal{L}(\mathcal{H}, \mathbb{C}).$$

(3) For $\varphi \in G^{\mathrm{loc}}_{\mathcal{A}}$, the constant term of h_t, defined by

$$F_t(x) = \lim_{z \to 0} h_t(x)(z) \tag{27}$$

is a one-parameter subgroup of $G_{\mathcal{A}} \cap \mathcal{L}(\mathcal{H}, \mathbb{C})$, the scalar-valued characters of $\mathcal{H}$.

PROOF. For any $\varphi \in G_{\mathcal{A}}$ one can write

$$\varphi^t = \varphi * h_t \tag{28}$$

with $h_t \in G_{\mathcal{A}}$. From (28), (24) and the definition of Ξ we immediately get

$$\varphi * \dot{h}_t = \varphi * h_t * z\Xi(\varphi * h_t).$$

Equation (26) then follows from the cocycle property (25). This proves the first assertion. Now take any character $\varphi \in G^{\mathrm{loc}}_{\mathcal{A}}$ with Birkhoff decomposition $\varphi = \varphi_-^{*-1} * \varphi_+$ and write the Birkhoff decomposition of φ^t:

$$\begin{aligned} \varphi^t &= (\varphi^t)_-^{*-1} * (\varphi^t)_+ \\ &= (\varphi_-)^{*-1} * (\varphi^t)_+ \\ &= (\varphi * \varphi_+^{*-1}) * (\varphi^t)_+ \\ &= \varphi * h_t, \end{aligned}$$

with h_t taking values in $\mathcal{A}_+$. Then $z\Xi(\varphi)$ also takes values in $\mathcal{A}_+$, as a consequence of equation (26) at $t = 0$. Conversely, suppose that $z\Xi(\varphi)$ takes values in $\mathcal{A}_+$. We show that h_t also takes values in $\mathcal{A}_+$ for any t, which immediately implies that φ belongs to $G^{\mathrm{loc}}_{\mathcal{A}}$.

For any $\gamma \in \mathfrak{g}_{\mathcal{A}}$, let us introduce the linear transformation U_γ of $\mathfrak{g}_{\mathcal{A}}$ defined by

$$U_\gamma(\delta) := \gamma * \delta + z\delta \circ Y.$$

If γ belongs to $\mathfrak{g}_{\mathcal{A}} \cap \mathcal{L}(\mathcal{H}, \mathcal{A}_+)$ then U_γ restricts to a linear transformation of $\mathfrak{g}_{\mathcal{A}} \cap \mathcal{L}(\mathcal{H}, \mathcal{A}_+)$.

LEMMA 2. *For any $\varphi \in G_{\mathcal{A}}$, $n \in \mathbb{N}$ we have*

$$z^n \varphi \circ Y^n = \varphi * U^n_{z\Xi(\varphi)}(e).$$

PROOF. The case $n = 0$ is obvious, $n = 1$ is just the definition of Ξ. We check thus by induction, using again the fact that composition on the right with Y is a derivation for the convolution product:

$$\begin{aligned} z^{n+1}\varphi \circ Y^{n+1} &= z(z^n\varphi \circ Y^n) \circ Y \\ &= z\big(\varphi * U^n_{z\Xi(\varphi)}(e)\big) \circ Y \\ &= z(\varphi \circ Y) * U^n_{z\Xi(\varphi)}(e) + z\varphi * \big(U^n_{z\Xi(\varphi)}(e) \circ Y\big) \\ &= \varphi * \big(z\Xi(\varphi) * U^n_{z\Xi(\varphi)}(e) + zU^n_{z\Xi(\varphi)}(e) \circ Y\big) \\ &= \varphi * U^{n+1}_{z\Xi(\varphi)}(e). \end{aligned}$$

□

Let us go back to the proof of Proposition 8. According to Lemma 2 we have for any t, at least formally,

$$\varphi^t = \varphi * \exp(tU_{z\Xi(\varphi)})(e). \tag{29}$$

We still have to fix the convergence of the exponential just above in the case when $z\Xi(\varphi)$ belongs to $\mathcal{L}(\mathcal{H}, \mathcal{A}_+)$. Let us consider the following decreasing bifiltration of $\mathcal{L}(\mathcal{H}, \mathcal{A}_+)$:

$$\mathcal{L}^{p,q}_+ = (z^q\mathcal{L}(\mathcal{H}, \mathcal{A}_+)) \cap \mathcal{L}^p,$$

where $\mathcal{L}^p$ is the set of those $\alpha \in \mathcal{L}(\mathcal{H}, \mathcal{A})$ such that $\alpha(x) = 0$ for any $x \in \mathcal{H}$ of degree $\leq p - 1$. In particular $\mathcal{L}^1 = \mathfrak{g}_0$. Considering the associated filtration,

$$\mathcal{L}^n_+ = \sum_{p+q=n} \mathcal{L}^{p,q}_+,$$

we see that for any $\gamma \in \mathfrak{g}_0 \cap \mathcal{L}(\mathcal{H}, \mathcal{A}_+)$ the transformation U_γ increases the filtration by 1, i.e.,

$$U_\gamma(\mathcal{L}^n_+) \subset \mathcal{L}^{n+1}_+.$$

The algebra $\mathcal{L}(\mathcal{H}, \mathcal{A}_+)$ is not complete with respect to the topology induced by this filtration, but the completion is $\mathcal{L}(\mathcal{H}, \widehat{\mathcal{A}_+})$, where $\widehat{\mathcal{A}_+} = \mathbb{C}[[z]]$ stands for the formal series. Hence the right-hand side of (29) is convergent in $\mathcal{L}(\mathcal{H}, \widehat{\mathcal{A}_+})$ with respect to this topology. Hence for any $\gamma \in \mathcal{L}(\mathcal{H}, \mathcal{A}_+)$ and for φ such that $z\Xi(\varphi) = \gamma$ we have $\varphi^t = \varphi * h_t$ with $h_t \in \mathcal{L}(\mathcal{H}, \widehat{\mathcal{A}_+})$ for any t. On the other hand we already know that h_t takes values in meromorphic functions for each t. So h_t belongs to $\mathcal{L}(\mathcal{H}, \mathcal{A}_+)$, which proves the first part of the second assertion. Equation (26) at $t = 0$ reads

$$z\Xi(\varphi) = \dot{h}(0) = \frac{d}{dt}_{|t=0}(\varphi^t)_+. \tag{30}$$

For $\varphi \in G^{\text{loc}}_{\mathcal{A}_-}$ we have, thanks to the property $\varphi(\operatorname{Ker}\varepsilon) \subset \mathcal{A}_-$,

$$\begin{aligned} h_t(x) = (\varphi^t)_+(x) &= (I - \pi)\Big(\varphi^t(x) + \sum_{(x)} \varphi^{*-1}(x')\varphi^t(x'')\Big) \\ &= t(I - \pi)\big(z|x|\varphi(x) + z\sum_{(x)} \varphi^{*-1}(x')\varphi(x'')|x''|\big) + O(t^2) \\ &= t\,\operatorname{Res}(\varphi \circ Y) + O(t^2), \end{aligned}$$

hence:

$$\dot{h}(0) = \operatorname{Res}\,(\varphi \circ Y). \tag{31}$$

From equations (24), (31) and the definition of Ξ we get

$$z\Xi(\varphi) = \text{Res } (\varphi \circ Y) \tag{32}$$

for any $\varphi \in G^{\text{loc}}_{\mathcal{A}_-}$, hence $z\widetilde{R}(\varphi) \in \mathfrak{g}^c$. Conversely let β in $\mathfrak{g}^c$. Consider $\psi = \Xi^{-1}(z^{-1}\beta)$. This element of $G_{\mathcal{A}}$ verifies, thanks to the definition of Ξ,

$$z\psi \circ Y = \psi * \beta.$$

Hence for any $x \in \operatorname{Ker}\varepsilon$ we have

$$z\psi(x) = \frac{1}{|x|}\Big(\beta(x) + \sum_{(x)} \psi(x')\beta(x'')\Big).$$

As $\beta(x)$ is a constant (as a function of the complex variable z) it is easily seen by induction on $|x|$ that the right-hand side evaluated at z has a limit when z tends to zero. Thus $\psi(x) \in \mathcal{A}_-$, and then

$$\psi = \Xi^{-1}\Big(\frac{1}{z}\beta\Big) \in G^{\text{loc}}_{\mathcal{A}_-},$$

which proves assertion (2).

Let us prove assertion (3): the equation $\varphi^t = \varphi * h_t$ together with $(\varphi^t)^s = \varphi^{t+s}$ yields

$$h_{s+t} = h_s * (h_t)^s. \tag{33}$$

Taking values at $z = 0$ immediately yields the one-parameter group property

$$F_{s+t} = F_s * F_t \tag{34}$$

thanks to the fact that the evaluation at $z = 0$ is an algebra morphism. □

We can now give a definition of the Beta function: for any $\varphi \in G^{\text{loc}}_{\mathcal{A}}$, the Beta function of φ is the generator of the one-parameter group F_t defined by equation (27) in Proposition 8. It is the element of the dual $\mathcal{H}^*$ defined by

$$\beta(\varphi) := \frac{d}{dt}\Big|_{t=0} F_t(x) \tag{35}$$

for any $x \in \mathcal{H}$.

PROPOSITION 9. *For any $\varphi \in G^{\text{loc}}_{\mathcal{A}}$ the Beta function of φ coincides with that of the negative part φ_-^{*-1} in the Birkhoff decomposition. It is given by any of the three expressions:*

$$\begin{aligned}\beta(\varphi) &= \text{Res } \Xi(\varphi)\\ &= \text{Res } (\varphi_-^{*-1} \circ Y)\\ &= -\text{Res } (\varphi_- \circ Y).\end{aligned}$$

PROOF. The third equality will be derived from the second by taking residues on both sides of the equation

$$0 = \Xi(e) = \Xi(\varphi_-) + \varphi_-^{*-1} * \Xi(\varphi_-^{*-1}) * \varphi_-,$$

which is a special instance of the cocycle formula (25). Suppose first $\varphi \in G^{\text{loc}}_{\mathcal{A}_-}$, hence $\varphi_-^{*-1} = \varphi$. Then $z\Xi(\varphi)$ is a constant according to assertion 2 of Proposition 8. The proposition then follows from equation (31) evaluated at $z = 0$, and equation

(32). Suppose now $\varphi \in G_{\mathcal{A}}^{\text{loc}}$, and consider its Birkhoff decomposition. As both components belong to $G_{\mathcal{A}}^{\text{loc}}$ we apply Proposition 8 to them. In particular we have

$$\begin{aligned}\varphi^t &= \varphi * h_t,\\ (\varphi_-^{*-1})^t &= \varphi_-^{*-1} * v_t,\\ (\varphi_+)^t &= \varphi_+ * w_t,\end{aligned}$$

and the equality $\varphi^t = (\varphi_-^{*-1})^t * (\varphi_+)^t$ yields

$$h_t = (\varphi_+)^{*-1} * v_t * \varphi_+ * w_t. \tag{36}$$

We denote by F_t, V_t, W_t the one-parameter groups obtained from h_t, v_t, w_t, respectively, by letting the complex variable z go to zero. It is clear that $\varphi^+|_{z=0} = e$, and similarly that W_t is the constant one-parameter group reduced to the co-unit ε. Hence equation (36) at $z = 0$ reduces to:

$$F_t = V_t, \tag{37}$$

hence the first assertion. the cocycle equation (25) applied to the Birkhoff decomposition reads

$$\Xi(\varphi) = \Xi(\varphi_+) + (\varphi_+)^{*-1} * \Xi(\varphi_-^{*-1}) * \varphi_+.$$

Taking residues of both sides yields

$$\text{Res } \Xi(\varphi) = \text{Res } \Xi(\varphi_-^{*-1}),$$

which ends the proof. □

The one-parameter group $F_t = V_t$ above is the *renormalization group* of φ [**9**].

REMARK 3. *As it is possible to reconstruct φ_- from $\beta(\varphi)$ using the explicit formula* (22) *above for Ξ^{-1}, the term φ_- (i.e. the divergence structure of φ) is uniquely determined by its residue.*

REMARK 4. *It would be interesting to define the renormalization group and the Beta function for other renormalization schemes and other target algebras $\mathcal{A}$. A first step in that direction can be found in* [**17**].

4. Rota–Baxter and dendriform algebras

We are interested in abstract versions of identities (10) and (19) fulfilled by the counterterm character φ_-. The general algebraic context is given by Rota–Baxter (associative) algebras of weight θ, which are themselves dendriform algebras. We first briefly recall the definition of Rota–Baxter (RB) algebra and its most important properties. For more details we refer the reader to the classical papers [**1, 2, 6, 45, 46**], as well as for instance to the references [**15, 16**].

4.1. From Rota-Baxter to dendriform. Let A be an associative not necessarily unital nor commutative algebra with $R \in \text{End}(A)$. We call a tuple (A, R) a *Rota–Baxter algebra* of weight $\theta \in k$ if R satisfies the *Rota–Baxter relation*

$$R(x)R(y) = R\big(R(x)y + xR(y) + \theta xy\big). \tag{38}$$

Note that the operator P of Section 2.5 is an idempotent Rota–Baxter operator. Its weight is thus $\theta = -1$. Changing R to $R' := \mu R$, $\mu \in k$, gives rise to a RB algebra of weight $\theta' := \mu\theta$, so that a change in the θ parameter can always be achieved, at least as long as weight non-zero RB algebras are considered.

Let us recall some classical examples of RB algebras. First, consider the integration by parts rule for the Riemann integral map. Let $A := C(\mathbb{R})$ be the ring of continuous real functions with pointwise product. The indefinite Riemann integral can be seen as a linear map on A:

$$I : A \to A, \qquad I(f)(x) := \int_0^x f(t)\,dt. \tag{39}$$

Then, integration by parts for the Riemann integral can be written compactly as

$$I(f)(x)I(g)(x) = I\big(I(f)g\big)(x) + I\big(fI(g)\big)(x), \tag{40}$$

dually to the classical Leibniz rule for derivations. Hence, we found our first example of a weight zero Rota–Baxter map. Correspondingly, on a suitable class of functions, we define the following Riemann summation operators:

$$R_\theta(f)(x) := \sum_{n=1}^{[x/\theta]-1} \theta f(n\theta) \qquad \text{and} \qquad R'_\theta(f)(x) := \sum_{n=1}^{[x/\theta]} \theta f(n\theta). \tag{41}$$

We observe readily that

$$\begin{aligned}\Big(\sum_{n=1}^{[x/\theta]} \theta f(n\theta)\Big)\Big(\sum_{m=1}^{[x/\theta]} \theta g(m\theta)\Big) &= \Big(\sum_{n>m=1}^{[x/\theta]} + \sum_{m>n=1}^{[x/\theta]} + \sum_{m=n=1}^{[x/\theta]}\Big)\theta^2 f(n\theta)g(m\theta)\\ &= \sum_{m=1}^{[x/\theta]} \theta^2\Big(\sum_{k=1}^{m} f(k\theta)\Big)g(m\theta) + \sum_{n=1}^{[x/\theta]} \theta^2\Big(\sum_{k=1}^{n} g(k\theta)\Big)f(n\theta) - \sum_{n=1}^{[x/\theta]} \theta^2 f(n\theta)g(n\theta)\\ &= R'_\theta\big(R'_\theta(f)g\big)(x) + R'_\theta\big(fR'_\theta(g)\big)(x) + \theta R'_\theta(fg)(x).\end{aligned} \tag{42}$$

Similarly for the map R_θ except that the diagonal, omitted, must be added instead of subtracted. Hence, the Riemann summation maps R_θ and R'_θ satisfy the weight θ and the weight $-\theta$ Rota–Baxter relation, respectively.

PROPOSITION 10. *Let (A, R) be a Rota–Baxter algebra. The map $\tilde{R} = -\theta id_A - R$ is a Rota–Baxter map of weight θ on A. The images of R and $\tilde{R}$, $A_\mp \subseteq A$, respectively, are subalgebras in A.*

The following Proposition follows directly from the Rota–Baxter relation:

PROPOSITION 11. *The vector space underlying A equipped with the product*

$$x *_\theta y := R(x)y + xR(y) + \theta xy \tag{43}$$

is again a Rota–Baxter algebra of weight θ with Rota–Baxter map R.

We denote it by (A_θ, R) and call it the *double Rota–Baxter algebra*. The Rota–Baxter map R becomes a (not necessarily unital even if A is unital) algebra homomorphism from the algebra A_θ to A. The result in Proposition 11 is best understood in the dendriform setting which we introduce now. A *dendriform algebra* [**32**] over a field k is a k-vector space A endowed with two bilinear operations $\prec$ and $\succ$ subject to the three axioms below:

$$(a \prec b) \prec c = a \prec (b * c), \quad (a \succ b) \prec c = a \succ (b \prec c), \quad a \succ (b \succ c) = (a * b) \succ c,$$

where $a * b$ stands for $a \prec b + a \succ b$. These axioms easily yield associativity for the law $*$. The bilinear operations $\rhd$ and $\lhd$ defined by

$$a \rhd b := a \succ b - b \prec a, \qquad a \lhd b := a \prec b - b \succ a \tag{44}$$

are left pre-Lie and right pre-Lie, respectively, which means that we have

$$(a \rhd b) \rhd c - a \rhd (b \rhd c) = (b \rhd a) \rhd c - b \rhd (a \rhd c), \tag{45}$$

$$(a \lhd b) \lhd c - a \lhd (b \lhd c) = (a \lhd c) \lhd b - a \lhd (c \lhd b). \tag{46}$$

The associative operation $*$ and the pre-Lie operations $\rhd$, $\lhd$ all define the same Lie bracket:

$$[a,b] := a * b - b * a = a \rhd b - b \rhd a = a \lhd b - b \lhd a. \tag{47}$$

PROPOSITION 12. **[11]** *Any Rota–Baxter algebra gives rise to two dendriform algebra structures given by*

$$a \prec b := aR(b) + \theta ab = -a\tilde{R}(b), \qquad a \succ b := R(a)b, \tag{48}$$

$$a \prec' b := aR(b), \qquad a \succ' b := R(a)b + \theta ab = -\tilde{R}(a)b. \tag{49}$$

The associated associative product $*$ is given for both structures by $a * b = aR(b) + R(a)b + \theta ab$ and thus coincides with the double Rota–Baxter product (43).

REMARK 5. **[11]** *In fact, by splitting again the binary operation $\prec$ (or alternatively $\succ'$), any Rota–Baxter algebra is* tri-dendriform **[34]**, *in the sense that the Rota–Baxter structure yields three binary operations $<, \diamond$ and $>$ subject to axioms refining the axioms of dendriform algebras. The three binary operations are defined by $a < b = aR(b)$, $a \diamond b = \theta ab$ and $a > b = R(a)b$. Choosing to put the operation $\diamond$ to the $<$ or $>$ side gives rise to the two dendriform structures above.*

Let $\overline{A} = A \oplus k.\mathbf{1}$ be our dendriform algebra augmented by a unit $\mathbf{1}$:

$$a \prec \mathbf{1} := a =: \mathbf{1} \succ a \qquad \mathbf{1} \prec a := 0 =: a \succ \mathbf{1}, \tag{50}$$

implying $a * \mathbf{1} = \mathbf{1} * a = a$. Note that $\mathbf{1} * \mathbf{1} = \mathbf{1}$, but that $\mathbf{1} \prec \mathbf{1}$ and $\mathbf{1} \succ \mathbf{1}$ are not defined **[44]**, **[4]**. We recursively define the following set of elements of $\overline{A}[[t]]$ for a fixed $x \in A$:

$$\begin{aligned}
w_{\prec}^{(0)}(x) &= w_{\succ}^{(0)}(x) = \mathbf{1},\\
w_{\prec}^{(n)}(x) &:= x \prec \big(w_{\prec}^{(n-1)}(x)\big),\\
w_{\succ}^{(n)}(x) &:= \big(w_{\succ}^{(n-1)}(x)\big) \succ x.
\end{aligned}$$

We also define the following set of iterated left and right pre-Lie products (44). For $n > 0$, let $a_1, \ldots, a_n \in A$:

$$\ell^{(n)}(a_1, \ldots, a_n) := \Big(\cdots\big((a_1 \rhd a_2) \rhd a_3\big) \cdots \rhd a_{n-1}\Big) \rhd a_n \tag{51}$$

$$r^{(n)}(a_1, \ldots, a_n) := a_1 \lhd \Big(a_2 \lhd \big(a_3 \lhd \cdots (a_{n-1} \lhd a_n)\big) \cdots\Big). \tag{52}$$

For a fixed single element $a \in A$ we can write more compactly for $n > 0$,

$$\ell^{(n+1)}(a) = \big(\ell^{(n)}(a)\big) \rhd a \quad \text{and} \quad r^{(n+1)}(a) = a \lhd \big(r^{(n)}(a)\big) \tag{53}$$

and $\ell^{(1)}(a) := a =: r^{(1)}(a)$. We have the following theorem **[23, 17]**.

THEOREM 2. *We have*

$$w_{\succ}^{(n)}(a) = \sum_{\substack{i_1 + \cdots + i_k = n\\ i_1, \ldots, i_k > 0}} \frac{\ell^{(i_1)}(a) * \cdots * \ell^{(i_k)}(a)}{i_1(i_1 + i_2) \cdots (i_1 + \cdots + i_k)},$$

$$w_{\prec}^{(n)}(a) = \sum_{\substack{i_1+\cdots+i_k=n\\ i_1,\ldots,i_k>0}} \frac{r^{(i_k)}(a)*\cdots*r^{(i_1)}(a)}{i_1(i_1+i_2)\cdots(i_1+\cdots+i_k)}.$$

PROOF. The free unital dendriform algebra with one generator a is naturally endowed with a connected graded cocommutative Hopf algebra structure. It has been shown in [**23**] that the associated Dynkin operator D verifies

$$D\big(w_{\succ}^{(n)}(a)\big) = \ell^{(n)}(a), \qquad D\big(w_{\prec}^{(n)}(a)\big) = r^{(n)}(a). \tag{54}$$

This result then comes from the formula (22). □

These identities nicely show how the dendriform pre-Lie and associative products fit together. This will become even more evident in the following: we are interested in the solutions X and Y in $\overline{A}[[t]]$ of the following two equations:

$$X = \mathbf{1} + ta \prec X, \qquad Y = \mathbf{1} - Y \succ ta. \tag{55}$$

Formal solutions to (55) are given by:

$$X = \sum_{n\ge 0} t^n w_{\prec}^{(n)}(a) \qquad \text{resp.} \qquad Y = \sum_{n\ge 0} (-t)^n w_{\succ}^{(n)}(a).$$

Let us introduce the following operators in A, where a is any element of A:

$$L_{\prec}[a](b) := a \prec b \quad L_{\succ}[a](b) := a \succ b \quad R_{\prec}[a](b) := b \prec a \quad R_{\succ}[a](b) := b \succ a$$
$$L_{\lhd}[a](b) := a \lhd b \quad L_{\rhd}[a](b) := a \rhd b \quad R_{\lhd}[a](b) := b \lhd a \quad R_{\rhd}[a](b) := b \rhd a.$$

We have recently obtained the following *pre-Lie Magnus expansion* [**19**]:

THEOREM 3. *Let $\Omega' := \Omega'(ta)$, $a \in A$, be the element of $tA[[t]]$ such that $X = \exp^*(\Omega')$ and $Y = \exp^*(-\Omega')$, where X and Y are the solutions of the two equations (55), respectively. This element obeys the following recursive equation:*

$$\Omega'(ta) = \frac{R_{\lhd}[\Omega']}{1-\exp(-R_{\lhd}[\Omega'])}(ta) = \sum_{m\ge 0} (-1)^m \frac{B_m}{m!} R_{\lhd}[\Omega']^m(ta), \tag{56}$$

or alternatively

$$\Omega'(ta) = \frac{L_{\rhd}[\Omega']}{\exp(L_{\rhd}[\Omega'])-1}(ta) = \sum_{m\ge 0} \frac{B_m}{m!} L_{\rhd}[\Omega']^m(ta), \tag{57}$$

where the B_l's are the Bernoulli numbers.

Recall that the Bernoulli numbers are defined via the generating series:

$$\frac{z}{\exp(z)-1} = \sum_{m\ge 0} \frac{B_m}{m!} z^m = 1 - \frac{1}{2}z + \frac{1}{12}z^2 - \frac{1}{720}z^4 + \cdots,$$

and observe that $B_{2m+3} = 0$, $m \ge 0$.

4.2. Non-commutative Bohnenblust-Spitzer formulas. Let n be a positive integer, and let $\mathcal{OP}_n$ be the set of ordered partitions of $\{1,\ldots,n\}$, i.e. sequences $(\pi_1,\ldots,\pi_k)$ of disjoint subsets (*blocks*) whose union is $\{1,\ldots,n\}$. We denote by $\mathcal{OP}_n^k$ the set of ordered partitions of $\{1,\ldots,n\}$ with k blocks. Let us introduce for any $\pi \in \mathcal{OP}_n^k$ the coefficient

$$\omega(\pi) = \frac{1}{|\pi_1|(|\pi_1|+|\pi_2|)\cdots(|\pi_1|+|\pi_2|+\cdots+|\pi_k|)}.$$

THEOREM 4. *Let $a_1, \ldots, a_n$ be elements in a dendriform algebra A. For any subset $E = \{j_1, \ldots, j_m\}$ of $\{1, \ldots, n\}$ let $\mathfrak{l}(E) \in A$ defined by*

$$\mathfrak{l}(E) := \sum_{\sigma \in S_m} \mathfrak{l}^{(m)}(a_{j_{\sigma_1}}, \ldots, a_{j_{\sigma_m}}).$$

We have

$$\sum_{\sigma \in S_n} \Big(...(a_{\sigma_1} \succ a_{\sigma_2}) \succ \cdots a_{\sigma_{n-1}} \Big) \succ a_{\sigma_n} = \sum_{k \geq 1} \sum_{\pi \in \mathcal{OP}_n^k} \omega(\pi) \mathfrak{l}(\pi_1) * \cdots * \mathfrak{l}(\pi_k).$$

See [**22**] where this identity is settled in the Rota–Baxter setting, see also [**18**]. The proof in the dendriform context is entirely similar. Another expression for the left-hand side can be obtained [**23**]. For any permutation $\sigma \in S_n$ we define the element $T_\sigma(a_1, \ldots, a_n)$ as follows: define first the subset $E_\sigma \subset \{1, \ldots, n\}$ by $k \in E_\sigma$ if and only if $\sigma_{k+1} > \sigma_j$ for any $j \leq k$. We write E_σ in the increasing order

$$1 \leq k_1 < \cdots < k_p \leq n - 1.$$

Then we set

$$T_\sigma(a_1, \ldots, a_n) := \ell^{(k_1)}(a_{\sigma_1}, \ldots, a_{\sigma_{k_1}}) * \cdots * \ell^{(n-k_p)}(a_{\sigma_{k_p+1}}, \ldots, a_{\sigma_n}) \tag{58}$$

There are $p+1$ packets separated by p stars in the right-hand side of the expression (58) above, and the parentheses are set to the left inside each packet. Following [**31**] it is convenient to write a permutation by putting a vertical bar after each element of E_σ. For example for the permutation $\sigma = (3261457)$ inside S_7 we have $E_\sigma = \{2, 6\}$. Putting the vertical bars,

$$\sigma = (32|6145|7),$$

we see that the corresponding element in A will then be

$$\begin{aligned} T_\sigma(a_1, \ldots, a_7) &= \ell^{(2)}(a_3, a_2) * \ell^{(4)}(a_6, a_1, a_4, a_5) * \ell^{(1)}(a_7) \\ &= (a_3 \rhd a_2) * \Big(\big((a_6 \rhd a_1) \rhd a_4 \big) \rhd a_5 \Big) * a_7. \end{aligned}$$

THEOREM 5. *For any $a_1, \ldots, a_n$ in the dendriform algebra A the following identity holds:*

$$\sum_{\sigma \in S_n} \big(...(a_{\sigma_1} \succ a_{\sigma_2}) \succ \cdots \big) \succ a_{\sigma_n} = \sum_{\sigma \in S_n} T_\sigma(a_1, \ldots, a_n). \tag{59}$$

A q-analog of this identity has been proved by J-C. Novelli and J-Y. Thibon [**41**].

References

[1] F. V. ATKINSON, *Some aspects of Baxter's functional equation*, J. Math. Anal. Appl. **7**, 1-30 (1963).

[2] G. BAXTER, *An analytic problem whose solution follows from a simple algebraic identity*, Pacific J. Math. **10**, 731-742 (1960).

[3] N. N. BOGOLIUBOV, O. S. PARASIUK, *On the multiplication of causal functions in the quantum theory of fields*, Acta Math. **97**, 227-266 (1957).

[4] J. COLLINS, *Renormalization*, Cambridge monographs in Math. Physics, Cambridge (1984).

[5] J. CARIÑENA, K. EBRAHIMI-FARD, H. FIGUEROA, J. M. GRACIA-BONDÍA, *Hopf algebras in dynamical systems theory*, Int. J. of Geom. Meth. in Mod. Phys. **4**, 577-646 (2007).

[6] P. CARTIER, *On the structure of free Baxter algebras*, Adv. Math. **9**, 253-265 (1972).

[7] A. CONNES, D. KREIMER, *Hopf algebras, Renormalization and Noncommutative Geometry*, Comm. Math. Phys. **199**, 203-242 (1998).

[8] A. CONNES, D. KREIMER, *Renormalization in Quantum Field Theory and the Riemann–Hilbert problem I: the Hopf algebra structure of graphs and the main theorem*, Comm. Math. Phys. **210**, 249-273 (2000).

[9] A. CONNES, D. KREIMER, *Renormalization in Quantum Field Theory and the Riemann–Hilbert problem II: the β-function, diffeomorphisms and the renormalization group*, Comm. Math. Phys. **216**, 215-241 (2001).

[10] A. CONNES, M. MARCOLLI, *Noncommutative Geometry, Quantum Fields and Motives*, Colloquium Publications, Amer. Math. Soc. (2007).

[11] K. EBRAHIMI-FARD, *Loday-type algebras and the Rota–Baxter relation*, Letters in Mathematical Physics **61**, 139-147 (2002).

[12] K. EBRAHIMI-FARD, L. GUO, D. KREIMER, *Spitzer's identity and the algebraic Birkhoff decomposition in pQFT*, J. Phys. A: Math. Gen. **37**, 11036-11052 (2004).

[13] K. EBRAHIMI-FARD, L. GUO, D. KREIMER, *Integrable renormalization II: the general case*, Ann. H. Poincaré **6**, 369-395 (2005).

[14] K. EBRAHIMI-FARD, L. GUO, *Matrix representation of renormalization in perturbative quantum field theory*, arXiv:hep-th/0508155 (2005).

[15] K. EBRAHIMI-FARD, D. KREIMER, *Hopf algebra approach to Feynman diagram calculations*, J. Phys. **A 38**), R385–R406 (2005).

[16] K. EBRAHIMI-FARD, L. GUO, D. MANCHON, *Birkhoff type decompositions and the Baker–Campbell-Hausdorff recursion*, Comm. Math. Phys. **267**, 821-845 (2006).

[17] K. EBRAHIMI-FARD, J. GRACIA-BONDÍA, F. PATRAS, *A Lie theoretic approach to renormalization*, Comm. Math. Phys. **276**, 519-549 (2007).

[18] K. EBRAHIMI-FARD, J. M. GRACIA-BONDÍA, F. PATRAS, *Rota–Baxter algebras and new combinatorial identities*, Letters in Mathematical Physics **81**, 61-75 (2007).

[19] K. EBRAHIMI-FARD, D. MANCHON, *A Magnus- and Fer-type formula in dendriform algebras*, Foundations of Computational Mathematics **9**, Number 3, 295-316 (2009).

[20] K. EBRAHIMI-FARD, L. GUO, J. M. GRACIA-BONDÍA, J. VÀRILLY, *Combinatorics of renormalization as matrix calculus*, Phys. Lett. **B 632**, 6552–6558 (2006).

[21] K. EBRAHIMI-FARD, D. MANCHON, *On matrix differential equations in the Hopf algebra of renormalization*, Adv. Theor. Math. Phys. **10**, 879-913 (2006).

[22] K. EBRAHIMI-FARD, D. MANCHON, F. PATRAS, *A noncommutative Bohnenblust-Spitzer identity for Rota-Baxter algebras solves Bogoliubov's recursion*, Journal of Noncommutative Geometry **3**, Issue 2, 181-222 (2009).

[23] K. EBRAHIMI-FARD, D. MANCHON, F. PATRAS, *New identities in dendriform algebras*, Journal of Algebra **320**, 708-727 (2008).

[24] H. FIGUEROA, J.M. GRACIA-BONDÍA, *Combinatorial Hopf algebras in Quantum Field Theory I*, Reviews of Mathematical Physics **17**, 881-976 (2005).

[25] I. M. GELFAND, D. KROB, A. LASCOUX, B. LECLERC, V. RETAKH, J.-Y. THIBON, *Noncommutative symmetric functions*, Adv. Math. **112**, 218-348 (1995.

[26] G. GREEN, *Researches on the Vibrations of Pendulums in Fluid Media*, Royal Society of Edinburgh Transactions, 315-324 (1836).

[27] K. HEPP, *Proof of the Bogoliubov–Parasiuk theorem on renormalization*, Comm. Math. Phys. **2**, 301–326 (1966).

[28] G. 'T HOOFT, M. VELTMAN, *Regularization and renormalization of gauge fields*, Nucl. Phys. **B44** 189-213 (1972).

[29] S. A. JONI, G-C. ROTA, *Coalgebras and bialgebras in combinatorics*, Stud. Appl. Math **61**, 93-139 (1979).

[30] D. KREIMER, *On the Hopf algebra structure of perturbative quantum field theories*, Adv. Theor. Math. Phys. **2**, 303-334 (1998).

[31] C. S. LAM, *Decomposition of time-ordered products and path-ordered exponentials*, J. Math. Phys. **39**, 5543–5558 (1998).

[32] J.-L. LODAY, *Dialgebras*, Lect. Notes Math. **1763**, 7-66 Springer (Berlin, 2001).

[33] J-L. LODAY, M. RONCO, *Hopf algebra of the planar binary trees*, Adv. Math. **139**, 293-309 (1998).

[34] J.-L. LODAY, M. RONCO, *Trialgebras and families of polytopes*, in *Homotopy Theory: Relations with Algebraic Geometry, Group Cohomology, and Algebraic K-theory*, Contemporary Mathematics **346**, 369–398 (2004).

[35] W. MAGNUS, *On the exponential solution of differential equations for a linear operator*, Commun. Pure Appl. Math. **7**, 649-673 (1954).

[36] D. MANCHON, *L'algèbre de Hopf bitensorielle*, Comm. Algebra **25**, 1537–1551 (1997).

[37] D. MANCHON, *Hopf algebras and renormalisation*, Handbook of Algebra,Vol. **5** (M. Hazewinkel ed.), 365-427 (2008).

[38] D MANCHON, *Quelques aspects combinatoires de la renormalisation*, Gazette des Mathématiciens, **119**, 5-15 (2009).

[39] S. MONTGOMERY, *Some remarks on filtrations of Hopf algebras*, Comm. Algebra **21**, 999-1007 (2003).

[40] B. MIELNIK, J. PLEBAŃSKI, *Combinatorial approach to Baker–Campbell–Hausdorff exponents*, Ann. Inst. Henri Poincaré **A XII**, 215-254 (1970).

[41] J-C. NOVELLI, J-Y. THIBON, *A one-parameter family of dendriform identities*, European Journal of Combinatorics, electronic version (12 pp., 2008) arXiv:0709.3235.

[42] F. PATRAS, CHR. REUTENAUER, *On Dynkin and Klyachko idempotents in graded bialgebras*, Adv. Appl. Math. **28**, 560-579 (2002).

[43] CH. REUTENAUER, *Free Lie algebras*, Oxford University Press, Oxford 1993.

[44] M. RONCO, *Primitive elements in a free dendriform algebra*, Contemp. Math. **207**, 245-263 (2000).

[45] G.-C. ROTA, *Baxter algebras and combinatorial identities. I, II.*, Bull. Amer. Math. Soc. **75**, 325-329 and 330-334(1969).

[46] G.-C. ROTA, D. A. SMITH, *Fluctuation theory and Baxter algebras*, Instituto Nazionale di Alta Matematica **IX**, 179-201(1972).

[47] F. SPITZER, *A combinatorial lemma and its application to probability theory*, Trans. Amer. Math. Soc. **82**, 323-339 (1956).

[48] M. E. SWEEDLER, *Hopf algebras*, Benjamin, New-York, 1969.

[49] R. S. STRICHARTZ, *The Campbell-Baker-Hausdorff-Dynkin formula and solutions of differential equations*, J. Func. Anal. **72**, 320-345 (1987).

[50] V. S. VARADARAJAN, *Lie groups, Lie algebras and their representations*, Springer Verlag, 1984.

[51] W. ZIMMERMANN, *Convergence of Bogoliubov's method of renormalization in momentum space*, Comm. Math. Phys. **15**, 208-234 (1969).

UNIVERSITÉ BLAISE PASCAL, C.N.R.S.-UMR 6620, 63177 AUBIÈRE, FRANCE
E-mail address: manchon@math.univ-bpclermont.fr
URL: http://math.univ-bpclermont.fr/~manchon/

Research Articles

Clay Mathematics Proceedings
Volume 12, 2010

Fredholm Realizations of Elliptic Symbols on Manifolds with Boundary II: Fibered Boundary

Pierre Albin and Richard Melrose

ABSTRACT. We consider two calculi of pseudodifferential operators on manifolds with fibered boundary: Mazzeo's edge calculus, which has as local model the operators associated to products of closed manifolds with asymptotically hyperbolic spaces, and the ϕ calculus of Mazzeo and the second author, which is similarly modeled on products of closed manifolds with asymptotically Euclidean spaces. We construct an adiabatic calculus of operators interpolating between them, and use this to compute the 'smooth' K-theory groups of the edge calculus, determine the existence of Fredholm quantizations of elliptic symbols, and establish a families index theorem in K-theory.

Introduction

If the boundary of a manifold is the total space of a fibration

$$Z \text{ ---- } \partial X \xrightarrow{\ \Phi\ } Y \tag{1}$$

one can quantize an invertible symbol

$$\sigma \in \mathcal{C}^\infty(S^*X; \hom(\pi^*E, \pi^*F))$$

as an elliptic pseudodifferential operator in the Φ-b or edge calculus, $\Psi^0_{\Phi\text{-b}}(X; E, F)$, introduced in [**8**] or alternately as an elliptic operator in the Φ-c or ϕ calculus, $\Psi^0_{\Phi\text{-c}}(X; E, F)$, introduced in [**9**]. As on a closed manifold, either of these operators will induce a bounded operator acting between natural L^2-spaces of sections but, in contrast to closed manifolds, these operators need not be Fredholm.

A well-known result of Atiyah and Bott [**2**] established that for a differential operator on a manifold with boundary X to admit local elliptic boundary conditions it is necessary and sufficient for the K-theory class of its symbol $[\sigma] \in K_c(T^*X)$ to map to zero under the natural map

$$K_c(T^*X) \to K^1_c(T^*\partial X).$$

2010 *Mathematics Subject Classification.* Primary 58J20, Secondary 19K56, 58J32, 58J40.
The first author was partially supported by an NSF postdoctoral fellowship.

At one extreme, when $Z = \{\text{pt}\}$, the Φ-b calculus is the zero calculus of [**7**] and the Φ-c calculus is the scattering calculus of [**12**], and for these the vanishing of the Atiyah-Bott obstruction is equivalent to the existence of a Fredholm quantization of σ. At the other extreme, when $Y = \{\text{pt}\}$, the Φ-b calculus is the b-calculus of [**11**] and the Φ-c calculus is the cusp calculus of [**9**], and in either case there is no obstruction to finding a Fredholm realization of an elliptic symbol σ. The case of general ϕ is intermediate between these two extremes. Indeed, it was established in [**14**] that one can use Φ and the families index to induce a map $K_c(T^*X) \to K^1_c(T^*Y)$ (see §4) and that σ has a Fredholm quantization in the Φ-c calculus if and only if

$$[\sigma] \in \ker \left(K_c(T^*X) \to K^1_c(T^*Y) \right).$$

In this paper the corresponding result for the edge calculus is established. We point out that another extension of the Atiyah-Bott obstruction, to a class of operators on stratified manifolds, is discussed in [**15**].

To this end we consider the even 'smooth K-theory' group of the edge calculus $\mathcal{K}_{\Phi\text{-b}}(X)$ consisting of equivalence classes of Fredholm edge operators under the relations generated by bundle stabilization, bundle isomorphisms, and smooth homotopy (see §1). This has a natural subgroup $\mathcal{K}_{\Phi\text{-b},-\infty}(X)$ of equivalence classes with principal symbol the identity, and there are analogous odd smooth K-theory groups defined by suspension.

The corresponding groups are defined and identified for the Φ-c calculus in [**14**]. The analysis of these groups for the Φ-c calculus is simpler for two related reasons. The first is that the Φ-c calculus is 'asymptotically normally commutative' whereas the Φ-b calculus is 'asymptotically normally non-commutative'. More precisely, the behavior of a Φ-c operator near the boundary is modeled by a family of operators, parametrized by Y, acting on the Lie group $\mathbb{R}^{h+1}$ times the closed manifold Z. On the other hand, the behavior of a Φ-b operator near the boundary is modeled by a family of operators, parametrized by Y, acting on the Lie group $\mathbb{R}^+ \ltimes \mathbb{R}^h$ times the closed manifold Z. To see the effect of this difference on the analysis of these calculi, one can compare the relative simplicity of identifying $\mathcal{K}_{\text{sc}}(X)$ in [**14**, §2] versus the corresponding identification of $\mathcal{K}_0(X)$ in [**1**].

The other related simplification is that the Φ-c calculus admits a smooth functional calculus, while the Φ-b calculus does not. This means that for the Φ-c calculus one can study the smooth K-theory groups using much the same constructions one would use to study the K-theory of its C^*-algebra (see for instance the constructions used in [**14**, §4] to define KK-classes). One could remove this difficulty by passing to a C^*-closure of the Φ-b calculus, but at the considerable cost of losing the smooth structure.

Instead, we will study the smooth K-theory groups of the Φ-b calculus by constructing an 'adiabatic' calculus of pseudodifferential operators interpolating between the Φ-b and Φ-c calculi. This induces maps, labeled ad, between their smooth K-theory groups. We work more generally in the context of a fibration $X - M \xrightarrow{\phi} B$ where the fibers are manifolds with fibered boundaries, thus altogether we have

(2)

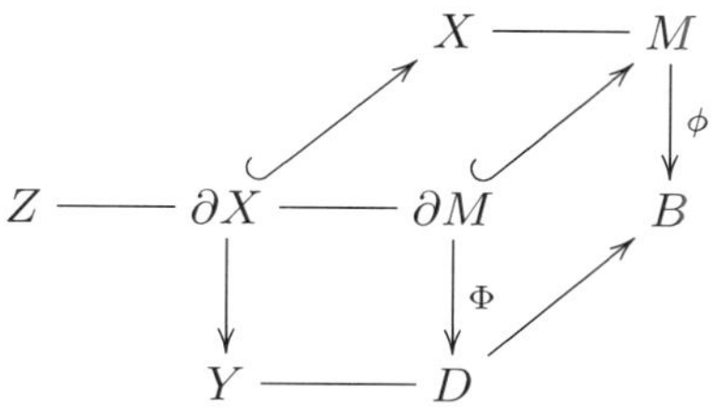

Using our analysis of the smooth K-theory groups of the zero calculus in [**1**], we establish the following theorem.

THEOREM 1. *The maps*

$$\mathcal{K}_{\Phi\text{-c}}(\phi) \xrightarrow{\text{ad}} \mathcal{K}_{\Phi\text{-b}}(\phi), \quad \mathcal{K}^1_{\Phi\text{-c}}(\phi) \xrightarrow{\text{ad}} \mathcal{K}^1_{\Phi\text{-b}}(\phi)$$

are isomorphisms, and restrict to isomorphisms

$$\mathcal{K}_{\Phi\text{-c},-\infty}(\phi) \xrightarrow{\text{ad}} \mathcal{K}_{\Phi\text{-b},-\infty}(\phi), \quad \mathcal{K}^1_{\Phi\text{-c},-\infty}(\phi) \xrightarrow{\text{ad}} \mathcal{K}^1_{\Phi\text{-b},-\infty}(\phi).$$

Furthermore, these groups fit into a commutative diagram

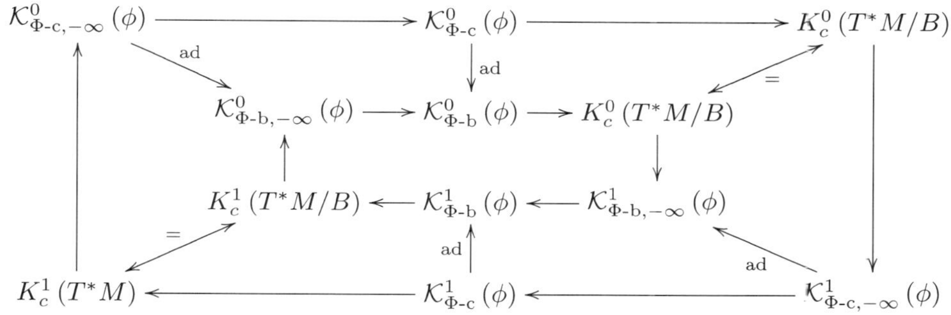

wherein the inner and outer six term sequences are exact.

In [**14**], the groups $\mathcal{K}^q_{\Phi\text{-c}}(\phi)$ are identified in terms of the KK-theory of the $C(B)$-module

$$\mathcal{C}_\Phi(M) = \{f \in \mathcal{C}(M) : f\big|_{\partial M} \in \Phi^*\mathcal{C}(D)\}$$

and the $\mathcal{K}^q_{\Phi\text{-c},-\infty}(\phi)$ groups are identified with the K-theory of the vertical cotangent bundle T^*D/B. Below, we establish the analogue of the latter directly and deduce the analogue of the former from Theorem 1.

COROLLARY 2. *There are natural isomorphisms*

$$\mathcal{K}^q_{\Phi\text{-b}}(\phi) \cong KK^q_B(\mathcal{C}_\Phi(M), \mathcal{C}(B)), \text{ and } \mathcal{K}^q_{\Phi\text{-b},-\infty}(\phi) \cong K^q_c(T^*D/B).$$

From the six-term exact sequence and this corollary the topological obstruction to realizing an elliptic symbol via a family of Fredholm Φ-b operators can be deduced.

COROLLARY 3. *An elliptic symbol* $\sigma \in \mathcal{C}^\infty(S^*M/B; \hom(\pi^*E, \pi^*F))$ *can be quantized as a Fredholm family of* Φ-b *operators if and only if its K-theory class satisfies*

$$[\sigma] \in \ker\left(K_c(T^*M/B) \to K^1_c(T^*D/B)\right).$$

Finally, in [**14**] the second author and Frédéric Rochon defined a topological index map $\mathcal{K}_{\Phi\text{-c}}(\phi) \to K(B)$ and showed that it coincided with the analytic index. Thus the group $\mathcal{K}_{\Phi\text{-b}}(\phi)$ inherits a topological index map from its isomorphism with $\mathcal{K}_{\Phi\text{-c}}(\phi)$ and, since the adiabatic limit commutes with the analytic index, the following K-theoretic index theorem follows.

COROLLARY 4. *The analytic and topological indices coincide as maps*

$$\mathcal{K}_{\Phi\text{-b}}(\phi) \to K(B).$$

In Section 1, we review the Φ-b and Φ-c algebras of pseudodifferential operators and the definition of the smooth K-theory groups. In Section 2, we set up the adiabatic calculus and prove that it is closed under composition in an appendix. In Section 3 we use an excision lemma and the analysis of the smooth K-theory of the zero calculus from [**1**] to identify the groups $\mathcal{K}^q_{\Phi\text{-b},-\infty}(\phi)$. Finally, in Section 4, we establish the six term exact sequence and finish the proof of Theorem 1.

1. Φ-b and Φ-c algebras of pseudodifferential operators

We recall some of the main features of these calculi and refer the reader to, e.g., [**11**], [**8**], and [**9**] for more details. For the moment we restrict attention to the case of a single operator (i.e., $B = \{\text{pt}\}$).

We start by describing the vector fields that generate the differential operators in the two calculi. Let $\{x, y_1, \dots, y_h, z_1, \dots, z_v\}$ be local coordinates near the boundary with y_i lifted from the base under the fibration (1) and the z_i vertical. Here x is a boundary defining function, i.e., a non-negative function on $\overline{X}$ with $\{x = 0\} = \partial X$ and $dx \neq 0$ on the boundary. The fibered cusp structure depends mildly on this choice.

The Lie algebra, $\mathcal{V}_{\Phi\text{-b}}$, of vector fields tangent to the fibers of the fibration over the boundary is locally spanned by the vector fields

$$\{x\partial_x, x\partial_{y_1}, \dots, x\partial_{y_h}, \partial_{z_1}, \dots \partial_{z_v}\}.$$

Fibered boundary differential operators are polynomials in these vector fields. That is, any $P \in \text{Diff}^k_{\Phi\text{-b}}(X)$ can be written locally as

$$P = \sum_{j+|\alpha|+|\beta|\leq k} a_{j,\alpha,\beta}(x,y,z)\,(x\partial_x)^j\,(x\partial_y)^\alpha\,(\partial_z)^\beta\,.$$

There is a vector bundle ${}^{\Phi\text{-b}}TX$, the Φ-b tangent bundle, whose space of smooth sections is precisely $\mathcal{V}_{\Phi\text{-b}}$. It plays the role of the usual tangent bundle in the study of the Φ-b calculus. For instance, Φ-b one-forms are elements of the dual bundle, ${}^{\Phi\text{-b}}T^*X$, and a Φ-b metric is a metric on ${}^{\Phi\text{-b}}TX$, e.g. locally

$$g_{\Phi\text{-b}} = \frac{dx^2}{x^2} + \frac{\Phi^* g_Y}{x^2} + g_Z. \tag{1.1}$$

'Extreme' cases of fibered boundary calculi are the b-calculus (where Y is a point) and the 0-calculus (where Z is a point). The b-calculus models non-compact manifolds with a cylindrical end. It was used in [**11**] to prove the APS-index theorem. The 0-calculus models non-compact manifolds that are asymptotically hyperbolic. It has applications to conformal geometry through the Fefferman-Graham construction and to physics (e.g. holography) through the AdS/CFT correspondence [**5**].

There is also a Φ-c tangent bundle, whose space of sections $\mathcal{V}_{\Phi\text{-c}}$ is locally spanned by

$$\{x^2\partial_x, x\partial_{y_1}, \ldots, x\partial_{y_h}, \partial_{z_1}, \ldots \partial_{z_v}\}.$$

Thus a fibered cusp differential operator $P \in \mathrm{Diff}^k_{\Phi\text{-c}}(X)$ can be written locally as

$$P = \sum_{j+|\alpha|+|\beta|\leq m} a_{j,\alpha,\beta}(x,y,z)\left(x^2\partial_x\right)^j\left(x\partial_y\right)^\alpha\left(\partial_z\right)^\beta.$$

The 'extreme' cases of fibered cusp calculi are known as the cusp-calculus (where Y is a point) and the scattering-calculus (where Z is a point). The cusp-calculus models the same geometric situation (asymptotically cylindrical manifolds) as the b-calculus, and indeed these calculi are very closely related (see [**1**, §3]). The scattering-calculus models non-compact, asymptotically locally Euclidean manifolds. Indeed, if one compactifies $\mathbb{R}^n$ radially to a half-sphere, the metric near the boundary takes the form

$$\frac{dx^2}{x^4} + \frac{h_x}{x^2},$$

and so defines a metric on the scattering tangent bundle.

As an illustration of the difference between the Φ-b and Φ-c calculi we point out that the scattering Lie algebra is asymptotically commutative in the sense that

$$[\mathcal{V}_{sc}, \mathcal{V}_{sc}] \subset x\mathcal{V}_{sc},$$

while the sub-bundle of the zero tangent bundle ${}^0TX$ spanned by commutators of zero vector fields is non-trivial over the boundary. This asymptotic commutativity of horizontal vector fields in the Φ-c calculus lies behind the simplifications over the Φ-b calculus.

As on a closed manifold, certain interesting operations (e.g. powers, parametrices, or inverses) require passing to a larger calculus of pseudodifferential operators. Pseudodifferential operators mapping sections of a bundle E to sections of a bundle F are denoted $\Psi^*_{\Phi\text{-b}}(X;E,F)$ and $\Psi^*_{\Phi\text{-c}}(X;E,F)$ respectively. Operators in these calculi act by means of distributional integral kernels as in the Schwartz kernel theorem, i.e.

$$Pf(\zeta) = \int_X \mathcal{K}_P(\zeta,\zeta')f(\zeta').$$

These integral kernels have singularities along the diagonal and when $\zeta, \zeta' \in \partial X$. The latter can be resolved by lifting the kernel to an appropriate blown-up space, denoted respectively $X^2_{\Phi\text{-b}}$ and $X^2_{\Phi\text{-c}}$. We will describe these spaces below, in §2, as part of the construction of the adiabatic calculus.

These pseudodifferential calculi each possess two symbol maps. On a closed manifold, the highest order terms in the local expression of a differential operator define invariantly a function on the cosphere bundle of the manifold. This same construction yields the principal symbol of a differential (and more generally pseudodifferential) Φ-b operator as a function on the Φ-b cosphere bundle, i.e., the bundle of unit vectors in ${}^{\Phi\text{-b}}T^*X$. This interior symbol fits into the short exact sequence,

$$\Psi^{k-1}_{\Phi\text{-b}}(X;E,F) \hookrightarrow \Psi^k_{\Phi\text{-b}}(X;E,F) \xrightarrow{\sigma} C^\infty\left({}^{\Phi\text{-b}}S^*X; (N^*X)^k \otimes \hom(\pi^*E, \pi^*F)\right),$$

where $(N^*X)^k$ denotes a line bundle whose sections have homogeneity k and π is the projection map ${}^{\Phi\text{-b}}S^*X \to X$. The interior symbol in the fibered cusp calculus works in much the same way, and the corresponding sequence

$$\Psi^{k-1}_{\Phi\text{-c}}(X;E,F) \hookrightarrow \Psi^{k}_{\Phi\text{-c}}(X;E,F) \xrightarrow{\sigma} C^\infty\left({}^{\Phi\text{-c}}S^*X; (N^*X)^k \otimes \hom(\pi^*E,\pi^*F)\right),$$

is also exact. The interior symbol is used to define ellipticity: an operator is elliptic if and only if its interior symbol is invertible.

The second symbol map, known as the normal operator, models the behavior of the operator near the boundary. Following [**3**], at any point $p \in Y$ let $\mathcal{I}_{\Phi^{-1}(p)} \subset \mathcal{V}_{\Phi\text{-b}}$ be the subspace of Φ-b vector fields that vanish along the fiber $\Phi^{-1}(p)$ (as Φ-b vector fields). These vector fields form an ideal and the quotient ${}^{\Phi\text{-b}}T_pX$ is a Lie algebra. The projection

$$\mathcal{V}_{\Phi\text{-b}} \to {}^{\Phi\text{-b}}T_pX$$

lifts to a map of enveloping algebras

$$\mathrm{Diff}^k_{\Phi\text{-b}}(X) \to \mathcal{D}\left({}^{\Phi\text{-b}}T_pX\right),$$

and the image of $P \in \mathrm{Diff}^k_{\Phi\text{-b}}(X)$ is known as the normal operator of P at p, $N_{\Phi\text{-b},p}(P)$. The normal operator extends to pseudodifferential operators and can be realized either globally (its kernel is obtained by restricting the kernel of P to a certain boundary face in $X^2_{\Phi\text{-b}}$) or locally by means of an appropriate rescaling as follows.

Assume that we have chosen a product neighborhood of the boundary and, in a neighborhood $\mathcal{U}_Y$ of a point $p \in Y$, we have chosen a local trivialization of Φ, so that $\mathcal{U} \subset X$ looks like

$$\mathcal{U} \cong [0,\varepsilon) \times \mathcal{U}_Y \times Z \xrightarrow{\Phi} \mathcal{U}_Y, \tag{1.2}$$

with coordinates $\{x\}$, $\{y_i\}$, and $\{z_i\}$ on the respective factors, and a corresponding decomposition of the tangent bundle

$${}^{\Phi\text{-b}}TX = {}^{\Phi\text{-b}}N\partial X \oplus V\partial X.$$

Using (1.2) we can define the dilation

$$(x, y_i, z_j) \overset{M_\delta}{\mapsto} (\delta x, \delta y_i, z_j).$$

If $V \in \mathcal{V}_{\Phi\text{-b}}$ then in local coordinates

$$V_\delta = \left(M_\delta^{-1}\right)^* V$$

is a smooth vector field defined in a neighborhood of zero (increasing as δ decreases). The map

$$\mathcal{V}_{\Phi\text{-b}} \ni V \mapsto \lim_{\delta\to 0} V_\delta \in T({}^{\Phi\text{-b}}N\partial Z) \oplus V\partial Z$$

has kernel $\mathcal{I}_{p,\Phi\text{-b}}$ and realizes ${}^{\Phi\text{-b}}T_pZ$ as a Lie algebra of smooth vector fields on T_pZ. More generally, Let $L_\mathcal{U}$ be a chart at p mapping into N_p^+Y, and define the normal operator of $P \in \Psi^0_{\Phi\text{-b}}(X)$ at p to be

$$N_{p,\Phi\text{-b}}(P)\,u = \lim_{\delta\to 0} (M_\delta)^* L_\mathcal{U}^* P \left(L_\mathcal{U}^{-1}\right)^* \left(M_{\frac{1}{\delta}}\right)^* u.$$

The Lie group structure on ${}^{\Phi\text{-b}}N\partial X$ at the point $p \in Y$ is that of $\mathbb{R}^+ \ltimes \mathbb{R}^h$, i.e.

$$(s,u)\cdot(s',u') = (ss', u + su').$$

The kernel of the normal operator at the point $p \in Y$ is invariant with respect to this action, so the normal operator acts by convolution in the associated variables.

We refer to this as a non-commutative suspension, and denote these operators by $\Psi^*_{N\mathrm{sus}}\left({}^{\Phi\text{-b}}N\partial X; E, F\right)$. They fit into a short exact sequence

$$x\Psi^k_{\Phi\text{-c}}(X;E,F) \hookrightarrow \Psi^k_{\Phi\text{-c}}(X;E,F) \xrightarrow{N_{\Phi\text{-b}}} \Psi^k_{N\mathrm{sus}}\left({}^{\Phi\text{-b}}N\partial X; E, F\right).$$

The same construction yields a model operator at every point $p \in Y$ for the fibered cusp calculus. In this case the Lie group is commutative and isomorphic to $\mathbb{R} \times \mathbb{R}^h$; the resulting calculus is referred to as the suspended calculus. The corresponding short exact sequence is

$$x\Psi^k_{\Phi\text{-c}}(X;E,F) \hookrightarrow \Psi^k_{\Phi\text{-c}}(X;E,F) \xrightarrow{N_{\Phi\text{-c}}} \Psi^k_{\mathrm{sus}}\left({}^{\Phi\text{-c}}N\partial X; E, F\right).$$

The volume form of a Φ-b metric (e.g. (1.1)), together with Hermitian metrics on E and F, defines a space of L^2 sections. An operator $P \in \Psi^k_{\Phi\text{-b}}(X;E,F)$ acts linearly on L^2 sections of E; boundedly if $k \leq 0$. There is a corresponding scale of Φ-b Sobolev spaces, $H^s_{\Phi\text{-b}}(X;E)$, and an element $P \in \Psi^k_{\Phi\text{-b}}(X;E,F)$ defines a bounded linear operator

$$H^s_{\Phi\text{-b}}(X;E) \xrightarrow{P} H^{s-k}_{\Phi\text{-b}}(X;F).$$

This operator is Fredholm if and only if both symbol maps $\sigma(P)$ and $N_{\Phi\text{-b}}(P)$ are invertible operators, in which case P is said to be *fully elliptic*.

In order to carry out our analysis below, we will need a good understanding of the normal operators of elements of $\Psi_0^{-\infty}$. A construction from [**8**] and [**6**] realizes normal operators of the zero calculus as families of 'b,c-calculus' on the interval, referred to as *reduced normal operators*. Operators in this calculus have b-behavior at one end of the interval and cusp-behavior at the other end. The reduced normal operator of an operator in $\Psi_0^{-\infty}(X;E)$ lies inside the space

$$\Psi^{-\infty,-\infty}_{b,c}([0,1], \pi^*E)$$

of operators with b-behavior of order $-\infty$ near 0 and which vanish to infinite order near 1. In the following result from [**6**], π denotes the canonical projection $S^*\partial X \to \partial X$.

PROPOSITION 1.1 ([**6**], Prop. 4.4.1). *The reduced normal operators of elements of $\Psi_0^{-\infty}(X;E)$ are precisely those functions*

$$\begin{aligned}\mathcal{N}(y',\eta;\tau,\rho) \in C^\infty(S^*\partial X)\,\widehat{\otimes}_\pi \dot{C}^\infty\left([-1,1],\Omega^{1/2}\right)\widehat{\otimes}_\pi \mathcal{S}\left(\mathbb{R}_+, {}^b\Omega^{1/2}\right)\\ \otimes_{C^\infty(\widetilde{\mathcal{I}}^2)} C^\infty\left(\widetilde{\mathcal{I}}^2, \beta^*\left(\mathrm{Hom}\left(\pi^*E \otimes {}^{b,c}\Omega^{-1/2}\right)\right)\right)\end{aligned} \tag{1.3}$$

with $\mathcal{N}(y',\eta;\cdot,\rho)$ the lift of a density on $T^*\partial X$, and which for each $\eta \in S^*\partial X$ extend to an element of $\Psi^{-\infty,-\infty}_{b,c}([0,1],\pi^*E)$ in such a way that

$$\mathcal{N} \in C^\infty\left(S^*\partial X, \Psi^{-\infty,-\infty}_{b,c}([0,1],\pi^*E)\right). \tag{1.4}$$

Remark. Notice that the fact that $\mathcal{N}(y',\eta;\cdot,\rho)$ is the lift of a density on $T^*\partial X$ implies in particular that the b-normal operator is a family defined on $S^*\partial X$ but actually only depending on ∂X.

Next, we define K-theory groups of Φ-b operators. Following [**14**, Definition 2], define

$$A_{\Phi\text{-b}}(M;E,F) = \{(\sigma(A), \mathcal{N}(A)) : A \in \Psi^0_{\Phi\text{-b}}(M;E,F) \text{ Fredholm}\}$$

so that $\mathcal{K}^0_{\Phi\text{-b}}(M)$ consists of equivalence classes of elements in $A_{\Phi\text{-b}}(M;E,F)$, where two elements are equivalent if there is a finite chain consisting of the following:

$$(1.5)\quad (\sigma,N)\in A_{\Phi\text{-b}}(M;E,F)\sim(\sigma',N')\in A_{\Phi\text{-b}}(M;E',F')$$

if there exists a bundle isomorphisms $\gamma_E : E\to E'$ and $\gamma_F : F\to F'$ such that $\sigma=\gamma_F^{-1}\circ\sigma'\circ\gamma_E$ and $\mathcal{N}=\gamma_F^{-1}\circ\mathcal{N}'\circ\gamma_E$,

$$(1.6)\quad (\sigma,N)\in A_{\Phi\text{-b}}(M;E,F)\sim\left(\widetilde{\sigma},\widetilde{N}\right)\in A_{\Phi\text{-b}}(M;E,F)$$

if there exists a homotopy of Fredholm operators A_t in $\Psi^0_{\Phi\text{-b}}(M;E,F)$, with $(\sigma,N)=(\sigma(A_0),\mathcal{N}(A_0))$ and $\left(\widetilde{\sigma},\widetilde{N}\right)=(\sigma(A_1),\mathcal{N}(A_1))$, and

$$(1.7)\quad (\sigma,N)\in A_{\Phi\text{-b}}(M;E,F)\sim(\sigma\oplus\mathrm{Id}_G,N\oplus\mathrm{Id}_G)\in A_{\Phi\text{-b}}(M;E\oplus G,F\oplus G).$$

Similarly, we define $\mathcal{K}^1_{\Phi\text{-b}}(M)$ as equivalence classes of elements in the space of based loops

$$\Omega A_{\Phi\text{-b}}(M;E,F)=\{s\in C^\infty\left(\mathbb{S}^1,A_{\Phi\text{-b}}(M;E,F)\right):s(1)=\mathrm{Id}\},$$

where the equivalences are finite chains of (1.5), (1.6), (1.7) with bundle transformations and homotopies required to be the identity at $1\in\mathbb{S}^1$.

In the same way, we can describe $\mathcal{K}^0_{\Phi\text{-b},-\infty}(M)$ and $\mathcal{K}^1_{\Phi\text{-b},-\infty}(M)$ as equivalence classes of elements in

$$A_{\Phi\text{-b},-\infty}(M;E,F)=\{N(\mathrm{Id}+A):A\in\Psi^{-\infty}_{\Phi\text{-b}}(M;E,F),\mathrm{Id}+A\ \text{ Fredholm}\},$$

and

$$\Omega A_{\Phi\text{-b},-\infty}(M;E,F)=\{s\in C^\infty\left(\mathbb{S}^1,A_{\Phi\text{-b},-\infty}(M;E,F)\right):s(1)=\mathrm{Id}\}$$

respectively.

2. Adiabatic limit of edge to ϕ calculi

For a compact manifold with boundary, with a specified fibration of the boundary, we show that there is an adiabatic limit construction passing from the fibered boundary (for $\varepsilon>0$) calculus to the fibered cusp calculus in the limit. Every Fredholm (i.e., totally elliptic) operator in the limiting calculi occurs in a totally elliptic family in the adiabatic calculus, so with constant index. This allows the K-theory of the two algebras to be identified and the (families) index map for one to be reduced to that of the other.

Let X be a compact manifold with boundary, with boundary fibration $Z \longrightarrow \partial X \twoheadrightarrow Y$. We construct a resolution of $X^2\times[0,\varepsilon_0]$ which carries the adiabatic calculus. First blow up the corner and consider

$$X^2_{(1)}=\left[X^2\times[0,\varepsilon_0];(\partial X)^2\times\{0\}\right]\xrightarrow{\beta_{(1)}}X^2\times[0,\varepsilon_0].$$

Then blow up the two sides to get

$$X^2_{(2)}=\left[X^2_{(1)};X\times\partial X\times\{0\};\partial X\times X\times\{0\}\right]\xrightarrow{\beta_{(2)}}X^2_{(1)}.$$

The lifts of these manifolds are disjoint in $X^2_{(1)}$ and commutativity of blow-ups shows that there is a smooth map, which is in fact a b-fibration, to the rescaled single space

$$X_\varepsilon=[X\times[0,\varepsilon_0];\partial X\times\{0\}]$$

in either factor:

$$X^2_{(2)} \overset{\pi_{L,\varepsilon}}{\underset{\pi_{R,\varepsilon}}{\rightrightarrows}} X_\varepsilon.$$

This leads to the action of the adiabatic operators on $\mathcal{C}^\infty(X_\varepsilon)$.

The choice of the fibered-cusp structure on X corresponds to the singling out of a class of boundary defining functions which all induce the same trivialization of the normal bundle to the boundary along the fibres of Φ. Let x be such a boundary defining function, on the left factor of X in X^2 and let x' denote the same function on the right factor. The hypersurface $x = x'$ is smooth near $(\partial X)^2 \subset X^2$ and lifts to be smooth near the front face of $X^2_{(2)}$, $\mathrm{ff}\left(X^2_{(2)}\right)$, i.e. the lift of $(\partial X)^2 \times \{0\}$. In fact it meets the front face in a smooth hypersurface; we denote by $\mathfrak{F}$ its intersection with the fibre diagonal

$$\mathfrak{F} \subset \mathrm{ff}\left(X^2_{(2)}\right).$$

In fact it is already smooth in $X^2_{(1)}$ and does not meet the side faces blown up to produce $X^2_{(2)}$.

Note that $\mathfrak{F}$ is not a p-submanifold because of its intersection with the lift of $(\partial X)^2 \times [0, \varepsilon_0]$ where it meets the boundary of the fibre diagonal

$$\mathfrak{B} \subset X^2_{(2)}$$

really of course the lift of the fibre diagonal inside $(\partial X)^2 \times [0, \varepsilon_0]$ to $X^2_{(2)}$. The blow up of $\mathfrak{B}$ resolves $\mathfrak{F}$ to a p-submanifold, so we can set

$$X^2_{\varepsilon\Phi} = \left[X^2_{(2)}; \mathfrak{B}; \mathfrak{F}\right].$$

We will also use $\mathfrak{B}$ and $\mathfrak{F}$ to denote the boundary hypersurfaces of $X^2_{\varepsilon\Phi}$ that result from blowing up these submanifolds.

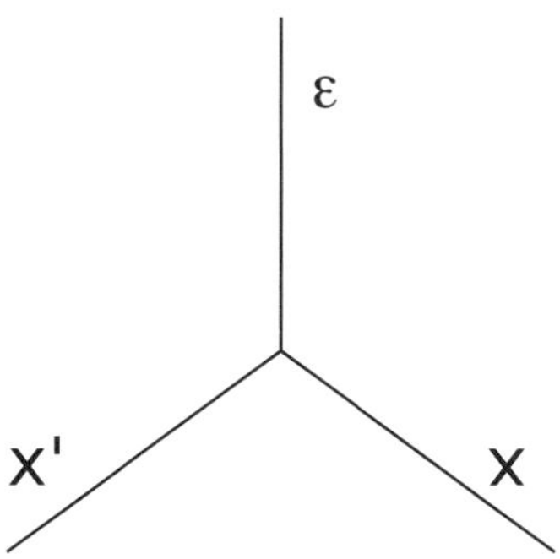

FIGURE 1. $X^2 \times [0, \varepsilon_0]$

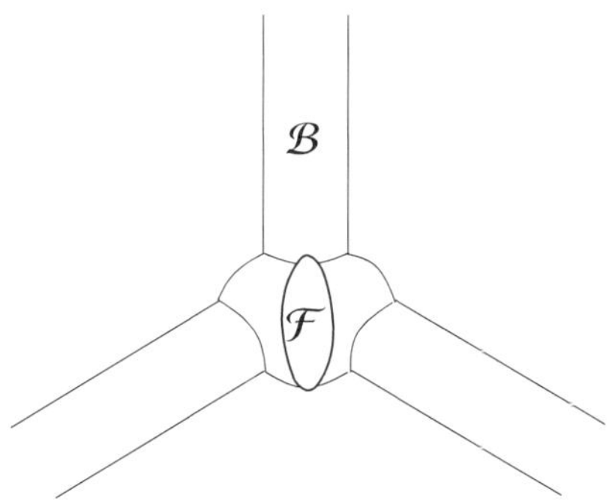

FIGURE 2. $X^2_{\varepsilon\Phi}$

PROPOSITION 2.1. *The diagonal,* $\mathrm{diag}_{\varepsilon\Phi} = \mathrm{diag} \times [0, \varepsilon_0] \subset X^2 \times [0, \varepsilon_0]$, *lifts to an interior p-submanifold of* $X^2_{\varepsilon\Phi}$ *and the projections lift to b-fibrations*

$$X^2_{\varepsilon\Phi} \overset{\pi_L}{\underset{\pi_R}{\rightrightarrows}} X_\varepsilon$$

which are transverse to the lifted diagonal.

PROOF. To show that the diagonal lifts to an interior p-submanifold, it is only necessary to consider it near the new boundary faces. Let $\{x, y_1, \dots, y_h, z_1, \dots, z_k\}$ be local coordinates near the boundary on the left factor of X^2 as in §1, and $\{x', y_1', \dots, y_h', z_1', \dots, z_k'\}$ a corresponding set of coordinates on the right factor of X^2. Then local coordinates on $X^2_{\varepsilon\Phi}$ near $\mathfrak{B} \cap \operatorname{diag}_{\varepsilon\Phi}$ are given by

$$s = \frac{x}{x'}, x', u_i = \frac{y_i - y_i'}{x'}, y_i', z_j, z_j'.$$

In these coordinates, $\operatorname{diag}_{\varepsilon\Phi} = \{s = 1, u_i = 0, z_j = z_j'\}$, hence $\operatorname{diag}_{\varepsilon\Phi}$ is a p-submanifold near $\mathfrak{B}$. In the corresponding local coordinates near $\mathfrak{F} \cap \operatorname{diag}_{\varepsilon\Phi}$,

$$S = \frac{s-1}{x'}, x', u_i, y_i', z_j, z_j',$$

the diagonal is given by $\operatorname{diag}_{\varepsilon\Phi} = \{S = 0, u_i = 0, z_j = z_j'\}$ and is again a p-submanifold. Next, to see that the maps π_R and π_L lift to b-fibrations, note that they do not map any boundary hypersurface to a corner and are fibrations in the interior of each boundary face. Finally, they are transverse to the lifted diagonal from the local coordinate description of $\operatorname{diag}_{\varepsilon\Phi}$. □

For any $\mathbb{Z}_2$-graded vector bundle $\mathbb{E} = (E_+, E_-)$ over X the space $\Psi^k_{\varepsilon\Phi}(X; \mathbb{E})$ of adiabatic-Φ pseudodifferential operators on X, between sections of E_+ and sections of E_-, is defined to be

$$\Psi^k_{\varepsilon\Phi}(X; E_+, E_-) = \big\{A \in I^{k-\frac{1}{4}}\left(X^2_{\varepsilon\Phi}, \operatorname{Diag} \times [0, \varepsilon_0]; \operatorname{Hom}(E_+, E_-) \otimes \Omega_{\varepsilon\Phi}\right) :$$
$$A \equiv 0 \text{ at the lifts of } \partial X \times X \times [0, \varepsilon_0], X \times \partial X \times [0, \varepsilon_0], \partial X \times X \times \{0\}, X \times \partial X \times \{0\}\big\}.$$

PROPOSITION 2.2. *These kernels define continuous linear operators*

$$\mathcal{C}^\infty(X_\varepsilon; E_+) \to \mathcal{C}^\infty(X_\varepsilon; E_-).$$

PROOF. This follows from the push-forward and pull-back theorems of [**10**] together with proposition 2.1. Indeed, given a section $f \in \mathcal{C}^\infty\left(X_\varepsilon, {}^{\varepsilon\Phi}\Omega^{1/2}\right)$ and an operator $A \in \Psi^k_{\varepsilon\Phi}\left(X; {}^{\varepsilon\Phi}\Omega^{1/2}\right)$ we have

$$Af = (\pi_L)_* \left(A \cdot \pi_R^* f\right).$$

Since A vanishes to infinite order at all faces of $X^2_{\varepsilon\Phi}$ not meeting the diagonal, and each boundary face meeting the diagonal maps down to a unique boundary face in X_ε it follows directly that smooth functions are mapped to smooth functions. In the same way, if $\mathcal{E}$ is a smooth index set for X_ε and $\mathcal{A}^{\mathcal{E}}$ is the associated set of polyhomogeneous functions then

$$\mathcal{A}^{\mathcal{E}} \xrightarrow{A} \mathcal{A}^{\mathcal{E}}.$$

The same argument holds with bundle coefficients with only notational differences. □

We will give a similar geometric proof of the composition

$$\Psi^k_{\varepsilon\Phi}(X; G, H) \circ \Psi^{k'}_{\varepsilon\Phi}(X; E, G) \subset \Psi^{k+k'}_{\varepsilon\Phi}(X; E, H).$$

in the appendix (Proposition A.2), after we describe the corresponding triple space.

Operators in the adiabatic calculus, along with the usual interior symbol, have four 'normal operators' from restricting their kernels to the different boundary faces. These restrictions in turn have natural interpretations as operators in simpler calculi derived from the suspended and non-commutative suspended calculi in which the normal operators of the fibered cusp and fibered boundary calculi respectively lie. All together, we have five surjective homomorphisms

$$\begin{pmatrix} \sigma \\ N_{\mathfrak{B}} \\ N_{\mathfrak{F}} \\ R_0 \\ R_1 \end{pmatrix} : \Psi^{m,k}_{\Phi\text{-a}}(X;\mathbb{E}) \longrightarrow \begin{pmatrix} S^m({}^{\Phi\text{-a}}T^*X, \hom(\mathbb{E})) \\ \mathcal{C}^\infty\left([0,1]_\varepsilon, \Psi^k_{\Phi\text{-Nsus}}(X;\mathbb{E})\right) \\ \mathcal{C}^\infty\left([-1,1]_\tau, \Psi^k_{\Phi\text{-sus}}(X;\mathbb{E})\right) \\ \Psi^m_{\Phi\text{-c}}(X;\mathbb{E}) \\ \Psi^m_{\Phi\text{-b}}(X;\mathbb{E}) \end{pmatrix}$$

with range respectively the homogeneous sections, the non-commutative Φ-suspended calculus, the Φ-suspended calculus, the fibered cusp and fibered boundary calculi respectively.

The model adiabatic calculus, where $N_{\mathfrak{F}}$ takes values, is a bundle of suspended pseudodifferential algebras on $Y \times (0,1)$. Namely, if $\mathrm{ff}(X_\varepsilon)$ denotes the lift of $\partial X \times \{0\}$ to X_ε, then the adiabatic front face of $X^2_{\varepsilon\Phi}$ as a bundle over $Y \times [-1,1] = \mathrm{ff}\,(X_\varepsilon)/Z$ (where Z is the fiber of Φ) with fiber over $\{y\} \times \{\frac{x-\varepsilon}{x+\varepsilon}\}$ the product of $RC(W) \times Z_y \times Z_y$ with $RC(W)$ the radial compactification of the subspace

$$W = \ker\left({}^{\varepsilon\Phi}TX_\varepsilon \hookrightarrow TX_\varepsilon\right).$$

It is useful to see how the product on the fibers evolves as $\varepsilon \to 0$. For simplicity assume we are in the case $Z = \{\mathrm{pt}\}$. In projective coordinates away from $\varepsilon = 0$,

$$(x, y, s, u, \varepsilon) = \left(x, y, \frac{x'}{x}, \frac{y'-y}{x}, \varepsilon\right),$$

the product on the fiber of the zero front face over the point $(0, y_0, \varepsilon)$ is given by

$$(s, u) \cdot (s', u') = (ss', u + su').$$

In the perhaps less natural coordinates

$$(x, y, S, U, \varepsilon) = \left(x, y, \frac{x'-x}{x}, \frac{y'-y}{x}, \varepsilon\right)$$

this product takes the form

$$(S, U) \cdot (S', U') = (S + S' + SS', U + U' + SU').$$

Near the adiabatic front face we can use coordinates

$$(x, y, \varepsilon, \tau, S_\varepsilon, U_\varepsilon) = \left(x, y, \varepsilon, \frac{x-\varepsilon}{x+\varepsilon}, \frac{x'-x}{x\,(x+\varepsilon)}, \frac{y'-y}{x}\right)$$

and obtain a family of products over the point $(0, y_0, 0, \tau_0)$

$$(S_\varepsilon, U) \cdot (S'_\varepsilon, U') = (S_\varepsilon + S'_\varepsilon + \varepsilon S_\varepsilon S'_\varepsilon, U + U' + \varepsilon S_\varepsilon U').$$

Note that if $\varepsilon = 1$ this product coincides with that on the zero front face, while at the adiabatic front face ($\varepsilon = 0$) it coincides with the product on the scattering front face.

Note that this construction extends to families: If $X - M \xrightarrow{\phi} B$ is a fibration with X as above, then carrying out the above blow-ups on $M \times_B M$ (the fiber product of M with itself) we obtain a fibration

$$\begin{array}{ccc} X^2_{\varepsilon\Phi} & \text{———} & M^2_{\varepsilon\Phi} \\ & & \downarrow \\ & & B \end{array}$$

which carries the kernels of families of adiabatic operators, $\Psi^{*}_{\varepsilon\Phi}(M/B;\mathbb{E})$.

PROPOSITION 2.3 (cf. [**14**], Proposition C.1). *If $P \in \Psi^{m}_{\varepsilon\Phi}(M/B;\mathbb{E})$ and the normal operators $\sigma(P)$, $N_{\mathfrak{B}}(P)$ and $N_{\mathfrak{F}}(P)$ are invertible then $R_0(P)$ and $R_1(P)$ are both fully elliptic and have the same families index.*

PROOF. The invertibility of $\sigma(P)$ allows the construction of a parametrix $Q_0 \in \Psi^{-m}_{\varepsilon\Phi}(M/B;E_-,E_+)$ with residues

$$\operatorname{Id} - Q_0 P \in \Psi^{-\infty}_{\varepsilon\Phi}(M/B;E_+), \qquad \operatorname{Id} - PQ_0 \in \Psi^{-\infty}_{\varepsilon\Phi}(M/B;E_-).$$

Each invertible normal operator permits this to be refined to a parametrix with residues vanishing to infinite order at the corresponding boundary face. Thus the hypothesis of the proposition permit us to find $Q \in \Psi^{-m}_{\varepsilon\Phi}(M/B;\mathbb{E}_-)$ with residues

$$\operatorname{Id} - QP \in x^{\infty}\Psi^{-\infty}_{\varepsilon\Phi}(M/B;E_+), \qquad \operatorname{Id} - PQ \in x^{\infty}\Psi^{-\infty}_{\varepsilon\Phi}(M/B;E_-).$$

The induced equations for $R_0(P)$ and $R_1(P)$ show that these operators are fully elliptic hence Fredholm.

The residues above are families of compact operators over $[0,1]$ in the intersection of the fibered cusp and fibered boundary calculi, hence it is possible to stabilize. That is, we can find an operator

$$A \in x^{\infty}\Psi^{-\infty}_{\varepsilon\Phi}(M/B;\mathbb{E}) = \mathcal{C}^{\infty}\left([0,1]_{\varepsilon}, \dot{\Psi}^{-\infty}(M/B;\mathbb{E})\right)$$

such that the null space of $P + A$ is in $\dot{\mathcal{C}}^{\infty}(M/B;E_+)$ and forms a trivial smooth bundle over $[0,1] \times B$ (see [**14**, Lemma 1.1]). It follows that the index bundles for $R_0(P)$ and $R_1(P)$ coincide in $K(B)$. □

THEOREM 2.4. *The adiabatic construction above induces a natural map*

$$\mathcal{K}^{*}_{\Phi\text{-c}}(\phi) \to \mathcal{K}^{*}_{\Phi\text{-b}}(\phi),$$

which restricts to

$$\mathcal{K}^{*}_{\Phi\text{-c},-\infty}(\phi) \to \mathcal{K}^{*}_{\Phi\text{-b},-\infty}(\phi),$$

and fits into the commutative diagram

$$\begin{array}{ccccc} & & \mathcal{K}^{*}_{\Phi\text{-c}}(\phi) & & \\ & \swarrow^{\operatorname{ind}_a} & \downarrow & \searrow^{\sigma} & \\ K^{*}(B) & \xleftarrow{\operatorname{ind}_a} & \mathcal{K}^{*}_{\Phi\text{-b}}(\phi) & \xrightarrow{\sigma} & K_c(T^{*}(X/B)) \end{array}$$

PROOF. Given a fibered cusp pseudodifferential operator, $P \in \Psi^{0}_{\Phi\text{-c}}(M/B;\mathbb{E})$ we can construct an element of $\Psi^{0}_{\varepsilon\phi}(M/B;\mathbb{E})$ by following the construction in [**14**, Proposition 8]. First the kernel of P gives us the normal operator of a putative $P(\varepsilon)$ at the (lift of) $\varepsilon = 0$. This fixes the boundary value for the face $\mathfrak{F}$, which as we have just seen is fibered over the lifted variable

$$\tau = \frac{x - \varepsilon}{x + \varepsilon} \in [-1,1],$$

with fiber given by the front face in the fibered cusp calculus. Knowing the value at $\tau = 1$, we extend to the rest of $\mathfrak{F}$ to be independent of τ. This fixes the kernel at the boundary of $\mathfrak{B}$ and we simply extend it as a conormal distribution to the diagonal and vanishing to infinite order at the other boundary faces. Having defined a kernel consistently on the boundary of $M^2_{\varepsilon\phi}$ we can choose an extension which is conormal to the diagonal in the interior, thus obtaining an element $P(\varepsilon) \in \Psi^0_{\varepsilon\Phi}(M/B;\mathbb{E})$.

If P is Fredholm then the normal operators $\sigma(P(\varepsilon))$ and $N_{\mathfrak{F}}(P(\varepsilon))$ are invertible as is $N_{\mathfrak{B}}(P(\varepsilon))$ for small enough ε, say $\varepsilon \in [0,\delta]$. Of course we can arrange for $N_{\mathfrak{B}}(P(\varepsilon))$ to be invertible for all ε, for instance by rescaling $[0,\delta]$ to $[0,1]$ or by perturbing $P(\varepsilon)$ with a family $A(\varepsilon)$ such that $N_{\mathfrak{B}}(A(\varepsilon))$ is a finite rank smoothing perturbation vanishing at $\varepsilon = 0$ and making the family $N_{\mathfrak{B}}(P(\varepsilon))$ invertible (as in [**14**, Remark 1.2]).

The invertibility of these normal operators allows us to apply Proposition 2.3 and conclude that $R_1(P(E))$ is fully elliptic. Thus, in terms of the maps

$$\begin{array}{ccccc} & & \Psi^0_{\varepsilon\Phi}(M/B;\mathbb{E}) & & \\ & \overset{R_0}{\swarrow} & & \overset{R_1}{\searrow} & \\ \Psi^0_{\Phi\text{-c}}(M/B;\mathbb{E}) & & & & \Psi^0_{\Phi\text{-b}}(M/B;\mathbb{E}) \end{array}$$

we see that we can lift each fully elliptic family P in the Φ-c calculus via R_0 to a family $P(\varepsilon)$ in the adiabatic calculus such that $R_1(P(\varepsilon))$ is a fully elliptic family in the Φ-b calculus. Notice that the choice of δ and extension to $\varepsilon = 1$ do not change the homotopy class of $R_1(P(\varepsilon))$. Furthermore, we can always extend a homotopy of P to a homotopy of $P(\varepsilon)$ as well as the other equivalence relations defining the K-groups (stabilization and bundle isomorphism), so we have a map

$$\mathcal{K}^*_{\Phi\text{-c}}(\phi) \ni [P] \xrightarrow{\text{ad}} [R_1(P(\varepsilon))] \in \mathcal{K}^*_{\Phi\text{-b}}(\phi),$$

well-defined independently of choices.

From the construction it is clear that perturbations of the identity by a smoothing operator are preserved as is the interior symbol, and from Proposition 2.3 so is the families index. $\square$

3. Fredholm perturbations of the identity

In this section, we prove that the map

$$\mathcal{K}^*_{\Phi\text{-c},-\infty}(\phi) \to \mathcal{K}^*_{\Phi\text{-b},-\infty}(\phi),$$

induced by the adiabatic construction in Theorem 2.4 is an isomorphism.

We first recall the excision argument from [**14**] and reduce to the case $\Phi = \text{Id}$, i.e., the scattering and zero calculi. We start by replacing M with a collar neighborhood of its boundary, $\partial M \times [0,1]_x$. The K-theory groups we are considering are made up of equivalence classes of normal operators, so there is no loss in restricting attention to the quantizations that are supported in this neighborhood, and reduce to the identity at $x = 1$.

The fibration extends off the boundary into the entire neighborhood $\partial M \times [0,1]_x$, so we have the simpler situation

$$\begin{array}{ccccc} Z \times [0,1] & \text{——} & \partial X \times [0,1] & \text{——} & \partial M \times [0,1] \\ & & \downarrow & & \downarrow \\ & & Y \times [0,1] & \text{——} & D \times [0,1] \\ & & & & \downarrow{\scriptstyle \widetilde{\phi}} \\ & & & & B \end{array}$$

where furthermore we have the following reduction.

LEMMA 3.1. *There are 'excision' isomorphisms*

$$\mathcal{K}^*_{\Phi\text{-c},-\infty}(\phi) \cong \mathcal{K}^*_{\mathrm{sc},-\infty}(\widetilde{\phi}) \text{ and } \mathcal{K}^*_{\Phi\text{-b},-\infty}(\phi) \cong \mathcal{K}^*_{0,-\infty}(\widetilde{\phi})$$

PROOF. Recall [**16**, Lemma 6.3], [**14**, proof of Prop. 3.1] that given

$$b \in \Psi^{-\infty}_{\Phi\text{-sus}}(\partial M/D; E)$$

such that $\mathrm{Id} + b$ is invertible, we can assume that b acts on a finite-rank sub-bundle W of $\mathcal{C}^\infty(\partial M/D)$ pulled-back to $T^*(D/B) \times \mathbb{R}$ and then think of b as the boundary family of a family in $\Psi^{-\infty}_{\Phi\text{-c}}(M/B; E)$ or in $\Psi^{-\infty}_{\mathrm{sc}}(D \times [0,1]/B; W)$. Doing the same to fibered boundary operators we obtain the isomorphisms above. □

Hence, it suffices to show that the adiabatic homomorphism between $\mathcal{K}^*_{\mathrm{sc},-\infty}(\widetilde{\phi})$ and $\mathcal{K}^*_{0,-\infty}(\widetilde{\phi})$ is an isomorphism. Both the kernel of an operator in $\Psi^{-\infty}_{\mathrm{sc}}(M; E)$ and the kernel of an operator in $\Psi^{-\infty}_{0}(M; E)$ are, near their respective front faces, translation invariant in the 'horizontal' directions. We can apply the Fourier transform in these directions and restrict to the front face, and obtain in either case an element of

$$\mathcal{S}(T^*\partial M)\, \widehat{\otimes}_\pi \dot{C}^\infty([-1,1])\, \widehat{\otimes}_\pi \mathrm{Hom}(E). \tag{3.1}$$

In the scattering case this is easily recognized as defining a class in $K^{-1}_c(T^*\partial M \times \mathbb{R}) \cong K^{-2}_c(T^*\partial M)$ and respecting the product structure. We show in the following theorem that the zero case defines the same class in $K^0(T^*\partial M)$. The coincidence of the classes defined by the commutative (scattering) product with that defined by the non-commutative (zero) product is a manifestation of Bott periodicity (cf. [**4**, Proposition 9.9]).

THEOREM 3.2. *The map $\mathcal{K}^*_{\mathrm{sc},-\infty}(\widetilde{\phi}) \to \mathcal{K}^*_{0,-\infty}(\widetilde{\phi})$ induced by the adiabatic calculus is an isomorphism and fits into the commutative diagram*

$$\begin{array}{ccc} \mathcal{K}^*_{\mathrm{sc},-\infty}(\phi) & \xrightarrow{\cong} & K^*_c(T^*\partial M/B) \\ \downarrow & \nearrow{\scriptstyle \cong} & \\ \mathcal{K}^*_{0,-\infty}(\phi) & & \end{array}$$

Before proving this theorem, we quickly review the relevant descriptions of the topological K-theory groups from [**13**]. Recall that for any manifold X we can describe $K^1(X)$ as stable homotopy classes of maps

$$K^1(X) = \lim_{\to}[X; \mathrm{GL}(N)],$$

so also [**13**, Proposition 3],

$$K^1(X) = \left[X; \dot{G}^{-\infty}(V;E)\right], \tag{3.2}$$

where V is any compact manifold with boundary and

$$\dot{G}^{-\infty}(V;E) = \{\mathrm{Id} + A : A \in \dot{\Psi}^{-\infty}(X;E) \text{ and } (\mathrm{Id} + A) \text{ is invertible}\}.$$

Similarly, we can define $K^0(X)$ from $K^1(X)$ by suspension, i.e.,

$$K^0(X) = \varinjlim \left[X; C^\infty\left((\mathbb{S}^1, 1); (\mathrm{GL}(N), \mathrm{Id})\right)\right]$$

and this can also be thought of as [**13**, Proposition 4]

$$K^0(X) = \left[X; G^{-\infty}_{\mathrm{sus}}(U;E)\right], \tag{3.3}$$

with U any closed manifold of positive dimension and, using $\mathcal{S}$ to dencte Schwartz functions,

$$G^{-\infty}_{\mathrm{sus}}(U;E) = \{\mathrm{Id} + A : A \in \mathcal{S}(\mathbb{R}; \Psi^{-\infty}(U;E)) \text{ and } (\mathrm{Id} + A) \text{ is invertible}\}.$$

For non-compact manifolds the same definitions apply, but the families of operators are required to be equal to the identity outside a compact set. Homotopies are also required to be the identity outside a compact set, however, in either case, the compact set is not fixed.

PROOF (OF THEOREM 3.2). Recall that in [**1**] it is shown that the groups $\mathcal{K}^q_{0,-\infty}(\phi)$ are canonically isomorphic to the groups $K^q_c(T^*\partial M/B)$. So, for each $\varepsilon_0 > 0$, we have an isomorphism into $K^0_c(T^*\partial M/B)$ consistent with the product on the $\varepsilon = \varepsilon_0$ slice of the adiabatic calculus. The smoothness of the product in the adiabatic calculus shows that, for any Fredholm operator $\mathrm{Id} + A \in \Psi^{-\infty}_{\varepsilon\phi}$, we obtain for each $\varepsilon_0 > 0$ the same class in $K^0_c(T^*\partial M)$.

Thus we have two homomorphisms from $\mathcal{K}^*_{\mathrm{sc},-\infty}(\phi)$ into $K^0_c(T^*\partial M)$. The first is the isomorphism given by identifying (3.1) with an element of $K^{-1}_c(T^*\partial M \times \mathbb{R}) \cong K^{-2}_c(T^*\partial M)$. The second comes from extending the given Fredholm scattering operator into a Fredholm adiabatic operator and applying the isomorphism of [**1**]. The first can be represented by a map

$$\begin{aligned} [T^*\partial M \times \mathbb{R}; \mathrm{GL}_N(\mathbb{C})] &= [S^*\partial M \times \mathbb{R}_+ \times \mathbb{R}; \mathrm{GL}_N(\mathbb{C})]_0 \\ &= [S^*\partial M; \mathcal{S}(\mathbb{R}_+ \times \mathbb{R}) \otimes \mathrm{GL}_N(\mathbb{C})]_0 \end{aligned} \tag{3.4}$$

with the zero subscript denoting that the restriction to zero in the $\mathbb{R}_+$ factor depends only on the base variable in $S^*\partial M$. The second homomorphism comes from quantizing the product in (3.4) into the b,c-calculus with a parameter ε:

$$[S^*\partial M; \mathcal{S}(\mathbb{R}_+ \times \mathbb{R}) \otimes \mathrm{GL}_N(\mathbb{C})]_0 \xrightarrow{q_\varepsilon} \left[S^*\partial M; \Psi^{-\infty,-\infty}_{\mathrm{b,c}}\left([0,1]; \mathbb{C}^N\right)\right]_0.$$

When $\varepsilon = 0$ we have the commutative product and for positive ε, the non-commutative (cf. [**4**, Proposition 9.9, Theorem 9.5]). Hence the isomorphism between $\mathcal{K}^0_{0,-\infty}(\phi)$ and $K^0_c(T^*\partial M)$ is taken into the isomorphism between $\mathcal{K}^0_{\mathrm{sc},-\infty}(\phi)$ and $K^0_c(T^*\partial M)$ by the adiabatic limit. □

4. Six term exact sequence

The Atiyah-Singer Index Theorem induces natural maps

$$K_c^q(T^*M/B) \ni [\sigma] \xrightarrow{I_q} [\operatorname{ind}_{AS} r_{\partial M}^* \sigma] \in K_c^{q+1}(T^*D/B)$$

and, together with the identifications $K_c^{q+1}(T^*D/B) \cong \mathcal{K}_{\Phi\text{-b},-\infty}^{q+1}(\phi)$, these are maps $K_c^q(T^*M/B) \to \mathcal{K}_{\Phi\text{-b},-\infty}^{q+1}(\phi)$. Just as in [**14**, Thm. 6.3], they fit into a diagram

$$\begin{array}{ccccc} \mathcal{K}_{\Phi\text{-b},-\infty}^{0}(\phi) & \xrightarrow{i_0} & \mathcal{K}_{\Phi\text{-b}}^{0}(\phi) & \xrightarrow{\sigma_0} & K_c^0(T^*M/B) \\ \uparrow{\scriptstyle I_1} & & & & \downarrow{\scriptstyle I_0} \\ K_c^1(T^*M/B) & \xleftarrow[\sigma_1]{} & \mathcal{K}_{\Phi\text{-b}}^{1}(\phi) & \xleftarrow[i_1]{} & \mathcal{K}_{\Phi\text{-b},-\infty}^{1}(\phi) \end{array} \tag{4.1}$$

with the maps i_* and σ_* induced by inclusion and the principal interior symbol, which we now show is exact.

PROPOSITION 4.1. *The diagram* (4.1) *is exact.*

PROOF. While exactness at $\mathcal{K}_{\Phi\text{-b}}^*(\phi)$ follows directly from the definitions, to establish exactness elsewhere we give a different description of the maps I_q.

Given a class $[\sigma_t] \in K_c^1(T^*M/B)$, it is shown in [**14**, Proposition 6.2] that there is a family of operators

$$\begin{gathered} P_t \in \Psi_{\Phi\text{-c}}^0(M/B;\mathbb{C}^N) \text{ and Fredholm, such that} \\ \sigma(P_t) = \sigma_t, \quad P_0 = \operatorname{Id}, \quad P_1 \in (\operatorname{Id} + \Psi_{\Phi\text{-c}}^{-\infty}(M/B;\mathbb{C}^N)) \text{ and Fredholm} \end{gathered}$$

and the identification $\mathcal{K}_{\Phi\text{-c},-\infty}^0(\phi) \cong K_c^0(T^*D/B)$ takes $[P_1]$ to $[\operatorname{ind}_{AS} r_{\partial M}^* \sigma]$. As in the proof of Theorem 2.4, we can lift the family P_t to a family of operators in the adiabatic calculus $P_t(\varepsilon)$ such that $\widetilde{P}_t = R_1(P_t(\varepsilon))$ satisfies

$$\begin{gathered} \widetilde{P}_t \in \Psi_{\Phi\text{-b}}^0(M/B;\mathbb{C}^N) \text{ and Fredholm, such that} \\ \sigma(\widetilde{P}_t) = \sigma_t, \quad \widetilde{P}_0 = \operatorname{Id}, \quad \widetilde{P}_1 \in (\operatorname{Id} + \Psi_{\Phi\text{-b}}^{-\infty}(M/B;\mathbb{C}^N)) \text{ and Fredholm} \end{gathered}$$

and, from Theorem 3.2, the identification $\mathcal{K}_{\Phi\text{-b},-\infty}(\phi) \cong K_c(T^*D/B)$ takes $[\widetilde{P}_1]$ to $[\operatorname{ind}_{AS} r_{\partial M}^* \sigma]$. Thus we can think of I_1 as being $[\sigma_t] \mapsto [\widetilde{P}_1]$.

An element $[\sigma_t]$ is in the null space of I_1 if $\widetilde{P}_1 \sim \operatorname{Id}$, i.e., precisely when $[\sigma_t]$ is in the image of the map $K_{\Phi\text{-b}}^1(\phi) \xrightarrow{\sigma_1} K_c^1(T^*M/B)$. On the other hand the homotopy $\widetilde{P}_t$ shows that $[\widetilde{P}_1]$ is automatically trivial as an element $K_{\Phi\text{-b}}^0(\phi)$, that is, $\operatorname{Im} I_1 \subseteq \operatorname{null} i_0$, and conversely, if $[Q] \in \operatorname{null} i_0$, there is (after stabilization) a homotopy of operators from the identity to Q, and the symbol of this homotopy defines an element in $K_c^1(T^*M/B)$ that is sent by I_1 to $[Q]$. Thus we see that the diagram is exact at $K_c^1(T^*M/B)$ and at $\mathcal{K}_{\Phi\text{-b}}^1(\phi)$. Just as in [**14**, Theorem 6.3] the same arguments prove exactness at $K_c^1(T^*M/B)$ and $\mathcal{K}_{\Phi\text{-b}}^1(\phi)$ using Bott periodicity. □

As explained in the proof of this proposition, the diagrams

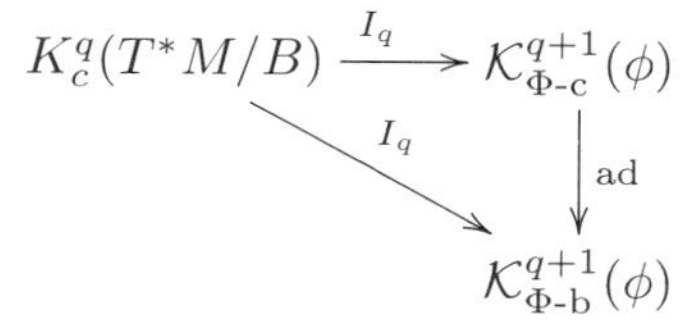

are commutative. Hence the six term exact sequences for the smooth K-theory of Φ-c and Φ-b operators make up a commutative diagram:

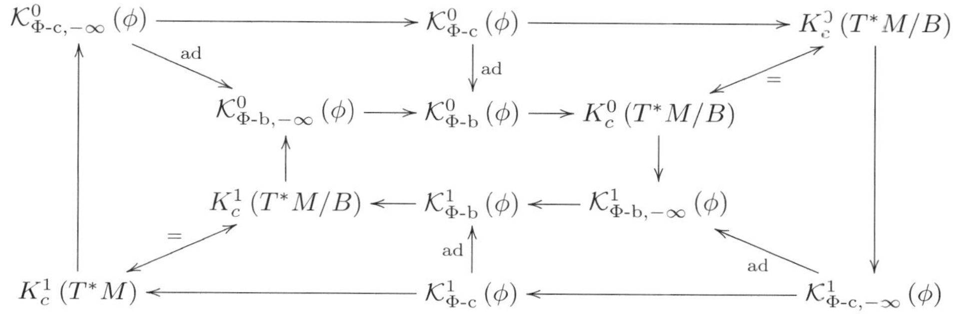

and, since we have already shown that $\mathcal{K}^q_{\Phi\text{-c}}(\phi) \xrightarrow{\text{ad}} \mathcal{K}^q_{\Phi\text{-b}}(\phi)$ is an isomorphism, the five-lemma implies the following.

THEOREM 4.2. *The adiabatic calculus induces an isomorphism of the smooth K-theory groups of the* Φ-c *calculus and those of the* Φ-b *calculus,*

$$\mathcal{K}^q_{\Phi\text{-c}}(\phi) \underset{\cong}{\xrightarrow{\text{ad}}} \mathcal{K}^q_{\Phi\text{-b}}(\phi).$$

Appendix A. Triple adiabatic space

In this section we prove the composition formula for the adiabatic calculus. The kernel of the composition is given by an appropriate interpretation of

$$\mathcal{K}_{A\circ B}(\zeta,\zeta',\varepsilon) = \int_{M^2_{\varepsilon\phi}} \mathcal{K}_A(\zeta,\zeta'',\varepsilon)\,\mathcal{K}_B(\zeta'',\zeta',\varepsilon).$$

Our method is geometric, we construct a space $M^3_{\varepsilon\phi}$ with three b-fibrations 'first', 'second', and 'composite' down to $M^2_{\varepsilon\phi}$. The kernel of the composite is then correctly given by

$$(\pi_C)_*\left(\pi_F^* A \cdot \pi_S^* B\right).$$

The basic triple space is $M^3 \times [0,\epsilon_0]$ where $\epsilon_0 > 0$ and M is a compact manifold with boundary and a specified fibration $\phi : \partial M \longrightarrow Y$ of the boundary. We first have to perform the 'boundary adiabatic' blow ups to be able to map back to the single and double spaces

$$\begin{gathered}
M^3_{(1)} = [M^3 \times [0,\epsilon_0], A_T, A_F, A_C, A_S, A_R, A_M, A_L],\\
A_T = (\partial M)^3 \times \{0\},\ A_F = M \times \partial M \times \partial M \times \{0\},\\
A_S = \partial M \times \partial M \times M \times \{0\},\ A_C = \partial M \times M \times \partial M \times \{0\},\\
A_R = M \times M \times \partial M \times \{0\},\ A_M = M \times \partial M \times M \times \{0\},\\
A_L = \partial M \times M \times M \times \{0\}.
\end{gathered} \tag{A.1}$$

Then consider the lifts of the triple and double fiber diagonals, lying within the lifts of the faces $(\partial M)^3 \times [0,\epsilon_0]$ and $(\partial M)^2 \times M \times [0,\epsilon_0]$ and its cyclic images. We can denote these as $\mathfrak{B}_T$, the triple boundary fibred diagonal and $\mathfrak{B}_S$, $\mathfrak{B}_C$ and $\mathfrak{B}_F$, the double fibered diagonals. These are all p-submanifolds, meeting in the standard way for diagonals so we may define

$$M^3_{(2)} = [M^3_{(1)}, \mathfrak{B}_T, \mathfrak{B}_S, \mathfrak{B}_C, \mathfrak{B}_F]. \tag{A.2}$$

The double and triple fibered-cusp diagonal faces are not p-submanifolds in $M_{(1)}$ but lift to be p-submanifolds, which we denote $\mathfrak{F}_T$, $\mathfrak{F}'_F$, $\mathfrak{F}'_C$, $\mathfrak{F}'_S$ and $\mathfrak{F}''_F$, $\mathfrak{F}''_C$ and $\mathfrak{F}''_F$. Here, $\mathfrak{F}'_O$ is the lift of the corresponding double diagonal in the face produced by the blow-up of A_O, $O = S, C, F$ and $\mathfrak{F}''_O$ is the lift of the intersection of this submanifold under the blow-up of A_T. The intersection properties of these submanifolds is essentially the same as for the fibered-cusp (or indeed the cusp) setting itself. Thus we complete the definition by blowing up in 'the usual' order

$$M^3_{\epsilon\phi} = [M^3_{(2)}, \mathfrak{F}_T, \mathfrak{F}''_S, \mathfrak{F}''_C, \mathfrak{F}''_F, \mathfrak{F}'_S, \mathfrak{F}'_C, \mathfrak{F}'_F]. \tag{A.3}$$

PROPOSITION A.1. *Each of the projections, dropping one or two factors of M, lift to b-fibrations*

$$\begin{gathered}\begin{pmatrix}\pi_F\\ \pi_C\\ \pi_S\end{pmatrix} : M^3_{\epsilon\phi} \rightrightarrows M^2_{\epsilon\phi},\\ \begin{pmatrix}\pi_L\\ \pi_M\\ \pi_R\end{pmatrix} : M^3_{\epsilon\phi} \rightrightarrows M_\epsilon.\end{gathered} \tag{A.4}$$

PROOF. This is the usual commutativity of blow ups argument. Rearranging the 'fibred-cusp' faces works just as for the (fibered-) cusp calculus. After these triple and two double faces have been blown down, the corresponding fibered-boundary faces can be blown down, as can the pure adiabatic faces. □

PROPOSITION A.2.

$$\Psi^k_{\varepsilon\phi}(M; G, H) \circ \Psi^{k'}_{\varepsilon\phi}(M; E, G) \subset \Psi^{k+k'}_{\varepsilon\phi}(M; E, H)$$

PROOF. This follows from Proposition A.1 via the push-forward and pull-back theorems of [**10**]. Indeed, given two operators (acting on half-densities for simplicity) A and B the kernel of their composition is given by

$$A \circ B = (\pi_C)_* (\pi_S^* A \cdot \pi_F^* B).$$

As the maps are b-fibrations and transversal to the diagonals this yields a distribution conormal to the diagonal of the appropriate degree. Since the kernels of A and B vanish to infinite order at every face not meeting the diagonal, the product of their lifts will vanish to infinite order at every face not meeting the triple diagonal (as this is the intersection of the lift of the two diagonals). Hence only the faces coming from $\operatorname{diag}(M^3) \times \{0\}$, and the blow-ups of $\mathfrak{B}_T$ and $\mathfrak{F}_T$ potentially contribute to the push-forward. It follows that the resulting distribution is smooth down to the each of the boundary faces. □

References

[1] Pierre Albin and Richard B. Melrose, *Fredholm realizations of elliptic symbols on manifolds with boundary.*, J. Reine Angew. Math. **627** (2009), 155–181.

[2] M. F. Atiyah and R. Bott, *The index problem for manifolds with boundary*, Differential Analysis, Bombay Colloq., 1964, Oxford Univ. Press, London, 1964, pp. 175–186.

[3] Charles L. Epstein, Richard B. Melrose, and Gerardo A. Mendoza, *Resolvent of the Laplacian on strictly pseudoconvex domains.*, Acta Math. **167** (1991), no. 1-2, 1–106.

[4] ———, *The Heisenberg algebra, index theory and homology*, preprint, February 2000.

[5] C. Robin Graham, *Volume and area renormalizations for conformally compact Einstein metrics*, The Proceedings of the 19th Winter School "Geometry and Physics" (Srní, 1999), no. 63, 2000, pp. 31–42.

[6] Robert Lauter, *Pseudodifferential analysis on conformally compact spaces.*, Mem. Amer. Math. Soc. **163** (2003), no. 777, xvi+92pp.

[7] Rafe Mazzeo, *The Hodge cohomology of a conformally compact metric*, J. Differential Geom. **28** (1988), no. 2, 309–339.

[8] ———, *Elliptic theory of differential edge operators. I*, Comm. Partial Differential Equations **16** (1991), no. 10, 1615–1664.

[9] Rafe Mazzeo and Richard B. Melrose, *Pseudodifferential operators on manifolds with fibred boundaries*, Asian J. Math. **2** (1998), no. 4, 833–866, Mikio Sato: a great Japanese mathematician of the twentieth century.

[10] Richard B. Melrose, *Differential analysis on manifolds with corners*, http://www-math.mit.edu/~rbm/.

[11] ———, *The Atiyah-Patodi-Singer index theorem*, Research Notes in Mathematics, vol. 4, A K Peters Ltd., Wellesley, MA, 1993.

[12] ———, *Spectral and scattering theory for the Laplacian on asymptotically Euclidian spaces*, Spectral and scattering theory (Sanda, 1992), Lecture Notes in Pure and Appl. Math., vol. 161, Dekker, New York, 1994, pp. 85–130. MR MR1291640 (95k:58168)

[13] Richard B. Melrose and Frédéric Rochon, *Families index for pseudodifferential operators on manifolds with boundary*, Int. Math. Res. Not. (2004), no. 22, 1115–1141.

[14] ———, *Index in K-theory for families of fibred cusp operators*, K-Theory **37** (2006), no. 1-2, 25–104.

[15] V. E. Nazaĭkinskiĭ, A. Yu. Savin, and B. Yu. Sternin, *On the homotopy classification of elliptic operators on stratified manifolds*, Izv. Ross. Akad. Nauk Ser. Mat. **71** (2007), no. 6, 91–118.

[16] Frédéric Rochon, *Bott periodicity for fibred cusp operators*, J. Geom. Anal. **15** (2005), no. 4, 685–722.

Department of Mathematics, Massachusetts Institute of Technology
E-mail address: pierre@math.mit.edu
E-mail address: rbm@math.mit.edu

Clay Mathematics Proceedings
Volume 12, 2010

Inversion of Series and the Cohomology of the Moduli Spaces $\mathcal{M}_{0,n}^{\delta}$

Jonas Bergström and Francis Brown

ABSTRACT. For $n \geq 3$, let $\mathcal{M}_{0,n}$ denote the moduli space of genus 0 curves with n marked points, and $\overline{\mathcal{M}}_{0,n}$ its smooth compactification. A theorem due to Ginzburg, Kapranov and Getzler states that the inverse of the exponential generating series for the Poincaré polynomial of $H^\bullet(\mathcal{M}_{0,n})$ is given by the corresponding series for $H^\bullet(\overline{\mathcal{M}}_{0,n})$. In this paper, we prove that the inverse of the ordinary generating series for the Poincaré polynomial of $H^\bullet(\mathcal{M}_{0,n})$ is given by the corresponding series for $H^\bullet(\mathcal{M}_{0,n}^{\delta})$, where $\mathcal{M}_{0,n} \subset \mathcal{M}_{0,n}^{\delta} \subset \overline{\mathcal{M}}_{0,n}$ is a certain smooth affine scheme.

1. Introduction

For $n \geq 3$, let $\mathcal{M}_{0,n}$ be the moduli space, defined over $\mathbf{Z}$, of smooth n-pointed curves of genus zero, and let $\mathcal{M}_{0,n} \subset \overline{\mathcal{M}}_{0,n}$ denote its smooth compactification, due to Deligne-Mumford and Knudsen. In [**1**], an intermediary space $\mathcal{M}_{0,n}^{\delta}$, which satisfies

$$\mathcal{M}_{0,n} \subset \mathcal{M}_{0,n}^{\delta} \subset \overline{\mathcal{M}}_{0,n},$$

was defined in terms of explicit polynomial equations. It is a smooth affine scheme over $\mathbf{Z}$. The automorphism group of $\mathcal{M}_{0,n}$ is the symmetric group $\mathfrak{S}_n$ permuting the n marked points, and this gives rise to a decomposition (see [**1**]),

$$\overline{\mathcal{M}}_{0,n} = \bigcup_{\sigma \in \mathfrak{S}_n} \sigma(\mathcal{M}_{0,n}^{\delta}) \ .$$

Thus $\mathcal{M}_{0,n}^{\delta}$ defines a symmetric set of canonical affine charts for $\overline{\mathcal{M}}_{0,n}$.

In this note, we compute the dimensions

$$a_{n,i} := \dim_{\mathbf{Q}} H^i(\mathcal{M}_{0,n}^{\delta}; \mathbf{Q})$$

of the de Rham cohomology of $\mathcal{M}_{0,n}^{\delta}$ for all i and n. Our main result can be expressed in terms of generating series, as follows. If X is a smooth scheme over $\mathbf{Q}$ of dimension d, we will denote its compactly supported Euler characteristic (or rather, Poincaré polynomial) by:

$$e_c(X)(q) = \sum_{i=0}^{2d} (-1)^i \dim_{\mathbf{Q}} H_c^i(X; \mathbf{Q})\, q^i \ .$$

Furthermore, if $H^i(X)=0$ whenever $i>d$, we define the (usual) Euler characteristic as:

$$e(X)(q)=\sum_{i=0}^{d}(-1)^i \dim_{\mathbf{Q}} H^i(X;\mathbf{Q})\, q^{d-i}\ .$$

Consider the exponential generating series:

$$\begin{aligned} g(x) &:= x-\sum_{n=2}^{\infty} e(\mathcal{M}_{0,n+1})(q^2)\,\frac{x^n}{n!}\ ,\\ \overline{g}(x) &:= x+\sum_{n=2}^{\infty} e_c(\overline{\mathcal{M}}_{0,n+1})(q)\,\frac{x^n}{n!}\ . \end{aligned}$$

The following formula is due to Ginzburg-Kapranov ([**4**], theorem 3.3.2) and Getzler ([**3**], §5.8)

$$\overline{g}(g(x))=g(\overline{g}(x))=x\ . \tag{1}$$

In this paper, we will consider the ordinary generating series:

$$\begin{aligned} f(x) &:= x-\sum_{n=2}^{\infty} e(\mathcal{M}_{0,n+1})(q)\,x^n\ ,\\ f_\delta(x) &:= x+\sum_{n=2}^{\infty} e(\mathcal{M}^\delta_{0,n+1})(q)\,x^n\ . \end{aligned}$$

THEOREM 1.1. *The following inversion formula holds:*

$$f(f_\delta(x))=f_\delta(f(x))=x\ . \tag{2}$$

Using the well-known formula

$$e(\mathcal{M}_{0,n+1})(q)=\prod_{i=2}^{n-1}(q-i) \tag{3}$$

and the purity of $\mathcal{M}^\delta_{0,n+1}$, we deduce a recurrence relation for $e(\mathcal{M}^\delta_{0,n+1})$ from (2), and hence also for the Betti numbers $a_{n,i}$. The proof of equation (2) uses the fact that the coefficients in the Lagrange inversion formula are precisely given by the combinatorics of Stasheff polytopes, which in turn determine the structure of the mixed Tate motive underlying $\mathcal{M}^\delta_{0,n}$.

In the special case $q=0$, the series $f(x)$ reduces to $x-\sum_{n=2}^\infty (n-1)!\,x^n$, which is essentially the generating series for the operad $\mathfrak{Lie}$. Comparing equation (2) to Lemma 8 in [**10**] we find that the dimensions $a_{n,n-3}=H^{n-3}(\mathcal{M}^\delta_{0,n};\mathbf{Q})$ are precisely the numbers of prime generators for $\mathfrak{Lie}$. We expect that there should be an explicit bijection between $H^{n-3}(\mathcal{M}^\delta_{0,n};\mathbf{Q})$ and the set of prime generators described in the proof of Proposition 6 in [**10**], and, more generally, an operad-theoretic interpretation of equation (2) for all q.

REMARK 1.2. The numbers $a_{n,n-3}$ count the number of convergent period integrals on the moduli space $\mathcal{M}_{0,n}$ defined in [**2**], called 'cell-zeta values'. Specifically, there is a connected component X_δ of the set of real points $\mathcal{M}_{0,n}(\mathbf{R})\subset\mathcal{M}^\delta_{0,n}(\mathbf{R})$

whose closure $\overline{X}_\delta \subset \mathcal{M}_{0,n}^{\delta}(\mathbf{R})$ is a compact manifold with corners, and is combinatorially a Stasheff polytope. For any $\omega \in H^{n-3}(\mathcal{M}_{0,n}^{\delta};\mathbf{Q})$, one can consider the integral

$$I(\omega) = \int_{X_\delta} \omega \in \mathbf{R} ,$$

which is the period of a framed mixed Tate motive over $\mathbf{Z}$, see [**6**]. By a theorem in [**1**], the number $I(\omega)$ is a $\mathbf{Q}$-linear combination of multiple zeta values. For example, when $n = 5$, we have $a_{5,2} = 1$, and there is essentially a unique such integral. Identifying $\mathcal{M}_{0,5}$ with $\{(t_1,t_2) \in (\mathbb{P}^1\backslash\{0,1,\infty\})^2, t_1 \neq t_2\}$, we can write $I(\omega)$ as

$$\int_{0<t_1<t_2<1} \frac{dt_1 dt_2}{(1-t_1)t_2} = \zeta(2) .$$

This work was begun at Institut Mittag-Leffler, Sweden, during the year 2006-2007 on moduli spaces. We thank the institute for the hospitality.

2. Geometry of $\mathcal{M}_{0,n}^{\delta}$

We recall some geometric properties of $\mathcal{M}_{0,n}^{\delta}$ from [**1**]. The set of real points $\mathcal{M}_{0,n}(\mathbf{R})$ is not connected but has $n!/2n$ components, and they can be indexed by the set of dihedral structures[1] δ on the set $\{1,\ldots,n\}$. Let X_δ denote one such connected component. Its closure in the real moduli space

$$\overline{X}_\delta \subset \overline{\mathcal{M}}_{0,n}(\mathbf{R})$$

is a compact manifold with corners. The variety $\mathcal{M}_{0,n}^{\delta} \subset \overline{\mathcal{M}}_{0,n}$ is then defined to be the complement $\overline{\mathcal{M}}_{0,n}\backslash A_\delta$, where A_δ is the set of all irreducible divisors $D \subset \overline{\mathcal{M}}_{0,n}\backslash\mathcal{M}_{0,n}$ which do not meet the closed cell $\overline{X}_\delta$. Conversely, every irreducible divisor $D \subset \overline{\mathcal{M}}_{0,n}\backslash\mathcal{M}_{0,n}$ which does meet the closed cell $\overline{X}_\delta$, defines an irreducible divisor $D \cap \mathcal{M}_{0,n}^{\delta} \subset \mathcal{M}_{0,n}^{\delta}\backslash\mathcal{M}_{0,n}$. In the case $n = 4$, we have:

$$\mathcal{M}_{0,4} \cong \mathbb{P}^1\backslash\{0,1,\infty\} , \ \mathcal{M}_{0,4}^{\delta} \cong \mathbb{P}^1\backslash\{\infty\} , \ \overline{\mathcal{M}}_{0,4} \cong \mathbb{P}^1 ,$$

where X_δ is the open interval $(0,1)$ and $\overline{X}_\delta$ is the closed interval $[0,1]$.

In the case $n = 5$, one can take four points in general position in $\mathbb{P}^2$ and identify $\mathcal{M}_{0,5}$ with the complement of a configuration of six lines passing through each pair of points. The compactification $\overline{\mathcal{M}}_{0,5}$ is obtained by blowing up these four points, giving a total of ten boundary divisors. Picturing $\mathbb{P}^2$ minus the six lines one sees that the set of real points $\mathcal{M}_{0,5}(\mathbb{R})$ has exactly 12 connected components which are triangles. Choosing one of these components X_δ, and blowing up only the two points which meet X_δ yields a space in which the boundary divisors incident to X_δ form a pentagon. The space $\mathcal{M}_{0,5}^{\delta}$ is obtained by removing all divisors of $\overline{\mathcal{M}}_{0,5}$ except the pentagon which bounds $\overline{X}_\delta$. Thus we obtain twelve isomorphic varieties $\mathcal{M}_{0,5}^{\delta}$, one for each connected component of $\mathcal{M}_{0,5}(\mathbb{R})$.

In general, $\overline{X}_\delta \subseteq \mathcal{M}_{0,n}^{\delta}(\mathbb{R})$ has the combinatorial structure of a Stasheff polytope. Its faces of codimension k are in bijection with the set of decompositions of a regular n-gon into $k+1$ polygons (with at least 3 sides) by k non-intersecting

[1] A dihedral structure on a set S is an identification of the elements of S with the edges of an unoriented polygon, i.e., considered modulo dihedral symmetries.

chords. Suppose, for each $i \geq 1$, that there are $\lambda_i(D)$ polygons in a decomposition D which has $i+2$ sides. Then the corresponding face is

$$F_D \cong \prod_{i=1}^{n-2} \prod_{j=1}^{\lambda_i(D)} \overline{X}_{i+2} \ ,$$

and $\overline{X}_{i+2}$ has itself the combinatorial structure of a Stasheff i-polytope. Note that $\overline{X}_3$ and $\mathcal{M}_{0,3}$ are just points. Since a closed polytope is the disjoint union of its open faces, we deduce the following stratification for $\mathcal{M}_{0,n}^{\delta}$:

$$\mathcal{M}_{0,n}^{\delta} = \coprod_D i_D \Big(\prod_{i=1}^{n-2} \prod_{j=1}^{\lambda_i(D)} \mathcal{M}_{0,i+2} \Big) \ . \tag{4}$$

Here, the disjoint union is taken over all decompositions D of a regular n-gon, and i_D is the isomorphism which restricts to the inclusion of each face $F_D \hookrightarrow \overline{X}_{\delta}$. The empty dissection corresponds to the inclusion of the open stratum $\mathcal{M}_{0,n}$.

EXAMPLE 2.1. There are nine chords in a regular hexagon, six of which decompose it into a pentagon and trigon, and three of which decompose it into two tetragons. It then has 21 decompositions into three pieces (a tetragon and two triangles), and 14 into four triangles. Therefore equation (4) can be abbreviated:

$$\mathcal{M}_{0,6}^{\delta} = \mathcal{M}_{0,6} \cup \Big(6\, \mathcal{M}_{0,5} \cup 3\, \mathcal{M}_{0,4}^2 \Big) \cup 21\, \mathcal{M}_{0,4} \cup 14\, \mathcal{M}_{0,3} \ . \tag{5}$$

3. Purity

Since $\mathcal{M}_{0,n}^{\delta}$ is stratified by products of varieties $\mathcal{M}_{0,r}$, which are isomorphic to an affine complement of hyperplanes and therefore of Tate type, it follows that $H^i(\mathcal{M}_{0,n}^{\delta})$ defines an element in the category of mixed Tate motives over $\mathbf{Q}$. In fact, it was proved in [**1**] that $\mathcal{M}_{0,n}^{\delta}$ is smooth and affine, so it follows by a theorem due to Grothendieck that its cohomology is generated by global regular forms. Using the well-known fact that $H^i(\mathcal{M}_{0,n})$ is pure [**3**], it follows that the subspace $H^i(\mathcal{M}_{0,n}^{\delta})$ is also pure. We can therefore work inside the semisimple subcategory (or Grothendieck group) generated by pure Tate motives. We have that,

$$H^i(\mathcal{M}_{0,n}^{\delta}) \cong \mathbf{Q}(-i)^{a_{n,i}} \ . \tag{6}$$

The purity of the spaces $\mathcal{M}_{0,n}^{\delta}$ has the important consequence that we have an equality of Poincaré polynomials (i.e. not only of Euler characteristics),

$$e(\mathcal{M}_{0,n}^{\delta}) = \sum_D \Big(\prod_{i=1}^{n-2} \prod_{j=1}^{\lambda_i(D)} e(\mathcal{M}_{0,i+2}) \Big). \tag{7}$$

4. Decompositions of regular n-gons

If λ is a partition of a number, we define λ_i to be the number of times i appears in this partition. For each partition λ of $n-2$, we then define $P(\lambda)$ to be the number of choices of $-1+\sum_i \lambda_i$ non-intersecting chords of an n-regular polygon that gives rise, for each i, to λ_i subpolygons with $i+2$ sides. Thus, $P(\lambda)$ counts the number of decompositions of an n-gon of given combinatorial type. This number is found to be equal to (see Ex. 2.7.14 on p. 127 in [**5**]):

$$P(\lambda) = \frac{(n-2+\sum_i \lambda_i)!}{(n-1)!\prod_i(\lambda_i!)}. \tag{8}$$

Combining this result and (7) we find that,

$$e(\mathcal{M}_{0,n}^{\delta}) = \sum_{\lambda \vdash n-2} P(\lambda) \cdot \prod_{i=1}^{n-2} e(\mathcal{M}_{0,i+2})^{\lambda_i}. \tag{9}$$

Using equation (3) we can now compute the $a_{n,i}$'s for any i and n,

EXAMPLE 4.1. From Example (5), we have

$$e(\mathcal{M}_{0,6}^{\delta}) = (q-2)(q-3)(q-4) + 6(q-2)(q-3) + 3(q-2)^2 + 21(q-2) + 14\ ,$$

which reduces to $q^3 + 5q - 4$. In particular, $a_{6,3} = \dim_{\mathbf{Q}} H^3(\mathcal{M}_{0,6}^{\delta}, \mathbf{Q}) = 4$.

Clearly $a_{n,0} = 1$ for all n, and it is also easy to see that $a_{n,1} = 0$ for all n. In the following table we present the results for n from five to eleven.

	$a_{n,1}$	$a_{n,2}$	$a_{n,3}$	$a_{n,4}$	$a_{n,5}$	$a_{n,6}$	$a_{n,7}$	$a_{n,8}$
$\mathcal{M}_{0,5}^{\delta}$	0	1						
$\mathcal{M}_{0,6}^{\delta}$	0	5	4					
$\mathcal{M}_{0,7}^{\delta}$	0	15	28	22				
$\mathcal{M}_{0,8}^{\delta}$	0	35	112	206	144			
$\mathcal{M}_{0,9}^{\delta}$	0	70	336	1063	1704	1089		
$\mathcal{M}_{0,10}^{\delta}$	0	126	840	3999	10848	15709	9308	
$\mathcal{M}_{0,11}^{\delta}$	0	210	1848	12255	49368	119857	159412	88562

There are no entries above the diagonal, because $\mathcal{M}_{0,n}^{\delta}$ is affine. For small i, one can use (9) to write down explicit formulae for $a_{n,i}$ as a function of n, e.g.,

$$a_{n,2} = \binom{n-1}{4} \quad \text{and} \quad a_{n,3} = 4\binom{n}{6}.$$

Finally, setting $q = 0$ in (9) gives the following closed formula for the dimension $a_{n,l}$ of the middle-dimensional de Rham cohomology of $\mathcal{M}_{0,n}^{\delta}$, where $l := n-3$,

$$\dim_{\mathbf{Q}} H^l(\mathcal{M}_{0,n}^{\delta}; \mathbf{Q}) = \sum_{\lambda \vdash l+1} P(\lambda) \cdot \prod_{i=1}^{l+1} \Big((-1)^l (l+1)!\Big)^{\lambda_i}. \tag{10}$$

5. An inversion formula

PROOF OF THEOREM 1.1. The proof is immediate on comparing equation (9) with the combinatorial interpretation of Lagrange's formula for the inversion of series in one variable (see [**9**], equation (4.5.12), p. 412). More precisely, consider the formal power series:

$$u(x) = x - \sum_{i=2}^{\infty} u_i x^i$$

Lagrange's formula states that the formal solution to $v(u(x)) = x$ is given by

$$v(x) = x + \sum_{i=2}^{\infty} v_i x^i$$

where $v_2 = u_2$, $v_3 = 2u_2^2 + u_3$, $v_4 = 5u_2^3 + 5u_2u_3 + u_4$, and in general:

$$v_n = \sum_{\lambda \vdash n-1} P(\lambda) \cdot \prod_{i=1}^{n-1} u_{i+1}^{\lambda_i} , \qquad \text{for } n \geq 2 .$$

The theorem follows from (9) on setting $u_i = e(\mathcal{M}_{0,i+1})$. □

REMARK 5.1. There is a stratification of $\overline{\mathcal{M}}_{0,n}$ similar to the one described by (4) for $\mathcal{M}_{0,n}^{\delta}$, but where $P(\lambda)$ should be replaced by $T(\lambda)$, and where $T(\lambda)$ is the number of dual graphs of n-pointed stable curves of genus zero that has λ_i components with a sum of $i+2$ marked points and nodes. Now note that from the proof of theorem 1.1 and (1) it follows that $T(\lambda) = P(\lambda) \cdot (n-1)!/\prod_i (i+1)!^{\lambda_i}$.

In the special case when $q = 0$, we deduce the following corollary.

COROLLARY 5.2. *The generating series for* $\dim_{\mathbf{Q}} H^{n-3}(\mathcal{M}_{0,n}^{\delta}; \mathbf{Q})$ *is obtained by inverting the series*

$$\sum_{n=1}^{\infty} (n-1)!\, x^n = x + x^2 + 2x^3 + 6x^4 + \ldots$$

REMARK 5.3. The cohomology of $\mathcal{M}_{0,n}$ is a module over the symmetric group $\mathfrak{S}_n$ with n elements, whose representation theory can for instance be found in [**3**] or [**8**]. The dihedral subgroup D_{2n} which stablizes a dihedral ordering δ acts upon the affine space $\mathcal{M}_{0,n}^{\delta}$, and hence its cohomology. It therefore would be interesting to compute the character of this group action on $H^{\bullet}(\mathcal{M}_{0,n}^{\delta})$, and compare its equivariant generating series to the one obtained by restriction $\mathrm{Res}_{D_{2n}}^{\mathfrak{S}_n} H^{\bullet}(\mathcal{M}_{0,n})$.

6. A recurrence relation

Let us alter our series slightly and put $F(x) := -f(-x)$ and $F_\delta(x) := -f_\delta(-x)$. By theorem 1.1 we find that $F_\delta(F(x)) = F(F_\delta(x)) = x$. The series $F(x)$ is easily seen to satisfy the differential equation:

$$x^2 F'(x) = (F(x) - x)(xq + 1) .$$

By differentiating $F_\delta(F(x)) = x$, we have $F_\delta'(F(x))F'(x) = 1$. Substituting the previous expression for $F'(x)$ gives:

$$F_\delta'(F(x))(F(x) - x)(xq + 1) = x^2 .$$

By changing variables $y = F(x)$, where $F_\delta(y) = F_\delta(F(x)) = x$, we obtain:

$$F_\delta'(y)(y - F_\delta(y))(q\, F_\delta(y) + 1) = F_\delta(y)^2 .$$

Expanding out gives:

$$yF_\delta' - F_\delta F_\delta' - F_\delta^2 + qyF_\delta F_\delta' - qF_\delta^2 F_\delta' = 0 .$$

If we write

$$F_\delta(y) = \sum_{n=1}^{\infty} a_n y^n ,$$

then the coefficient of y^n is exactly:

$$na_n - \sum_{k+l=n+1} ka_k a_l - \sum_{k+l=n} a_k a_l + q \sum_{k+l=n} ka_k a_l - q \sum_{k+l+m=n+1} ka_k a_l a_m = 0 .$$

Decomposing the first sum $\sum_{k+l=n+1} ka_k a_l = (n+1)a_1 a_n + \sum_{k=2}^{n-1} ka_k a_{n+1-k}$, and using the fact that $a_1 = 1$, gives the recurrence relation:

$$a_n = - \sum_{k+l=n+1, k,l\geq 2} ka_k a_l + \sum_{k+l=n} (qk-1) a_k a_l - q \sum_{k+l+m=n+1} ka_k a_l a_m .$$

THEOREM 6.1. *The recurrence relation*

$$a_n = - \sum_{\substack{k+l=n+1\\ k,l\geq 2}} ka_k a_l + \sum_{k+l=n} (qk-1) a_k a_l - q \sum_{k+l+m=n+1} ka_k a_l a_m ,$$

with initial conditions $a_0 = 0$, $a_1 = 1$, has a unique solution given by

$$a_n = (-1)^{n+1} e(\mathcal{M}^{\delta}_{0,n+1}).$$

In the special case $q = 0$, we have the following corollary. Note that in theorem 9 of [**10**] there is an equivalent presentation of this recurrence relation.

COROLLARY 6.2. *The dimensions $b_n := \dim_{\mathbf{Q}} H^{n-2}(\mathcal{M}^{\delta}_{0,n+1}; \mathbf{Q})$ are the unique solutions to the recurrence relation:*

$$b_n = \sum_{\substack{k+l=n+1\\ k,l\geq 2}} k\, b_k b_l + \sum_{k+l=n} b_k b_l \ , \ for \quad n \geq 2 \ ,$$

with initial conditions $b_0 = 0, b_1 = -1$.

PROOF. Set $b_n = -a_n|_{q=0}$ in the previous theorem. □

References

[1] **F. C. S. Brown**: *Multiple zeta values and periods of moduli spaces $\mathfrak{M}_{0,n}$*, Ann. Scient. Éc. Norm. Sup., 4e série, t. 42, p. 373-491, (2009)

[2] **F. C. S. Brown, S. Carr, L. Schneps**: *The algebra of cell-zeta values*, to appear in *Compositio Math.*

[3] **E. Getzler**: *Operads and moduli spaces of genus 0 Riemann surfaces*, The moduli space of curves (Texel Island, 1994), 199–230, Progr. Math., 129.

[4] **V. Ginzburg, M. Kapranov**: *Koszul duality for operads*, Duke Math. J. 76 (1994), no. 1, 203–272.

[5] **I. P. Goulden, D. M. Jackson**: *Combinatorial enumeration*, Dover, NY (2004).

[6] **A. B. Goncharov, Y. I. Manin**: *Multiple ζ-motives and moduli spaces $\overline{\mathcal{M}}_{C,n}$*, Compositio Math. **140** (2004), 1-14.

[7] **N.M. Katz**: *Review of l-adic cohomology*, Motives (Seattle, WA, 1991), 21–30, Proc. Sympos. Pure Math., 55 (1994).

[8] **M. Kisin, G. I. Lehrer**: *Equivariant Poincaré polynomials and counting points over finite fields*, J. Algebra **247** (2002), 435–451.

[9] **P. Morse, H. Feshbach**: *Methods of Theoretical Physics*, Part I. New York: McGraw-Hill (1953).

[10] **P. Salvatore, R. Tauraso**: *The Operad Lie is Free*, J. Pure Applied Algebra 213 no. 2, 224-230 (2009).

Korteweg-de Vries Instituut, Universiteit van Amsterdam, Plantage Muidergracht 24, 1018 TV Amsterdam, The Netherlands.
E-mail address: o.l.j.bergstrom@uva.nl

CNRS and Institut Mathématiques de Jussieu. 175 rue du Chevaleret, 75013 Paris, France.
E-mail address: brown@math.jussieu.fr

Clay Mathematics Proceedings
Volume 12, 2010

C^*-Algebras in Tensor Categories

Peter Bouwknegt, Keith C. Hannabuss, and Varghese Mathai

ABSTRACT. We define and systematically study nonassociative C^*-algebras as C^*-algebras internal to a topological tensor category. We also offer a concrete approach to these C^*-algebras, as G-invariant, norm closed $*$-subalgebras of bounded operators on a G-Hilbert space, with deformed composition product. Our central results are those of stabilization and Takai duality for (twisted) crossed products in this context.

1. Introduction

In [**12**] we gave an account of some nonassociative algebras and their applications to T-duality, with a brief mention of the role of categories at the end. In this paper we will develop the theory more systematically from the category theoretic perspective. In particular we shall not need to assume that the groups are abelian, and will mostly work with general three-cocycles rather than antisymmetric tricharacters. We believe, however, that the results in this paper are of interest independent of our original motivation. Since writing [**12**] we have become aware of more of the large literature in this subject, for example, the work of Fröhlich, Fuchs, Runkel, Schweigert in conformal field theory [**22, 23, 24, 25, 26, 27**], and of Beggs, Majid and collaborators [**1, 2, 4, 5, 6, 14**]. Most of that work is algebraic in spirit, working with finite groups or finite dimensional Hopf algebras, whereas we are primarily interested in locally compact groups and C^*-algebras, which necessitate the development of a rather different set of techniques. The work of Nesterov and collaborators [**35, 36, 37, 38, 39**] does use vector groups, but has rather different aims and methods. In addition, there is a well-established theory of C^*- and W^*-categories, [**28, 34**], in which the algebras are generalised to morphisms in a suitable category, which means that they are automatically associative. One could generalise to weak higher categories, but the examples discussed in [**12**] suggest that one starts by looking at categories in which the algebras and their modules are objects, which is also more directly parallel to the algebraic cases already mentioned.

2010 *Mathematics Subject Classification.* Primary: 46L70 Secondary: 47L70, 46M15,18D10, 46L08, 46L55, 22D35, 46L85.

Key words and phrases. nonassociative, C^*-algebras, stabilization, Takai duality, topological tensor categories, twisted crossed products.

In Section 2 we introduce the tensor categories that we use and give some elementary examples, based on our earlier work in [**12**]. Within the category it is possible to define algebras and modules. The next three sections show how to obtain nonassociative algebras of twisted compact and bounded operators on a Hilbert space, and introduce Morita equivalence. We then link this to exterior equivalence in Section 6, which establishes that exterior equivalent twisted actions give rise to isomorphic twisted crossed product C*-algebras.

Section 7 is devoted to an extension of the nonassociative Takai duality proved in [**12**]. This is especially useful, because it provides a method of stabilising algebras. For example, an associative algebra with a very twisted group action has a nonassociative dual and double dual which admit ordinary untwisted group actions. The double dual is Morita equivalent to the original, and so one could replace the original associative algebra and twisted action by an equivalent nonassociative algebra and ordinary action.

The main result in Section 8 establishes the fact that twisted crossed products can be obtained by repeated ordinary crossed products, but with a possible modified automorphism action of the final subgroup, a result that goes a long way towards proving an analog of the Connes-Thom isomorphism theorem [**15, 16**] in our context, as briefly discussed in the final section.

A theorem of MacLane [**34**] asserts that every monoidal category can be made strict, that is, associative, but in general the functor which does this is quite complicated. However, in our category things are much simpler, and in Appendix A, it is shown that whenever a nonassociative algebra acts on a module, its multiplication can be modified to an associative multiplication. Examples of the strictification process are discussed there, in particular to the algebras of twisted compact and bounded operators which are defined to act on a module, but also have more serious implications for physics, where the algebras are generally represented by actions on modules. (On the other hand this does not mean that we can simply dismiss the nonassociativity, because there are known nonassociative algebras such as the octonions, cf. [**3**], which fit into our framework.) Appendix B gives a concrete approach to our nonassociative C*-algebras, as G-invariant, norm closed $*$-subalgebras of bounded operators on a G-Hilbert space, with composition product deformed by a 3-cocycle on G. In Appendix C, we revisit the construction of our nonassociative torus, via a geometric construction that realizes it as a nonassociative deformation of the C*-algebra of continuous functions on the torus. For completeness, we have summarized in Appendix D, the original motivation for this work, namely, T-duality in string theory.

Acknowledgments. KCH would like to thank E. Beggs and S. Majid for discussions at the "Noncommutative Geometry" program held at the Newton Institute (Cambridge) in 2006. VM would like to thank A. Connes for feedback and suggestions at the "Noncommutative Geometry" workshop held at Banff, Canada in 2006, and PB would like to thank the participants at the "Morgan-Phoa mathematics workshop" at the ANU for useful feedback. Part of this work was done while we were visiting the Erwin Schrödinger Institute in Vienna for the program on "Gerbes, Groupoids and Quantum Field Theory". KCH would also like to thank the University of Adelaide for hospitality during the intermediate stages of the project. Both PB and VM acknowledge financial support from the Australian Research Council.

2. Tensor categories and their algebras

As in [**29**] we use tensor category to mean just a monoidal category, without any of the other structures often assumed elsewhere. That is, a tensor category is a category in which associated to each pair of objects A and B there exists a product object $A \otimes B$, and there is an identity object $\mathbf{1}$, such that $\mathbf{1} \otimes A \cong A \cong A \otimes \mathbf{1}$, together with associator isomorphisms

$$\Phi = \Phi_{A,B,C} : A \otimes (B \otimes C) \to (A \otimes B) \otimes C \tag{2.1}$$

for any three objects A, B and C, satisfying the consistency pentagonal identity on quadruple products:

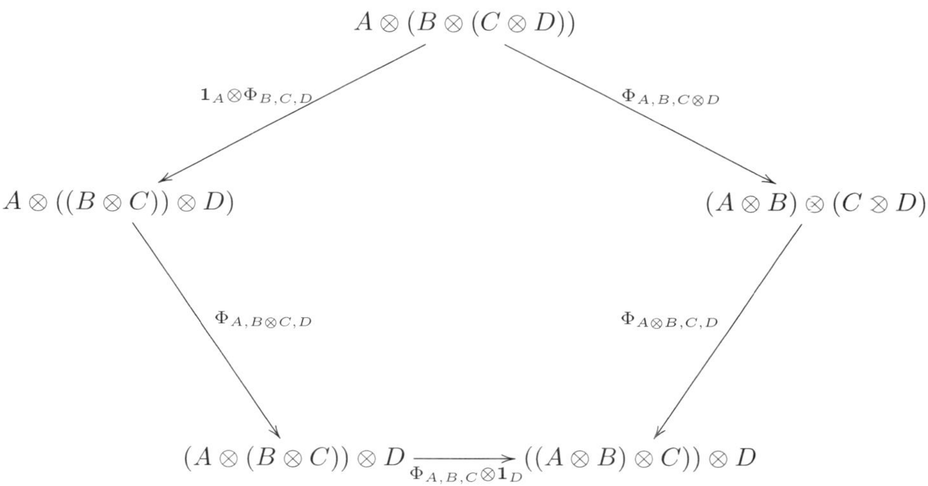

with each arrow the appropriate map Φ, and the triangle relation:

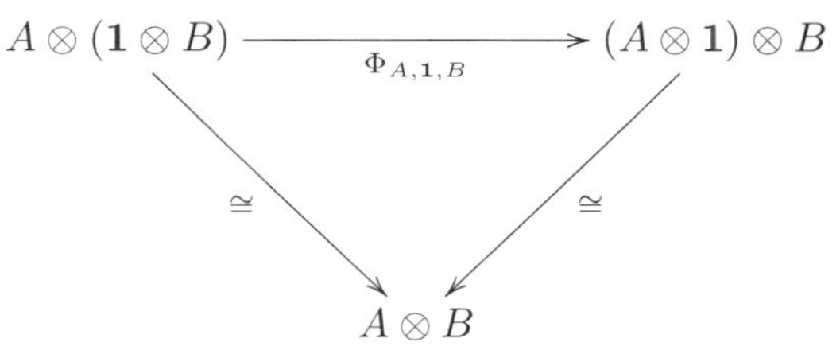

MacLane's coherence theorem ensures that these conditions are sufficient to guarantee consistency of all other rebracketings, [**30**]. (The theorem proceeds by showing that one can always take the category to be a strictly associative category. We prove this explicitly for our examples in Appendix A)

Module categories provide two standard examples of tensor categories with the obvious identification map $\Phi = \mathrm{id}$. One is the category of $\mathcal{R}$-modules and $\mathcal{R}$-morphisms, for $\mathcal{R}$ a commutative algebra over $\mathbb{C}$. The appropriate tensor product of objects A and B is $A \otimes_{\mathcal{R}} B$, and the identity object is $\mathcal{R}$ itself. Continuous trace algebras with spectrum S are $C_0(S)$-modules, and so can be studied within this category with $\mathcal{R} = C_0(S)$, though some care is needed in defining the appropriate tensor products in the case of topological algebras, but this can be done explicitly in this case, cf. [**40**, Sect 6.1].

A subtly different example is provided by the algebra of functions $H = C_0(\mathsf{G})$ on a separable locally compact group G. As well as being an algebra under pointwise multiplication it also has a comultiplication $\Delta : H \to H \otimes H$ taking a function $f \in C_0(\mathsf{G})$ to $(\Delta f)(x, y) = f(xy)$, and a counit $\epsilon : H \to \mathbb{C}$, which evaluates f at the group identity. This enables us to equip the category of H-modules with a tensor product $A \otimes B$ over $\mathbb{C}$, on which f acts as $\Delta(f)$, and the identity object being $\mathbb{C}$ with the trivial H-action of multiplication by $\epsilon(f)$. Writing the comultiplication in abbreviated Sweedler notation $\Delta f = f_{(1)} \otimes f_{(2)}$, the action on $A \otimes B$ is

$$f[a \star b] = f_{(1)}[a] \star f_{(2)}[b].$$

Since everything has been defined in terms of the comultiplication and counit of H, this clearly generalises to bialgebras, and even to quasi-bialgebras. If G is an abelian group, with Pontryagin dual group $\widehat{\mathsf{G}} = \mathrm{Hom}(\mathsf{G}, \mathsf{U}(1))$, then we can work with $C^*(\widehat{\mathsf{G}})$ instead of $C_0(\mathsf{G})$, and the tensor product action of $\xi \in \widehat{\mathsf{G}}$ is just $\xi \otimes \xi$. If the group $\widehat{\mathsf{G}}$ acts on S, the two examples can be combined in the category of modules for the crossed product, or transformation groupoid, algebra $C_0(S) \rtimes \widehat{\mathsf{G}}$, equipped with the tensor product over $C_0(S)$, and identity object $C_0(S)$.

Both examples use modules for a $*$-algebra, with $f^*(s) = \overline{f(s)}$ in $C_0(S)$ and $f^*(x) = \overline{f(x^{-1})}$ in $C_0(\mathsf{G})$, and the category contains conjugate objects A^*, having the same underlying set, but with the algebra action changed to that of f^* and conjugated scalar multiplication. For a an element of some object A and a^* the same element considered as an element of A^* we then have $f^*[a^*] = f[a]^*$. Working with $C^*(\widehat{\mathsf{G}})$ instead of $C_0(\mathsf{G})$ one has $\overline{\xi(x^{-1})} = \xi(x)$ so that $\xi^* = \xi$. This too can be generalised to quasi-Hopf algebras [**29**, Section XV.5] which provide such structure as do the coinvolutions in Kac C*-algebras. Conjugation $A \mapsto A^*$ preserves direct sums and gives a covariant functor. Another crucial property follows by noting that in $C_0(\mathsf{G})$ one has

$$\Delta(f^*)(x, y) = f^*(xy) = \overline{f(y^{-1}x^{-1})} = \overline{\Delta(f))(y^{-1}, x^{-1})} = (f^*_{(2)} \otimes f^*_{(1)})(x, y),$$

so that conjugation reverses the order of factors in the tensor product. There is a natural isomorphism between $(A \otimes B)^*$ and $B^* \otimes A^*$. Since preparing the first draft of this paper the preprint [**6**] has appeared and gives a systematic account of bar categories, of which these form one example. We refer the reader there for more detail.

Tensor categories have enough structure to define algebras and modules.

DEFINITION 2.1. An object $\mathcal{A}$ is an algebra (or monoid) in a tensor category if there is a morphism $\star : \mathcal{A} \otimes \mathcal{A} \to \mathcal{A}$ that is associative in the category, that is, $\star(\star \otimes \mathrm{id})\Phi = \star(\mathrm{id} \otimes \star)$ as maps $\mathcal{A} \otimes (\mathcal{A} \otimes \mathcal{A}) \to \mathcal{A}$. An algebra $\mathcal{A}$ in the category is called a $*$-algebra if the category has the above conjugation of objects and $\mathcal{A}^* = \mathcal{A}$. A left (respectively, right) module $\mathcal{M}$ for $\mathcal{A}$ is an object such that there is a morphism, which we also denote by $\star$, sending $\mathcal{A} \otimes \mathcal{M}$ to $\mathcal{M}$ (respectively, $\mathcal{M} \otimes \mathcal{A}$ to $\mathcal{M}$), and satisfying the usual composition law in the category, that is, for left modules $\star(\star \otimes \mathrm{id})\Phi = \star(\mathrm{id} \times \star)$ as maps $\mathcal{A} \otimes (\mathcal{A} \otimes \mathcal{M}) \to \mathcal{M}$. For brevity, the term module will mean a left module, unless otherwise specified. An $\mathcal{A}$-$\mathcal{B}$-bimodule $\mathcal{X}$ for two algebras $\mathcal{A}$ and $\mathcal{B}$ in the category is a left module for $\mathcal{A}$ and a right module for $\mathcal{B}$, with commuting actions (allowing for rebracketing given by an associator map).

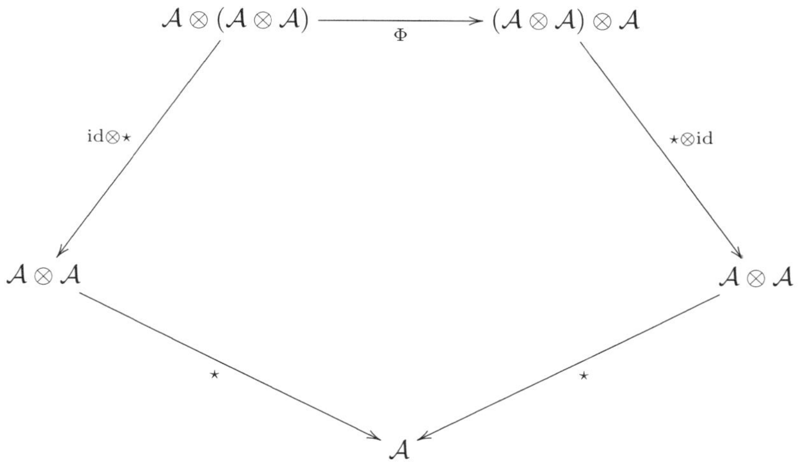

As an example we note that continuous trace algebras with spectrum S are algebras in the category of $C_0(S)$-modules introduced above. In fact $C_0(S)$ can be identified with a subalgebra of the centre $ZM(\mathcal{A})$ of the multiplier algebra of a continuous trace algebra $\mathcal{A}$, and so it also acts on all $\mathcal{A}$-modules, so that $\mathcal{A}$-modules in the ordinary sense are also $\mathcal{A}$-modules in the category. Similarly, continuous trace algebras with spectrum S on which $\widehat{\mathsf{G}}$ acts as automorphisms are algebras in the category of $C_0(S)\rtimes\widehat{\mathsf{G}}$-modules. Although we have so far taken Φ to be the usual identification map, there are other possibilities. Let G be a separable locally compact group with dual $\widehat{\mathsf{G}}$, and let $\phi\in C(\mathsf{G}\times\mathsf{G}\times\mathsf{G})$ be normalised to take the value 1 whenever any of its arguments is the identity $1\in\mathsf{G}$, and satisfy the pentagonal cocycle identity

$$\phi(x,y,z)\phi(x,yz,w)\phi(y,z,w)=\phi(xy,z,w)\phi(x,y,zw)\,. \tag{2.2}$$

Since $C(\mathsf{G})$ is the multiplier algebra of $C_0(\mathsf{G})$ it also acts on $C_0(\mathsf{G})$-modules. Unlike the algebraic case there are various module tensor products, and it is assumed that we have chosen one for which the action of $C(\mathsf{G})\otimes C(\mathsf{G})\otimes C(\mathsf{G})$ is defined.

Definition 2.2. The category $\mathcal{C}_{\mathsf{G}}(\phi)$ has for objects normed $C_0(\mathsf{G})$-modules, or equivalently normed $\widehat{\mathsf{G}}$-modules, and its morphisms are continuous linear maps commuting with the action. The tensor structure comes from taking the tensor product of modules with the tensor product action of the coproduct $(\Delta f)(x,y)=f(xy)$ for $f\in C(\mathsf{G})$, (or diagonal tensor product action $\xi\otimes\xi$ of $\xi\in\widehat{\mathsf{G}}$). For any three objects $\mathcal{A}$, $\mathcal{B}$ and $\mathcal{C}$ the associator map, $\Phi:\mathcal{A}\otimes(\mathcal{B}\otimes\mathcal{C})\to(\mathcal{A}\otimes\mathcal{B})\otimes\mathcal{C}$ is given by the action of $\phi\in C(\mathsf{G}\times\mathsf{G}\times\mathsf{G})\cong C(\mathsf{G})\otimes C(\mathsf{G})\otimes C(\mathsf{G})$. The identity object is the trivial one-dimensional module $\mathbb{C}$, on which $\widehat{\mathsf{G}}$ acts trivially, or, equivalently, $f\in C(\mathsf{G})$ multiplies by $f(1)$, where 1 is the identity element in G.

We could introduce a similar structure for $C_0(S)\rtimes\widehat{\mathsf{G}}$ modules.

As already mentioned, the algebras (or monoids) in $\mathcal{C}_{\mathsf{G}}(\phi)$ are objects $\mathcal{A}$ for which there is a product morphism $\mathcal{A}\otimes\mathcal{A}\to\mathcal{A}$, which we shall write as $a\otimes b\mapsto a\star b$. An algebra $\mathcal{A}$ can have a module $\mathcal{M}$, when there is a morphism $\mathcal{A}\otimes\mathcal{M}\to\mathcal{M}$, which we also write $a\otimes m\mapsto a\star m$, relying on the context to distinguish multiplications from actions. The category structure forces interesting compatibility conditions on algebras and modules.

PROPOSITION 2.3. *Let $\mathcal{A}$ be an algebra in the category $\mathcal{C}_{\mathsf{G}}(\phi)$. Then the group $\widehat{\mathsf{G}}$ acts on an algebra $\mathcal{A}$ by automorphisms, and the action of $\widehat{\mathsf{G}}$ on an $\mathcal{A}$-module M gives a covariant representation of $\mathcal{A}$ and $\widehat{\mathsf{G}}$ or, equivalently, a representation of the crossed product $\mathcal{A} \rtimes \widehat{\mathsf{G}}$.*

PROOF. For $a, b \in \mathcal{A}$, and $\xi \in \widehat{\mathsf{G}}$ we have $\xi[a] \star \xi[b] = \xi[a \star b]$, showing that the action of $\widehat{\mathsf{G}}$ is by endomorphisms, and since $\widehat{\mathsf{G}}$ is a group, these are invertible, and so automorphisms.

The morphism property gives $\xi[a] \star \xi[m] = \xi[a \star m]$, showing that the actions of $\mathcal{A}$ and $\widehat{\mathsf{G}}$ combine into a covariant representation of $(\mathcal{A}, \widehat{\mathsf{G}})$. Standard theory then tells us that this is equivalent to having an action of the crossed product $\mathcal{A} \rtimes \widehat{\mathsf{G}}$. □

We need to take care concerning the ordering of products where Φ sets up the appropriate associativity conditions. When $\mathcal{A}$ is an algebra in the category, and $\mathcal{M}$ is an $\mathcal{A}$-module, we shall simplify the notation by writing $\Phi(a \star (b \star m))$ for the composition of the maps $\Phi : \mathcal{A} \otimes (\mathcal{A} \otimes \mathcal{M}) \to (\mathcal{A} \otimes \mathcal{A}) \otimes \mathcal{M}$ and the multiplications and action

$$(\mathcal{A} \otimes \mathcal{A}) \otimes \mathcal{M} \to \mathcal{A} \otimes \mathcal{M} \to \mathcal{M},$$

applied to $a \otimes (b \otimes c)$, and similarly for triple products of algebra elements. We shall similarly abbreviate the notation for other operations involving tensor products.

The algebra of twisted compact operators $\mathcal{K}_\phi(L^2(G))$, introduced in [**12**], provides an example of an algebra in the category $\mathcal{C}_{\mathsf{G}}(\phi)$. There we assumed that ϕ was an antisymmetric tricharacter, that is, $\phi(x, y, z)$ is a character in each of its arguments for fixed values of the others, and is inverted by any transposition of its arguments. This is sufficient to ensure that it satisfies the cocycle identity, but at the expense of slightly more complicated formulae the cocycle condition usually suffices, for example the product of two integral kernels is defined by

$$(k_1 \star k_2)(x, z) = \int \phi(xy^{-1}, yz^{-1}, z) k_1(x, y) k_2(y, z)\, dydz\,. \tag{2.3}$$

The action of $f \in C(\mathsf{G})$ on $\mathcal{K}_\phi(L^2(\mathsf{G}))$ multiplies the kernel $K(x, y)$ by $f(xy^{-1})$, and one checks that this defines an automorphism using the identity $(\Delta f)(xy^{-1}, yz^{-1}) = f(xz^{-1})$. Moreover, as may be readily checked using the pentagonal identity:

$$\begin{aligned} &((k_1 \star k_2) \star k_3))(x, w) \qquad\qquad (2.4)\\ &= \int \phi(xy^{-1}, yz^{-1}, z)\phi(xz^{-1}, zw^{-1}, w) k_1(x, y) k_2(y, z) k_3(z, w)\, dydz \\ &= \int \phi(xy^{-1}, yz^{-1}, zw^{-1})\phi(xy^{-1}, yw^{-1}, w)\phi(yz^{-1}, zw^{-1}, w) \\ &\quad \times k_1(x, y) k_2(y, z) k_3(z, w)\, dydz \\ &= \Phi(k_1 \star (k_2 \star k_3))(x, w)\,. \qquad\qquad (2.5) \end{aligned}$$

In Appendix B the twisted compact operators $\mathcal{K}_\phi(L^2(\mathsf{G}))$, introduced in [**12**], and the twisted bounded operators $\mathcal{B}_\phi(L^2(\mathsf{G}))$ are systematically defined and studied, providing examples of C*-algebras in the category $\mathcal{C}_{\mathsf{G}}(\phi)$. In fact, any norm closed, G-invariant $\star$-subalgebra of $\mathcal{B}_\phi(L^2(\mathsf{G}))$ gives an example of a C*-algebra in the category. The reworking of the twisted compact operators and twisted bounded operators with a general cocycle ϕ shows that the arguments of [**12**] need not depend on ϕ being a tricharacter, although there may be simplifications when it is.

PROPOSITION 2.4. *Let $\lambda : A \otimes (B \otimes C) \to \mathbb{C}$ be a morphism in $\mathcal{C}_{\mathsf{G}}(\phi)$. If ϕ is a tricharacter then $\lambda(\Phi(a \otimes (b \otimes c))) = \lambda(a \otimes (b \otimes c))$, for all $a \in A$, $b \in B$ and $c \in C$. The same applies for a morphism $\lambda : A \otimes (B \otimes C) \to A \otimes \mathbb{C} \to A$ which factors through $\mathbb{C}$.*

PROOF. The fact that λ is a morphism means that for any $f \in C_0(\mathsf{G})$

$$\epsilon(f)\lambda\Phi = \lambda(\mathrm{id} \otimes \Delta)\Delta(f))\Phi = \lambda(f_{(1)} \otimes (f_{(2)} \otimes f_{(3)}))\phi\,,$$

or

$$\lambda\epsilon(f)\Phi = \lambda(f_{(1)} \otimes (f_{(2)} \otimes f_{(3)}))\phi\,,$$

Thus the effect of $(f_{(1)} \otimes (f_{(2)} \otimes f_{(3)}))\phi(x,y,z) = f(x(yz))\phi(x,y,z)$ is just the same as $\epsilon(f)\phi(x,y,z) = f(1)\phi(x,y,z)$. In other words, the effect of $\lambda\Phi$ is concentrated where $xyz = 1$. However, when ϕ is an antisymmetric tricharacter, it takes the value 1 when arguments are repeated, and so

$$\phi(x,y,z) = \phi(x,y,x)\phi(x,y,y)\phi(x,y,z) = \phi(x,y,xyz) = 1.$$

In other words $\lambda\Phi = \lambda$. A similar argument applied to $B \otimes C$ covers the case when λ factors through $\mathbb{C}$.

We conclude by remarking that in the abbreviated notation introduced above, one would write $\lambda(\Phi(a \otimes (b \otimes c))) = \lambda(a \otimes (b \otimes c))$. □

More generally $C(\mathsf{G})$ can be replaced by other algebras. To give the desired structure we require at least a comultiplication Δ to define a tensor product action, a linear functional ϵ satisfying $\epsilon(h_1)h_2 = h = h_1\epsilon(h_2)$ defining the action on the identity object, and a three-cocycle Φ in the multiplier algebra $\mathcal{M}(\mathsf{H} \otimes \mathsf{H} \otimes \mathsf{H})$, consistent with these. The pentagonal cocyle condition is

$$(\Delta \otimes \mathrm{id} \otimes \mathrm{id})(\Phi)(\mathrm{id} \otimes \mathrm{id} \otimes \Delta)(\Phi) = (\Phi \otimes \mathrm{id})(\mathrm{id} \otimes \Delta \otimes \mathrm{id})(\Phi)(\mathrm{id} \otimes \Phi)\,. \tag{2.6}$$

Consistency of the associativity rebracketing with the action of H on tensor products of modules requires

$$(\Delta \otimes \mathrm{id})\Delta(h) = \mathrm{ad}\Phi(\mathrm{id} \otimes \Delta)\Delta(h)\,, \tag{2.7}$$

whilst consistency with the action on the identity object means that ϵ contracted with the middle part of Φ gives the identity. These are precisely the conditions satisfied by a quasi-bialgebra [**29**, Section XV.1].

DEFINITION 2.5. The objects in the category $\mathcal{C}^{\mathsf{H}}(\Phi)$ are H-modules, and the morphisms are linear H-endomorphisms, with the associator map given by the action of $\Phi \in \mathsf{H}\otimes\mathsf{H}\otimes\mathsf{H}$. The action on tensor products is given by the comultiplication and the trivial object is $\mathbb{C}$ with the action given by the counit.

The quasi-bialgebra version of Proposition 2.3 asserts that H acts by automorphisms of an algebra $\mathcal{A}$ in the category, that is, $h[a \star b] = h_{(1)}[a] \star h_{(2)}[b]$, for all $a, b \in A$, and the actions of H and $\mathcal{A}$ on an A-module are covariant, so that $h[a \star m] = h_{(1)}[a] \star h_{(2)}[m]$, for all $m \in \mathcal{M}$, and one has a module for the crossed product $A \rtimes \mathsf{H}$.

We shall often use H to include the case of $C(\mathsf{G})$, though in the latter case there are extra analytic conditions. (In principle one might use Kac algebras, [**20, 21**], but that would require too big a digression.)

We shall outline our results for topological groups (more directly linked to our applications) and for quasi-Hopf algebras.

We shall work in tensor categories of modules for an appropriate locally compact abelian group (more directly linked to our applications) or for a quasi-Hopf algebra (which makes the algebraic structure particularly transparent), though it is probably possible to extend much of this to tensor categories of modules for Kac algebras [**20, 21**], which provide a natural framework for considerations of duality. Even locally compact groups present challenges beyond those present in the purely algebraic case of Hopf algebras, for example, the algebra $C_0(\mathsf{G})$ of compactly supported functions on G has no unit, since the constant function 1 is not compactly supported, whilst $L^\infty(\mathsf{G})$ has no counit since evaluation at the identity is not defined, and in neither case is there an antipode, though both have a coinvolution as Kac C*-algebras. One also requires modules which are not finitely generated.

3. Hilbert modules in tensor categories

The notion of the twisted kernels on $L^2(\mathsf{G})$ can be extended to more general Hilbert spaces. There are several equivalent characterisations of the bounded operators in a normal Hilbert space, and for our purposes the most useful approach is rather indirect.

Let $\mathcal{H}$ be a $\widehat{\mathsf{G}}$-module with an inner product $\langle\cdot,\cdot\rangle$. Since the inner product takes values in the trivial object $\mathbb{C}$, consistency with the action of $\widehat{\mathsf{G}}$ requires that $\langle\xi[\psi_1],\xi[\psi_2]\rangle = \xi[\langle\psi_1,\psi_2\rangle] = \langle\psi_1,\psi_2\rangle$, for all $\xi\in\widehat{\mathsf{G}}$ and $\psi_1,\ \psi_2\in\mathcal{H}$, so that the $\widehat{\mathsf{G}}$ action is unitary, or equivalently $\langle\psi_1, f[\psi_2]\rangle = \langle f^*[\psi_1],\psi_2\rangle$ for all $f\in C(\mathsf{G})$.

DEFINITION 3.1. An object $\mathcal{H}$ in $\mathcal{C}_{\mathsf{G}}(\phi)$ with an inner product $\langle\cdot,\cdot\rangle$ with respect to which the action of $\widehat{\mathsf{G}}$ is unitary (or, equivalently, consistent with the $*$-structure of $C(\mathsf{G})$) is called a pre-Hilbert space in $\mathcal{C}_{\mathsf{G}}(\phi)$. If it is complete in the norm topology it is called a Hilbert space in $\mathcal{C}_{\mathsf{G}}(\phi)$.

The unitarity of the action of $\widehat{\mathsf{G}}$ means that the inner product on $\mathcal{H}$ defines a morphism from the conjugate space $\mathcal{H}^*$ introduced earlier to the dual of $\mathcal{H}$ given by $\psi^* = \langle\psi,\cdot\rangle$. We can alternatively think of an inner product as a morphism $\mathcal{H}^*\otimes\mathcal{H}\to\mathbb{C}$, written $\psi_1^*\otimes\psi_2\mapsto\langle\psi_1,\psi_2\rangle$, which satisfies the positivity condition.

PROPOSITION 3.2. *In order that the map taking A to A^* be consistent with the associator isomorphism it is necessary and sufficient that*

$$\Phi(A\otimes(B\otimes C))^* = \Phi^{-1}((A\otimes(B\otimes C))^*)\,, \tag{3.1}$$

for all objects A, B and C, or, equivalently in $\mathcal{C}_{\mathsf{G}}(\phi)$, that the ϕ be unitary.

PROOF. We have

$$\begin{aligned}\Phi(A\otimes(B\otimes C))^* &= ((A\otimes B)\otimes C)^*\\ &= C^*\otimes(B^*\otimes A^*)\\ &= \Phi^{-1}((C^*\otimes B^*)\otimes A^*)\\ &= \Phi^{-1}((A\otimes(B\otimes C))^*),\end{aligned}$$

which amounts to saying that the function $\phi^* = \phi^{-1}$, so that ϕ is unitary. □

In ordinary Hilbert spaces bounded operators can be characterised as those that are adjointable, and this definition is easy to generalise.

DEFINITION 3.3. Modifying the usual definition we shall call a linear operator A on $\mathcal{H}$ adjointable if for all ψ_1 and $\psi_2 \in \mathcal{H}$ there is an operator $A^\star$ such that

$$\langle A \star \psi_1, \psi_2 \rangle = \Phi(\langle \psi_1, A^\star \star \psi_2 \rangle),$$

where the ordering on each side is that given by the order-reversing conjugation, that is, the two sides are images of $(\psi_1^* \otimes A^*) \otimes \psi_2$ and $\Phi(\psi_1^* \otimes (A^\star \otimes \psi_2))$ under the module action and inner product map.

We note that this definition is taken to mean that A and $A^\star$ are in a $*$-algebra with a module $\mathcal{H}$. The unitarity of the action means that the ordinary bounded operators are objects in the category with the action $\xi[A] = \xi \circ A \circ \xi^{-1}$. The equality of the images of $(\psi_1^* \otimes A^*) \otimes \psi_2$ and $\Phi(\psi_1^* \otimes (A^\star \otimes \psi_2))$ allows us to identify the adjoint $A^\star$ with the conjugate A^*. We note that the unique action of $\widehat{\mathsf{G}}$ consistent with the covariance property gives $\xi : A \mapsto \xi \circ A \circ \xi^{-1}$. When Φ is given by an antisymmetric tricharacter ϕ we may apply Proposition 2.4 to the morphism $\lambda(\psi_1 \otimes (A \otimes \psi_2)) = \langle \psi_1, A \star \psi_2 \rangle$ to deduce that Φ acts trivially and the condition for the adjoint reduces to $\langle A^\star \psi_1, \psi_2 \rangle = \langle \psi_1, A\psi_2 \rangle$, as usual. In this case the adjointable operators are therefore just the bounded operators on $\mathcal{H}$. For general ϕ one will have a natural generalisation of the bounded operators, the subject of the next section.

4. Twisted compact and twisted bounded operators

Regarding $\mathcal{H}$ as a right $\mathbb{C}$-module for the scalar multiplication action, Rieffel's method allows us to define the dual inner product $\langle\!\langle \cdot, \cdot \rangle\!\rangle$ such that

$$\langle\!\langle \psi_0, \psi_1 \rangle\!\rangle \star \psi_2 = \Phi(\psi_0 \star \langle \psi_1, \psi_2 \rangle).$$

(As usual when ϕ is an antisymmetric tricharacter, as in [**12**], the invariance of the inner product renders the Φ action trivial, so that $\Phi(\psi_0 \star \langle \psi_1, \psi_2 \rangle) = \psi_0 \star \langle \psi_1, \psi_2 \rangle$.)

In the associative case the dual inner products $\langle\!\langle \psi_0, \psi_1 \rangle\!\rangle$ span the algebra $\mathcal{K}(\mathcal{H})$ of compact operators, and in the nonassociative case we define the norm-closure of the span to be the twisted compact operators $\mathcal{K}_\phi(\mathcal{H})$. The dual inner product is not generally invariant under the action of $\widehat{\mathsf{G}}$. Indeed, we have

$$\langle\!\langle \xi[\psi_0], \xi[\psi_1] \rangle\!\rangle \star \xi[\psi_2] = \xi[\psi_0 \star \langle \psi_1, \psi_2 \rangle], \tag{4.1}$$

from which it follows that $\langle\!\langle \xi[\psi_0], \xi[\psi_1] \rangle\!\rangle = \xi \circ \langle\!\langle \psi_0, \psi_1 \rangle\!\rangle \circ \xi^{-1}$. This means that there is a new multiplication $\star$ on $\mathcal{K}_\phi(\mathcal{H})$, so that, for $K_1, K_2 \in \mathcal{K}(\mathcal{H})$, $\psi \in \mathcal{H}$,

$$(K_1 \star K_2) \star \psi = \Phi(K_1 \star (K_2 \star \psi)), \tag{4.2}$$

where the right hand side is just the iterated natural action of $\mathcal{K}_\phi(\mathcal{H})$ on $\mathcal{H}$.

LEMMA 4.1. *The compact operators in $\mathcal{K}_\phi(\mathcal{H})$ are automatically adjointable with $\langle\!\langle \psi_2, \psi_1 \rangle\!\rangle$ being the adjoint of $\langle\!\langle \psi_1, \psi_2 \rangle\!\rangle$.*

PROOF. This is proved by consideration of $\langle\!\langle \psi_0, \langle \psi_1, \psi_2 \rangle \star \psi_3 \rangle\!\rangle$ for $\psi_j \in \mathcal{H}$, $j = 0, 1, 2, 3$. To keep the orderings clearer we write $\mathcal{H}_j = \mathcal{H}$, $j = 0, \ldots, 3$, and think of $\psi_j \in \mathcal{H}_j$, so that the vectors involved in the inner product

$$\psi_0^* \otimes ((\psi_1 \otimes \psi_2^*) \otimes \psi_3) \in \mathcal{H}_0^* \otimes ((\mathcal{H}_1 \otimes \mathcal{H}_2^*) \otimes \mathcal{H}_3).$$

Rewriting $\langle\!\langle \psi_1, \psi_2 \rangle\!\rangle \star \psi_3 = \Phi(\psi_1 \star \langle \psi_2, \psi_3 \rangle)$, is just a rebracketing, and with three more such steps we can rebracket it as $(\mathcal{H}_0^* \otimes (\mathcal{H}_1 \otimes \mathcal{H}_2^*)) \otimes \mathcal{H}_3$, which differs from the original by a single rebracketing, and so by just one application of Φ (by the pentagonal identity). We also have

$$(\mathcal{H}_0^* \otimes (\mathcal{H}_1 \otimes \mathcal{H}_2^*)) \cong ((\mathcal{H}_1 \otimes \mathcal{H}_2^*)^* \otimes \mathcal{H}_0)^* \cong ((\mathcal{H}_2 \otimes \mathcal{H}_1^*) \otimes \mathcal{H}_0)^*,$$

from which we see that the adjoint of $\langle\!\langle \psi_1, \psi_2 \rangle\!\rangle$ is $\langle\!\langle \psi_2, \psi_1 \rangle\!\rangle$. □

The associative case suggests that the multiplier algebra of the twisted compact operators should be a twisted version of the bounded operators, but we can be more explicit.

DEFINITION 4.2. The twisted bounded operators $\mathcal{B}_\phi(\mathcal{H})$ are the adjointable operators, equipped with the $\widehat{\mathsf{G}}$ action given by $\xi \cdot A = \xi \circ A \circ \xi^{-1}$, and the twisted multiplication given, for $A, B \in \mathcal{B}_\phi(\mathcal{H})$ and $\psi \in \mathcal{H}$, by

$$(A \star B) \star \psi = \Phi(A \star (B \star \psi)) \,. \tag{4.3}$$

LEMMA 4.3. *If A and B are adjointable then so is $A \star B$ and $(A \star B)^* = B^* \star A^*$. Thus the adjointable operators with this multiplication form a (generally nonassociative) $*$-algebra of twisted bounded operators $\mathcal{B}_\phi(\mathcal{H})$. The twisted compact operators $\mathcal{K}_\phi(\mathcal{H})$ are an ideal in $\mathcal{B}_\phi(\mathcal{H})$ (and so, in particular, $\mathcal{K}_\phi(\mathcal{H})$ is a subalgebra). When ϕ is an antisymmetric tricharacter the twisted bounded operators are bounded operators but with a different multiplication. In that case, when $\mathcal{H} = L^2(\mathsf{G})$ with the multiplication action of $\widehat{\mathsf{G}}$, $\xi[\psi](x) = \xi(x)\psi(x)$, then $\mathcal{K}_\phi(\mathcal{H})$ is the algebra $\mathcal{K}_\phi(L^2(\mathsf{G}))$ defined in* [**12**].

PROOF. This time we consider $\langle \psi_1, (A \star B) \star \psi_2 \rangle$, which arises from

$$\psi_1^* \otimes ((A \otimes B) \otimes \psi_2) \in \mathcal{H}^* \otimes ((\mathcal{A}_A \otimes \mathcal{A}_B) \otimes \mathcal{H}),$$

where the indices on $\mathcal{A}$ just serve as reminders of where operator lives. Similar rebracketings to those in the previous lemma

$$\begin{aligned}\psi_1^* \otimes ((A \otimes B) \otimes \psi_2) \to \psi_1^* \otimes (A \otimes (B \otimes \psi_2)) \to (\psi_1^* \otimes A) \otimes (B \otimes \psi_2) \to \\ \to ((\psi_1^* \otimes A) \otimes B) \otimes \psi_2 \to (\psi_1^* \otimes (A \otimes B)) \otimes \psi_2 \,,\end{aligned}$$

give us an element of $(\mathcal{H}^* \otimes (\mathcal{A} \otimes \mathcal{A})) \otimes \mathcal{H}$. Moreover,

$$\mathcal{H}^* \otimes (\mathcal{A}_A \otimes \mathcal{A}_B) \cong ((\mathcal{A}_B^* \otimes \mathcal{A}_A^*) \otimes \mathcal{H})^*,$$

from which it follows that

$$\langle (B^* \star A^*) \star \psi_1, \psi_2 \rangle = \Phi(\langle \psi_1, (A \star B) \star \psi_2 \rangle),$$

so that $B^* \star A^* = (A \star B)^*$. This proves that $A \star B$ is adjointable, and allows us to form an algebra. In the associative case it is totally straightforward to see that $\mathcal{K}(\mathcal{H})$ is an ideal because, for any $A \in \mathcal{B}(\mathcal{H})$ and $\psi_0, \psi_1, \psi_2 \in \mathcal{H}$, we have

$$(A \star \langle\!\langle \psi_0, \psi_1 \rangle\!\rangle) \star \psi_2 = A \star (\psi_0 \langle \psi_1, \psi_2 \rangle) = \langle\!\langle A \star \psi_0, \psi_1 \rangle\!\rangle \star \psi_2,$$

showing that $A \star \langle\!\langle \psi_0, \psi_1 \rangle\!\rangle = \langle\!\langle A \star \psi_0, \psi_1 \rangle\!\rangle$. Similarly $\langle\!\langle \psi_0, \psi_1 \rangle\!\rangle \star A = \langle\!\langle \psi_0, A^* \star \psi_1 \rangle\!\rangle$, from which it is obvious that left and right multiplication preserve the generators of $\mathcal{K}(\mathcal{H})$. In the nonassociative case, there are several rebracketings involved but, thanks to the pentagonal identity these reduce to

$$\langle\!\langle A \star \psi_0, \psi_1 \rangle\!\rangle = \Phi(A \star \langle\!\langle \psi_0, \psi_1 \rangle\!\rangle).$$

The right hand side is

$$\int \phi(x,y,z)\overline{\xi(x)\eta(y)\zeta(z)}\langle\langle \xi[A]\star\eta[\psi_0],\zeta[\psi_1]\rangle\rangle\, dxdydz\, d\xi d\eta d\zeta,$$

and this is in the closure $\mathcal{K}_\phi(\mathcal{H})$ of the span of the dual inner product.

We have already seen that when ϕ is an antisymmetric tricharacter the adjointable operators are the same as usual, that is, the bounded operators. The argument of Proposition 2.4 applied to $\lambda(\psi_0\otimes(\psi_1\otimes\psi_2)) = \psi_0\langle\psi_1,\psi_2\rangle$ shows that Φ also disappears from the formula for the dual inner product in this case; for $\mathcal{H} = L^2(\mathsf{G})$ with the natural inner product we get the same rank one operators defined by the dual inner product as when $\phi = 1$, that is, the ordinary compact operators, which are the closure of the kernels $C_c(\mathsf{G}\times\mathsf{G})$, and only the multiplication changes. More precisely we see that

$$(\langle\langle\psi_0,\psi_1\rangle\rangle\star\psi_2)(x) = \Phi(\psi_0(x)\langle\psi_1,\psi_2\rangle) = \psi_0(x)\int_{\mathsf{G}}\phi(x,y,y)\overline{\psi_1(y)}\psi_2(y)\,dy,$$

so that for an antisymmetric tricharacter $\langle\langle\psi_0,\psi_1\rangle\rangle(x,y) = \psi_0(x)\overline{\psi_1(y)}$. It is straightforward to check that multiplication follows from the formula of [**12**, Sect 5]. □

COROLLARY 4.4. *The twisted bounded operators form a $*$-algebra in $\mathcal{C}_{\mathsf{G}}(\phi)$ with $A^\star$ identified with A^*.*

It is no coincidence that the same space $\mathcal{B}(\mathcal{H})$ can carry both associative and nonassociative multiplications, as one can see from the discussion of strictification in Appendix A.

We have thus shown that the twisted bounded operators are closed under twisted multiplication, without deriving any simple relationship between the norm of $A\star B$ and those of A and B. In particular we do not have the C*-algebra identity $\|A^\star\star A\| = \|A\|^2$. However, there is a simple substitute for this.

PROPOSITION 4.5. *For any A in the algebra of twisted bounded operators $\mathcal{B}_\phi(\mathcal{H})$, $A^\star\star A = 0$ if and only if $A = 0$.*

PROOF. It is clear that $A = 0$ implies $A^\star\star A = 0$. Conversely, for any ψ, $\langle A\psi, A\psi\rangle$ is obtained from $\langle\psi,(A^\star\star A)\psi\rangle$ by the appropriate actions of Φ. Since these are linear, when $A^\star\star A = 0$ we must also have $\langle A\psi, A\psi\rangle = 0$, and this forces $A = 0$. □

DEFINITION 4.6. A C*-algebra in $\mathcal{C}_{\mathsf{G}}(\phi)$ is a $*$-algebra which is $*$-isomorphic to a norm-closed $*$-subalgebra of $\mathcal{B}_\phi(\mathcal{H})$ for some Hilbert module $\mathcal{H}$. (Here $*$-isomorphisms are defined as usual except that they must also be $\mathcal{C}_{\mathsf{G}}(\phi)$-morphisms.)

5. Morita equivalence

In Section 2 we defined bimodules in a category, but we now want to study them in a little more detail.

DEFINITION 5.1. Let $\mathcal{X}$ be a right $\mathcal{A}$-module for $\mathcal{A}$ a C*-algebra in $\mathcal{C}_{\mathsf{G}}(\phi)$, and let $\mathcal{X}^*$ denote the conjugate left $\mathcal{A}$-module. An $\mathcal{A}$-valued inner product on $\mathcal{X}$ is a morphism $\mathcal{X}^*\times\mathcal{X}\to\mathcal{A}$, written $\psi_1^*\otimes\psi_2\mapsto\langle\psi_1,\psi_2\rangle_R$, such that for all ψ_1, $\psi_2\in\mathcal{X}$, and $b\in\mathcal{A}$,

(i) $\Phi(\langle\psi_1,\psi_2\star b\rangle_R) = \langle\psi_1,\psi_2\rangle_R\star b$, where Φ is given by the action of a cocycle ϕ on G;

(ii) $\langle\psi_1, \psi_1\rangle_R$ is positive in $\mathcal{A}$, in the sense that it can be written as a sum of elements of the form $b_j^\star \star b_j$ with $b_j \in \mathcal{A}$, and vanishes only when $\psi_1 = 0$. One calls $\mathcal{X}$ a (right) Hilbert $\mathcal{A}$-module.

REMARK. Assumptions (i) and (ii) are consistent because a series of rebracketings and uses of (i) gives

$$\langle\psi \star b, \psi \star b\rangle_R \to b^\star \star (\langle\psi, \psi\rangle_R \star b) = b^\star \star ((\sum b_j^\star \star b_j) \star b),$$

and undoing the various rebracketings takes us back to the obviously positive term

$$\sum_j (b_j \star b)^\star \star (b_j \star b).$$

This argument is slightly more delicate than one might realise, but as a morphism in the category the inner product must satisfy

$$f[\langle\psi_1, \psi_2\rangle_R] = \langle f_1^*[\psi_1], f_2[\psi_2]\rangle_R,$$

as do products $f[b_j^\star \star b_j] = f_1^*[b_j^\star] \star f_2[b_j]$, so that the various actions of ϕ on the inner product and on the algebra products really do match each other. This was the reason for taking this definition of positive rather than the other possibilities, such as having positive spectrum, which are equivalent in the associative case, but undefined or less useful here.

PROPOSITION 5.2. *For all ψ_1 and $\psi_2 \in \mathcal{X}$ we have $\langle\psi_1, \psi_2\rangle_R = \langle\psi_2, \psi_1\rangle_R^\star$.*

PROOF. The polarisation identity gives

$$\langle\psi_1, \psi_2\rangle_R = \sum_{r=0}^{3} i^{-r}\langle\psi_1 + i^r\psi_2, \psi_1 + i^r\psi_2\rangle = \sum_{r=0}^{3} i^r\langle i^r\psi_1 + \psi_2, i^r\psi_1 + \psi_2\rangle$$

and, since property (ii) tells that the inner product is real, this is the same as $\langle\psi_2, \psi_1\rangle_R^\star$. □

Obviously ordinary Hilbert spaces when $\mathcal{A} = \mathbb{C}$, and ordinary Hilbert C*-modules (when G is trivial), provide examples. Any C*-algebra $\mathcal{A}$, considered as an $\mathcal{A}$-$\mathcal{A}$-bimodule for the left and right multiplication actions, can be given the $\mathcal{A}$-valued inner product $\langle a_1, a_2\rangle_{\mathcal{A}} = a_1^* a_2$. This is certainly bilinear on $\mathcal{A}^* \times \mathcal{A}$ and for the action by automorphisms one has

$$f[\langle a_1, a_2\rangle_R] = f[a_1^* a_2] = f_1[a_1^*]f_2[a_2] = \langle f_1^*[a_1], f_2[a_2]\rangle_R.$$

The inner product is obviously positive, and Proposition 4.5 ensures that $a^\star \star a = 0$ if and only if $a = 0$. This example can be combined with a Hilbert space $\mathcal{H}$ to obtain $\mathcal{H} \otimes \mathcal{A}$ with the $\mathcal{A}$-valued inner product

$$\langle\psi_1 \otimes a_1, \psi_2 \otimes a_2\rangle_R = \langle\psi_1, \psi_2\rangle a_1^* a_2,$$

compatible with the (right) action of $\mathcal{A}$ by right multiplication.

One can similarly define $\mathcal{A}$-valued inner products for left $\mathcal{A}$-modules, either as the conjugate of the right $\mathcal{A}$-module $\mathcal{X}^*$, or directly as follows.

DEFINITION 5.3. Let $\mathcal{X}$ be a left $\mathcal{A}$-module for $\mathcal{A}$ a C*-algebra in $\mathcal{C}_\mathsf{G}(\phi)$, and let $\mathcal{X}^*$ denote the conjugate left $\mathcal{A}$-bimodule. An $\mathcal{A}$-valued inner product on $\mathcal{X}$ is a morphism $\mathcal{X} \times \mathcal{X}^* \to \mathcal{A}$, written $\psi_1^* \otimes \psi_2 \mapsto \langle\psi_1, \psi_2\rangle_L$, such that for all ψ_1, $\psi_2 \in \mathcal{X}$, and $a \in \mathcal{A}$,

(i) $\Phi_1(\langle a \star \psi_1, \psi_2\rangle_L) = a \star \langle \psi_1, \psi_2\rangle_L$, where Φ_1 is given by the action of a cocycle ϕ_1 on G_1;
(ii) $\langle \psi_1, \psi_1\rangle_L$ is positive in $\mathcal{A}$, in the sense that it can be written as a sum of elements of the form $a_j^\star \star a_j$ with $a_j \in \mathcal{A}$, and vanishes only when $\psi_1 = 0$. One calls $\mathcal{X}$ a (left) Hilbert $\mathcal{A}$-module.

As with the $\mathcal{A}$-valued inner product a polarisation argument shows that $\langle \psi_1, \psi_2\rangle_L = \langle \psi_2, \psi_1\rangle_L^\star$.

Returning to right Hilbert C*-modules, the next task is to study the algebra of adjointable operators a on $\mathcal{X}$ which commute with the action of $C(\mathsf{G})$ and admit an adjoint $a^\star$ satisfying

$$\langle a \star \psi, \theta\rangle_R = \Phi\langle \psi, a^\star \star \theta\rangle_R\,. \tag{5.1}$$

Examples are provided by rank one operators

$$\langle \psi_0, \psi_1\rangle_L : \psi_2 \mapsto \Phi[\psi_0 \star \langle \psi_1, \psi_2\rangle_R],$$

where $\Phi = \Phi_1 \times \Phi_2$.

All this suggests an abstraction of this structure into the idea of an imprimitivity $\mathcal{A}_1$-$\mathcal{A}_2$-module $\mathcal{X}$.

DEFINITION 5.4. Let $\mathcal{X}$ be an $\mathcal{A}_1$-$\mathcal{A}_2$-bimodule, for $\mathcal{A}_j$ a C*-algebra in $\mathcal{C}_{\mathsf{G}}(\phi)$, $j = 1, 2$, such that the actions of $(\widehat{\mathsf{G}}, \mathcal{A}_1)$ and $(\widehat{\mathsf{G}}, \mathcal{A}_2)$are mutually commuting, and each generates the commutant of the other. Let $\mathcal{X}$ have $\mathcal{A}_1$ and $\mathcal{A}_2$-valued inner products, such that each algebra is adjointable in the inner product associated with the other:

$$\langle a \star \psi_1, \psi_2\rangle_R = \Phi\langle \psi_1, a^\star \star \psi_2\rangle_R, \qquad \langle \psi_1, \psi_2 \star b\rangle_L = \Phi\langle \psi_1 \star b^\star, \psi_2\rangle_L\,. \tag{5.2}$$

In addition one asks that the inner products are full in the sense that their images are dense in $\mathcal{A}_1$ and $\mathcal{A}_2$, respectively, and linked by the imprimitivity condition that each is dual to the other,

$$\langle \psi_0, \psi_1\rangle_L \star \psi_2 = \Phi(\psi_0 \star \langle \psi_1, \psi_2\rangle_R)\,. \tag{5.3}$$

Then $\mathcal{X}$ is said to be an imprimitivity bimodule for $\mathcal{A}_1$ and $\mathcal{A}_2$.

We have defined bimodules within a single category $\mathcal{C}_{\mathsf{G}}(\phi)$, but this is easily extended to cover bimodules $\mathcal{X}$ for C*-algebras $\mathcal{A}_1$ and $\mathcal{A}_2$ in different tensor categories, $\mathcal{C}_{\mathsf{G}_1}(\phi_1)$ and $\mathcal{C}_{\mathsf{G}_2}(\phi_2)$, (where G_1, G_2 are separable locally compact abelian groups with three-cocycles ϕ_1 and ϕ_2). We take $\mathsf{G} = \mathsf{G}_1 \times \mathsf{G}_2$ with the product cocycle $\phi = \phi_1 \times \phi_2$, and then $\mathcal{C}_{\mathsf{G}_1}(\phi_1)$ forms a subcategory of $\mathcal{C}_{\mathsf{G}}(\phi)$ on which $C(\mathsf{G}_2)$ acts trivially (by its counit, ϵ_2), and similarly with indices reversed. Within $\mathcal{C}_{\mathsf{G}}(\phi)$ there is no problem in taking an $\mathcal{A}_1$-$\mathcal{A}_2$-bimodule $\mathcal{X}$, which as a left module is in $\mathcal{C}_{\mathsf{G}_1}(\phi_1)$, and as a right module in $\mathcal{C}_{\mathsf{G}_2}(\phi_2)$. In particular, when G_2 is trivial, one can use the bimodule to compare modules for twisted and untwisted algebras (cf. [**12**]).

As with Hilbert spaces, when ϕ_1 and ϕ_2 are antisymmetric tricharacters the rebracketing which links the dual inner products is independent of Φ.

LEMMA 5.5. *In the case of $\mathcal{A}_j$ in category $\mathcal{C}_{\mathsf{G}_j}(\phi_j)$, described above, when ϕ_1 and ϕ_2 are antisymmetric tricharacters the inner products are linked by*

$$\langle \psi_0, \psi_1\rangle_L \star \psi_2 = \psi_0 \star \langle \psi_1, \psi_2\rangle_R\,. \tag{5.4}$$

PROOF. This follows from Proposition 2.4, since the inner product $\psi_1^* \otimes \psi_2 \mapsto \langle \psi_1, \psi_2 \rangle_R$ is a $C_0(\mathsf{G}_1)$-morphism, so that Φ_1 acts trivially on the terms involving this, and $\psi_1 \otimes \psi_2^* \mapsto \langle \psi_1, \psi_2 \rangle_L$ is a $C_0(\mathsf{G}_2)$-morphism, so that Φ_2 acts trivially on the terms involving that. By rewriting the equation linking the inner products as

$$\Phi_2^{-1}(\langle \psi_0, \psi_1 \rangle_L \star \psi_2) = \Phi_1(\psi_0 \star \langle \psi_1, \psi_2 \rangle_R),$$

we see that each Φ_j acts trivially, so that

$$\langle \psi_0, \psi_1 \rangle_L \star \psi_2 = \psi_0 \star \langle \psi_1, \psi_2 \rangle_R.$$

□

DEFINITION 5.6. Two C*-algebras $\mathcal{A}_1$ and $\mathcal{A}_2$ are said to be Morita equivalent if there exists an imprimitivity $\mathcal{A}_1$-$\mathcal{A}_2$-bimodule.

THEOREM 5.7. *Morita equivalence is an equivalence relation.*

PROOF. Reflexivity follows by using $\mathcal{A}$ with the inner product $\langle a_1, a_2 \rangle = a_1^* a_2$ as an imprimitivity $\mathcal{A}$-$\mathcal{A}$-bimodule. Symmetry follows by replacing an $\mathcal{A}$-$\mathcal{B}$-bimodule $\mathcal{X}$, by the conjugate $\mathcal{B}$-$\mathcal{A}$-bimodule $\mathcal{X}^*$, equipped with the dual inner product.

To prove transitivity, suppose that $\mathcal{X}$ and $\mathcal{Y}$ are imprimitivity $\mathcal{A}$-$\mathcal{B}$- and $\mathcal{B}$-$\mathcal{C}$-bimodules, respectively, and set $\mathcal{Z} = \mathcal{X} \otimes_{\mathcal{B}} \mathcal{Y}$. Certainly $\mathcal{Z}$ is an $\mathcal{A}$-$\mathcal{C}$-bimodule, and it may be equipped with dual inner products as follows. The right inner product is obtained from the following composition of maps

$$(\mathcal{Y}^* \otimes \mathcal{X}^*) \otimes (\mathcal{X} \otimes \mathcal{Y}) \to \mathcal{Y}^* \otimes (\mathcal{X}^* \otimes (\mathcal{X} \otimes \mathcal{Y})) \to \mathcal{Y}^* \otimes ((\mathcal{X}^* \otimes \mathcal{X}) \otimes \mathcal{Y}) \to \mathcal{Y}^* \otimes (\mathcal{B} \otimes \mathcal{Y}) \\ \to \mathcal{Y}^* \otimes \mathcal{Y} \to \mathcal{C}\,,$$

where the first two maps are given by the appropriate associator Φ, the third is the right inner product on $\mathcal{X}$, the next is the action of $\mathcal{B}$ on $\mathcal{Y}$, and the last is the right inner product on Y. The end result is $\langle y_1, \langle x_1, x_2 \rangle_R \star y_2 \rangle_R$, and this actually factors through $\mathcal{X} \otimes_{\mathcal{B}} \mathcal{Y}$. For example, the associator gives a map

$$\langle x_1, x_2 \star b \rangle_R \star y_2 \to \langle x_1, x_2 \rangle_R \star (b \star y_2)$$

so that the composition depends only on $x_2 \otimes_{\mathcal{B}} y_2$. That will also apply for the other argument and so one really has a map $\mathcal{Z}^* \otimes \mathcal{Z} \to \mathcal{C}$. The map is a morphism, because

$$\begin{aligned} \langle f_{(1)}^*[y_1], \langle f_{(2)}^*[x_1], f_{(3)}[x_1] \rangle_R \star f_{(4)}[y_2] \rangle_R &= \langle f_{(1)}^*[y_1], f_{(2)}[\langle x_1, x_2 \rangle_R] \star f_{(3)}[y_2] \rangle_R \\ &= \langle f_{(1)}^*[y_1], f_{(2)}[\langle x_1, x_2 \rangle_R \star y_2] \rangle_R \\ &= f[\langle y_1, \langle x_1, x_2 \rangle_R \star y_2 \rangle_R]. \end{aligned}$$

The positivity follows from the positivity of the inner products on $\mathcal{X}$ and $\mathcal{Y}$, exploiting the fact that exactly the same rebracketings occur for the inner products and for sums of the form $\sum_j b_j^* \star b_j$.

The left inner product $\mathcal{Z} \otimes \mathcal{Z}^* \to \mathcal{A}$ is similarly obtained from the composition of the following maps:

$$(\mathcal{X} \otimes \mathcal{Y}) \otimes (\mathcal{Y}^* \otimes \mathcal{X}^*) \to \mathcal{X} \otimes (\mathcal{Y} \otimes (\mathcal{Y}^* \otimes \mathcal{X}^*)) \to \mathcal{X} \otimes ((\mathcal{Y} \otimes \mathcal{Y}^*) \otimes \mathcal{X}^*) \to \mathcal{X} \otimes (\mathcal{B} \otimes \mathcal{X}^*) \\ \to \mathcal{X} \otimes \mathcal{X}^* \to \mathcal{A}\,,$$

and its properties similarly checked. □

We saw at the end of the previous section that $\mathcal{X} = \mathcal{H}$ is an imprimitivity $\mathcal{K}_\phi(\mathcal{H})$-$\mathbb{C}$-bimodule.

THEOREM 5.8. *The twisted compact operators $\mathcal{K}_\phi(\mathcal{H})$ are Morita equivalent to $\mathbb{C}$, with $\mathcal{H}$ providing the bimodule which gives the equivalence, and shows that $\mathcal{K}_\phi(\mathcal{H})$ has trivial representation theory.*

Now $\mathcal{K}_{\phi_j}(\mathcal{H}_j)$ is Morita equivalent to $\mathbb{C}$ via the bimodule $\mathcal{H}_j$ for any Hilbert space $\mathcal{H}_j$ and twistings ϕ_j, so, by the symmetry and transitivity of the equivalence, $\mathcal{K}_{\phi_1}(\mathcal{H}_1)$ and $\mathcal{K}_{\phi_2}(\mathcal{H}_2)$ are Morita equivalent via the bimodule $\mathcal{H}_1 \otimes_{\mathbb{C}} \mathcal{H}_2^* = \mathcal{H}_1 \otimes \mathcal{H}_2^*$.

6. Exterior equivalence for nonassociative algebras

The nonassociative algebras appear in [**12**] as twisted crossed products of ordinary associative algebras, when one had to lift an outer automorphism. In this section we return to that situation.

In the case of a Dixmier-Douady class on a principal T-bundle E described by a de Rham form along the fibres, there is a homomorphism from T to the outer automorphism group of an algebra $\mathcal{A}$ with spectrum E. This is lifted to a map $\alpha : \mathsf{T} \to \mathrm{Aut}(\mathcal{A})$ with

$$\alpha_x \alpha_y = \mathrm{ad}(u(x, y))\alpha_{xy}\,, \tag{6.1}$$

and

$$u(x, y)u(xy, z) = \phi(x, y, z)\alpha_x[u(y, z)]u(x, yz)\,. \tag{6.2}$$

Any other lifting β would have the form

$$\beta_x[a] = \mathrm{ad}(w_x)\alpha_x[a]\,, \tag{6.3}$$

for suitable $w_x \in \mathcal{A}$, and

$$\beta_x \beta_y = \mathrm{ad}(v(x, y))\beta_{xy}\,. \tag{6.4}$$

For trivial u and v this is just the usual exterior equivalence.

LEMMA 6.1. *If $\beta_x = \mathrm{ad}(w_x)\alpha_x$ then $\beta_x\beta_y = \mathrm{ad}(v(x, y))\beta_{xy}$ with*

$$v(x, y) = c(x, y)w_x\alpha_x[w_y]u(x, y)w_{xy}^{-1}\,, \tag{6.5}$$

for some central $c(x, y)$, and this is the most general form of v.

PROOF. For consistency, we must have

$$\begin{aligned} \mathrm{ad}(v(x, y)w_{xy})\alpha_{xy}[a] &= \mathrm{ad}(w_x)\alpha_x[\mathrm{ad}(w_y)\alpha_y[a]] \\ &= \mathrm{ad}(w_x\alpha_x[w_y])[\alpha_x\alpha_y[a]] \\ &= \mathrm{ad}(w_x\alpha_x[w_y]u(x, y))[\alpha_{xy}[a]], \end{aligned}$$

so that we must have

$$v(x, y)w_{xy} = c(x, y)w_x\alpha_x[w_y]u(x, y)\,,$$

for some central $c(x, y)$. □

For genuine representations when u and v are identically 1, this reduces to the usual requirement for exterior equivalence,

$$w_{xy} = w_x\alpha_x[w_y]\,. \tag{6.6}$$

We can now prove the following result well known in the algebraic context.

LEMMA 6.2. *Different liftings of an outer automorphism give cohomologous cocycles ϕ, that is, cocycles differing by a coboundary*

$$(dc)(x,y,z) = \frac{c(x,y)c(xy,z)}{c(x,yz)c(y,z)}\,. \tag{6.7}$$

PROOF. In general, we can now calculate that

$$\begin{aligned}
&v(x,y)v(xy,z)\\
&= c(x,y)c(xy,z)w_x\alpha_x[w_y]u(x,y)w_{xy}^{-1}w_{xy}\alpha_{xy}[w_z]u(xy,z)w_{xyz}^{-1}\\
&= c(x,y)c(xy,z)w_x\alpha_x[w_y]u(x,y)\alpha_{xy}[w_z]u(xy,z)w_{xyz}^{-1}\\
&= c(x,y)c(xy,z)w_x\alpha_x[w_y]\alpha_x\alpha_y[w_z]u(x,y)u(xy,z)w_{xyz}^{-1}\\
&= \phi(x,y,z)c(y,z)c(x,yz)w_x\alpha_x[w_y\alpha_y[w_z]u(y,z)]u(x,yz)w_{xyz}^{-1}\\
&= (\phi.dc)(x,y,z)c(y,z)c(x,yz)w_x\alpha_x[w_y\alpha_y[w_z]u(y,z)w_{yz}^{-1}]\alpha_x[w_{yz}]u(x,yz)w_{xyz}^{-1}\\
&= (\phi.dc)(x,y,z)w_x\alpha_x[v(y,z)]w_x^{-1}v(x,yz)\\
&= (\phi.dc)(x,y,z)\beta_x[v(y,z)]v(x,yz),
\end{aligned}$$

showing that u and v have cohomologous cocycles ϕ and $\phi.dc$. □

THEOREM 6.3. *The crossed product algebras $\mathcal{A}\rtimes_{\alpha,u}\mathsf{G}$ and $\mathcal{A}\rtimes_{\beta,v}\mathsf{G}$, with $\beta_x = \mathrm{ad}(w_x)\alpha_x$ and $v(x,y)w_{xy} = w_x\alpha_x[w_y]u(x,y)$, are isomorphic.*

PROOF. The product of $f,g\in\mathcal{A}\rtimes_{\beta,v}\mathsf{G}$ is given by

$$\begin{aligned}
(f\star_{\beta,v}g)(x) &= \int f(y)\beta_y[g(y^{-1}x)]v(y,y^{-1}x)\,dy\\
&= \int f(y)w_y\alpha_y[g(y^{-1}x)]w_y^{-1}v(y,y^{-1}x)\,dy\\
&= \int f(y)w_y\alpha_y[g(y^{-1}x)w_{y^{-1}x}]u(y,y^{-1}x)w_x^{-1}\,dy,
\end{aligned}$$

so that we may set $f_w(x) = f(x)w_x$ to get

$$(f\star_{\beta\,v}g)_w(x) = (f_w\star_{\alpha,u}g_w)(x),$$

and similar calculations on f^* confirm that $f\mapsto f_w$ is the required isomorphism. □

So, up to isomorphism, the twisted crossed product depends only on the outer automorphism group, and a matching multiplier. The choice of v does matter, since even when $u=1$ the crossed product is not usually isomorphic to a twisted crossed product. In fact, if we take $\phi = dc$, the composition of twisted compact operators is given by

$$(K_1\star K_2)(x,z) \tag{6.8}$$

$$\begin{aligned}
&= \int K_1(x,y)K_2(y,z)dc(xy^{-1},yz^{-1},z)\,dy \qquad (6.9)\\
&= \int K_1(x,y)K_2(y,z)c(xy^{-1},yz^{-1})c(xz^{-1},z)c(xy^{-1},y)^{-1}c(yz^{-1},z)^{-1}\,dy\\
&= c(xz^{-1},z)\int K_1(x,y)c(xy^{-1},y)^{-1}K_2(y,z)c(yz^{-1},z)^{-1}c(xy^{-1},yz^{-1})\,dy\,.
\end{aligned}$$

This can be rewritten in terms of $K_1^c(x,y) = K_1(x,y)c(xy^{-1},y)^{-1}$ as

$$(K_1 \star K_2)^c(x,z) = \int K_1^c(x,y)K_2^c(y,z)c(xy^{-1},yz^{-1})\,dy\,, \tag{6.10}$$

showing that the new multiplication is still twisted. Indeed, it is obtained by applying the untwisted multiplication to the image of $K_1 \otimes K_2$ under the action of $c \in C(\mathsf{G}) \otimes C(\mathsf{G})$. Even such slightly deformed products can have very different differential calculi, as investigated by Majid and collaborators (particularly the recent preprint [**14**]).

We now define two automorphism groups with algebra-valued cocycles (α, u) and (β, v) to be exterior equivalent if

$$\beta_x = \mathrm{ad}(w_x)\alpha_x, \qquad \text{and} \qquad v(x,y) = w_x\alpha_x(w_y)u(x,y)w_{xy}^{-1}\,. \tag{6.11}$$

These can be rewritten as

$$\alpha_x = \mathrm{ad}(w_x^{-1})\beta_x, \qquad \text{and} \qquad u(x,y) = \beta_x(w_y^{-1})w_x^{-1}v(x,y)w_{xy}\,, \tag{6.12}$$

demonstrating the reflexivity of the equivalence, and similarly the product of the algebra elements gives transitivity.

We now can easily rephrase (and shorten) the derivation of Packer–Raeburn equivalence [**12**, Cor 4.2]. The cocycle equation for u can be written as

$$u(x,y)u(xy,y^{-1}x^{-1}z) = \phi(x,y,y^{-1}x^{-1}z)\alpha_x[u(y,y^{-1}x^{-1}z)]u(x,x^{-1}z).$$

We now work in $\mathcal{A} \otimes \mathcal{K}(L^2(\mathsf{G}))$ which acts on $a \in \mathcal{A} \otimes L^2(\mathsf{G})$. We define

$$(w_x a)(z) = u(x,x^{-1}z)a(x^{-1}z),$$

and $\beta_x = \mathrm{ad}(w_x)\alpha_x$ and use the above version of the cocycle identity to show that (α, u) is exterior equivalent to (β, v) with $(v(x,y)a)(z) = \phi(x,y,y^{-1}x^{-1}z)a(z)$.

7. The general duality result

In this section we generalise the construction at the end of Section 6 to general C*-algebras with twisted actions.

THEOREM 7.1. *Let $\mathcal{B}$ be a C*-algebra on which the group* G *acts by twisted automorphisms β_g, with twisting given by $v(x,y)$, satisfying the deformed cocycle condition with tricharacter obstruction ϕ. The twisted crossed product $(\mathcal{B} \rtimes_{\beta,v} \mathsf{G}) \rtimes \widehat{\mathsf{G}}$ is isomorphic to the algebra of $\mathcal{B}$- valued twisted kernels $\mathcal{B} \otimes \mathcal{K}_{\overline{\phi}}(L^2(\mathsf{G}))$ with the product*

$$(k_1 \star k_2)(w,z) = \int_{\mathsf{G}} k_1(w,u)k_2(u,z)\phi(wu^{-1},uz^{-1},z)^{-1}\,du\,. \tag{7.1}$$

The double dual action on $(\mathcal{B} \rtimes_{\beta,v} \mathsf{G}) \rtimes \widehat{\mathsf{G}}$ is equivalent to

$$(\widehat{\widehat{\beta}}_y k_F)(w,z) = \phi(wz^{-1},z,y)\mathrm{ad}(V_y)^{-1}[\beta_y[k_F(wy,zy)]]\,, \tag{7.2}$$

which is the product of the original action β_y and a twisted adjoint action of

$$V_y(z) = \frac{\phi(y,z^{-1},z)}{\phi(yz^{-1},zy^{-1},y)}v(y,z^{-1})\,, \tag{7.3}$$

on $\mathcal{B}$-valued kernels.

PROOF. The twisted crossed product $\mathcal{B} \rtimes_{\beta,v} \mathsf{G}$ consists of $\mathcal{B}$-valued functions on G with product

$$(f * g)(x) = \int_{\mathsf{G}} f(y)\beta_y[g(y^{-1}x)]v(y, y^{-1}x)\, dy,$$

and $(\mathcal{B} \rtimes_{\beta,v} \mathsf{G}) \rtimes \widehat{\mathsf{G}}$ consists of $\mathcal{B}$-valued functions on $\mathsf{G} \times \widehat{\mathsf{G}}$ with product

$$\begin{aligned}(F * G)(x, \xi) &= \int_{\widehat{G}} (F(., \eta)\widehat{\beta}_\eta[G(., \eta^{-1}\xi)])(x)\, d\eta \\ &= \int_{\widehat{G}\times G} F(y, \eta)\beta_y\widehat{\beta}_\eta[G(y^{-1}x, \eta^{-1}\xi)]v(y, y^{-1}x)\, dyd\eta \\ &= \int_{\widehat{G}\times G} F(y, \eta)\beta_y[\eta(y^{-1}x)G(y^{-1}x, \eta^{-1}\xi)])v(y, y^{-1}x)\, dyd\eta.\end{aligned}$$

We now Fourier transform with respect to the second argument, so that

$$\widehat{F}(x, z) = \int_{\widehat{G}} F(x, \xi)\xi(z)\, d\xi,$$

to get

$$\begin{aligned}(\widehat{F * G})(x, z) &= \int F(y, \eta)\beta_y[\eta(y^{-1}x)G(y^{-1}x, \eta^{-1}\xi)])v(y, y^{-1}x)\xi(z)\, dyd\eta d\xi \\ &= \int F(y, \eta)\eta(y^{-1}xz)\beta_y[G(y^{-1}x, \eta^{-1}\xi)(\eta^{-1}\xi)(z))]v(y, y^{-1}x)\, dyd\eta d\xi \\ &= \int_G \widehat{F}(y, y^{-1}xz)\beta_y[\widehat{G}(y^{-1}x, z)]v(y, y^{-1}x)\, dy.\end{aligned}$$

Next we introduce $\widehat{k}_F(w, z) = \beta_{w^{-1}}[\widehat{F}(wz^{-1}, z)]v(w^{-1}, wz^{-1})$ and by setting $w = xz$ in the last product formula, applying $\beta_{w^{-1}}$, and using the standard identities for v and ϕ, we obtain

$$\begin{aligned}&\widehat{k}_{F*G}(w, z) = \beta_{w^{-1}}[\widehat{F * G})(wz^{-1}, z)]v(w^{-1}, wz^{-1}) \\ &= \int_G \beta_{w^{-1}}[\widehat{F}(y, y^{-1}w)]\beta_{w^{-1}}\beta_y[\widehat{G}(y^{-1}wz^{-1}, z)]\beta_{w^{-1}}[v(y, y^{-1}wz^{-1})]v(w^{-1}, wz^{-1})\, dy \\ &= \int_G \beta_{w^{-1}}[\widehat{F}(y, y^{-1}w)]v(w^{-1}, y)\beta_{w^{-1}y}[\widehat{G}(y^{-1}wz^{-1}, z)] \\ &\qquad \times v(w^{-1}y, y^{-1}wz^{-1})\phi(w^{-1}, y, y^{-1}wz^{-1})\, dy \\ &= \int_G \widehat{k}_F(w, y^{-1}w)\widehat{k}_G(y^{-1}w, z)\phi(w^{-1}, y, y^{-1}wz^{-1})\, dy \\ &= \int_G \widehat{k}_F(w, u)\widehat{k}_G(u, z)\phi(w^{-1}, wu^{-1}, uz^{-1})\, du.\end{aligned}$$

Now, by the cocycle identity

$$\phi(w^{-1}, wu^{-1}, uz^{-1}) = \frac{\phi(u^{-1}, uz^{-1}, z)\phi(w^{-1}, wu^{-1}, u)}{\phi(wu^{-1}, uz^{-1}, z)\phi(w^{-1}, wz^{-1}, z)},$$

from which it follows that with $k_F(w, u) = \widehat{k}_F(w, u)\phi(w^{-1}, wu^{-1}, u)$ we have

$$k_{F*G}(w, z) = \int_G k_F(w, u)k_G(u, z)\phi(wu^{-1}, uz^{-1}, z)^{-1}\, du.$$

Thus we have an isomorphism with $\mathcal{B}$-valued twisted kernels $\mathcal{B} \otimes \mathcal{K}_{\overline{\phi}}(L^2(\mathsf{G}))$.

We can also compute that the double dual (untwisted) action of $y \in G$ takes $F(x,\xi)$ to $\xi(y)F(x,\xi)$. Integrating against $\xi(z)$ we see that this takes $\widehat{F}(x,z)$ to $\widehat{F}(x,zy)$, and so $\widehat{k}_F$ to

$$\begin{aligned}
(\widehat{\widehat{\beta}}_y\widehat{k}_F)(w,z) &= \beta_{w^{-1}}[\widehat{F}(wz^{-1},zy)]v(w^{-1},wz^{-1})\\
&= \beta_{y(wy)^{-1})}[\widehat{F}((wy)(zy)^{-1},zy)]v(w^{-1},wz^{-1})\\
&= \mathrm{ad}(v(y,(wy)^{-1}))^{-1}\beta_y\beta_{(wy)^{-1}}[\widehat{F}((wy)(zy)^{-1},zy)]v(w^{-1},wz^{-1})\\
&= v(y,(wy)^{-1})^{-1}\beta_y[\widehat{k}_F(wy,zy)v((wy)^{-1},wz^{-1})^{-1}]v(y,(wy)^{-1})v(w^{-1},wz^{-1})\\
&= v(y,(wy)^{-1})^{-1}\beta_y[\widehat{k}_F(wy,zy)]\beta_y[v((wy)^{-1},wz^{-1})]^{-1}v(y,(wy)^{-1})v(w^{-1},wz^{-1})\\
&= v(y,(wy)^{-1})^{-1}\beta_y[\widehat{k}_F(wy,zy)]\phi(y,(wy)^{-1},wz^{-1})^{-1}v(y,(zy)^{-1}).
\end{aligned}$$

This in turn, with a couple of applications of the pentagonal identity, gives the following expression for $(\widehat{\widehat{\beta}}_y k_F)(w,z)$:

$$\begin{aligned}
&\phi(w^{-1},wz^{-1},z)\phi(y^{-1}w^{-1},wz^{-1},zy)^{-1}\phi(y,(wy)^{-1},wz^{-1})^{-1}v(y,(wy)^{-1})^{-1}\\
&\qquad\times \beta_y[k_F(wy,zy)]v(y,(zy)^{-1})\\
&= \frac{\phi(w^{-1},wz^{-1},z)}{\phi(w^{-1},wz^{-1},zy)}\frac{\phi(y,y^{-1}z^{-1},zy)}{\phi(y,y^{-1}w^{-1},wy)}v(y,(wy)^{-1})^{-1}\beta_y[k_F(wy,zy)]v(y,(zy)^{-1})\\
&= \phi(wz^{-1},z,y)\frac{\phi(w^{-1},w,y)}{\phi(y,y^{-1}w^{-1},wy)}v(y,(wy)^{-1})^{-1}\beta_y[k_F(wy,zy)]v(y,(zy)^{-1})\\
&\qquad\times \frac{\phi(y,y^{-1}z^{-1},zy)}{\phi(z^{-1},z,y)} = \phi(wz^{-1},z,y)\mathrm{ad}(V_y)^{-1}\beta_y[k_F(wy,zy)],
\end{aligned}$$

where V_y is defined in the Theorem. This is the product of the original action β_y, and a twisted action on $\mathcal{B}$-valued kernels, which combines the original action β_y with an adjoint action of V_y, and an action on kernels of the type discussed in [**12**, Sect 5]. (Two actions α and β linked by an inner automorphism $\mathrm{ad}\, u$ can be removed by an exterior equivalence.) $\square$

As a consequence of this result we can, when convenient, replace an algebra $\mathcal{B}$ with twisted action by a (stable) nonassociative algebra $\mathcal{B}\otimes\mathcal{K}_\phi(L^2(\mathsf{G}))$ with an ordinary action.

Since the arguments are now quite general they could also be used to show that the third dual is isomorphic to the first dual tensored with ordinary compact operators.

8. Twisted and repeated crossed products

The standard Takai duality theorem can be seen from a different perspective by identifying the repeated crossed product $(\mathcal{A}\rtimes\mathsf{G})\rtimes\widehat{\mathsf{G}}$ with the twisted crossed product $\mathcal{A}\rtimes(\mathsf{G}\times_\sigma\widehat{\mathsf{G}})$, where the twisting is done by the Mackey multiplier σ on $\mathsf{G}\times\widehat{\mathsf{G}}$ given by $\sigma((y,\eta),(x,\xi)) = \eta(x)$. (For $\mathsf{G}=\mathbb{R}$, every multiplier is equivalent to a Mackey multiplier.) In fact the crossed product $\mathcal{A}\rtimes\mathsf{G}$ consists of $\mathcal{A}$-valued functions a, b, on G with product

$$(a*b)(x) = \int_{\mathsf{G}} a(y)\alpha_y[b(y^{-1}x)]\,dy\,, \tag{8.1}$$

with $\widehat{\mathsf{G}}$-action $(\widehat{\alpha}_\xi[a])(x) = \overline{\xi(x)}a(x)$, and the repeated crossed product similarly consists of functions from $\mathsf{G} \times_\sigma \widehat{\mathsf{G}}$ to $\mathcal{A}$ with product

$$(a * b)(x, \xi) = \int a(y, \eta)\overline{\eta(y^{-1}x)}\alpha_y[b(y^{-1}x, \eta^{-1}\xi)]\, dy d\eta\,. \tag{8.2}$$

Now a twisted crossed product given by a projective action β of $\mathsf{G} \times_\sigma \widehat{\mathsf{G}}$ would have product

$$(a * b)(x, \xi) = \int a(y, \eta)\beta_{(y,\eta)}[b(y^{-1}x, \eta^{-1}\xi)]\overline{\sigma((y, \eta), (y^{-1}x, \eta^{-1}\xi))}\, dy d\eta\,, \tag{8.3}$$

and we see that these match if $\beta_{(y,\eta)} = \alpha_y$ and $\sigma((y, \eta), (x, \xi)) = \eta(x)$. The duality theorem then follows from Green's results on imprimitivity algebras and the fact that the twisted group algebra of $\mathsf{G} \times_\sigma \widehat{\mathsf{G}}$ is essentially the algebra of compact operators on $L^2(\mathsf{G})$.

This equivalence between a twisted crossed product and a repeated (untwisted) crossed product has an analogue when the twisting is done with a three-cocycle, which can be exploited in reverse, to reinterpret the twisted crossed product as a repeated crossed product. We take two exterior equivalent twisted automorphism groups (β, v) and $(\widehat{\beta}, \widehat{v})$ of a C*-algebra $\mathcal{A}$:

$$\widehat{\beta}_x = \mathrm{ad}(w_x)\beta_x, \qquad \text{and} \qquad \widehat{v}(x, y) = w_x\beta_x(w_y)v(x, y)w_{xy}^{-1}\,. \tag{8.4}$$

Both v and $\widehat{v}$ define the same three-cocycle ϕ, which we shall assume to be an antisymmetric tricharacter, so that when trivial it is identically 1.

THEOREM 8.1. *For a separable locally compact abelian group* G *and a stable* C^**-algebra* $\mathcal{A}$, *let* $\beta : \mathsf{G} \to \mathrm{Aut}(\mathcal{A})$ *be a twisted homomorphism with* $\beta_\xi\beta_\eta = \mathrm{ad}(v(\xi, \eta))\beta_{\xi\eta}$, *for all* $\xi, \eta \in \mathsf{G}$. *Suppose that* $\mathsf{G} = \mathsf{G}_1 \times \mathsf{G}_2$, *where* G_j *has trivial Moore cohomology* $H^3(\mathsf{G}_j, \mathsf{U}(1))$. *Suppose that the corresponding three-cocycle* ϕ *is identically 1 on* G_1 *and is also 1 when two of its arguments are in* G_2 *(this being true for antisymmetric tricharacters on* $\mathbb{R}^2 \times \mathbb{R}$*). Then we may take* v *to be trivial on the subgroups* G_j *and* (β, v) *is exterior equivalent to* $(\widehat{\beta}, \widehat{v})$, *where*

$$\widehat{\beta}_{xX} = \beta_x\beta_X, \qquad \text{and} \qquad \widehat{v}(xX, yY) = \beta_x[\widetilde{v}(X, y)] \tag{8.5}$$

where lower case letters denote elements of G_1 *and capitals elements of* G_2, *and* $\widetilde{v}(X, y) = v(X, y)v(y, X)^{-1}$. *The cocycle* $\widetilde{v}$ *satisfies*

$$\begin{aligned} \widetilde{v}(XY, z) &= \beta_X[\widetilde{v}(Y, z)]\widetilde{v}(X, z), \qquad \text{and} \\ \widetilde{v}(X, yz) &= \phi(X, y, z)^{-3}\widetilde{v}(X, y)\beta_y[\widetilde{v}(X, z)]\,, \end{aligned} \tag{8.6}$$

and $\widehat{v}$ *defines the three-cocycle* $\varphi(xX, yY, zZ) = \phi(X, y, z)^3$.

PROOF. Since $\phi = 1$ on the subgroups, the stabilisation theorem ([**12**, Cor4.2] and below) tells us that we may also take v to be 1 on the subgroups. It follows from the definitions that $\widehat{\beta}_{xX} = \mathrm{ad}(v(x, X))\beta_{xX}$, so that we could take $w_{xX} = v(x, X)$. In fact it is more convenient to make a slightly different choice since we also have

$$\beta_\xi\beta_\eta = v(\xi, \eta)\beta_{\xi\eta} = \mathrm{ad}(\widetilde{v}(\xi, \eta))\beta_\eta\beta_\xi,$$

from which we deduce

$$\begin{aligned} (\beta_x\beta_X)(\beta_y\beta_Y) &= \beta_x(\beta_X\beta_y)\beta_Y \\ &= \beta_x\mathrm{ad}(\widetilde{v}(X, y))\beta_y\beta_X\beta_Y \end{aligned}$$

$$= \mathrm{ad}(\beta_x[\widetilde{v}(X,y)])\mathrm{ad}(v(x,y))\beta_{xy}\mathrm{ad}(v(X,Y))\beta_{XY}\,.$$

Since the cocycle $\widehat{v}$ is determined up to scalars, and $v(x,y)=1$, $v(X,Y)=1$ on the subgroups, this gives us the required form $\widehat{v}(xX,yY)=\beta_x[\widetilde{v}(X,y)]$. Alternatively we can see that up to scalar factors $w_{xX}\beta_{xX}[w_{yY}]v(xX,yY)w^{-1}_{xyXY}$ is

$$\beta_x[\widetilde{v}(X,y)]v(x,y)\beta_{xy}][v(X,Y)] = \beta_x[\widetilde{v}(X,y)]$$

using the triviality of the restriction of v to the subgroups. Now, we also have

$$\begin{aligned}
v(XY,z) &= \phi(X,Y,z)v(X,Y)^{-1}\beta_X[v(Y,z)]v(X,Yz)\\
&= \phi(X,Y,z)v(X,Y)^{-1}\beta_X[\widetilde{v}(Y,z)]\phi(X,z,Y)^{-1}v(X,z)v(Xz,Y)\\
&= \phi(X,Y,z)\phi(X,z,Y)^{-1}v(X,Y)^{-1}\\
&\qquad\times \beta_X[\widetilde{v}(Y,z)]\widetilde{v}(X,z)\phi(z,X,Y)\beta_z[v(X,Y)]v(z,XY)\,,
\end{aligned}$$

and using the triviality of v on subgroups, and the fact that ϕ must be trivial whenever two of its arguments are in G_2, this reduces to

$$\widetilde{v}(XY,z) = \beta_X[\widetilde{v}(Y,z)]\widetilde{v}(X,z).$$

Similarly, we have

$$\begin{aligned}
\beta_X[v(y,z)]v(X,yz) &= \phi(X,y,z)^{-1}v(X,y)v(Xy,z)\\
&= \phi(X,y,z)^{-1}\phi(y,X,z)\widetilde{v}(X,y)\beta_y[v(X,z)]v(y,Xz)\\
&= \phi(X,y,z)^{-1}\phi(y,X,z)\widetilde{v}(X,y)\beta_y[\widetilde{v}(X,z)]\phi(y,z,X)^{-1}v(y,z)v(yz,X)\,,
\end{aligned}$$

and using the triviality of v on subgroups as well as the antisymmetry of ϕ this reduces to

$$\widetilde{v}(X,yz) = \phi(X,y,z)^{-3}\widetilde{v}(X,y)\beta_y[\widetilde{v}(X,z)].$$

We then have

$$\begin{aligned}
&\widehat{v}(xX,yY)\widehat{v}(xyXY,zZ)\widehat{v}(xX,yzYZ)^{-1}\\
&= \beta_x[\widetilde{v}(X,y)]\beta_{xy}[\widetilde{v}(XY,z)]\beta_x[\widetilde{v}(X,yz)]^{-1}\\
&= \beta_x[\widetilde{v}(X,y)\beta_y[\widetilde{v}(XY,z)]\widetilde{v}(X,yz)^{-1}]\\
&= \phi(X,y,z)^{-3}\beta_x[\widetilde{v}(X,y)\beta_y[\widetilde{v}(XY,z)\widetilde{v}(X,z)^{-1}\widetilde{v}(X,y)^{-1}]\\
&= \phi(X,y,z)^{-3}\beta_x[\widetilde{v}(X,y)\beta_y[\beta_X[\widetilde{v}(Y,z)]\widetilde{v}(X,y)^{-1}]\\
&= \phi(X,y,z)^{-3}\beta_x\beta_X[\beta_y[\widetilde{v}(Y,z)]]\\
&= \phi(X,y,z)^{-3}\beta_x\beta_X[\widehat{v}(yY,zZ)]\,,
\end{aligned}$$

showing that a cocycle identity holds for $\widehat{v}$. The corresponding three-cocycle $\varphi(xX,yY,zZ)=\phi(X,y,z)^3$ is not antisymmetric, but its antisymmetrisation,

$$\begin{aligned}
[\varphi(xX,yY,zZ)\varphi(yY,zZ,xX)\varphi(zZ,xX,yY)]^{\frac{1}{3}} &= \phi(X,y,z)\phi(Y,z,x)\phi(Z,x,y)\\
&= \phi(xX,yY,zZ),
\end{aligned}$$

is just the original cocycle ϕ. □

THEOREM 8.2. *Writing $f_Y(y)=f(yY)$, the crossed product algebra $\mathcal{A}\rtimes_{\widehat{\beta},\widehat{v}}G \sim (\mathcal{A}\rtimes \mathsf{G}_1)\rtimes_{\widetilde{\beta}}\mathsf{G}_2$ has twisted convolution product*

$$(f*g)_X = \int_{\mathsf{G}_2} f_Y *_1 \widetilde{\beta}_Y[g_{Y^{-1}X}]\,dY\,, \tag{8.7}$$

where $\widetilde{\beta}_X[f](y) = \beta_X[f(y)]\widetilde{v}(X,y)$, *and* $*_2$ *denotes the convolution product on* $\mathcal{A} \rtimes \mathsf{G}_1$. *The map* $X \to \widetilde{\beta}_X$ *is a group homomorphism, and* $\widetilde{\beta}_X$ *gives an isomorphism between twisted crossed products for multipliers* σ *and* $\phi_X\sigma$, *where* $\phi_X(y,z) = \phi(X,y,z)$.

PROOF. We calculate that

$$\begin{aligned}(f*g)(xX) &= \int f(yY)\beta_y\beta_Y[g(y^{-1}xY^{-1}X)\beta_y[\widetilde{v}(Y,y^{-1}x)]\,dydY\\ &= \int f(yY)\beta_y[\beta_Y[g(y^{-1}xY^{-1}X)]\widetilde{v}(Y,y^{-1}x)]\,dydY\\ &= \int f(yY)\beta_y[\widetilde{\beta}_Y[g(y^{-1}xY^{-1}X)]]\,dydY\,,\end{aligned}$$

so that

$$(f*g)_X = \int f_Y *_1 \widetilde{\beta}_Y[g_{Y^{-1}X}]\,dY.$$

We can then check that

$$\begin{aligned}\widetilde{\beta}_X\widetilde{\beta}_Y[f](z) &= \beta_X[\beta_Y[f(z)]\widetilde{v}(Y,z)]\widetilde{v}(X,z)\\ &= \beta_{XY}[f(z)]\beta_X[\widetilde{v}(Y,z)]\widetilde{v}(X,z)\\ &= \beta_{XY}[f(z)]\widetilde{v}(XY,z)\\ &= \widetilde{\beta}_{XY}[f](z)\,.\end{aligned}$$

Next consider a twisted crossed product with $U(1)$-valued multiplier σ

$$\begin{aligned}(\widetilde{\beta}_X[f]*\widetilde{\beta}_X[g])(x) &= \int \widetilde{\beta}_X[f](y)\beta_y[\widetilde{\beta}_X[g](y^{-1}z)]\sigma(y,y^{-1}x)\,dy\\ &= \int \beta_X[f(y)]\widetilde{v}(X,y)\beta_y[\beta_X[g(y^{-1}z)]\widetilde{v}(X,y^{-1}z)]\sigma(y,y^{-1}x)\,dy\\ &= \int \beta_X[f(y)]\beta_X[\beta_y[g(y^{-1}x)]]\widetilde{v}(X,y)\beta_y[\widetilde{v}(X,y^{-1}x)]\sigma(y,y^{-1}x)\,dy\\ &= \beta_X[\int f(y)\beta_y[g(y^{-1}x)]\varphi(X,y,y^{-1}x)\sigma(y,y^{-1}x)\,dy]\widetilde{v}(X,x)\,,\end{aligned}$$

which differs from $\widetilde{\beta}_X[f*g](x)$ by the insertion of $\varphi(X,y,y^{-1}x)$ in the convolution integral, changing the multiplier σ to $\varphi(X,\cdot,\cdot)\sigma$. □

This result tells us that the twisted crossed product multiplication for $\mathcal{A}\rtimes\mathsf{G}$ can be obtained by doing repeated crossed products but with a modified automorphism action of the final subgroup. The result is still nonassociative because one still has the three-cocycle φ. There is no inconsistency because $\widetilde{\beta}_X$ does not act as automorphisms of a crossed product $\mathcal{A}\rtimes\mathsf{G}_1$. We shall now show how to modify things to get a more useful result.

Suppose now that we replace the algebra $\mathcal{A}$ by an algebra $\mathcal{B}$ which admits not only an action of G but also a compatible action of $C(\mathsf{G}_2) = C(\mathsf{G}/\mathsf{G}_1)$. (This would be automatic for an algebra induced to G from a subgroup N.)

THEOREM 8.3. *Let* $\mathcal{B}$ *be a* C^**-algebra admitting an action* β *of the abelian group* $\mathsf{G} = \mathsf{G}_1\times\mathsf{G}_2$ *with multiplier* v *and* ϕ *satisfying the conditions of the previous theorems and also a compatible action of* $C(\mathsf{G}_2)$. *Then* $\mathcal{B}\rtimes_{\beta,v}\mathsf{G}$ *is stably equivalent* *to an ordinary repeated crossed product* $(\mathcal{B} \rtimes_{\widehat{\beta}_1} \mathsf{G}_1) \rtimes_{\widehat{\beta}_2} \mathsf{G}_2$, *where* $\widehat{\beta}_j$ *is the restriction of* $\widehat{\beta}$ *to* G_j.

PROOF. We have already seen that, by replacing (β, v) by $(\widehat{\beta}, \widehat{v})$, the multipliers on the subgroups are trivial and that when one splits the crossed product into a repeated crossed product the first part is just an ordinary crossed product with the action $\widehat{\beta}_1$. Now, by the Packer–Raeburn trick, this is stably equivalent to a projective crossed product with the $C(\mathsf{G}_2)$-valued multiplier $\varphi_\bullet(y,z)$ defined as the function $X \mapsto \varphi(X,y,z)$, The advantage is that $\widetilde{\beta}_X$ is now an automorphism, since when we put $\sigma = \varphi_\bullet$ we have

$$\begin{aligned}
(\widetilde{\beta}_X[f] *_{\varphi_\bullet} \widetilde{\beta}_X[g])(x) &= \int \widetilde{\beta}_X[f](y)\beta_y[\widetilde{\beta}_X[g](y^{-1}z)]\varphi_\bullet(y,y^{-1}x)\,dy \\
&= \int \beta_X[f(y)]\widetilde{v}(X,y)\beta_y[\beta_X[g(y^{-1}z)]\widetilde{v}(X,y^{-1}z)]\varphi_\bullet(y,y^{-1}x)\,dy \\
&= \int \beta_X[f(y)]\beta_X[\beta_y[g(y^{-1}x)]]\widetilde{v}(X,y)\beta_y[\widetilde{v}(X,y^{-1}x)]\varphi_\bullet(y,y^{-1}x)\,dy \\
&= \int \beta_X[f(y)]\beta_X[\beta_y[g(y^{-1}x)]]\varphi_\bullet(y,y^{-1}x)\varphi(X,y,y^{-1}x)\,dy]\widetilde{v}(X,x) \\
&= \beta_X[\int f(y)\beta_y[g(y^{-1}x)]\varphi_\bullet(y,y^{-1}x)\,dy]\widetilde{v}(X,x) = \widetilde{\beta}_X[f *_{\varphi_\bullet} g]\,,
\end{aligned}$$

where $\varphi_Z(y,z)\varphi(X,y,z) = \varphi_{ZX}(y,z) = \widehat{\beta}_X(\varphi_Z(y,z)\widehat{\beta}_X^{-1}$ follows from compatibility of the actions. □

This result takes us quite a long way towards proving the analogue of the Connes-Thom isomorphism in our context. We take $\mathsf{G} = \mathbb{R}^3$, with the subgroups $\mathsf{G}_1 = \mathbb{R}^2$, and $\mathsf{G}_2 = \mathbb{R}$. In [**12**] the algebra has the form $\mathcal{B} = \mathrm{ind}_{\mathbf{Z}^3}^{\mathbb{R}^3}\mathcal{A}$, and so has an action of $\mathbb{R}^3$. As already noted the Moore cohomology group H^3 is trivial on the subgroups $\mathsf{G}_1 = \mathbb{R}^2$ and $\mathsf{G}_2 = \mathbb{R}$. Using Theorem 7.1 we express the twisted crossed product $\mathcal{B} \rtimes \mathbb{R}^3$ as a repeated crossed product $(\mathcal{B} \rtimes \mathbb{R}^2) \rtimes \mathbb{R}$. The standard Connes-Thom isomorphism tells us that $K_*(\mathcal{B} \rtimes \mathbb{R}^2) \cong K_*(\mathcal{B})$, so that the first crossed product does not change the K-theory. The second crossed product with G_2 is more problematic because the group does not act as automorphisms. However, it is stably equivalent to the case where one does have automorphisms. In that stably equivalent algebra the ordinary Connes-Thom theorem then asserts that the K-theory of the crossed product $(\mathcal{B} \rtimes \mathbb{R}^2) \rtimes \mathbb{R}$ is the same as that of $\mathcal{B} \rtimes \mathbb{R}^2$ and so of $\mathcal{B}$, apart from a shift of 1 in degree. Superficially this appears to have proved the desired generalisation of the Connes-Thom theorem to the twisted algebra $\mathcal{B} \rtimes_{\beta,v} \mathbb{R}^3$, but the missing ingredient is to check that the stabilisation equivalence valid for ordinary K-theory is also consistent with the definition of K-theory in the category $\mathcal{C}_{\mathsf{G}}(\phi)$.

9. Some remarks and speculations related to K-theory

Here we give an application, to a version of K-theory, of Takai duality in our context. For $\mathcal{A}$ an algebra in the category $\mathcal{C}_{\mathsf{G}}(\phi)$ we can define the ring $K_0(\mathcal{A}, \Phi)$ of stable equivalence classes of projective finite rank $\mathcal{A}$-modules in $\mathcal{C}_{\mathsf{G}}(\phi)$. We have already seen that these modules are also $\mathcal{A} \rtimes \widehat{\mathsf{G}}$-modules, and it follows from Proposition A.2 that these are also finite rank and projective. There is thus a natural identification of $K_0(\mathcal{A}, \Phi)$ with the stable equivalences of finite rank projective $\mathcal{A} \rtimes \widehat{\mathsf{G}}$-modules. There is a caveat that $K_0(\mathcal{A}, 1)$ does not reduce to the K-theory of the G-algebra $\mathcal{A}$, since the finite projective $\mathcal{A}$-modules used in the definition are also expected to be G-invariant. The reason is that only G-invariant projections are well defined in the category $\mathcal{C}_{\mathsf{G}}(\phi)$, cf. below. In fact, $K_0(\mathcal{A}, 1) \cong K_0(\mathcal{A} \rtimes \widehat{\mathsf{G}})$, and for $\mathsf{G} \cong \mathbb{R}^d$ Connes-Thom isomorphism theorem gives $K_0(\mathcal{A} \rtimes \widehat{\mathsf{G}}) \cong K_d(\mathcal{A})$, so that $K_0(\mathcal{A}, 1) \cong K_d(\mathcal{A})$.

We actually want to apply this to a C*-algebra of the form $\mathcal{A} = \mathcal{B} \rtimes \mathsf{G}$, with $\mathcal{B}$ associative, to find $K_0(\mathcal{B} \rtimes \mathsf{G}, \Phi)$ in terms of equivalence classes of $(\mathcal{B} \rtimes \mathsf{G}) \rtimes \widehat{\mathsf{G}}$-modules. By Theorem 7.1, one has $(\mathcal{B} \rtimes \mathsf{G}) \rtimes \widehat{\mathsf{G}} \cong \mathcal{B} \otimes \mathcal{K}_{\overline{\phi}}(L^2(\mathsf{G}))$, which by Appendix A, can be strictified to the associative C*-algebra $\mathcal{B} \otimes \mathcal{K}(L^2(\mathsf{G}))$. (There is a natural correspondence between the modules since the nonassociative effect of Φ appears only for repeated actions, and the action itself can be defined in the same way for both cases. The strictification functor thus preserves the properties of being finite rank and projective.) This associative C*-algebra is Morita equivalent to $\mathcal{B}$ itself, so that the stable equivalence classes of finite rank projective modules for $(\mathcal{B} \rtimes \mathsf{G}) \rtimes \widehat{\mathsf{G}}$ are in natural bijective correspondence with those for $\mathcal{B}$, that is, with $K_0(\mathcal{B} \otimes \mathcal{K}(L^2(\mathsf{G})))$. In other words $K_0(\mathcal{B} \rtimes \mathsf{G}, \Phi) \cong K_0(\mathcal{B} \otimes \mathcal{K}(L^2(\mathsf{G})))$.

At first sight a degree change appears to be missing, but this is a consequence of the way we have defined $K_0(\mathcal{A}, \Phi)$, as previously explained. The morphisms in the category $\mathcal{C}_{\mathsf{G}}(\phi)$ with associator ϕ are G-maps, and so projective modules for an algebra $\mathcal{A}$ in this category are submodules of free modules defined by G-invariant projections E. Such modules can also be thought of as submodules of free modules defined by idempotents e in a matrix algebra $M_n(\mathcal{A})$, so that $Ev = v \star e$. We then see that

$$v \star g[e] = g(g^{-1}v \star e) = g(Eg^{-1}v) = Ev = v \star e\,, \tag{9.1}$$

so that $e = g[e]$ is invariant under the action of G.

To properly define the K-theory of a C*-algebra in the category $\mathcal{C}_{\mathsf{G}}(\phi)$, we need to consider an enveloping category that includes not just G-morphisms. One such candidate is the Karoubian enveloping category, and will be considered in a future work. We also plan to investigate Tannakian duality and its consequences in our context. More precisely, let G denote the Euclidean group, and ϕ the 3-cocycle on it as in the text. Then our tensor category $\mathcal{C}_{\mathsf{G}}(\phi)$ is just a twisted representation category, $\mathrm{Rep}_c(\mathsf{G}, \phi)$, where the subscript is a reminder that we take topology into consideration. Then the dual tensor category consists of continuous, tensorial functors $\mathcal{F} : \mathrm{Rep}_c(\mathsf{G}, \phi) \to \mathcal{B}(V)$, where $\mathcal{B}(V)$ denotes bounded operators on the Hilbert space V. Here V varies, and tensorial means respecting the structures in the tensor category. The dual tensor category is a tensor category denoted by $\mathrm{Rep}_c(\mathsf{G}, \phi)'$, and the putative analog of Tannakian duality in this context would say that $\mathrm{Rep}_c(\mathsf{G}, \phi)$ and $\mathrm{Rep}_c(\mathsf{G}, \phi)''$ are equivalent tensor categories. The consequences of Tannakian duality applied to C*-algebras within $\mathcal{C}_{\mathsf{G}}(\phi) = \mathrm{Rep}_c(\mathsf{G}, \phi)$ will also be explored.

Appendix A. Equivalence to a strict category

MacLane showed that every monoidal category is equivalent to a strict category in which the structure maps such as Φ are the obvious identity maps. The general construction is described and applied to the category $\mathcal{C}_{\mathsf{G}}(\phi)$ in [**1, 2**]. We shall give a simpler alternative construction which works in this case, and it is one of those cases where the structure is more transparent in the more general category $\mathcal{C}^{\mathsf{H}}(\Phi)$ of modules for a Hopf algebra H, although we shall apply it to $\mathsf{H} = C^*(\widehat{\mathsf{G}}) \sim C_0(\mathsf{G})$. (It is to some extent motivated by our observation that $\mathcal{A}$-modules in the category $\mathcal{C}_{\mathsf{G}}(\phi)$ are automatically modules for the crossed product $\mathcal{A} \rtimes \widehat{\mathsf{G}}$.)

The algebra H is an H-H-bimodule under the left and right multiplication actions, and so, in particular it is an object in $\mathcal{C}^{\mathsf{H}}(\Phi)$, though it is not an algebra in the category, since its multiplication is associative in the strict sense. We shall exploit this dual role of H to construct the functor to a strict monoidal category.

Consider the functor F which takes each object $\mathcal{A}$ of $\mathcal{C}^{\mathsf{H}}(\Phi)$ to the algebraic tensor product $F(\mathcal{A}) = \mathcal{A} \otimes \mathsf{H}$, and each morphism $T \in \hom(\mathcal{A},\mathcal{B})$ to $F(T) \in \hom(F(\mathcal{A}), F(\mathcal{B}))$ which sends $a \otimes h$ to $T(a) \otimes h$. (When $\mathsf{H} = C_0(\mathsf{G})$ we can use $F(\mathcal{A}) = C(\mathsf{G},\mathcal{A})$ as a more convenient alternative.) Since H is a bimodule $F(\mathcal{A})$ is an H-H-bimodule, with the right multiplication action by H, and the left comultiplication action, given in Sweedler notation by

$$h \cdot (a \otimes k) = \Delta h(a \otimes k) = h_{(1)}[a] \otimes h_{(2)}k.$$

With these actions we may take the new tensor product operation to be $\mathcal{A} \otimes_F \mathcal{B} = \mathcal{A} \otimes_{\mathsf{H}} \mathcal{B}$ (or $\mathcal{A} \otimes_{C_0(\mathsf{G})} \mathcal{B}$), for which the new identity object is $F(\mathbb{C}) = \mathsf{H}$ (or $C_0(\mathsf{G})$) itself.

THEOREM A.1.

(i) *For $\mathcal{A} \in \mathcal{C}^{\mathsf{H}}(\Phi)$ set $F(\mathcal{A}) = \mathcal{A}\otimes\mathsf{H}$ and let the tensor product $F(\mathcal{A})\otimes_{\mathsf{H}}F(\mathcal{B})$ be the quotient of $(\mathcal{A} \otimes \mathsf{H}) \otimes (\mathcal{B} \otimes \mathsf{H})$ by the equivalence relation that $(a \cdot h) \otimes b \sim a \otimes (h \cdot b)$, for all $a \in F(\mathcal{A})$, $b \in F(\mathcal{B})$ and $h \in \mathsf{H}$. Then*
$$F(\mathcal{A}) \otimes_{\mathsf{H}} F(\mathcal{B}) \cong F(\mathcal{A} \otimes \mathcal{B}).$$

(ii) *For $\mathcal{A} \in \mathcal{C}_{\mathsf{G}}(\phi)$ set $F(\mathcal{A}) = C(\mathsf{G},\mathcal{A})$ and let the tensor product $[F(\mathcal{A})\otimes_{C_0(\mathsf{G})} F(\mathcal{B})]$ be the quotient of $C(\mathsf{G},\mathcal{A})\otimes C(\mathsf{G},\mathcal{B})$ by the equivalence relation that $(a \cdot h) \otimes b \sim a \otimes (h \cdot b)$, for all $a \in F(\mathcal{A})$, $b \in F(\mathcal{B})$ and $h \in C_0(\mathsf{G})$. Then*
$$F(\mathcal{A}) \otimes_{C_0(\mathsf{G})} F(\mathcal{B}) \cong F(\mathcal{A} \otimes \mathcal{B}).$$

In each case set $F(T) = T \otimes \mathrm{id}$ for a morphism T. Then F defines a functor between tensor categories. (The associator in each case is just Φ tensored with the identity.)

PROOF. We have

$$F(\mathcal{A}) \otimes_{\mathsf{H}} F(\mathcal{B}) = (\mathcal{A} \otimes \mathsf{H}) \otimes_{\mathsf{H}} (\mathcal{B} \otimes \mathsf{H}) \cong (\mathcal{A} \otimes \mathcal{B}) \otimes \mathsf{H} = F(\mathcal{A} \otimes \mathcal{B}),$$

giving the result and showing consistency with the previous tensor product. (In Sweedler notation we have the isomorphism is given explicitly by $(a\otimes h)\otimes_H (b\otimes k) \mapsto (a \otimes \Delta(h)(b \otimes k)) = (a \otimes (h_{(1)}b) \otimes (h_{(2)}k).)$ Most of the rest is easily checked. In particular, we find that

$$\begin{aligned}((a \otimes h) \otimes_{\mathsf{H}} (b \otimes k)) \otimes_{\mathsf{H}} (c \otimes l) &= ((a \otimes h_{(1)}[b]) \otimes (h_{(2)}k_{(1)}[c])) \otimes h_{(3)}k_{(2)}l \\ &= \Phi[(a \otimes (h_{(1)}[b] \otimes h_{(2)}k_{(1)}[c]))] \otimes h_{(3)}k_{(2)}l \\ &= (\Phi \otimes \mathrm{id})[(a \otimes h) \otimes_{\mathsf{H}} [(b \otimes k) \otimes_{\mathsf{H}} (c \otimes l)]] .\end{aligned}$$

□

For future reference we also note the connection with the crossed product.

PROPOSITION A.2. *If $\mathcal{A}$ is an algebra with multiplication $\star$ then $F(\mathcal{A})$ can be given the multiplication defined by the maps*

$$F(\mathcal{A}) \otimes_{\mathsf{H}} F(\mathcal{A}) = F(\mathcal{A}\otimes\mathcal{A}) \cong (\mathcal{A}\otimes\mathcal{A})\otimes\mathsf{H} \longrightarrow (\mathcal{A}\otimes\mathsf{H}) = F(\mathcal{A}),$$

where the arrow denotes the map $\star\otimes 1$. When $\mathsf{H} = C_0(\mathsf{G})$, $F(\mathcal{A})$ equipped with this multiplication is just the crossed product $\mathcal{A}\rtimes\mathsf{G}$. If $\mathcal{M}$ is a module for $\mathcal{A}$ then $F(\mathcal{M})$ is a module for $F(\mathcal{A})$, and $(\mathrm{id}\otimes\epsilon)F(\mathcal{M})$ can be identified with $\mathcal{M}$ regarded as a $\mathcal{A}\rtimes\widehat{\mathsf{G}}$-module.

PROOF. We have

$$(a\otimes h)\otimes(b\otimes k) = (a\otimes h_{(1)}[b])\otimes h_{(2)}k \mapsto (a\star h_{(1)}[b])\otimes h_{(2)}k,$$

which is the crossed product multiplication in $\mathcal{A}\rtimes\mathsf{G}$, when $\mathsf{H} = C_0(\mathsf{G})$. Replacing b by an element of $\mathcal{M}$ and applying ϵ (which is a counit and a multiplicative homomorphism) we have

$$\begin{aligned}(\mathrm{id}\otimes\epsilon)[(a\otimes h)\star(m\otimes k)] &= (a\otimes h_{(1)}[m])\epsilon(h_{(2)}k)\\ &= (a\star h_{(1)}[m])\otimes\epsilon(h_{(2)})\epsilon(k)\\ &= (a\star h[m])\otimes\epsilon(k),\end{aligned}$$

as required. □

The advantage of expressing the crossed product action in terms of F is that the functor respects direct sums and maps the free $\mathcal{A}$-module $\mathcal{A}^n$ to the free $\mathcal{A}\rtimes\widehat{\mathsf{G}}$-module $(\mathcal{A}\rtimes\widehat{\mathsf{G}})^n$. If $\mathcal{M}$ is a finite rank projective module defined by $e:\mathcal{A}^n\to\mathcal{M}$, then $F(\mathcal{M})$ is a finite rank projective module defined by $(\mathcal{A}\rtimes\widehat{\mathsf{G}})^n\to F(\mathcal{M})$, and, taking the image under $\mathrm{id}\otimes\epsilon$, we see that $\mathcal{M}$ is a finite rank projective $\mathcal{A}\rtimes\widehat{\mathsf{G}}$-module.

We now introduce a new tensor product $F(\mathcal{A})\circ F(\mathcal{B})$, such that for $a\in F(\mathcal{A})$, $b\in F(\mathcal{B})$, and $c\in F(\mathcal{C})$, we have

$$(a\circ b)\otimes_{\mathsf{H}} c = a\otimes_{\mathsf{H}}(b\otimes_{\mathsf{H}} c).$$

This can be done explicitly using Φ, since

$$(a\circ b)\otimes_{\mathsf{H}} c = \Phi^{-1}((a\otimes_{\mathsf{H}} b)\otimes_{\mathsf{H}} c),$$

and Φ^{-1} is given by the left action of an element $\phi^{-1}\in\mathsf{H}\otimes\mathsf{H}\otimes\mathsf{H}$ on $\mathcal{A}\otimes\mathcal{B}\otimes\mathcal{C}$. For convenience we shall write ϕ^{-1} in a Sweedler type notation as $\phi^{-1} = \phi'\otimes\phi''\otimes\phi'''$, but this is neither intended to imply that ϕ is decomposable nor that this is in the range of $(\Delta\otimes 1)\Delta$. We then have

$$(a\circ b)\otimes_{\mathsf{H}} c = (\phi'\cdot a\otimes_{\mathsf{H}}\phi''\cdot b)\otimes_{\mathsf{H}}\phi'''c = (\phi'\cdot a\otimes_{\mathsf{H}}\phi''\cdot b)\cdot\phi'''\otimes_{\mathsf{H}} c\,,$$

or, formally,

$$a\circ b = (\phi'\cdot a\otimes_{\mathsf{H}}\phi''\cdot b)\cdot\phi'''\in F(\mathcal{A})\otimes_{\mathsf{H}} F(\mathcal{B}).$$

In fact, this expression makes perfectly good sense (in the multiplier algebra if not in the original algebra), and can be used as a definition of $a\circ b$. (This would not have been true in the original setting with $a\in\mathcal{A}$ and $b\in\mathcal{B}$, and the original tensor product, since there was only a left action and the argument would have led to $a\circ b = (\phi'\cdot a\otimes_{\mathsf{H}}\phi''\cdot b)\otimes\phi'''\in(\mathcal{A}\otimes\mathcal{B})\otimes\mathsf{H}$, not in $\mathcal{A}\otimes\mathcal{B}$. Whether it makes sense in the original algebra or in some slightly extended algebra depends on the detail of the situation. The case when $\mathsf{H} = C_0(\mathsf{G})$, $F(\mathcal{A}) = C(\mathsf{G},\mathcal{A})$ gives a large enough algebra to work whilst the algebraic tensor product would generally be too small.)

Another useful way of thinking about the product is that $F(\mathcal{A}\otimes\mathcal{B}) \sim [(\mathcal{A}\otimes\mathcal{B})\otimes\mathsf{H}]$, and Φ^{-1} maps this to $\mathcal{A}\otimes(\mathcal{B}\otimes\mathsf{H}) \sim \mathcal{A}\otimes F(\mathcal{B}) \sim F(\mathcal{A})\otimes_{\mathsf{H}} F(\mathcal{B})$.

As a consequence of the definition, for $d \in F(\mathcal{D})$

$$\begin{aligned}(a\circ b)\circ c)\otimes d &= (a\circ b)\otimes(c\otimes d)\\ &= a\otimes(b\otimes(c\otimes d))\\ &= a\otimes((b\circ c)\otimes d)\\ &= (a\circ(b\circ c))\otimes d,\end{aligned}$$

and taking $\mathcal{D}=\mathsf{H}$ we see that we now have strict associativity. We note that if H has a unit 1 the original algebra $\mathcal{A}$ can be identified with the subalgebra $\mathcal{A}\otimes 1 \subseteq \mathcal{A}\otimes\mathsf{H}$, (or in the case of $\mathsf{H}=C_0(\mathsf{G})$ with the constant functions in the algebra $C(\mathsf{G},\mathcal{A})$). These are not closed under the new tensor product $\circ$.

The new tensor product now carries over to products on algebra, so that $\star$ is replaced by a new product $a * b = \psi_z[a\star b]$, where $\phi_z(x,y) = \varphi(x,y,z)$ acts as an element of $C(\mathsf{G})\otimes C(\mathsf{G})$, The associativity of this may now be checked by either of the above calculations, and similarly for actions of algebras on modules. In summary, the action of an algebra $\mathcal{A}$ on a $C_0(\mathsf{G})$-module can always be replaced by an action of the crossed product $\mathcal{A}\rtimes\widehat{\mathsf{G}}$, which a Fourier transform identifies with $C(\mathsf{G},\mathcal{A})$, which means that for modules one always works with the objects whose multiplication can be made associative. This result clarifies the paradox of the relevance of nonassociative algebras when algebras of operators are always associative, for we see that in representation the action of nonassociative algebras can always be replaced by an associative action if one so wishes. The situation is rather similar to that of projective representations of groups, where, for any multiplier σ on a group G, a projective σ-representation of G can be obtained from an ordinary representation of its central extension G^σ. Nonetheless, there are situations in which G and σ appear naturally or linking a number of different situations, so that although G^σ is technically useful, it does not really capture the essence of the situation. The study of the canonical commutation relations as a projective representation of a vector group provides a good example. The central extension has representations allowing all possible values of Planck's constant, and so misses an important feature of the physical situation.

In addition to the left and right actions of H, when H is a Hopf algebra there is also a right coaction of H on $F(\mathcal{A}) = \mathcal{A}\otimes\mathsf{H}$ given by

$$\mathrm{id}\otimes\Delta : \mathcal{A}\otimes\mathsf{H}\mapsto(\mathcal{A}\otimes\mathsf{H})\otimes\mathsf{H},$$

and similarly for $C_0(\mathsf{G})$. There is no rebracketing problem since H is the algebra whose modules give the objects rather than an object itself, and the fact that this is a coaction follows from the coassociativity of Δ. To check compatibility with the tensor product structure, we note that

$$\begin{aligned}(a\otimes h)\otimes_{\mathsf{H}}(b\otimes k) &= (a\otimes h_{(1)})\otimes_{\mathsf{H}}(b\otimes h_{(2)}k)\\ &\mapsto [(a\otimes h_{(1)})\otimes_{\mathsf{H}}(b\otimes h_{(2)}k_{(1)})]\otimes h_{(3)}k_{(2)}\\ &= [(a\otimes h_{(1)})\otimes_{\mathsf{H}}(b\otimes k_{(1)})]\otimes h_{(2)}k_{(2)},\end{aligned}$$

showing direct compatibility with the coaction on the individual factors. (With a little modification of the functor F it is possible to work with quasi-Hopf algebras too.)

If we instead consider the dual action of θ in the dual Hopf algebra H^* given in Sweedler notation by

$$\theta[a \otimes h] = \theta(h_{(2)})a \otimes h_{(1)},$$

where $\theta(h_{(2)})$ denotes the pairing of H^* with H, then this shows that one has the correct action on tensor products. We can therefore regard F as a functor from $\mathcal{C}^{\mathsf{H}}(\Phi)$ to $\mathcal{C}^{\mathsf{H}^*}(1)$ (or from $\mathcal{C}_{\mathsf{G}}(\phi)$ to $\mathcal{C}_{\widehat{\mathsf{G}}}(1)$).

A.1. Example: Strictification of the twisted compact operators. The easiest way to see how the strictifcation works is to consider an example, such as the algebra $\mathcal{K}_\phi(L^2(\mathsf{G}))$, which maps to $F(\mathcal{K}_\phi(L^2(\mathsf{G}))) = C(\mathsf{G}, \mathcal{K}_\phi(L^2(\mathsf{G})))$. We first note that the product of $a \otimes h$ and $b \otimes k$ in $\mathcal{A} \otimes C_0(\mathsf{G})$ is given by

$$((a \otimes h) \star (b \otimes k))(x) = (a \star h_{(1)}[b]) \otimes (h_{(2)}k)(x) = (a \star h_{(1)}[b]) \otimes h_{(2)}(x)k(x).$$

Now, defining $h_x(u) = h(ux) = (\Delta h)(u, x) = h_{(1)}(u)h_{(2)}(x)$, we see that the product can be rewritten as

$$((a \otimes h) \star (b \otimes k))(x) = (a \star h_x[b])k(x).$$

Writing $k(x)(v, w) = k(v, w; x)$ for the $\mathcal{K}_\phi(L^2(\mathsf{G}))$-valued function on G, the action of a function h_x is just multiplication by $h_x(vw^{-1}) = h(vw^{-1}x)$ The product of two such kernel-valued functions is therefore

$$(k_1 \star k_2)(u, w; x) = \int k_1(u, v; vw^{-1}x)k_2(v, w; x)\phi(uv^{-1}, vw^{-1}, w)\, dv\,.$$

This can be rewritten as

$$(k_1 \star k_2)(u, w; wx) = \int k_1(u, v; vx)k_2(v, w; wx)\phi(uv^{-1}, vw^{-1}, w)\, dv\,,$$

so that $k_j \mapsto k_j'(u, w; x) = k_j(u, w; wx)$ gives an isomorphism with $C_0(\mathsf{G}) \otimes \mathcal{K}(L^2(\mathsf{G}))$ equipped with componentwise multiplication.

According to the general prescription we can turn this into an associative product by applying Φ^{-1}, that is, multiplying by $\phi(uv^{-1}, vw^{-1}, wx)^{-1}$, to get

$$\begin{aligned}(k_1 * k_2)&(u, w; wx)\\ &= \int k_1(u, v; vx)k_2(v, w; wx)\phi(uv^{-1}, vw^{-1}, wx)^{-1}\phi(uv^{-1}, vw^{-1}, w)\, dv\,,\end{aligned}$$

The cocycle identity tells us that

$$\phi(uw^{-1}, w, x)\phi(uv^{-1}, vw^{-1}, wx) = \phi(uv^{-1}, vw^{-1}, w)\phi(uv^{-1}, v, x)\phi(vw^{-1}, w, x)\,,$$

so that we can rewrite the product as

$$\begin{aligned}(k_1 * k_2)&(u, w; wx)\phi(uw^{-1}, w, x)^{-1}\\ &= \int k_1(u, v; vx)k_2(v, w; wx)\phi(uv^{-1}, v, x)^{-1}\phi(vw^{-1}, w, x)^{-1}\, dv\,.\end{aligned}$$

Setting $k^F(u, w; x) = k(u, w, wx)\phi(uw^{-1}, w, x)^{-1}$ gives

$$(k_1 * k_2)^F(u, w, x) = \int k_1^F(u, v; x)k_2^F(v, w; x)\, dv\,.$$

showing that the new product is isomorphic to the usual product on $C(\mathsf{G}, \mathcal{K}(L^2(\mathsf{G})))$ $\cong C_0(\mathsf{G}) \otimes \mathcal{K}(L^2(\mathsf{G}))$, which is certainly associative.

For future reference we note that the same argument applies to algebra-valued compact operators $\mathcal{K}_\phi(L^2(\mathsf{G})) \otimes \mathcal{A}$ when $\mathcal{A}$ is associative, the only change being that the product of k_j in the composition formula must be interpreted as a product in the algebra rather than of scalars.

This example shows that quite different nonassociative algebras can have the same associative version, since for any cocycle ϕ, $\mathcal{K}_\phi(L^2(\mathsf{G}))$ has $C(\mathsf{G}, \mathcal{K}(L^2(\mathsf{G})))$ as its associative version (including the case of trivial ϕ, when the algebra is associative from the start).

This is important in clarifying the duality theorem [**12**, 9.2], where it was shown that the double dual of $\mathcal{B} = u\text{-ind}_{\mathsf{N}}^{\mathsf{G}}(\mathcal{A})$ is $\mathcal{K}_\phi(L^2(\mathsf{G})) \otimes \mathcal{B}$, since the associative version $\mathcal{K}(L^2(\mathsf{G})) \otimes \mathcal{B}$ is stably equivalent to $\mathcal{B}$, as usual. This certainly means that all the representations (that is, the modules) correspond naturally with those of $\mathcal{B}$. We summarise this in a theorem.

THEOREM A.3. *The dual $\widehat{\mathcal{B}} = \mathcal{B} \rtimes_{\beta,v} \mathsf{G}$ of the algebra $\mathcal{B} = u\text{-}ind_N^{\mathsf{G}}(\mathcal{A})$ is a nonassociative algebra with a natural action of $\widehat{\mathsf{G}}$, and the double dual $\widehat{\mathcal{B}} \rtimes \widehat{\mathsf{G}}$ can be given the associative product $\mathcal{K}(L^2(\mathsf{G})) \otimes u\text{-}ind_N^{\mathsf{G}}(\mathcal{A})$.*

We conclude by noting that the octonions $\mathbb{O}$ can be constructed from $\mathbb{R}$ using $\mathsf{G} = \mathbb{Z}_2 \times \mathbb{Z}_2 \times \mathbb{Z}_2$ with the cocycle

$$\phi(\mathbf{a}, \mathbf{b}, \mathbf{c}) = (-1)^{[\mathbf{a},\mathbf{b},\mathbf{c}]} = (-1)^{\mathbf{a}\cdot(\mathbf{b}\times\mathbf{c})}, \tag{A.1}$$

where G is identified with $\{0,1\}^3 \subset \mathbb{R}^3$. Thus the same procedure can be used to give an associative version of $C(\mathbb{Z}_2 \times \mathbb{Z}_2 \times \mathbb{Z}_2, \mathbb{O})$, as a 64-dimensional algebra over $\mathbb{R}$.

Appendix B. Nonassociative bounded operators, tempered distributions & a concrete approach to nonassociative C^*-algebras

We begin with an illustrative example. Let $\mathsf{G} = \mathbb{R}^n$, and consider the space of all bounded operators $\mathcal{B}(L^2(\mathsf{G}))$ on the Hilbert space $L^2(\mathsf{G})$. We begin by showing that $T \in \mathcal{B}(L^2(\mathsf{G}))$ determines a unique tempered distribution k_T on G^2. That is, there is a canonical embedding, $\mathcal{B}(L^2(\mathsf{G})) \hookrightarrow \mathcal{S}'(\mathsf{G}^2)$. This embedding will be frequently used, for instance to give the algebra $\mathcal{B}(L^2(\mathsf{G}))$ a nonassociative product, which has the advantage of being rather explicit. Later, we will also determine other closely related results. Recall that the Sobolev spaces $H^s(\mathsf{G})$, $s \in \mathbb{R}$, are defined as follows: the Fourier transform on Schwartz functions on G is a topological isomorphism, $\widehat{\ }: \mathcal{S}(\mathsf{G}) \to \mathcal{S}(\mathsf{G})$, where we identify G with its Pontryagin dual group. It extends uniquely to an isometry on square integrable functions on G, $\widehat{\ }: L^2(\mathsf{G}) \to L^2(\mathsf{G})$. Moreover, by duality, the Fourier transform extends to be a topological isomorphism on tempered distributions on G, $\widehat{\ }: \mathcal{S}'(\mathsf{G}) \to \mathcal{S}'(\mathsf{G})$. Then for $s \in \mathbb{R}$, define $H^s(\mathsf{G})$ to be the Hilbert space of all tempered distributions Q such that $(1+|\xi|^2)^{s/2}\widehat{Q}(\xi)$ is in $L^2(\mathsf{G})$, with inner product $\langle Q_1, Q_2\rangle_s = \langle (1+|\xi|^2)^{s/2}\widehat{Q}_1(\xi), (1+|\xi|^2)^{s/2}\widehat{Q}_2(\xi)\rangle_0$, where $\langle\ ,\ \rangle_0$ denotes the inner product on $L^2(\mathsf{G})$.

The following are some basic properties of Sobolev spaces, which are established in any basic reference on distribution theory. For $s < t$, $H^t(\mathsf{G}) \subset H^s(\mathsf{G})$ and moreover the inclusion map $H^t(\mathsf{G}) \hookrightarrow H^s(\mathsf{G})$ is continuous. Also one has $\mathcal{S}(\mathsf{G}) = \bigcap_{s\in\mathbb{R}} H^s(\mathsf{G})$, $\mathcal{S}'(\mathsf{G}) = \bigcup_{s\in\mathbb{R}} H^s(\mathsf{G})$ and the inclusions $\iota_s\colon \mathcal{S}(\mathsf{G}) \hookrightarrow H^s(\mathsf{G})$ and $\kappa_s\colon H^s(\mathsf{G}) \hookrightarrow \mathcal{S}'(\mathsf{G})$ are continuous for any $s \in \mathbb{R}$. The renowned Schwartz kernel theorem says that a continuous linear operator $T\colon \mathcal{S}(\mathsf{G}) \to \mathcal{S}'(\mathsf{G})$ determines a unique tempered distribution k_T on G^2, and conversely.

LEMMA B.1. *There is a canonical embedding,*

$$\mathcal{B}(L^2(\mathsf{G})) \hookrightarrow \mathcal{S}'(\mathsf{G}^2)\,, \tag{B.1}$$

whose image is contained in the subspace of composable tempered distributions.

PROOF. Suppose that $T \in \mathcal{B}(L^2(\mathsf{G}))$. Then in the notation above, the composition

$$\kappa_0 \circ T \circ \iota_0 \colon \mathcal{S}(\mathsf{G}) \to \mathcal{S}'(\mathsf{G})\,, \tag{B.2}$$

is a continuous linear operator. By the Schwartz kernel theorem, it determines a unique tempered distribution $k_T \in \mathcal{S}'(\mathsf{G}^2)$. Suppose now that $S \in \mathcal{B}(L^2(\mathsf{G}))$. Then $ST \in \mathcal{B}(L^2(\mathsf{G}))$ and

$$k_{ST}(x,y) = \int_{z\in\mathsf{G}} k_S(x,z)k_T(z,y)\;dz\,, \tag{B.3}$$

where $\int_{z\in\mathsf{G}} dz$ denotes the distributional pairing. □

We can now define a new product on $\mathcal{B}(L^2(\mathsf{G}))$ making it into a nonassociative C*-algebra.

DEFINITION B.2. Let $\phi \in C(\mathsf{G}\times\mathsf{G}\times\mathsf{G})$ be an antisymmetric tricharacter on G. For $S, T \in \mathcal{B}(L^2(\mathsf{G}))$, define the tempered distribution $k_{S\star T} \in \mathcal{S}'(\mathsf{G}^2)$ by the formula

$$k_{S\star T}(x,y) = \int_{z\in\mathsf{G}} k_S(x,z)k_T(z,y)\phi(x,y,z)\;dz\,. \tag{B.4}$$

Then for all $\xi, \psi \in L^2(\mathsf{G})$, the linear operator $S\star T$ given by the prescription

$$\langle \xi, S\star T\psi\rangle_0 = \int_{x,y\in\mathsf{G}} k_{S\star T}(x,y)\bar{\xi}(x)\psi(y)\;dxdy\,, \tag{B.5}$$

defines a bounded linear operator in $\mathcal{B}(L^2(\mathsf{G}))$, which follows from the earlier observation that $S\star T$ is an adjointable operator.

This extends the definition in [**12**] of twisted compact operators $\mathcal{K}_\phi(L^2(\mathsf{G}))$. Then by §4, $\star$ defines a nonassociative product on $\mathcal{B}(L^2(\mathsf{G}))$ which agrees with the nonassociative product on the twisted compact operators, which will be justified in what follows. We denote by $\mathcal{B}_\phi(L^2(\mathsf{G}))$ the space $\mathcal{B}(L^2(\mathsf{G}))$ endowed with the nonassociative product $\star$, and call it the algebra of twisted bounded operators.

There is an involution $k_{S^*}(x,y) = \overline{k_S(y,x)}$ for all $S \in \mathcal{B}_\phi(L^2(\mathsf{G}))$, and the norm on $\mathcal{B}_\phi(L^2(\mathsf{G}))$ is the usual operator norm. The following are obvious from the definition: $\forall\,\lambda\in\mathbb{C}$, $\forall\,S_1,S_2\in\mathcal{B}_\phi(L^2(\mathsf{G}))$,

$$\begin{aligned} (S_1+S_2)^* &= S_1^*+S_2^*\,,\\ (\lambda S_1)^* &= \bar{\lambda}S_1^*\,,\\ S_1^{**} &= S_1\,. \end{aligned} \tag{B.6}$$

The following lemma can be proved as in Section 5 in [**12**].

LEMMA B.3. $\forall\,S_1,S_2\in\mathcal{B}_\phi(L^2(\mathsf{G}))$,

$$(S_1\star S_2)^* = S_2^*\star S_1^*\,. \tag{B.7}$$

What appears to be missing for the deformed bounded operators $\mathcal{B}_\phi(L^2(\mathsf{G}))$ is the so called C*-identity,

$$||S_1^* \star S_1|| = ||S_1^* S_1|| = ||S_1||^2. \tag{B.8}$$

However, we will continue to call $\mathcal{B}_\phi(L^2(\mathsf{G}))$ a nonassociative C*-algebra and this prompts the following definition of a general class of nonassociative C*-algebras.

DEFINITION B.4. A nonassociative C^*-subalgebra $\mathcal{A}$ of $\mathcal{B}_\phi(L^2(\mathsf{G}))$, is defined to be a G-invariant, $\star$-subalgebra of $\mathcal{B}_\phi(L^2(\mathsf{G}))$ that is closed under taking adjoints and also closed in the operator norm topology.

In particular, such an $\mathcal{A}$ satisfies the identities in equations (B.6) and (B.7). Examples include the algebra of twisted bounded operators $\mathcal{B}_\phi(L^2(\mathsf{G}))$ and the algebra of twisted compact operators $\mathcal{K}_\phi(L^2(\mathsf{G}))$. The following two propositions can be proved as in Section 5 of [**12**].

PROPOSITION B.5. *The group* G *acts on the twisted algebra of bounded operators* $\mathcal{B}_\phi(L^2(\mathsf{G}))$ *by natural* $*$*-automorphisms*

$$\theta_x[k](z,w) = \phi(x,z,w)k(zx,wx), \tag{B.9}$$

and $\theta_x\theta_y = \mathrm{ad}(\sigma(x,y))\theta_{xy}$, *where* $\mathrm{ad}(\sigma(x,y))[k](z,w) = \phi(x,y,z)k(z,w)\phi(x,y,w)^{-1}$ *comes from the multiplier* $\sigma(x,y)(v) = \phi(x,y,v)$.

PROPOSITION B.6. $\mathcal{B}_\phi(L^2(\mathsf{G}))$ *is a continuous deformation of* $\mathcal{B}(L^2(\mathsf{G}))$.

Appendix C. Nonassociative crossed products and nonassociative tori

C.1. Nonassociative tori – revisited. Here will present a slightly different, more geometric, approach to the definition of the nonassociative torus as defined in [**12**], which in fact generalizes the construction there, and also realizes it as a nonassociative deformation of the algebra continuous functions on the torus. We begin with a general construction, and later specialize to the case when M is the torus.

Basic Setup: *Let* M *be a compact manifold with fundamental group* Γ, *and* $\widetilde{M}$ *its universal cover. Assume for simplicity that* $\widetilde{M}$ *is contractible, that is* $M = B\Gamma$ *is the classifying space of* Γ. *In that case we have an isomorphism* $H^n(M,\mathbb{Z}) \cong H^n(\Gamma,\mathbb{Z})$, *due to Eilenberg and MacLane* [**17, 18**].

A large class of examples of manifolds M that satisfy the hypotheses of the Basic Setup are locally symmetric spaces $M = \Gamma\backslash\mathsf{G}/\mathsf{K}$, where G is a Lie group, K a maximal compact subgroup of G, Γ a discrete, torsion-free cocompact subgroup of G, since in this case $\widetilde{M} = \mathsf{G}/\mathsf{K}$ is a contractible manifold. This includes tori and hyperbolic manifolds in particular.

The isomorphism $H^n(M,\mathbb{R}) \to H^n(\Gamma,\mathbb{R})$ can be explicitly constructed by making use of the double complex $(C^{p,q}(\widetilde{M},\Gamma);\delta,d)$, with

$$C^{p,q}(\widetilde{M},\Gamma) = C^p(\Gamma,\Omega^q(\widetilde{M})) = \{f : \Gamma^{\otimes p} \to \Omega^q(\widetilde{M})\}, \tag{C.1}$$

where we think of q-forms on $\widetilde{M}$ as a left Γ-module through the action $\gamma\cdot\omega = (\gamma^*)^{-1}\omega$. The differential $d : C^{p,q}(\widetilde{M},\Gamma) \to C^{p,q+1}(\widetilde{M},\Gamma)$ is the de Rham differential on $\Omega(\widetilde{M})$, hence this complex is acyclic since $\widetilde{M}$ is contractible. The differential $\delta : C^{p,q}(\widetilde{M},\Gamma) \to C^{p+1,q}(\widetilde{M},\Gamma)$ is given by

(C.2)
$$(\delta f)_{\gamma_1,\dots,\gamma_{p+1}} = \gamma_1\cdot f_{\gamma_2,\dots,\gamma_{p+1}} - f_{\gamma_1\gamma_2,\dots,\gamma_{p+1}} + (-1)^p f_{\gamma_1,\dots,\gamma_p\gamma_{p+1}} + (-1)^{p+1} f_{\gamma_1,\dots,\gamma_p}.$$

Its cohomology $H^p(\Gamma, \Omega^q(\widetilde{M}))$ is known as the group cohomology with coefficients in the Γ-module $\Omega(\widetilde{M})$. The map $H(M,\mathbb{R}) \to H(\Gamma,\mathbb{R})$ is now obtained by 'zigzagging' through this double complex, much in the same way as for the Čech-de Rham complex. We will illustrate the procedure in the case of our interest, i.e. how to explicitly associate to a closed degree 3 differential form H on M a $\mathsf{U}(1)$-valued 3-cocycle σ on the discrete group Γ.

First we let $\widetilde{H}$ denote the lift of H to $\widetilde{M}$. Now $\widetilde{H} = dB$, where B is a 2-form on $\widetilde{M}$, i.e. $B \in C^{0,2}(\widetilde{M},\Gamma)$. Since $\widetilde{H}$ is Γ-invariant, it follows that for all $\gamma \in \Gamma$, we have $0 = \gamma \cdot \widetilde{H} - \widetilde{H} = d(\gamma \cdot B - B) = d\delta B$, so that $\gamma \cdot B - B$ is a closed 2-form on $\widetilde{M}$. By hypothesis, it follows that

$$(\delta B)_\gamma = \gamma \cdot B - B = dA_\gamma\,, \tag{C.3}$$

for some 1-form A_γ on $\widetilde{M}$, i.e. $A \in C^{1,1}(\widetilde{M},\Gamma)$. Then by (C.3), it follows that the following identity holds for all $\beta, \gamma \in \Gamma$:

$$d(\beta \cdot A_\gamma - A_{\beta\gamma} + A_\beta) = d\delta A = \delta dA = \delta^2 B = 0\,.$$

By hypothesis, it follows that

$$\beta \cdot A_\gamma - A_{\beta\gamma} + A_\beta = (\delta A)_{\beta,\gamma} = df_{\beta,\gamma}$$

for some smooth function $f_{\beta,\gamma}$ on $\widetilde{M}$, that is, $f \in C^{2,0}(\widetilde{M},\Gamma)$. Continuing, one computes that,

$$d(\alpha \cdot f_{\beta,\gamma} - f_{\alpha\beta,\gamma} + f_{\alpha,\beta\gamma} - f_{\alpha,\beta}) = d\delta f = \delta^2 A = 0\,.$$

Therefore

$$\alpha \cdot f_{\beta,\gamma} - f_{\alpha\beta,\gamma} + f_{\alpha,\beta\gamma} - f_{\alpha,\beta} = (\delta f)_{\alpha,\beta,\gamma} = c(\alpha,\beta,\gamma)\,,$$

where $c(\alpha,\beta,\gamma)$ is a constant. That is, $c : \Gamma \times \Gamma \times \Gamma \to \mathbb{R}$ is a 3-cocycle on Γ, and we can set for all $t \in \mathbb{R}$,

$$\sigma_t(\alpha,\beta,\gamma) = \exp\left(itc(\alpha,\beta,\gamma)\right)\,. \tag{C.4}$$

Then $\sigma_t(\alpha,\beta,\gamma)$ is a $\mathsf{U}(1)$-valued 3-cochain on γ, satisfying the pentagonal identity,

$$\sigma_t(\alpha,\beta,\gamma)\sigma_t(\alpha,\beta\gamma,\delta)\sigma_t(\beta,\gamma,\delta) = \sigma_t(\alpha\beta,\gamma,\delta)\sigma_t(\alpha,\beta,\gamma\delta)\,, \tag{C.5}$$

for all $\alpha,\beta,\gamma,\delta \in \Gamma$. That is, $\sigma_t(\alpha,\beta,\gamma)$ is actually a $\mathsf{U}(1)$-valued 3-cocycle on Γ.

It is convenient to normalize the function $f_{\alpha,\beta}$ such that $f_{\alpha,\beta}(x_0) = 0$ for all $\alpha,\beta \in \Gamma$ and for some $x_0 \in \widetilde{M}$. Then the formula for the $\mathsf{U}(1)$-valued 3-cocycle on Γ simplifies to,

$$\sigma_t(\alpha,\beta,\gamma) = \exp\left(it(\alpha^{*-1} f_{\beta,\gamma}(x_0))\right)\,.$$

Consider the unitary operator $u_t(\beta,\gamma)$ acting on $L^2(\Gamma)$ given by

$$(u_t(\beta,\gamma)\psi)(\alpha) = \exp(it\alpha^{*-1} f_{\beta,\gamma}(x_0))\psi(\alpha) = \sigma_t(\alpha,\beta,\gamma)\psi(\alpha).$$

We easily see that

$$\sigma_t(\alpha,\beta,\gamma)u_t(\alpha,\beta)u_t(\alpha\beta,\gamma) = \xi_\alpha[u_t(\beta,\gamma)]u_t(\alpha,\beta\gamma)$$

where $\xi_\alpha = \mathrm{ad}(\rho(\alpha))$ and $(\rho(\alpha)\psi)(g) = \psi(g\alpha)$ is the right regular representation. One can define, analogous to what was done in [**12**], a twisted convolution product and adjoint on $C(\Gamma,\mathcal{K})$, $\mathcal{K} = \mathcal{K}(L^2(\Gamma))$, by

$$(f *_t g)(x) = \sum_{y\in\Gamma} f(y)\xi_y[g(y^{-1}x)]u_t(y,y^{-1}x)\,, \tag{C.6}$$

and

$$f^*(x) = u_t(x, x^{-1})^{-1}\xi_x[f(x^{-1})]^* . \tag{C.7}$$

The operator norm completion is the nonassociative twisted crossed product C*-algebra

$$C^*(\mathcal{K}, \Gamma, \sigma_t) = \mathcal{K}(L^2(\Gamma)) \rtimes_{\xi, u_t} \Gamma .$$

where as before, $\sigma_t(\alpha, \beta, \gamma)$ is a $\mathsf{U}(1)$-valued 3-cocycle on Γ as above. This construction extends easily to the case when $\mathcal{K}$ is replaced by a general Γ-C*-algebra $\mathcal{A}$, giving rise to a nonassociative C*-algebra denoted by $C^*(\mathcal{A}, \Gamma, \sigma_t) = \mathcal{A} \rtimes_{\xi, u_t} \Gamma$.

In the special case when $M = \mathbb{T}^n$ is a torus, we get the nonassociative torus $A_{\sigma_t}(n)$. Now $A_{\sigma_0}(n)$ is just the ordinary crossed product $\mathcal{K}(L^2(\mathbb{Z}^n)) \rtimes \mathbb{Z}^n$ where $\mathbb{Z}^n$ acts on $\mathcal{K}(L^2(\mathbb{Z}^n))$ via the the adjoint of the left regular representation. By the Stabilization theorem, and using the Fourier transform,

$$\mathcal{K}(L^2(\mathbb{Z}^n)) \rtimes \mathbb{Z}^n \cong C^*(\mathbb{Z}^n) \otimes \mathcal{K}(L^2(\mathbb{Z}^n)) \cong C(\mathbb{T}^n) \otimes \mathcal{K}(L^2(\mathbb{Z}^n)).$$

This then indicates why $A_{\sigma_t}(n)$ is a nonassociative deformation of the ordinary torus $\mathbb{T}^n$ for $t \neq 0$.

EXAMPLE. As an explicit example, consider the 3-torus $M = \mathbb{T}^3$. We have $\widetilde{M} = \mathbb{R}^3$ and $\Gamma = \mathbb{Z}^3$. Let us take, for $H \in H^3(M, \mathbb{R})$, k times the volume form on M (i.e. k times the image in de Rham cohomology of the generator of $H^3(M, \mathbb{Z}) \cong \mathbb{Z}$). Its lift to $\widetilde{M}$ is explicitly given by

$$\widetilde{H} = k\, dx_1 \wedge dx_2 \wedge dx_3 ,$$

where (x_1, x_2, x_3) are standard coordinates on $\widetilde{M} = \mathbb{R}^3$. Let us denote elements of $\Gamma = \mathbb{Z}^3$ by $\mathbf{n} = (n_1, n_2, n_3)$. Going through the procedure above, we see that a representative of this 3-form in group cohomology is given by $c(\mathbf{l}, \mathbf{m}, \mathbf{n}) = k\, l_1 m_2 n_3$. However, by making different choices for $B, A_{\mathbf{n}}$, etc., specifically

$$\begin{aligned} B &= \tfrac{1}{3}k\,(x_1\, dx_2 \wedge dx_3 + \text{cycl}) , \\ A_{\mathbf{n}} &= \tfrac{1}{6}k\ (n_1(x_2 dx_3 - x_3 dx_2) + \text{cycl}) , \\ f_{\mathbf{m},\mathbf{n}} &= \tfrac{1}{6}k\ (m_1(n_2 x_3 - n_3 x_2) + \text{cycl}) , \end{aligned}$$

we can also construct a completely antisymmetric representative, namely

$$c(\mathbf{l}, \mathbf{m}, \mathbf{n}) = \tfrac{1}{6}k\, \mathbf{l} \cdot (\mathbf{m} \times \mathbf{n}) . \tag{C.8}$$

It is this representative which gives rise to an antisymmetric tricharacter σ_t on Γ. Note that the image of an integer cohomology class in $H^3(M, \mathbb{R})$ is not necessarily integer-value, but in this example rather ends up in $\frac{1}{6}\mathbb{Z}$. This example explains and corrects a discrepancy between our earlier paper [**12**] and the physical interpretation of our nonassociative 3-torus in the context of open string theory by Ellwood and Hashimoto (cf. Eqn. (5.14) in [**19**]).

C.2. Factors of automorphy and continuous trace C*-algebras. We next study principal PU-bundles P and associated bundles of compact operators $\mathcal{K}_P$ and their sections over manifolds M that satisfy the assumptions of the Basic Setup of Appendix C.1. Let $\widetilde{P}$ denote the lift of P to $\widetilde{M}$. Since $H^3(\widetilde{M}) = 0$, it follows that $\widetilde{P}$ is trivializable, i.e. $\widetilde{P} \cong \widetilde{M} \times \mathsf{PU}$. Having fixed this isomorphism, we can define a continuous map $j : \Gamma \times \widetilde{M} \to \mathsf{PU} = \mathrm{Aut}(\mathcal{K})$ by the following commutative diagram,

$$\begin{array}{ccc} \mathcal{K} = (\mathcal{K}_{\widetilde{P}})_x & \xrightarrow{\ j(\gamma,x)\ } & \mathcal{K} = (\mathcal{K}_{\widetilde{P}})_{\gamma\cdot x} \\ & {\scriptstyle p}\searrow \quad \swarrow{\scriptstyle p} & \\ & (\mathcal{K}_P)_{p(x)} & \end{array} \tag{C.9}$$

Then

$$j(\gamma_1\gamma_2, x)^{-1} j(\gamma_1, \gamma_2 x) j(\gamma_2, x) = 1. \tag{C.10}$$

is a factor of automorphy for the bundle $\mathcal{K}_P$. Conversely, given a continuous map $j : \Gamma \times \widetilde{M} \to \mathsf{PU} = \mathrm{Aut}(\mathcal{K})$ satisfying (C.10), we can define a bundle of compact operators,

$$\mathcal{K}_j = (\widetilde{M} \times \mathcal{K})/\Gamma \to M \tag{C.11}$$

where $\gamma \cdot (x, \xi) = (\gamma \cdot x, j(\gamma, x)\xi)$ for $\gamma \in \Gamma$ and $(x, \xi) \in \widetilde{M} \times \mathcal{K}$.

Given any two algebra bundles of compact operators, $\mathcal{K}_j, \mathcal{K}_{j'}$ over M with factors of automorphy j, j' respectively, and an isomorphism $\phi : \mathcal{K}_j \longrightarrow \mathcal{K}_{j'}$, we get an induced isomorphism

$$\widetilde{\phi} : \widetilde{M} \times \mathcal{K} = \widetilde{\mathcal{K}}_j \longrightarrow \mathcal{K}_{j'} = \widetilde{M} \times \mathcal{K} \tag{C.12}$$

given by $\widetilde{\phi}(x, \xi) = (x, u(x)\xi)$, where $u : \widetilde{M} \to \mathsf{PU}$ is continuous. Since $\widetilde{\phi}$ commutes with the action of Γ, $\gamma \cdot \widetilde{\phi}(x, \xi) = \widetilde{\phi}(\gamma(x, \xi))$, we deduce that $(\gamma \cdot x, j'(\gamma, x)u(x)\xi) = (\gamma \cdot x, u(\gamma \cdot x) j(\gamma, x)\xi)$. Therefore

$$j'(\gamma, x) = u(\gamma \cdot x) j(\gamma, x) u(x)^{-1} \tag{C.13}$$

for all $x \in \widetilde{M}$ and $\gamma \in \Gamma$. Conversely, two factors of automorphy j, j' give rise to isomorphic algebra bundles $\mathcal{K}_j, \mathcal{K}_{j'}$ of compact operators if they are related by (C.13) for some continuous function $u : \widetilde{M} \to \mathsf{PU}$.

Therefore continuous sections of $\mathcal{K}_P$ can be viewed as continuous maps $f \in C(\widetilde{M}, \mathcal{K})$ satisfying the property,

$$f(\gamma \cdot x) = j(\gamma, x) f(x), \qquad \forall \gamma \in \Gamma,\ x \in \widetilde{M}. \tag{C.14}$$

For example, $f(x) := \sum_{\gamma \in \Gamma} j(\gamma, x)^{-1} F(\gamma \cdot x)$ converges uniformly on compact subsets of $\widetilde{M}$ whenever $F : \widetilde{M} \to \mathcal{K}$ is a compactly supported continuous function, and satisfies (C.14), therefore defining a continuous section of $\mathcal{K}_P$.

We would like to write the Dixmier-Douady invariant of the algebra bundle of compact operators $\mathcal{K}_P$ over M in terms of the factors of automorphy. There is no obstruction to lifting the factor of automorphy $j : \Gamma \times \widetilde{M} \to \mathsf{PU} = \mathrm{Aut}(\mathcal{K})$ to $\widehat{j} : \Gamma \times \widetilde{M} \to \mathsf{U}$, because of our assumptions on $\widetilde{M}$. However the cocycle condition (C.10) has to be modified,

$$\widehat{j}(\gamma_1\gamma_2, x)^{-1} \widehat{j}(\gamma_1, \gamma_2 x) \widehat{j}(\gamma_2, x) = \tau(\gamma_1, \gamma_2, x)\,, \tag{C.15}$$

where $\tau : \Gamma \times \Gamma \times \widetilde{M} \to \mathsf{U}(1)$. There is no obstruction to lifting $\tau : \Gamma \times \Gamma \times \widetilde{M} \to \mathsf{U}(1)$ to $\widehat{\tau} : \Gamma \times \Gamma \times \widetilde{M} \to \mathbb{R}$, however the cocycle condition satified by τ has to be modified to $\delta\widehat{\tau}(\gamma_1,\gamma_2,\gamma_3) = \eta(\gamma_1,\gamma_2,\gamma_3)$, where $\eta : \Gamma\times\Gamma\times\Gamma \to \mathbb{Z}$ is a $\mathbb{Z}$-valued 3-cocycle on Γ. One can show $DD(P) = \delta([\tau']) = [\eta]$. Thus, given a principal PU bundle P on M, we have derived a cohomology class $[\eta] \in H^3(\Gamma,\mathbb{Z}) \cong H^3(M,\mathbb{Z})$ which is by standard arguments independent of the choices made. The relation with the previous discussion is that $[\eta] = [c]$.

To see that the converse is true, notice that τ can be viewed as a continuous map $\tau' : \Gamma\times\Gamma \to C(\widetilde{M},\mathsf{U}(1))$, which is easily verified to be a $C(\widetilde{M},\mathsf{U}(1))$-valued 2-cocycle on Γ. Recall from standard group cohomology theory that equivalence classes of extensions of a group Γ by an abelian group $C(\widetilde{M},\mathsf{U}(1))$ on which Γ acts is in bijective correspondence correspondence with the group cohomology with coefficients, $H^2(\Gamma, C(\widetilde{M},\mathsf{U}(1)))$. We will first show that possible extensions $\widetilde{\Gamma}$ of Γ by $C(\widetilde{M},\mathsf{U}(1))$ are in bijective correspondence with elements of $H^3(M,\mathbb{Z})$ called the Dixmier-Douady invariant, and we will also compute $DD(P) \in H^3(M,\mathbb{Z})$ in our case. Now there is an exact sequence of abelian groups,

$$0 \to \mathbb{Z} \to C(\widetilde{M},\mathbb{R}) \to C(\widetilde{M},\mathsf{U}(1)) \to 0\,. \tag{C.16}$$

This leads to a change of coefficients long exact sequence,

$$\cdots \to H^2(\Gamma, C(\widetilde{M},\mathbb{R})) \to H^2(\Gamma, C(\widetilde{M},\mathsf{U}(1))) \xrightarrow{\delta} H^3(\Gamma,\mathbb{Z}) \to H^3(\Gamma, C(\widetilde{M},\mathbb{R})) \to \cdot \tag{C.17}$$

Since Γ acts freely on $\widetilde{M}$ and $C(\widetilde{M},\mathbb{R})$ is an induced module, it follows that $H^j(\Gamma, C(\widetilde{M},\mathbb{R})) = 0$ for all $j > 0$. Therefore $H^j(\Gamma, C(\widetilde{M},\mathsf{U}(1))) \cong H^{j+1}(\Gamma,\mathbb{Z}) = H^{j+1}(M,\mathbb{Z})$ for all $j > 0$, and in particular for $j = 2$ as claimed. In particular, since $[\tau'] \in H^2(\Gamma, C(\widetilde{M},\mathsf{U}(1)))$, we see that $DD(P) = \delta([\tau']) = [\eta] \in H^3(\Gamma,\mathbb{Z}) = H^3(M,\mathbb{Z})$ is the Dixmier-Douady invariant of P.

We next explain how the data $(H, B, A_\gamma, f_{\alpha,\beta})$ also determine a bundle gerbe in a natural way. The bundle gerbe consists of the submersion $\widetilde{M} \to M$. Then the fibered product $\widetilde{M}^{[2]}$ is equivariantly isomorphic to $\Gamma\times\widetilde{M}$. Under this identification, the two projection maps $\pi_i : \widetilde{M}^{[2]} \to \widetilde{M}$, $i = 1,2$ become the action $\mu : \Gamma\times\widetilde{M} \to \widetilde{M}$ of Γ on $\widetilde{M}$ and the projection $p : \widetilde{M}\times\Gamma \to \widetilde{M}$, respectively. Then A_γ defines a connection on the trivial line bundle $\mathcal{L}_\gamma \to \{\gamma\}\times\widetilde{M}$ whose curvature is dA_γ. The choice of curving is B, satisfying the equation $\mu^*B - p^*B = dA$, which on the sheet $\{\gamma\}\times\widetilde{M}$ reduces to $\gamma^*B - B = dA_\gamma$. The 3-curvature $dB = \widetilde{H}$ is the lift of the closed 3-form H on M. What is surprising is that H need not have integral periods!

Appendix D. Motivation from T-duality in String Theory

For completeness we have summarized the original motivation for this work, namely T-duality in string theory, in this appendix. We believe, however, that the results in this paper are of interest independent of our original motivation.

T-duality, also known as target space duality, plays an important role in string theory and has been the subject of intense study for many years. In its most basic form, T-duality relates a string theory compactified on a circle of radius R, to a string theory compactified on the dual circle of radius $1/R$ by the interchange of the string momentum and winding numbers. T-duality can be generalized to locally defined circles (principal circle bundles, circle fibrations), higher rank torus bundles or fibrations, and in the presence of a background H-flux which is represented by a closed, integral Čech 3-cocycle H on the spacetime manifold Y. It is closely related to mirror symmetry through the SYZ-mechanism.

An amazing feature of T-duality is that it can relate topologically distinct spacetime manifolds by the interchange of topological characteristic classes with components of the H-flux. Specifically, denoting by $(Y, [H])$ the pair of an (isomorphism class of) principal circle bundle $\pi : Y \to X$, characterized by the first Chern class $[F] \in H^2(X, \mathbb{Z})$ of its associated line bundle, and an H-flux $[H] \in H^3(Y, \mathbb{Z})$, the T-dual again turns out to be a pair $(\widehat{Y}, [\widehat{H}])$, where the principal circle bundles

$$\begin{array}{ccc} \mathbb{T} & \longrightarrow & Y \\ & & \downarrow{\scriptstyle \pi} \\ & & X \end{array}, \qquad \begin{array}{ccc} \mathbb{T} & \longrightarrow & \widehat{Y} \\ & & \downarrow{\scriptstyle \widehat{\pi}} \\ & & X \end{array}$$

are related by $[\widehat{F}] = \pi_*[H]$, $[F] = \widehat{\pi}_*[\widehat{H}]$, such that on the correspondence space

$$\begin{array}{ccccc} & & Y \times_X \widehat{Y} & & \\ & \swarrow^{p} & & \searrow^{\widehat{p}} & \\ Y & & & & \widehat{Y} \\ & \searrow_{\pi} & & \swarrow_{\widehat{\pi}} & \\ & & X & & \end{array}$$

we have $p^*[H] - \widehat{p}^*[\widehat{H}] = 0$ [**8, 9**].

In earlier papers we have argued that the twisted K-theory $K^\bullet(Y, [H])$ (see, e.g., [**7**]) classifies charges of D-branes on Y in the background of H-flux $[H]$ [**13**], and indeed, as a consistency check, one can prove that T-duality gives an isomorphism of twisted K-theory (and the closely related twisted cohomology $H^\bullet(Y, [H])$ by means of the twisted Chern character ch_H) [**8**]

$$\begin{array}{ccc} K^\bullet(Y, [H]) & \xrightarrow{T_!} & K^{\bullet+1}(\widehat{Y}, [\widehat{H}]) \\ \downarrow{\scriptstyle ch_H} & & \downarrow{\scriptstyle ch_{\widehat{H}}} \\ H^\bullet(Y, [H]) & \xrightarrow{T_*} & H^{\bullet+1}(\widehat{Y}, [\widehat{H}]) \end{array}$$

The above considerations were generalized to principal torus bundles in [**10, 11**].

Since the projective unitary group of an infinite dimensional Hilbert space $\mathsf{PU}(\mathcal{H})$ is a model for $K(\mathbb{Z}, 2)$, we can 'geometrize' the H-flux in terms of an (isomorphism class of) principal $\mathsf{PU}(\mathcal{H})$-bundle P over Y. We can reformulate the discussion of T-duality above in terms of noncommutative geometry as follows. The space of continuous sections vanishing at infinity, $\mathcal{A} = C_0(Y, \mathcal{E})$, of the associated algebra bundle of compact operator $\mathcal{K}$ on the Hilbert space $\mathcal{E} = P \times_{\mathsf{PU}(\mathcal{H})} \mathcal{K}$. $\mathcal{A}$ is a stable, continuous-trace, C*-algebra with spectrum Y, and has the property that it is locally Morita equivalent to continuous functions on Y. Thus the H-flux has the effect of making spacetime noncommutative. The K-theory of $\mathcal{A}$ is just the twisted K-theory $K^\bullet(Y, [H])$. The $\mathbb{T}$-action on Y lifts essentially uniquely to an $\mathbb{R}$-action on $\mathcal{A}$. In this context T-duality is the operation of taking the crossed product $\mathcal{A} \rtimes \mathbb{R}$, which turns out to be another continuous trace algebra associated to $(\widehat{Y}, [\widehat{H}])$ as above. A fundamental property of T-duality is that when applied twice, yields the original algebra $\mathcal{A}$, and the reason that it works in this case is due to Takai duality. The isomorphism of the D-brane charges in twisted K-theory is, in this context, due to the Connes-Thom isomorphism. These methods have been generalized to principal torus bundles by Mathai and Rosenberg [**31, 32, 33**], however novel features arise. First of all the $\mathbb{T}^n$-action on the principal torus bundle Y need not always lift to an $\mathbb{R}^n$-action on $\mathcal{A}$. Even if it does, this lift need not be unique. Secondly, the crossed product $\mathcal{A} \rtimes \mathbb{R}^n$ need not be a continuous-trace algebra, but rather, it could be a continuous field of noncommutative tori [**31**], and necessary and sufficient conditions are given when these T-duals occur. More generally, as argued in [**10**], when the $\mathbb{T}^n$-action on the principal torus bundle Y does not lift to an $\mathbb{R}^n$-action on $\mathcal{A}$, one has to leave the category of C^*-algebras in order to be able to define a "twisted" lift. The associator in this case is the restriction of the H-flux H to the torus fibre of Y, and the "twisted" crossed product is defined to be the T-dual. The fibres of the T-dual are noncommutative, nonassociative tori. That this is a proper definition of T-duality is due to our results which show that the analogs of Takai duality and the Connes-Thom isomorphism hold in this context. Thus an appropriate context to describe nonassociative algebras that arise as T-duals of spacetimes with background flux, such as nonassociative tori, is that of C*-algebras in tensor categories, which is the subject of this paper.

References

[1] Akrami, SE, Majid, S, Braided cyclic cocycles and nonassociative geometry, J. Math Phys. 45 (2004), no. 10, 3883–3911, [arXiv:math.QA/0406005].

[2] Albuquerque, H, Majid, S, New approach to octonions and Cayley algebras. Nonassociative algebra and its applications (Sao Paulo, 1998), 1–7, Lecture Notes in Pure and Appl. Math., 211, Dekker, New York, 2000, [arXiv:math.QA/9810037].

[3] Baez, JC, The octonions, Bull. Amer. Math. Soc. (N.S.) 39 (2002), no. 2, 145–205. Errata, Bull. Amer. Math. Soc. (N.S.) 42 (2005), no. 2, 213.

[4] Beggs, EJ, Majid, S, Poisson-Lie T-duality for quasitriangular Lie bialgebras, Comm. Math. Phys. 220 (2001), no. 3, 455–488, [arXiv:math.QA/9906040].

[5] Beggs, EJ, Majid, S, Quantization by cochain twists and nonassociative differentials, [arXiv:math.QA/0506450].

[6] Beggs, EJ, Majid, S, Bar categories and star operations, Algebras and Representation Theory, to appear, [arXiv:math.QA/0701008].

[7] Bouwknegt, P, Carey, A, Mathai, V, Murray M, Stevenson, D, Twisted K-theory and K-theory of bundle gerbes, Comm. Math. Phys. 228 (2002) 17-45, [arXiv:hep-th/0106194].

[8] Bouwknegt, P, Evslin, J, Mathai, M, T-duality: Topology change from H-flux, Commun. Math. Phys. 249 (2004) 383-415, [arXiv:hep-th/0306062].

[9] Bouwknegt, P, Evslin, J, Mathai, M, On the topology and H-flux of T-dual manifolds, Phys. Rev. Lett. 92 (2004) 181601, [arXiv:hep-th/0312052].

[10] Bouwknegt, P, Hannabuss, K, Mathai, V, T-duality for principal torus bundles, J. High Energy Phys. 03 (2004) 018, [arXiv:hep-th/0312284].

[11] Bouwknegt, P, Hannabuss, K, Mathai, V, T-duality for principal torus bundles and dimensionally reduced Gysin sequences, Adv. Theor. Math. Phys. 9 (2005) 749-773, [arXiv:hep-th/0412268].

[12] Bouwknegt, P, Hannabuss, KC, Mathai, V, Nonassociative tori and applications to T-duality, Comm. Math. Phys. 264 (2006) 41–69, [arXiv:hep-th/0412092].

[13] Bouwknegt, P, Mathai, V, D-branes, B-fields and twisted K-theory, J. High Energy Phys. 03 (2000) 007, [arXiv:hep-th/0002023].

[14] Brain, S, Majid, S, Quantisation of twistor theory by cocycle twist, Comm. Math. Phys. 284 (2008) 713–774, [arXiv:math.QA/0701893].
[15] Connes, A, An analogue of the Thom isomorphism for crossed products of a C^*-algebra by an action of $\mathbb{R}$, Adv. in Math. 39, no. 1 (1981), 31–55.
[16] Connes, A, Noncommutative Geometry, (New York, Academic Press, 1994).
[17] Eilenberg, S, MacLane, S, Relations between homotopy and homology groups, Proc. Nat. Acad. Sc. USA 29 (1943), 155–158.
[18] Eilenberg, S, MacLane, S, Relations between homotopy and homology groups of spaces, Ann. Math. 46 (1945) 480–509.
[19] Ellwood, I, Hashimoto, A, Effective descriptions of branes on non-geometric tori, J. High Energy Phys. 12 (2006) 025, [arXiv:hep-th/0607135].
[20] Enock, M, Schwartz, J-M, Kac algebras and duality of locally compact groups. With a preface by Alain Connes. With a postface by Adrian Ocneanu. Springer-Verlag, Berlin, 1992.
[21] Enock, M, Vallin, J-M, C^*-algèbres de Kac et algèbres de Kac, Proc. London Math. Soc. (3) 66 (1993), no. 3, 619–650.
[22] Fjelstad, J, Fuchs, J, Runkel, I, Schweigert, C, TFT construction of RCFT correlators V: Proof of modular invariance and factorisation, Theor. Appl. Categ. 16 (2006) 342–433, [arXiv:hep-th/0503194].
[23] Fröhlich, J, Fuchs, J, Runkel, I, Schweigert, C, Algebras in tensor categories and coset conformal field theories, Proceedings of the 36th International Symposium Ahrenshoop on the Theory of Elementary Particles. Fortschr. Phys. 52 (2004), no. 6-7, 672–677.
[24] Fuchs, J, Runkel, I, Schweigert, C, TFT construction of RCFT correlators. I. Partition functions, Nuclear Phys. B 646 (2002), no. 3, 353–497, [arXiv:hep-th/0204148].
[25] Fuchs, J, Runkel, I, Schweigert, C, TFT construction of RCFT correlators. II. Unoriented world sheets, Nuclear Phys. B 678 (2004), no 3, 511–637, [arXiv:hep-th/0306164].
[26] Fuchs, J, Runkel, I, Schweigert, C, TFT construction of RCFT correlators. III. Simple currents, Nuclear Phys. B 694 (2004), no. 3, 277–353, [arXiv:hep-th/0403157].
[27] Fuchs, J, Runkel, I, Schweigert, C, TFT construction of RCFT correlators. IV. Structure constants and correlation functions, Nuclear Phys. B 715 (2005) 539–638, [arXiv:hep-th/0412290].
[28] Ghez, P, Lima, R, Roberts, JE, W^*-categories, Pacific J. Math. 120 (1985), no. 1, 79–109.
[29] Kassel, Christian Quantum groups. Graduate Texts in Mathematics, 155, Springer-Verlag, New York, 1995. xii+531 pp. ISBN: 0-387-94370-6
[30] MacLane, S, Categories for the working mathematician, Graduate Texts in Mathematics, Vol. 5. Springer-Verlag, New York-Berlin, 1971. ix+262 pp.
[31] Mathai, V, Rosenberg, J, T-duality for torus bundles via noncommutative topology, Comm. Math. Phys. 253 (2005) 705-721, [arXiv:hep-th/0401168].
[32] Mathai, V, Rosenberg, J, On mysteriously missing T-duals, H-flux and the T-duality group, in "Differential Geometry and Physics", editors Mo-Lin Ge and Weiping Zhang, Nankai Tracts in Mathematics, Volume 10, World Scientific 2006, pages 350-358, [arXiv:hep-th/0409073].
[33] Mathai, V, Rosenberg, J, T-duality for torus bundles with H-fluxes via noncommutative topology, II: the high-dimensional case and the T-duality group, Adv. Theor. Math. Phys. 10 (2006) 123-158, [arXiv:/hep-th/0508084].
[34] Mitchener, PD, C*-categories, groupoid actions, equivariant KK-theory, and the Baum-Connes conjecture, J. Funct. Anal. 214 (2004), no. 1, 1–39, [arXiv:math.KT/0204291].
[35] Nesterov, AI, Three-cocycles, nonassociative gauge transformations and Dirac's monopole, Phys. Lett. A 328 (2004), no. 2-3, 110–115, [arXiv:hep-th/0406073].
[36] Nesterov, AI, Towards nonassociative gauge theories. Fourth Workshop on Gravitation and Mathematical Physics (Chapala, 2001), Rev. Mexicana Fs. 49 (2003), suppl. 2, 52–57.
[37] Nesterov, AI, Principal loop bundles: toward nonassociative gauge theories. Clifford algebras and their applications (Ixtapa, 1999), Internat. J. Theoret. Phys. 40 (2001), no. 1, 339–350.
[38] Nesterov, AI, Sabinin, LV, Nonassociative geometry: towards discrete structure of spacetime. Phys. Rev. D (3) 62 (2000), no. 8, 081501, 5 pp, [arXiv:hep-th/0010159].
[39] Nesterov, AI, Principal Q-bundles. Nonassociative algebra and its applications (Sao Paulo, 1998), 247–257, Lecture Notes in Pure and Appl. Math., 211, Dekker, New York, 2000.
[40] Raeburn, I, Williams, DP, Morita equivalence and continuous-trace C^*-algebras, Mathematical Surveys and Monographs, **60**. American Mathematical Society, Providence, RI, 1998.

Department of Theoretical Physics, Research School of Physics and Engineering, and Department of Mathematics, Mathematical Sciences Institute, The Australian National University, Canberra, ACT 0200, Australia
E-mail address: peter.bouwknegt@anu.edu.au

Mathematical Institute, 24-29 St. Giles', Oxford, OX1 3LB, and Balliol College, Oxford, OX1 3BJ, England
E-mail address: kch@balliol.oxford.ac.uk

Department of Pure Mathematics, University of Adelaide, Adelaide, SA 5005, Australia
E-mail address: mathai.varghese@adelaide.edu.au

Clay Mathematics Proceedings
Volume 12, 2010

Structural Relations of Harmonic Sums and Mellin Transforms at Weight $w = 6$

Johannes Blümlein

ABSTRACT. We derive the structural relations between nested harmonic sums and the corresponding Mellin transforms of Nielsen integrals and harmonic polylogarithms at weight $\mathsf{w} = 6$. They emerge in the calculations of massless single–scale quantities in QED and QCD, such as anomalous dimensions and Wilson coefficients, to 3– and 4–loop order. We consider the set of the multiple harmonic sums at weight six without index $\{-1\}$. This restriction is sufficient for all known physical cases. The structural relations supplement the algebraic relations, due to the shuffle product between harmonic sums, studied earlier. The original number of 486 possible harmonic sums contributing at weight $\mathsf{w} = 6$ reduces to 99 sums with no index $\{-1\}$. Algebraic and structural relations lead to a further reduction to 20 basic functions. These functions supplement the set of 15 basic functions up to weight $\mathsf{w} = 5$ derived formerly. We outline an algorithm to obtain the analytic representation of the basic sums in the complex plane.

1. Introduction

Inclusive and semi-inclusive scattering cross sections in Quantum Field Theories such as Quantum Electrodynamics (QED) and Quantum Chromodynamics (QCD) at higher loop order can be expressed in terms of special classes of fundamental numbers and functions. Zero scale quantities, like the loop-expansion coefficients for renormalized couplings and masses in massless field theories, are given by special numbers, which are the multiple ζ-values [**1, 2**] in the known orders. At higher orders and in the massive case other quantities will also contribute [**3**]. The next class of interest is comprised by the single scale quantities to which the anomalous dimensions and Wilson coefficients belong [**4–6**], as also other hard scattering cross sections which are differential in one variable $z = \hat{L}/L$ given by the ratio of two Lorentz invariants with support $z \in [0, 1]$. A natural way to study these quantities consists in representing them in Mellin space by performing the integral transform

$$\mathbf{M}\left[f(z)\right](N) = \int_0^1 dz\ z^N\ f(z)\ . \tag{1.1}$$

2010 *Mathematics Subject Classification.* Primary: 33B30, Secondary: 81T15, 81V05.
Key words and phrases. Harmonic Sums, Special Relations, Mellin Transforms.

In the light-cone expansion [**7**] these quantities naturally emerge as moments for physical reasons with $N \in \mathbf{N}$. Their mathematical representation is obtained in terms of nested harmonic sums [**8–10**]

$$S_{b,\vec{a}}(N) = \sum_{k=1}^{N} \frac{(\text{sign}b)^k}{k^{|b|}} S_{\vec{a}}(k), \;\; S_{\emptyset}(k) = 1 \;, \tag{1.2}$$

which form a common language. This is the main reason to adopt this prescription also for other quantities of this kind. The harmonic sums lead to the multiple ζ-values in the limit $N \to \infty$ for $b \neq 1$. In the latter case the harmonic sums diverge. [1] To obtain a representation which is as compact as possible we seek to find all relations between the harmonic sums. There are two classes of relations :
i) the algebraic relations, cf. [**11**]. They are due to the index set of the harmonic sums only and result from their quasi–shuffle algebra [**12**].
ii) the structural relations. These relations depend on the other properties of the harmonic sums. One sub-class refers to relations which are obtained by considering harmonic sums at N and integer multiples or fractions of N, which leads to a continuation of $N \in \mathbf{Q}$. Harmonic sums can be represented in terms of Mellin integrals of harmonic polylogarithms $H_{\vec{a}}(z)$ weighted by $1/(1 \pm z)$ [**13**], which belong to the Poincaré–iterated integrals [**14**]. [2] The Mellin integrals are valid for $N \in \mathbf{R}, N \geq N_0$. From these representations integration-by-parts relations can be derived. Furthermore, there is a large number of differentiation relations

$$\frac{d^l}{dN^l} \mathbf{M}\left[f(z)\right](N) = \mathbf{M}\left[\ln^l(z) f(z)\right](N) \;. \tag{1.3}$$

We analyzed a wide class of physical single scale massless processes and those containing a single mass scale at two and three loops [**4–6**] in the past, which led to the same set of basic harmonic sums and, related to it, basic Mellin transforms. As in the case of zero scale quantities, this points to a unique representation, which is generally process independent and is rather related to the topological structure of the contributing Feynman integrals. The representation in terms of harmonic sums is usually more compact than a corresponding representation by harmonic polylogarithms, since *i*) Mellin convolutions emerge as simple products; *ii*) harmonic polylogarithms are multiple integrals, which are usually not reducible to more compact analytic representations. The latter one requires to solve (part of) these integrals analytically. In the case of harmonic sums the analytic continuation of their argument N to complex values has to be performed to apply them in physics problems. As outlined in Ref. [**16–18**] this is possible since harmonic sums can be represented in terms of factorial series [**19**] up to known algebraic terms. Harmonic sums turn out to be meromorphic functions with single poles at the non-positive integers. One may derive their asymptotic representation analytically and they obey recursion relations for complex arguments N. Due to this their unique representation is given in the complex plane.

In the present paper we derive the structural relations of the weight $\mathsf{w} = 6$ harmonic sums extending earlier work on the structural relations of harmonic sums

---

[1]Due to the algebraic relations [**11**] of the harmonic sums one may show that this divergence is at most of $O(\ln^m(N))$, where m is the number of indices equal to one at the beginning of the index set.

[2]Generalized polylogarithms and Z-sums were considered in [**15**].

up to weight $\mathsf{w} = 5$ [**18**]. [3] The paper is organized as follows. In Sections 2–6 we derive the structural relations of the harmonic sums of weight $\mathsf{w} = 6$ of depth 2 to 6 for the harmonic sums not containing the index $\{-1\}$. The restriction to this class of functions is valid in the massless case at least to three-loop order and in the massive case to two-loop order. In Section 7 we summarize the set of basic functions chosen. The principal method to derive the analytic continuation of the harmonic sums to complex values of N is outlined in Section 8 in an example. Section 9 contains the conclusions. Some useful integrals are summarized in the appendix.

2. Twofold Sums

The following $\mathsf{w} = 6$ two–fold sums occur : $S_{\pm 5,1}(N), S_{\pm 4,\pm 2}(N), S_{-3,3}(N)$ and $S_{3,3}(N), S_{-3,-3}(N)$. The latter sums are related to single harmonic sums through Euler's relation.

$$S_{a,b}(N) + S_{b,a}(N) = S_a(N)S_b(N) + S_{a\wedge b}(N) \ , \tag{2.1}$$

with $a \wedge b = \mathrm{sign}(a) \cdot \mathrm{sign}(b)(|a| + |b|)$. For the former six sums we only consider the algebraically irreducible cases. In Ref. [**18**] the basic functions, which determine the harmonic sums without index $\{-1\}$ through their Mellin transform, up to $\mathsf{w} = 5$ were found :

$$\mathsf{w} = 1 : 1/(x-1) \tag{2.2}$$

$$\mathsf{w} = 2 : \ \ln(1+x)/(x+1) \tag{2.3}$$

$$\mathsf{w} = 3 : \mathrm{Li}_2(x)/(x \pm 1) \tag{2.4}$$

$$\mathsf{w} = 4 : \mathrm{Li}_3(x)/(x+1), \qquad S_{1,2}(x)/(x \pm 1) \tag{2.5}$$

$$\begin{aligned} \mathsf{w} = 5 : \mathrm{Li}_4(x)/(x \pm 1), \qquad & S_{1,3}(x)/(x \pm 1), \qquad S_{2,2}(x)/(x \pm 1), \\ \mathrm{Li}_2^2(x)/(x \pm 1), \qquad & [\ln(x)S_{1,2}(-x) - \mathrm{Li}_2^2(-x)/2]/(x \pm 1) \end{aligned} \tag{2.6}$$

In the following we determine the corresponding basic functions for $\mathsf{w} = 6$.

In case of the double sums we show that they all can be related to

$$\mathbf{M}\left[\frac{\mathrm{Li}_5(x)}{1+x}\right](N) \tag{2.7}$$

up to derivatives of basic functions of lower degree and polynomials of known harmonic sums. The representation of $S_{\pm 5,1}(N)$ read:

$$\begin{aligned} S_{5,1}(N) = \mathbf{M}\left[\left(\frac{\mathrm{Li}_5(x)}{x-1}\right)_+\right](N) - S_1(N)\zeta(5) + S_2(N)\zeta(4) \\ - S_3(N)\zeta(3) + S_4(N)\zeta(2) \end{aligned} \tag{2.8}$$

$$\begin{aligned} S_{-5,1}(N) = (-1)^N \mathbf{M}\left[\frac{\mathrm{Li}_5(x)}{1+x}\right](N) + \frac{15}{16}\zeta(5)\ln(2) - s_6 - S_{-1}(N)\zeta(5) \\ + S_{-2}(N)\zeta(4) - S_{-3}(N)\zeta(3) + S_{-4}(N)\zeta(2) \ , \end{aligned} \tag{2.9}$$

with

$$\int_0^1 dx g(x)[f(x)]_+ = \int_0^1 dx[g(x) - g(1)]f(x) \tag{2.10}$$

[3] We correct a typographical error in [**18**]. The bracket in (5.10) has to close before $\zeta(k)$.

and

$$s_6 = \frac{15}{16}\ln(2)\zeta(5) + \int_0^1 dz \frac{\mathrm{Li}_5(z)}{1+z} \tag{2.11}$$

being one of the basic constants at weight $\mathsf{w} = 6$. For the determination of the constants in the alternating case we use the tables associated to Ref. [**10**]. To express one of the sums given below we also give a second representation of $S_{-5,1}(N)$,

(2.12)

$$\begin{aligned} S_{-5,1}(N) &= S_{-5}(N)S_1(N) + S_{-6}(N) \\ &+ (-1)^{(N+1)}\,\mathbf{M}\left[\frac{\mathrm{Li}_5(-x) - \ln(x)\mathrm{Li}_4(-x) + \ln^2(x)\mathrm{Li}_3(-x)/2}{x+1}\right](N) \\ &+ (-1)^{(N+1)}\,\mathbf{M}\left[\frac{-\ln^3(x)\mathrm{Li}_2(-x)/6 - \ln^4(x)\ln(1+x)/24}{x+1}\right](N) \\ &- \frac{15}{16}\zeta(5)\left[S_{-1}(N) - S_1(N)\right] - \frac{23}{70}\zeta(2)^3 + \frac{3}{4}\zeta(3)^2 + \frac{23}{8}\zeta(5)\ln(2) - s_6\ . \end{aligned}$$

The other two-fold sums are

(2.13)

$$\begin{aligned} S_{-4,-2}(N) &= -\,\mathbf{M}\left[\left(\frac{4\mathrm{Li}_5(-x) - \ln(x)\mathrm{Li}_4(-x)}{x-1}\right)_+\right](N) \\ &+ \frac{1}{2}\zeta(2)\left[S_4(N) - S_{-4}(N)\right] - \frac{3}{2}\zeta(3)S_3(N) + \frac{21}{8}\zeta(4)S_2(N) \\ &- \frac{15}{4}\zeta(5)S_1(N) \end{aligned}$$

$$\begin{aligned} S_{-4,2}(N) &= (-1)^N\,\mathbf{M}\left[\frac{4\mathrm{Li}_5(x) - \mathrm{Li}_4(x)\ln(x)}{1+x}\right](N) + 2\zeta(3)S_{-3}(N) \\ &- 3\zeta(4)S_{-2}(N) + 4\zeta(5)S_{-1}(N) + \frac{239}{840}\zeta(2)^3 - \frac{3}{4}\zeta(3)^2 \\ &- \frac{15}{4}\zeta(5)\ln(2) + 4s_6 \end{aligned} \tag{2.14}$$

$$\begin{aligned} S_{4,-2}(N) &= \frac{1}{2}\zeta(2)\left[S_{-4}(N) - S_4(N)\right] - \frac{3}{2}\zeta(3)S_{-3}(N) + \frac{21}{8}\zeta(4)S_{-2}(N) \\ &- \frac{15}{4}\zeta(5)S_{-1}(N) + (-1)^{N+1}\,\mathbf{M}\left[\frac{4\mathrm{Li}_5(-x) - \ln(x)\mathrm{Li}_4(-x)}{1+x}\right](N) \\ &- \frac{313}{840}\zeta(2)^3 + \frac{15}{16}\zeta(3)^2 + 4\zeta(5)\ln(2) - 4s_6 \end{aligned} \tag{2.15}$$

(2.16)

$$\begin{aligned} S_{4,2}(N) &= -\,\mathbf{M}\left[\left(\frac{4\mathrm{Li}_5(x) - \ln(x)\mathrm{Li}_4(x)}{x-1}\right)_+\right](N) + 2\zeta(3)S_3(N) - 3\zeta(4)S_2(N) \\ &+ 4\zeta(5)S_1(N) \end{aligned}$$

(2.17)
$$S_{-3,-3}(N) = \mathbf{M}\left[\frac{6\mathrm{Li}_5(-x) - 3\ln(x)\mathrm{Li}_4(-x) + \ln^2(x)\mathrm{Li}_3(-x)/2}{x-1}\right](N) - \frac{3}{4}\zeta(3)\left[S_{-3}(N) - S_3(N)\right] - \frac{21}{8}\zeta(4)S_2(N) + \frac{45}{8}\zeta(5)S_1(N) = \frac{1}{2}\left[S_{-3}^2(N) + S_6(N)\right]$$

(2.18)
$$S_{-3,3}(N) = 3\zeta(4)S_{-2}(N) - 6\zeta(5)S_{-1}(N) + (-1)^{N+1}6\,\mathbf{M}\left[\left(\frac{S_{1,4}(1-x) - \zeta(5)}{1+x}\right)_+\right](N) + (-1)^{N+1}\mathbf{M}\left[\left(\frac{3\ln(x)\left[S_{1,3}(1-x) - \zeta(4)\right] + \ln^2(x)\left[S_{1,2}(1-x) - \zeta(3)\right]/2}{1+x}\right)_+\right](N) - \frac{271}{280}\zeta(2)^3 + \frac{81}{32}\zeta(3)^2 + \frac{45}{8}\zeta(5)\ln(2) - 6s_6$$

(2.19)
$$S_{3,3}(N) = 3\zeta(4)S_2(N) - 6\zeta(5)S_1(N) - 6\,\mathbf{M}\left[\left(\frac{S_{1,4}(1-x) - \zeta(5)}{x-1}\right)_+\right](N) - \mathbf{M}\left[\left(\frac{\ln^2(x)\left[S_{1,2}(1-x) - \zeta(3)\right]/2 + 3\ln(x)\left[S_{1,3}(1-x) - \zeta(4)\right]}{x-1}\right)_+\right](N) = \frac{1}{2}\left[S_3^2(N) + S_6(N)\right] .$$

In the above relations Nielsen integrals, [**20**], given by

(2.20)
$$S_{p,n}(x) = \frac{(-1)^{p+n+1}}{(p-1)!n!}\int_0^1 \frac{dz}{z}\ln^{p-1}(z)\ln^n(1-xz)$$

occur. The corresponding functions $S_{1,k}(1-x)$ are given by

(2.21)
$$S_{1,2}(1-x) = -\mathrm{Li}_3(x) + \log(x)\mathrm{Li}_2(x) + \frac{1}{2}\log(1-x)\log^2(x) + \zeta(3)$$

(2.22)
$$S_{1,3}(1-x) = -\mathrm{Li}_4(x) + \log(x)\mathrm{Li}_3(x) - \frac{1}{2}\log^2(x)\mathrm{Li}_2(x) - \frac{1}{6}\log^3(x)\log(1-x) + \zeta(4)$$

(2.23)
$$S_{1,4}(1-x) = -\mathrm{Li}_5(x) + \ln(x)\mathrm{Li}_4(x) - \frac{1}{2}\ln^2(x)\mathrm{Li}_3(x) + \frac{1}{6}\ln^3(x)\mathrm{Li}_2(x) + \frac{1}{24}\ln^4(x)\ln(1-x) + \zeta(5) .$$

They are used to express the respective sums in terms of the Mellin transforms of basic functions and their derivatives w.r.t. N.

The algebraic relation for $S_{3,3}(N)$ can be used to express $\mathbf{M}[(\mathrm{Li}_5(x)/(x-1))_+](N)$. The Mellin transform in $S_{-3,-3}(N)$ allows one to express $S_{-4,-2}(N)$

and $S_{-4,2}(N)$ through (2.12). $S_{4,2}(N)$ and $S_{-3,3}(N)$ do not contain new Mellin transforms. Therefore the only non-trivial Mellin transform needed to express the double sums at $\mathsf{w} = 6$ is $\mathbf{M}[\mathrm{Li}_5(x)/(1+x)](N)$.

In some of the harmonic sums Mellin transforms of the type

$$\frac{\mathrm{Li}_k(-x)}{x \pm 1}\ . \tag{2.24}$$

contribute. For odd values of $k = 2l+1$ the harmonic sums $S_{1,-(k-1)}(N)$, $S_{-(k-1),1}(N)$ and $S_{-l,-l}(N)$ allow one to substitute the Mellin transforms of these functions in terms of Mellin transforms of basic functions and derivatives thereof.

For even values of k this argument applies to $\mathbf{M}[\mathrm{Li}_k(-x)/(1+x)](N)$ but not to $\mathbf{M}[\mathrm{Li}_k(-x)/(1+x)](N)$. In the latter case one may use the relation

$$\frac{1}{2^{k-2}}\frac{\mathrm{Li}_k(x^2)}{1-x^2} = \frac{\mathrm{Li}_k(x)}{1-x} + \frac{\mathrm{Li}_k(x)}{1+x} + \frac{\mathrm{Li}_k(-x)}{1-x} + \frac{\mathrm{Li}_k(x)}{1+x}\ . \tag{2.25}$$

Since in massless quantum field-theoretic calculations both denominators occur, one may apply this decomposition based on the first two cyclotomic polynomials, cf. [**21**], and the relation between $\mathrm{Li}_k(x^2)$ and $\mathrm{Li}_k(\pm x)$, [**22**]. The corresponding Mellin transforms also require half–integer arguments. In more general situations other cyclotomic polynomials might emerge. The relation

(2.26)
$$\begin{aligned}\frac{1}{2^{k-1}}\mathbf{M}\left[\left(\frac{\mathrm{Li}_k(x^2)}{x^2-1}\right)_+\right]\left(\frac{N-1}{2}\right) &= \mathbf{M}\left[\left(\frac{\mathrm{Li}_k(x)}{x-1}\right)_+\right](N)\\ &+ \mathbf{M}\left[\left(\frac{\mathrm{Li}_k(x)}{x+1}\right)_+\right](N) + \mathbf{M}\left[\left(\frac{\mathrm{Li}_k(-x)}{x-1}\right)_+\right](N)\\ &+ \mathbf{M}\left[\left(\frac{\mathrm{Li}_k(-x)}{x+1}\right)_+\right](N) - \int_0^1 dx \frac{\mathrm{Li}_k(x^2)}{1+x}\end{aligned}$$

determines $\mathbf{M}[\mathrm{Li}_k(-x)/(1+x)](N)$. For $k = 2, 4$ the last integral in (2.26) is given by

$$\int_0^1 dx \frac{\mathrm{Li}_2(x^2)}{1+x} \quad = \quad \zeta(2)\ln(2) - \frac{3}{4}\zeta(3) \tag{2.27}$$

$$\int_0^1 dx \frac{\mathrm{Li}_4(x^2)}{1+x} \quad = \quad \frac{2}{5}\ln(2)\zeta(2)^2 + 3\zeta(2)\zeta(3) - \frac{25}{4}\zeta(5)\ . \tag{2.28}$$

The corresponding relations for $\mathbf{M}[\mathrm{Li}_k(-x)/(1+x)](N)$ are :

(2.29)
$$\begin{aligned}\mathbf{M}\left[\frac{\mathrm{Li}_2(-x)}{x+1}\right](N) &= -\frac{1}{2}\mathbf{M}\left[\left(\frac{\mathrm{Li}_2(x)}{x-1}\right)_+\right]\left(\frac{N-1}{2}\right) + \mathbf{M}\left[\left(\frac{\mathrm{Li}_2(x)}{x-1}\right)_+\right](N)\\ &+ \mathbf{M}\left[\left(\frac{\mathrm{Li}_2(-x)}{x-1}\right)_+\right](N) - \mathbf{M}\left[\frac{\mathrm{Li}_2(x)}{x+1}\right](N)\\ &+ \frac{3}{8}\zeta(3) - \frac{1}{2}\zeta(2)\ln(2)\end{aligned}$$

(2.30)

$$\begin{aligned}\mathbf{M}\left[\frac{\mathrm{Li}_4(-x)}{x+1}\right](N) &= -\frac{1}{8}\mathbf{M}\left[\left(\frac{\mathrm{Li}_4(x)}{x-1}\right)_+\right]\left(\frac{N-1}{2}\right) + \mathbf{M}\left[\left(\frac{\mathrm{Li}_4(x)}{x-1}\right)_+\right](N) \\ &+ \mathbf{M}\left[\left(\frac{\mathrm{Li}_4(-x)}{x-1}\right)_+\right](N) - \mathbf{M}\left[\frac{\mathrm{Li}_4(x)}{x+1}\right](N) \\ &- \frac{1}{20}\zeta(2)^2\ln(2) - \frac{3}{8}\zeta(2)\zeta(3) + \frac{25}{32}\zeta(5)\ .\end{aligned}$$

In the case of $\mathsf{w} = 6$ these relations do not lead to a further reduction of basic functions but are required at lower weights, cf. [**18**].

3. Threefold Sums

The triple sums are :

(3.1)

$$\begin{aligned}S_{4,1,1}(N) &= -\mathbf{M}\left[\left(\frac{S_{3,2}(x)}{x-1}\right)_+\right](N) + S_1(N)(2\zeta(5) - \zeta(2)\zeta(3)) - \frac{\zeta(4)}{4}S_2(N) \\ &+ \zeta(3)S_3(N)\end{aligned}$$

(3.2)

$$\begin{aligned}S_{-4,1,1}(N) &= (-1)^{N+1}\mathbf{M}\left[\frac{S_{3,2}(x)}{1+x}\right](N) + (2\zeta(5) - \zeta(2)\zeta(3))S_{-1}(N) \\ &- \frac{\zeta(4)}{4}S_{-2}(N) + \zeta(3)S_{-3}(N) + \frac{71}{840}\zeta(2)^3 + \frac{1}{8}\zeta(3)^2 - \frac{29}{32}\zeta(5)\ln(2) \\ &- \zeta(2)\zeta(3)\ln(2) + \frac{3}{2}s_6\end{aligned}$$

(3.3)

$$\begin{aligned}S_{-3,-2,1}(N) &= \mathbf{M}\left[\frac{H_{0,0,-1,0,1}(x)}{x-1}\right](N) + \zeta(2)S_{-3,-1}(N) + [S_{-3}(N) - S_3(N)] \\ &\times \left[\zeta(2)\ln(2) - \frac{5}{8}\zeta(3)\right] + \frac{3}{40}\zeta(2)^2 S_2(N) \\ &- \left(\frac{9}{4}\zeta(2)\zeta(3) - \frac{67}{16}\zeta(5)\right)S_1(N)\end{aligned}$$

(3.4)

$$\begin{aligned}S_{-2,-3,1}(N) &= S_{-2}(N)S_{-3,1}(N) + S_{5,1}(N) + S_{-3,-3}(N) - S_{-3,1,-2} \\ &- S_{-3,-2,1}\end{aligned}$$

(3.5)

$$\begin{aligned}S_{1,-2,-3}(N) &= S_{-3}(N)S_{1,-2}(N) + S_{1,5}(N) - S_1(N)S_{-3,-2}(N) - S_{-3,-3}(N) \\ &+ S_{-3,-2,1}(N)\end{aligned}$$

(3.6)

$$\begin{aligned}S_{1,-3,-2}(N) &= S_1(N)S_{-3,-2}(N) + S_{-4,-2}(N) \\ &+ S_{-3,-3}(N) - S_{-3,1,-2}(N) - S_{-3,-2,1}(N)\end{aligned}$$

(3.7)

$$S_{-2,1,-3}(N) = S_{-3,-3}(N) - S_{-3}(N)S_{1,-2}(N) - S_{1,5}(N) + S_1(N)S_{-2,-3}(N) - S_{-2}(N)S_{-3,1}(N) - S_{5,1}(N) + S_{-2,-4}(N) + S_1(N)S_{-3,-2}(N) + S_{-3,1,-2}(N)$$

(3.8)

$$S_{-3,1,-2}(N) = \mathbf{M}\left[\left(\frac{A_1(-x)/2 + S_{3,2}(-x) - S_{2,2}(-x)\ln(x)}{x-1}\right)_+\right](N) - \frac{1}{2}\zeta(2)\left[S_{-3,1}(N) - S_{-3,-1}(N)\right] - \left[\frac{1}{8}\zeta(3) - \frac{1}{2}\zeta(2)\ln(2)\right]\left[S_{-3}(N) - S_3(N)\right] + \frac{1}{8}\zeta(2)^2 S_2(N) + \left[\frac{23}{16}\zeta(5) - \frac{7}{8}\zeta(2)\zeta(3)\right]S_1(N)$$

(3.9)

$$S_{-3,2,1}(N) = (-1)^N\mathbf{M}\left[\frac{2S_{3,2}(x) - A_1(x)/2}{x+1}\right](N) + \zeta(2)S_{-3,1}(N) - \frac{3}{4}\zeta(4)S_{-2}(N) - \left(\frac{11}{2}\zeta(5) - 3\zeta(2)\zeta(3)\right)S_{-1}(N) + \frac{23}{168}\zeta(2)^3 + \frac{59}{64}\zeta(3)^2 + \frac{41}{32}\zeta(5)\ln(2) + \frac{1}{2}\zeta(2)^2\ln^2(2) + \frac{5}{4}\zeta(2)\zeta(3)\ln(2) - \frac{1}{12}\zeta(2)\ln^4(2) - 2\zeta(2)\mathrm{Li}_4\left(\frac{1}{2}\right) - \frac{7}{2}s_6$$

(3.10)

$$S_{2,-3,1}(N) = (-1)^N\mathbf{M}\left[\frac{H_{0,-1,0,0,1}(x)}{1+x}\right](N) - \left(\frac{83}{16}\zeta(5) - \frac{21}{8}\zeta(2)\zeta(3)\right)S_{-1}(N) + \zeta(2)S_{2,-2}(N) - \zeta(3)S_{2,-1}(N) + \left[-\frac{3}{5}\zeta(2)^2 + 2\mathrm{Li}_4\left(\frac{1}{2}\right) + \frac{3}{4}\zeta(3)\ln(2) - \frac{1}{2}\zeta(2)\ln^2(2) + \frac{1}{12}\ln^4(2)\right] \times\left[S_2(N) - S_{-2}(N)\right] - \frac{7}{8}\zeta(2)\zeta(3)\ln(2) - \frac{1}{6}\zeta(2)\ln^4(2) - 4\zeta(2)\mathrm{Li}_4\left(\frac{1}{2}\right) + \zeta(2)^2\ln^2(2) + \frac{11}{12}\zeta(2)^3 + \frac{87}{64}\zeta(3)^2 - \frac{11}{32}\zeta(5)\ln(2) - \frac{5}{2}s_6$$

(3.11)

$$S_{1,2,-3}(N) = S_{-3}(N)S_{1,2}(N) + S_{1,-5}(N) - S_1(N)S_{-3,2}(N) - S_{-3,3}(N) + S_{-3,2,1}(N)$$

(3.12)

$$S_{1,-3,2}(N) = -S_2(N)S_{-3,1}(N) - S_{-5,1}(N) + S_{2,-3,1}(N) + S_1(N)S_{-3,2}(N) + S_{-4,2}(N)$$

(3.13)

$$S_{2,1,-3}(N) = S_{3,-3}(N) - S_{-3}(N)S_{1,2}(N) - S_{1,-5}(N) + S_1(N)S_{2,-3}(N) - S_{2,-3,1}(N) + S_{2,-4}(N) + S_1(N)S_{-3,2}(N) + S_{-3,3}(N) - S_{-3,2,1}(N)$$

(3.14)

$$S_{-3,1,2}(N) = S_2(N)S_{-3,1}(N) + S_{-5,1}(N) + S_{-3,3}(N) - S_{2,-3,1}(N) - S_{-3,2,1}(N)$$

(3.15)

$$S_{-2,3,1}(N) = (-1)^N \mathbf{M}\left[\frac{(3/2)A_1(x) - \mathrm{Li}_2(x)\mathrm{Li}_3(x)}{x+1}\right](N)$$
$$+ \zeta(2)S_{-2,2}(N) - \zeta(3)S_{-2,1}(N)$$
$$+ \left(\frac{9}{2}\zeta(5) - 2\zeta(2)\zeta(3)\right)S_{-1}(N) - \frac{13}{12}\zeta(2)^3 - \frac{43}{32}\zeta(3)^2 + \frac{51}{32}\zeta(5)\ln(2)$$
$$- \zeta(2)^2\ln^2(2) + \frac{3}{2}\zeta(2)\zeta(3)\ln(2) + \frac{1}{6}\zeta(2)\ln^4(2) + 4\zeta(2)\mathrm{Li}_4\left(\frac{1}{2}\right) + \frac{3}{2}s_6$$

(3.16)

$$S_{3,-2,1}(N) = S_{3,-3}(N) - S_{3,1,-2}(N) - S_{-2,3,1}(N) + S_{-2}(N)S_{3,1}(N) + S_{-5,1}(N)$$

(3.17)

$$S_{1,-2,3}(N) = S_{-3,3}(N) - S_3(N)S_{-2,1}(N) - S_{-2,4}(N) + S_{-2}(N)S_{1,3}(N)$$
$$+ S_{1,-5}(N) + 2S_{-2}(N)S_{3,1}(N) + S_{3,-3}(N) - S_1(N)S_{3,-2}(N)$$
$$- S_{4,-2}(N) + S_{-5,1}(N) - S_{3,1,-2}(N) - S_{-2,3,1}(N)$$

(3.18)

$$S_{1,3,-2}(N) = -S_{-2}(N)S_{3,1}(N) - S_{-5,1}(N) + S_{-2,3,1}(N) + S_1(N)S_{3,-2}(N)$$
$$+ S_{4,-2}(N)$$

(3.19)

$$S_{-2,1,3}(N) = S_3(N)S_{-2,1}(N) + S_{-2,4}(N) - S_{-2}(N)S_{3,1}(N)$$
$$- S_{3,-3}(N) + S_{3,1,-2}(N)$$

(3.20)

$$S_{3,1,-2}(N) = (-1)^N \mathbf{M}\left[\frac{A_1(-x)/2 + S_{3,2}(-x) - \ln(x)S_{2,2}(-x)}{1+x}\right](N)$$
$$- \frac{1}{2}\zeta(2)\left[S_{3,1}(N) - S_{3,-1}(N)\right]$$
$$- \left[\frac{1}{8}\zeta(3) - \frac{1}{2}\zeta(2)\ln(2)\right]\left[S_3(N) - S_{-3}(N)\right]$$
$$+ \frac{1}{8}\zeta(2)^2 S_{-2}(N) + \left[\frac{23}{16}\zeta(5) - \frac{7}{8}\zeta(2)\zeta(3)\right]S_{-1}(N)$$
$$+ \frac{113}{560}\zeta(2)^3 - \frac{17}{64}\zeta(3)^2 - \frac{1}{2}\zeta(5)\ln(2) - \frac{7}{8}\zeta(2)\zeta(3)\ln(2) + s_6$$

(3.21)

$$S_{3,2,1}(N) = \mathbf{M}\left[\left(\frac{2S_{3,2}(x) - A_1(x)/2}{x-1}\right)_+\right](N)$$
$$+ \zeta(2)S_{3,1}(N) - \frac{3}{4}\zeta(4)S_2(N) - \left(\frac{11}{2}\zeta(5) - 3\zeta(2)\zeta(3)\right)S_1(N)$$

(3.22)

$$S_{2,3,1}(N) = \mathbf{M}\left[\left(\frac{(3/2)A_1(x) - \mathrm{Li}_2(x)\mathrm{Li}_3(x)}{x-1}\right)_+\right](N)$$
$$+ \zeta(2)S_{2,2}(N) - \zeta(3)S_{2,1}(N) + \left(\frac{9}{2}\zeta(5) - 2\zeta(2)\zeta(3)\right)S_1(N)$$

(3.23)

$$S_{1,2,3}(N) = S_3(N)S_{1,2}(N) + S_{1,5}(N) - S_1(N)S_{3,2}(N) - S_{3,3}(N) + S_{3,2,1}(N)$$

(3.24)

$$S_{2,1,3}(N) = 2S_{3,3}(N) - S_3(N)S_{1,2}(N) - S_{1,5}(N) + S_1(N)S_{2,3}(N) - S_{2,3,1}(N) + S_{2,4}(N) + S_1(N)S_{3,2}(N) - S_{3,2,1}(N)$$

(3.25)

$$S_{1,3,2}(N) = -S_2(N)S_{3,1}(N) - S_{5,1}(N) + S_{2,3,1}(N) + S_1(N)S_{3,2}(N) + S_{4,2}(N)$$

(3.26)

$$S_{3,1,2}(N) = S_2(N)S_{3,1}(N) + S_{5,1}(N) + S_{3,3}(N) - S_{2,3,1}(N) - S_{3,2,1}(N)$$

(3.27)

$$\begin{aligned} S_{2,2,2}(N) = &-\mathbf{M}\left[\left(\frac{2A_1(x) + \mathrm{Li}_2^2(x)\ln(x)/2 - 2S_{2,2}(x)\ln(x)}{x-1}\right)_+\right](N) \\ &- \mathbf{M}\left[\left(\frac{4S_{3,2}(x) - 2\mathrm{Li}_2(x)\mathrm{Li}_3(x)}{x-1}\right)_+\right](N) + 2\zeta(3)S_{2,1}(N) \\ &+ 2\left(\zeta(5) - \zeta(2)\zeta(3)\right)S_1(N) \\ = &\frac{1}{6}S_2^3(N) + \frac{1}{2}S_2(N)S_4(N) + \frac{1}{3}S_6(N) \end{aligned}$$

(3.28)

$$\begin{aligned} S_{-2,2,2}(N) = &(-1)^{(N+1)}\,\mathbf{M}\left[\frac{2A_1(x) + \mathrm{Li}_2^2(x)\ln(x)/2 - 2S_{2,2}(x)\ln(x)}{x+1}\right](N) \\ &+ (-1)^{(N+1)}\,\mathbf{M}\left[\frac{4S_{3,2}(x) - 2\mathrm{Li}_2(x)\mathrm{Li}_3(x)}{x+1}\right](N) \\ &+ 2\zeta(3)S_{-2,1}(N) + 2\left(\zeta(5) - \zeta(2)\zeta(3)\right)S_{-1}(N) \\ &- \frac{1}{4}\zeta(2)\zeta(3)\ln(2) + \frac{1}{12}\zeta(2)\ln^4(2) + 2\zeta(2)\mathrm{Li}\left(\frac{1}{2}\right) - \frac{1}{2}\zeta(2)^2\ln^2(2) \\ &+ \frac{37}{80}\zeta(2)^3 - 2\zeta(3)^2 - \frac{23}{4}\zeta(5)\ln(2) + 4s_6 \\ S_{2,-2,2}(N) = &-2S_{-2,2,2}(N) + S_2(N)S_{-2,2}(N) + S_{-4,2}(N) + S_{-2,4}(N) \end{aligned}$$

(3.29)

$$\begin{aligned} S_{2,2,-2}(N) = &S_{-2,2,2}(N) - \frac{1}{2}\left[S_2(N)S_{-2,2}(N) + S_{-4,2}(N) + S_{-2,4}(N)\right. \\ &\left. - S_2(N)S_{2,-2}(N) - S_{4,-2}(N) - S_{2,-4}(N)\right] \end{aligned}$$

(3.30)

$$\begin{aligned} S_{-2,2,-2}(N) = &\mathbf{M}\left[\left(\frac{-4S_{3,2}(-x) - \ln(x)\mathrm{Li}_2^2(-x)/2 + 2\ln(x)S_{2,2}(-x)}{x-1}\right)_+\right](N) \\ &+ \mathbf{M}\left[\left(\frac{2\mathrm{Li}_3(-x)\mathrm{Li}_2(-x) - 2A_1(-x)}{x-1}\right)_+\right](N) \\ &- \frac{3}{2}\zeta(3)S_{-2,-1}(N) + \frac{1}{2}\zeta(2)\left[S_{-2,-2}(N) - S_{-2,2}(N)\right] \\ &+ \left[-\frac{11}{8}\zeta(2)^2 + 4\mathrm{Li}_4\left(\frac{1}{2}\right) + 2\zeta(3)\ln(2) - \zeta(2)\ln^2(2) + \frac{1}{6}\ln^4(2)\right] \\ &\times\left[S_{-2}(N) - S_2(N)\right] + \left(\frac{11}{4}\zeta(2)\zeta(3) - \frac{23}{4}\zeta(5)\right)S_1(N) \end{aligned}$$

(3.31)
$$S_{2,-2,-2}(N) = \frac{1}{2}\left[-S_{-2,2,-2}(N) + S_{-2}(N)S_{2,-2}(N) + S_{-4,-2}(N) + S_{2,4}(N)\right]$$
(3.32)
$$S_{-2,-2,2}(N) = \frac{1}{2}\left[-S_{-2,2,-2}(N) + S_{-2}(N)S_{-2,2}(N) + S_{4,2}(N) + S_{-2,-4}(N)\right]$$
(3.33)
$$S_{-2,-2,-2}(N) = \frac{1}{6}S_{-2}^3(N) + \frac{1}{2}S_{-2}(N)S_4(N) + \frac{1}{3}S_{-6}(N)~.$$

There emerge numerator functions, which do not belong to the class of Nielsen integrals [4],

$$A_1(x) = \int_0^x \frac{dy}{y}\mathrm{Li}_2^2(y) \tag{3.34}$$
$$A_2(x) = \int_0^x \frac{dy}{y}\ln(1-y)S_{1,2}(y) \tag{3.35}$$
$$A_3(x) = \int_0^x \frac{dy}{y}[\mathrm{Li}_4(1-y) - \zeta(4)]~. \tag{3.36}$$

As seen in Eqs. (3.27), $(A_1(x)/(x-1))_+$ is not a basic function since its Mellin transform reduces to single harmonic sums and known Mellin transforms algebraically. Furthermore, some numerator functions are given by harmonic polylogarithms $H_{a_1,\ldots,a_k}(x)$, $a_i \in \{-1,0,+1\}$, which cannot be significantly reduced further. Harmonic polylogarithms are Poincaré–iterated integrals [**14**] over the alphabet $[f_0, f_1, f_{-1}] = [1/x, 1/(x-1), 1/(x+1)]$, [**13**], with

$$H_0(x) = \ln(x) \tag{3.37}$$
$$H_1(x) = -\ln(1-x) \tag{3.38}$$
$$H_{-1}(x) = \ln(1+x) \tag{3.39}$$

and

$$H_{a,\vec{b}}(x) = \int_0^x dy~f_a(y)H_{\vec{b}}(y)~. \tag{3.40}$$

4. Fourfold Sums

The quadruple–index sums are :

(4.1)
$$\begin{aligned}S_{-3,1,1,1}(N) &= (-1)^N\,\mathbf{M}\left[\frac{S_{2,3}(x)}{x+1}\right](N) + \zeta(4)S_{-2}(N) - (2\zeta(5) - \zeta(2)\zeta(3))S_{-1}(N)\\ &+ \frac{1}{8}\zeta(2)\zeta(3)\ln(2) - \frac{1}{6}\zeta(2)\ln^4(2) - \zeta(2)\mathrm{Li}_4\left(\frac{1}{2}\right) + \frac{1}{4}\zeta(2)^2\ln^2(2)\\ &- \frac{257}{840}\zeta(2)^3 + \frac{7}{24}\zeta(3)\ln^3(2) + \frac{41}{64}\zeta(3)^2 - \frac{33}{32}\zeta(5)\ln(2) + 2\ln(2)\mathrm{Li}_5\left(\frac{1}{2}\right)\\ &+ \ln^2(2)\mathrm{Li}_4\left(\frac{1}{2}\right) + \frac{1}{36}\ln^6(2) + 2\mathrm{Li}_6\left(\frac{1}{2}\right) - \frac{s_6}{2}\end{aligned}$$

[4]Note a misprint in Eq. (14), [**17**]. $\mathrm{Li}_4(y)$ should read $\mathrm{Li}_4(1-y)$.

(4.2)
$$S_{3,1,1,1}(N) = -\mathbf{M}\left[\left(\frac{S_{2,3}(x)}{x-1}\right)_+\right](N) + \zeta(4)S_2(N) - (2\zeta(5) - \zeta(2)\zeta(3))S_1(N)$$

(4.3)
$$\begin{aligned} S_{-2,2,1,1}(N) = (-1)^{N+1}\mathbf{M}\left[\frac{3S_{2,3}(x) + A_2(x)}{x+1}\right](N) + \zeta(3)S_{-2,1}(N) \\ + \left(\frac{11}{2}\zeta(5) - 3\zeta(2)\zeta(3)\right)S_{-1}(N) - \frac{5}{4}\zeta(2)\zeta(3)\ln(2) + \frac{1}{3}\zeta(2)\ln^4(2) \\ + 2\zeta(2)\mathrm{Li}_4\left(\frac{1}{2}\right) - \frac{1}{2}\zeta(2)^2\ln^2(2) + \frac{411}{560}\zeta(2)^3 \\ - \frac{7}{12}\zeta(3)\ln^3(2) - \frac{9}{8}\zeta(3)^2 + \frac{73}{64}\zeta(5)\ln(2) - 4\ln(2)\mathrm{Li}_5\left(\frac{1}{2}\right) \\ - 2\ln^2(2)\mathrm{Li}_4\left(\frac{1}{2}\right) - \frac{1}{18}\ln^6(2) - 4\mathrm{Li}_6\left(\frac{1}{2}\right) + \frac{9}{4}s_6 \end{aligned}$$

(4.4)
$$\begin{aligned} S_{2,-2,1,1}(N) = (-1)^{N+1}\mathbf{M}\left[\frac{H_{0,-1,0,1,1}(x)}{1+x}\right](N) + \zeta(3)S_{2,-1}(N) \\ + \left[\frac{11}{16}\zeta(2)\zeta(3) - \frac{41}{32}\zeta(5)\right]S_{-1}(N) \\ + \left[-\mathrm{Li}_4\left(\frac{1}{2}\right) + \frac{1}{8}\zeta(2)^2 + \frac{1}{8}\zeta(3)\ln(2) + \frac{1}{4}\zeta(2)\ln^2(2) - \frac{1}{24}\ln^4(2)\right] \\ \times [S_2(N) - S_{-2}(N)] - \frac{17}{16}\zeta(2)\zeta(3)\ln(2) \\ - \frac{1}{3}\zeta(2)\ln^4(2) - 2\zeta(2)\mathrm{Li}_4\left(\frac{1}{2}\right) \\ + \frac{1}{2}\zeta(2)^2\ln^2(2) - \frac{87}{280}\zeta(2)^3 + \frac{7}{12}\zeta(3)\ln^3(2) \\ + \frac{105}{128}\zeta(3)^2 - \frac{103}{32}\zeta(5)\ln(2) \\ + 4\ln(2)\mathrm{Li}_5\left(\frac{1}{2}\right) + 2\ln^2(2)\mathrm{Li}_4\left(\frac{1}{2}\right) + \frac{1}{18}\ln^6(2) + 4\mathrm{Li}_6\left(\frac{1}{2}\right) + s_6 \end{aligned}$$

(4.5)
$$\begin{aligned} S_{-2,1,1,2}(N) = \\ (-1)^N\mathbf{M}\left[\frac{A_3(x)}{1+x}\right](N) - (\zeta(2)\zeta(3) - 3\zeta(5))S_{-1}(N) - \zeta(3)S_{-2,1}(N) \\ + \zeta(2)S_{-2,1,1}(N) + \frac{5}{16}\zeta(2)\zeta(3)\ln(2) + \frac{1}{16}\zeta(2)\ln^4(2) + \frac{3}{2}\zeta(2)\mathrm{Li}_4\left(\frac{1}{2}\right) \\ - \frac{3}{8}\zeta(2)^2\ln^2(2) + \frac{11}{120}\zeta(2)^3 - \frac{27}{16}\zeta(3)^2 - \frac{59}{32}\zeta(5)\ln(2) + \frac{5}{2}s_6 \end{aligned}$$

(4.6)

$$
\begin{aligned}
S_{-2,-2,1,1}(N) = &-\mathbf{M}\left[\left(\frac{H_{0,-1,0,1,1}(x)}{x-1}\right)_+\right](N) + \zeta(3)S_{-2,-1}(N) \\
&+ \left(\frac{11}{16}\zeta(2)\zeta(3) - \frac{41}{32}\zeta(5)\right) S_1(N) \\
&+ \left(-\mathrm{Li}_4\left(\frac{1}{2}\right) + \frac{1}{8}\zeta(2)^2 + \frac{1}{8}\zeta(3)\ln(2) + \frac{1}{4}\zeta(2)\ln^2(2) - \frac{1}{24}\ln^4(2)\right) \\
&\times \left[S_{-2}(N) - S_2(N)\right]
\end{aligned}
$$

(4.7)
$$
\begin{aligned}
S_{2,2,1,1}(N) = &-\mathbf{M}\left[\left(\frac{3S_{2,3}(x) + A_2(x)}{x-1}\right)_+\right](N) + \zeta(3)S_{2,1}(N) \\
&+ \left(\frac{11}{2}\zeta(5) - 3\zeta(2)\zeta(3)\right) S_1(N)\ .
\end{aligned}
$$

Here, the harmonic polylogarithm $H_{0,-1,0,1,1}(x)$ is given by

(4.8)
$$
H_{0,-1,0,1,1}(x) = \int_0^x \frac{dy}{y} \int_0^y dz \frac{S_{1,2}(z)}{1+z}\ .
$$

We tested the above sum-relations containing harmonic polylogarithms in the Mellin transforms numerically using the code of Ref. [**23**].

5. Fivefold Sums

Two 5–fold sums contribute :

(5.1)

$$
S_{2,1,1,1,1}(N) = -\mathbf{M}\left[\left(\frac{S_{1,4}(x)}{x-1}\right)_+\right](N) + \zeta(5)S_1(N)
$$

(5.2)

$$
\begin{aligned}
S_{-2,1,1,1,1}(N) = &(-1)^{N+1}\mathbf{M}\left[\frac{S_{1,4}(x)}{1+x}\right](N) + \zeta(5)S_{-1}(N) \\
&+ \frac{7}{16}\zeta(2)\zeta(3)\ln(2) + \frac{1}{12}\zeta(2)\ln^4(2) + \frac{1}{2}\zeta(2)\mathrm{Li}_4\left(\frac{1}{2}\right) - \frac{1}{8}\zeta(2)^2\ln^2(2) \\
&- \frac{7}{48}\zeta(3)\ln^3(2) - \frac{49}{128}\zeta(3)^2 - \ln(2)\mathrm{Li}_5\left(\frac{1}{2}\right) - \frac{1}{2}\ln^2(2)\mathrm{Li}_4\left(\frac{1}{2}\right) \\
&- \frac{1}{72}\ln^6(2) - \mathrm{Li}_6\left(\frac{1}{2}\right) + \ln(2)\zeta(5)\ .
\end{aligned}
$$

All other sums can be traced back to these sums using algebraic relations [**11**]. The other Mellin transforms emerging in their representation were all calculated in Refs. [**9, 18**] before.

6. Sixfold Sums

Only one sixfold sum contributes at $\mathsf{w} = 6$, $S_{1,1,1,1,1,1}(N)$. This sum is completely reducible into a polynomial of single harmonic sums, cf. [**9**],

$$S_{\underbrace{1,\ldots,1}_{6}} = \frac{1}{720}S_1^6 + \frac{1}{48}S_2\,S_1^4 + \frac{1}{18}S_3\,S_1^3 + \frac{1}{8}S_4S_1^2 + \frac{1}{5}S_5\,S_1 + \frac{1}{16}S_1^2S_2^2 + \frac{1}{6}S_1\,S_2\,S_3 + \frac{1}{48}S_2^3 + \frac{1}{8}S_2\,S_4 + \frac{1}{18}S_3^2 + \frac{1}{6}S_6 \qquad (6.1)$$

7. The Basic Functions

In the following we summarize the basic functions the Mellin transforms of which represents the harmonic sums up to weight $\mathsf{w} = 6$ without those carrying an index $\{-1\}$. The corresponding sums of lower weight were determined in Refs. [**8,18,24**]. The 20 new functions are given by

$$\begin{aligned}(7.1)\qquad \mathsf{w}=6: &\ \mathrm{Li}_5(x)/(x+1), \quad S_{1,4}(x)/(x\pm 1), \quad S_{2,3}(x)/(x\pm1),\\ &S_{3,2}(x)/(x\pm1), \quad \mathrm{Li}_2(x)\mathrm{Li}_3(x)/(x\pm1),\\ &A_1(x)/(x+1), \quad A_2(x)/(x\pm1), \quad A_3(x)/(x+1)\\ &H_{0,-1,0,1,1}(x)/(x\pm1), \quad H_{0,0,-1,0,1}(x)/(x\pm1)\\ &[A_1(-x)+2S_{3,2}(-x)-2S_{2,2}(-x)\ln(x)]/(x\pm1)\\ &[A_1(-x)+2S_{3,2}(-x)-S_{2,2}(-x)\ln(x)+\mathrm{Li}_2^2(-x)\ln(x)/4\\ &-\mathrm{Li}_3(-x)\mathrm{Li}_2(-x)]/(x-1)\end{aligned}$$

and extend the set Eqs. (2.2–2.6). The algebraic relations allow one to express the initial set of 99 functions by 30 functions and the structural relations reduce the basis further to 20 functions.

8. Complex Analysis of Harmonic Sums

The anomalous dimensions and Wilson coefficients expressed in Mellin space allow simple representations of the scale evolution of single-scale observables, which are given by ordinary differential equations. The experimental measurement of the observables requires the representation in z–space. Therefore, one has to perform the analytic continuation of harmonic sums to complex values of N. Precise numerical representations for the analytic continuation of the basic functions up to weight $\mathsf{w} = 5$ were derived in [**25**] based on the `MINIMAX`-method [**26**]. One may even obtain corresponding representations for quite general functions $\Phi(z)$, $z \in [0,1]$, as worked out for the heavy flavor Wilson coefficients to 2-loop order in [**27**]. [5] For other effective parameterizations see [**28**].

Here we aim at exact representations. The inverse Mellin transforms are obtained by a contour integral around the singularities of the respective functions in the complex plane.

[5] For another proposal for the analytic continuation of harmonic sums to $N \in \mathbf{R}$, for which some simple examples were presented, cf. [**29**].

We traced back all the harmonic sums to Mellin transforms of basic functions $f_i(z)$,

(8.1)
$$F_i^-(N) = \int_0^1 dz f_i(z) \frac{z^N - 1}{z-1}, \qquad F_i^+(N) = \int_0^1 dz f_i(z) \frac{(-z)^N - 1}{z+1} .$$

Eqs. (8.1) imply the recursion relations

$$F_i^-(N+1) = -F_i^-(N) + \int_0^1 dz z^N f_i(z) , \tag{8.2}$$

$$F_i^+(N+1) = F_i^+(N) + (-1)^{N+1} \int_0^1 dz z^N f_i(z) . \tag{8.3}$$

The remaining integrals are simpler Mellin transforms, which correspond to harmonic sums of lower weight.

If the functions $f_i(z)/(z-1)$, $f_i(z)/(z+1)$ are analytic at $z=1$ the Mellin transforms (8.1) can be represented in terms of factorial series [**19**]. Not all basic functions chosen above have this property. A corresponding analytic relation replacing

$$f_i(z) \rightarrow f_i(1-z) \tag{8.4}$$

always exists. The additional terms are lower weight functions in N or are related to these by differentiation. We use this representation and consider the factorial series. Due to this both the pole–structure and the asymptotic relation for $|N| \rightarrow \infty$ are known. [6] The poles are located at the integers below a fixed value N_0. The recursion relations (8.1) are used to express the respective harmonic sums at any value $N \in \mathbf{C}$ except the poles.

Let us illustrate this representation in an example for the harmonic sum $S_{2,1,1,1,1}(N)$. The corresponding basic function is

$$\left(\frac{S_{1,4}(z)}{z-1}\right)_+ . \tag{8.5}$$

The recursion relation is given by

$$\mathbf{M}\left[\frac{S_{1,4}(z)}{z-1}\right](N+1) = \mathbf{M}\left[\left(\frac{S_{1,4}(z)}{z-1}\right)_+\right](N) + \mathbf{M}\left[S_{1,4}(z)\right](N) , \tag{8.6}$$

with

$$\mathbf{M}\left[S_{1,4}(z)\right](N) = \frac{1}{N+1}\left[\zeta(5) - \frac{1}{N+1} S_{1,1,1,1}(N)\right] , \tag{8.7}$$

cf. [**33**].

The numerator function possesses a branch–point at $z=1$. The contributions related to terms containing $\ln^k(1-z)/(z \pm 1)$ have to be subtracted explicitly due

---

[6]In [**30**] asymptotic relations for non-alternating harmonic sums to low orders in $1/N^k$ were derived. Our algorithm given below is free of these restrictions. The main ideas were presented in January 2004 [**31**], see also [**32**].

to their logarithmic growth (to a power) for $|N| \to \infty$. This is either possible using the relation $S_{1,4}(z)$ to $\mathrm{Li}_5(1-z)$

$$S_{1,4}(z) = -\mathrm{Li}_5(1-z) + \ln(1-z)\mathrm{Li}_4(1-z) - \frac{1}{2}\ln^2(1-z)\mathrm{Li}_3(1-z) + \frac{1}{6}\ln^3(1-z)\mathrm{Li}_2(1-z) + \frac{1}{24}\ln^4(1-z)\ln(z) + \zeta(5) \tag{8.8}$$

or considering harmonic sums, which are algebraically equivalent to the above and are related to a basic function which is regular at $z \to 1$. We will follow the latter way and use the algebraic relations [**11**] to express $S_{2,1,1,1,1}(N)$ afterwards,

$$\begin{aligned} S_{2,1,1,1,1} = S_{1,1,1,1,2} &+ \frac{1}{4}\Big[S_1 S_{2,1,1,1} + S_{3,1,1,1,} + S_{2,2,1,1} + S_{2,1,2,1} + S_{2,1,1,2}\Big] \\ &- \frac{1}{12}\Big[S_1 S_{1,2,1,1} + S_{2,2,1,1} + S_{1,3,1,1} + S_{1,2,1,2} - S_1 S_{1,1,2,1} - S_{2,1,2,1} \\ &- S_{1,1,3,1} - S_{1,1,2,2}\Big] \\ &- \frac{1}{4}\Big[S_1 S_{1,1,1,2} + S_{2,1,1,2} + S_{1,2,1,2} + S_{1,1,2,2} + S_{1,1,1,3}\Big] \end{aligned} \tag{8.9}$$

through known harmonic sums of lower weight. The latter sum obeys the representation

$$\begin{aligned} S_{1,1,1,1,2}(N) = -\mathbf{M}\left[\frac{\mathrm{Li}_5(1-x)}{1-x}\right](N) + \zeta(2)S_{1,1,1,1}(N) - \zeta(3)S_{1,1,1}(N) \\ + \zeta(4)S_{1,1}(N) - \zeta(5)S_1(N) + \zeta(6)\ . \end{aligned} \tag{8.10}$$

The function in the remaining Mellin transform is regular at $z = 1$ and can be represented in terms of a factorial series. The remainder terms in (8.10) are polynomials of single harmonic sums. Therefore the poles of $S_{1,1,1,1,2}(N)$, resp. $S_{2,1,1,1,1}(N)$, are located at the non-positive integers. Finally we need the asymptotic representations of $\mathbf{M}\left[\mathrm{Li}_5(1-x)/(1-x)\right](N)$,

$$
\begin{aligned}
\mathbf{M}\left[\frac{\mathrm{Li}_5(1-x)}{1-x}\right](z) \sim\ & \frac{1}{z} + \frac{1}{32z^2} - \frac{179}{7776}\frac{1}{z^3} + \frac{515}{41472}\frac{1}{z^4} - \frac{216383}{194400000}\frac{1}{z^5} \\
& - \frac{183781}{25920000}\frac{1}{z^6} + \frac{4644828197}{653456160000}\frac{1}{z^7} + \frac{153375307}{49787136000}\frac{1}{z^8} \\
& - \frac{371224706507}{25204737600000}\frac{1}{z^9} + \frac{959290541}{160030080000}\frac{1}{z^{10}} \\
& + \frac{575134377343021}{16913534146740000}\frac{1}{z^{11}} - \frac{14855426650259}{312400053504000}\frac{1}{z^{12}} \\
& - \frac{291066196744896915257 29}{31970282063722722720000 0}\frac{1}{z^{13}} + \frac{225456132288901603}{7886010795062400 00}\frac{1}{z^{14}} \\
& + \frac{263567702701300558681}{105396534276008976 0000}\frac{1}{z^{15}} - \frac{355061945309358701}{187184432058624000}\frac{1}{z^{16}} \\
& - \frac{14324775585473770544568437 33}{49882668989177092212144000 00}\frac{1}{z^{17}} \\
& + \frac{192140702840923335916939}{130281924583069459200 00}\frac{1}{z^{18}} \\
& - \frac{2027981189268747465011536794768001}{254294408120596135866406712880000}\frac{1}{z^{19}} + O\left(\frac{1}{z^{20}}\right)
\end{aligned}
\tag{8.11}
$$

The corresponding representations for all other harmonic sums of weight $\mathsf{w}=6$ will be given in a forthcoming paper.

9. Conclusions

We derived the basic functions spanning the nested harmonic (alternating) sums up to weight $\mathsf{w}=6$ with no index $\{-1\}$. This sub class governs the functions contributing to the massless single-scale quantities, like the anomalous dimensions and Wilson coefficients to 3-loop order in QED and QCD. The structural relations, unlike the index-based algebraic relations, depend on the quantity under investigation. Furthermore, the number of independent quantities changes with the set of relations considered. We limited the investigation to harmonic sums without indices $a=-1$, since they do not occur in physics applications up to $\mathsf{w}=6$. For this class we considered differentiation relations, square-argument relations, partial integration relations, and relations for the functions emerging in the integrands of the Mellin transforms associated to the harmonic sums to obtain the minimal set of 20 functions at w=6. More details on these relations and their derivation are given in Ref. [**18**]. We did not find further relations.

There are first indications that, in the massive case, even in the limit $Q^2 \gg m^2$ this class needs to be extended at 3-loop order, cf. [**34**]. Up to weight $\mathsf{w}=5$ all basic functions were given by polynomials of Nielsen integrals, Eq. (2.20), of argument x or $-x$ weighted by $1/(x\pm1)$. Although most of the basic functions at $\mathsf{w}=6$ share this property, some contain 1-dimensional integrals over polynomials of Nielsen integrals $A_i(\pm x)|_{1\ldots3}$ and more dimensional integrals, which are not reducible. This is generally expected and the cases up to $\mathsf{w}=5$ form an exception.

We outlined how the exact representation of the Mellin transforms of the basic functions can be obtained, generalizing effective numerical high-precision representations [**25, 27**]. Up to terms which can be determined algebraically the Mellin transforms of the basic functions are factorial series. The singularities of the Mellin transforms are located at the non-positive integers. They obey recursion relations for $N \to N+1$. The asymptotic representation of the Mellin transforms can be determined analytically. The basic Mellin transforms are thus generalizations of Euler's ψ-function and their derivatives, which describe the single harmonic sums. In a forthcoming publication, the formulae and relations given in the present paper, and those of lower weight [**18**] will be made available in computer algebra code.

Acknowledgment.
I would like to thank Steve Rosenberg for his invitation and the organization of a very interesting and stimulating workshop. For useful conversations I would like to thank J. Ablinger, Y. André, F. Brown, and C. Schneider. This paper was supported in part by the Clay Mathematics Institute, DFG Sonderforschungsbereich Transregio 9, Computergestützte Theoretische Physik, and by the European Commission MRTN HEPTOOLS under Contract No. MRTN-CT-2006-035505.

10. Appendix A: Useful Integrals

In this appendix we list useful constants and integrals.

$$\mathrm{Li}_k(1) = \zeta_k \tag{10.1}$$

$$S_{1,k}(1) = \zeta_{k+1} \tag{10.2}$$

$$S_{2,2}(1) = \frac{1}{10}\zeta(2)^2 \tag{10.3}$$

$$S_{3,2}(1) = 2\zeta(5) - \zeta(2)\zeta(3) \tag{10.4}$$

$$S_{3,2}(-1) = -\frac{29}{32}\zeta(5) + \frac{1}{2}\zeta(2)\zeta(3) \tag{10.5}$$

$$S_{2,3}(1) = 2\zeta(5) - \zeta(2)\zeta(3) \tag{10.6}$$

$$A_1(1) = -3\zeta(5) + 2\zeta(2)\zeta(3) \tag{10.7}$$

$$A_1(-1) = -\frac{17}{16}\zeta(5) + \frac{3}{4}\zeta(2)\zeta(3) \tag{10.8}$$

$$A_2(1) = -\frac{1}{2}\zeta(5) \tag{10.9}$$

$$A_3(1) = -3\zeta(5) + \zeta(2)\zeta(3) \tag{10.10}$$

$$\int_0^x dy \frac{\mathrm{Li}_3(-y)}{1+y} = \ln(1+x)\mathrm{Li}_3(-x) + \frac{1}{2}\mathrm{Li}_2^2(-x) \tag{10.11}$$

$$\begin{aligned}\int_0^x dy \frac{\ln(y)\mathrm{Li}_2(-y)}{1+y} &= \ln(1+x)\ln(x)\mathrm{Li}_2(-x) + \frac{1}{2}\mathrm{Li}_2^2(-x) \\ &\quad - 2S_{2,2}(-x) + 2\ln(x)S_{1,2}(-x)\end{aligned} \tag{10.12}$$

$$\int_0^x dy \frac{S_{1,2}(y)}{y-1} = \ln(1-x)S_{1,2}(x) + 3S_{1,3}(x) \tag{10.13}$$

$$\int_0^x \frac{dy}{y}\left[\mathrm{Li}_2(1-y) - \zeta(2)\right] = -2\mathrm{Li}_3(x) + \ln(x)\mathrm{Li}_2(x) \tag{10.14}$$

$$\int_0^x dy \frac{\mathrm{Li}_3(y)}{y-1} = \frac{1}{2}\mathrm{Li}_2^2(x) + \ln(1-x)\mathrm{Li}_3(x) \tag{10.15}$$

$$\int_0^x dy \frac{\ln(y)}{y-1} \mathrm{Li}_2(y) = \frac{1}{2}\mathrm{Li}_2^2(x) + \ln(x)\ln(1-x)\mathrm{Li}_2(x) - 2S_{2,2}(x) + 2\ln(x)S_{1,2}(x) \tag{10.16}$$

$$\int_0^x \frac{dy}{y} \ln(1-y)\mathrm{Li}_3(y) = -\mathrm{Li}_2(x)\mathrm{Li}_3(x) + A_1(x) \tag{10.17}$$

$$\int_0^x \frac{dy}{y} \mathrm{Li}_2(y)\ln(y)\ln(1-y) = -\frac{1}{2}\ln(x)\mathrm{Li}_2^2(x) + \frac{1}{2}A_1(x) \tag{10.18}$$

$$\int_0^x \frac{dy}{y} \mathrm{Li}_2(-y)\ln(y)\ln(1+y) = -\frac{1}{2}\ln(x)\mathrm{Li}_2^2(-x) + \frac{1}{2}A_1(-x) \tag{10.19}$$

$$\int_0^x \frac{dy}{y} \mathrm{Li}_3(-y)\ln(1+y) = -\mathrm{Li}_2(-x)\mathrm{Li}_3(-x) + A_1(-x) \tag{10.20}$$

$$\int_0^x \frac{dy}{y} \left[\mathrm{Li}_3(1-x) - \zeta(3)\right] = -S_{1,2}(x)\ln(x) + \frac{1}{2}\mathrm{Li}_2^2(x) - \zeta(2)\mathrm{Li}_2(x) \tag{10.21}$$

$$\int_0^x \frac{dy}{y} S_{1,2}(y)\ln(y) = S_{2,2}(x)\ln(x) - S_{3,2}(x) \tag{10.22}$$

$$\int_0^x \frac{dy}{y} S_{1,2}(-y)\ln(y) = S_{2,2}(-x)\ln(x) - S_{3,2}(-x) \tag{10.23}$$

References

[1] L. Euler, Novi Comm. Acad. Sci. Petropol. **20** (1775) 140;
D. Zagier, First European Congress of Mathematics, (Birkhäuser, Boston, 1994), Vol.II, p. 497;
For a recent study of relations between non-alternating resp. alternating multiple zeta values to weight w=24 resp. w=12 see : J. Blümlein, D. J. Broadhurst and J. A. M. Vermaseren, Comput. Phys. Commun. **181** (2010) 582 [arXiv:0907.2557 [math-ph]].

[2] A wide range of references on multiple zeta values and harmonic sums is maintained by M.E. Hoffman : `http://www.usna.edu/Users/math/meh/biblio.html`

[3] D. J. Broadhurst, Eur. Phys. J. C **8** (1999) 311 [arXiv:hep-th/9803091];
F. Brown, arxiv:0804.1660 [math.AG];
Y. André, these Proceedings.

[4] D. J. Gross and F. Wilczek, Phys. Rev. D **8** (1973) 3633; Phys. Rev. D **9** (1974) 980;
H. Georgi and H. D. Politzer, Phys. Rev. D **9** (1974) 416;
G. Altarelli and G. Parisi, Nucl. Phys. B **126** (1977) 298;
E. G. Floratos, D. A. Ross and C. T. Sachrajda, Nucl. Phys. B **129** (1977) 66 [Erratum-ibid. B **139** (1978) 545]; Nucl. Phys. B **152** (1979) 493;
G. Curci, W. Furmanski and R. Petronzio, Nucl. Phys. B **175** (1980) 27;
W. Furmanski and R. Petronzio, Phys. Lett. B **97** (1980) 437;
R. Hamberg and W. L. van Neerven, Nucl. Phys. B **379** (1992) 143;
R. Mertig and W. L. van Neerven, Z. Phys. C **70** (1996) 637 [arXiv:hep-ph/9506451];
W. Vogelsang, Nucl. Phys. B **475** (1996) 47 [arXiv:hep-ph/9603366];
S. Moch, J. A. M. Vermaseren and A. Vogt, Nucl. Phys. B **688** (2004) 101 [arXiv:hep-ph/0403192]; Nucl. Phys. B **691** (2004) 129 [arXiv:hep-ph/0404111].

[5] J. Blümlein, M. Kauers, S. Klein and C. Schneider, Comput. Phys. Commun. **180** (2009) 2143 [arXiv:0902.4091 [hep-ph]].

[6] E.B. Zijlstra and W.L. van Neerven, Phys. Lett. **B273** (1991) 476; Nucl. Phys. **B383** (1992) 525;
W.L. van Neerven and E.B. Zijlstra, Phys. Lett. **B272** (1991) 127;
D. I. Kazakov and A. V. Kotikov, Nucl. Phys. B **307** (1988) 721 [Erratum-ibid. B **345** (1990) 299];
J. Sánchez Guillén, J. Miramontes, M. Miramontes, G. Parente and O. A. Sampayo, Nucl. Phys. B **353** (1991) 337;
S. Moch and J. A. M. Vermaseren, Nucl. Phys. B **573** (2000) 853 [arXiv:hep-ph/9912355];

J. A. M. Vermaseren, A. Vogt and S. Moch, Nucl. Phys. B **724** (2005) 3 [arXiv:hep-ph/0504242];
M. Buza, Y. Matiounine, J. Smith, R. Migneron and W. L. van Neerven, Nucl. Phys. B **472** (1996) 611 [arXiv:hep-ph/9601302];
M. Buza, Y. Matiounine, J. Smith and W. L. van Neerven, Eur. Phys. J. C **1** (1998) 301 [arXiv:hep-ph/9612398];
J. Blümlein, A. De Freitas, W. L. van Neerven and S. Klein, Nucl. Phys. B **755** (2006) 272 [arXiv:hep-ph/0608024];
I. Bierenbaum, J. Blümlein, S. Klein and C. Schneider, Nucl. Phys. B **803** (2008) 1 [arXiv:0803.0273 [hep-ph]];
I. Bierenbaum, J. Blümlein and S. Klein, Nucl. Phys. B **780** (2007) 40 [arXiv:hep-ph/0703285]; Phys. Lett. B **648** (2007) 195 [arXiv:hep-ph/0702265].

[7] K.G. Wilson, Phys. Rev. **179** (1969) 1699;
R.A. Brandt and G. Preparata, Fortschr. Phys. **18** (1970) 249;
W. Zimmermann, Lect. on Elementary Particle Physics and Quantum Field Theory, Brandeis Summer Inst., Vol. 1, (MIT Press, Cambridge, 1970), p. 395;
Y. Frishman, Ann. Phys. **66** (1971) 373.

[8] A. González-Arroyo, C. Lopez and F. J. Yndurain, Nucl. Phys. B **153** (1979) 161.

[9] J. Blümlein and S. Kurth, Phys. Rev. D **60** (1999) 014018 [arXiv:hep-ph/9810241].

[10] J. A. M. Vermaseren, Int. J. Mod. Phys. A **14** (1999) 2037 [arXiv:hep-ph/9806280].

[11] J. Blümlein, Comput. Phys. Commun. **159** (2004) 19 [arXiv:hep-ph/0311046].

[12] M.E. Hoffman, J. Algebra, **194** (1997) 477.

[13] E. Remiddi and J. A. M. Vermaseren, Int. J. Mod. Phys. A **15** (2000) 725 [arXiv:hep-ph/9905237].

[14] H. Poincaré, Acta Math **4** (1884) 201;
J.A. Lappo-Danielevsky, Mémoirs sur la Théorie des Systèmes Différentielles Linéaires, (Chelsea Publishing Company, New York, 1953).

[15] A.B. Goncharov, Math. Res. Lett. **5** (1998) 497;
J.M. Borwein, D.M. Bradley, D.J. Broadhurst, and P. Lisonek Trans. Am. Math. Soc. **353** (2001) 907 [`math.CA/9910045`];
S. Moch, P. Uwer and S. Weinzierl, J. Math. Phys. **43** (2002) 3363 [arXiv:hep-ph/0110083].

[16] J. Blümlein and S. Klein, PoS ACAT (2007) 084 [arXiv:0706.2426 [hep-ph]].

[17] J. Blümlein, Nucl. Phys. Proc. Suppl. **183** (2008) 232 [arXiv:0807.0700 [math-ph]].

[18] J. Blümlein, Comput. Phys. Commun. **180** (2009) 2218 [arXiv:0901.3106 [hep-ph]].

[19] N. Nielsen, Handbuch der Theorie der Gammafunktion, (Teubner, Leipzig, 1906); reprinted by Chelsea Publishing Company, Bronx, NY, 1965;
E. Landau, Über die Grundlagen der Theorie der Fakultätenreihen, S.-Ber. math.-naturw. Kl. Bayerische Akad. Wiss. München, **36** (1906) 151.

[20] N. Nielsen, Der Eulersche Dilogarithmus und seine Verallgemeinerungen, Nova Acta Leopoldina, Vol. XC, Nr. 3, Halle, 1909;
K.S. Kölbig, SIAM J. Math. Anal. **16** (1986) 1232.

[21] S. Lang, Algebra, (Springer, Berlin, 2002).

[22] L. Lewin, Dilogarithms and Associated Functions, (Macdonald, London, 1958); Polylogarithms and Associated Functions, (North-Holland, New York, 1981);
A. Devoto and D.W. Duke, Riv. Nuovo Cimento **7** (1981) 1.

[23] T. Gehrmann and E. Remiddi, Comput. Phys. Commun. **141** (2001) 296 [arXiv:hep-ph/0107173].

[24] J. Blümlein and V. Ravindran, Nucl. Phys. B **716** (2005) 128 [arXiv:hep-ph/0501178]; Nucl. Phys. B **749** (2006) 1 [arXiv:hep-ph/0604019].
J. Blümlein and S. Moch, in preparation.

[25] J. Blümlein, Comput. Phys. Commun. **133** (2000) 76 [arXiv:hep-ph/0003100];
J. Blümlein and S. O. Moch, Phys. Lett. B **614** (2005) 53 [arXiv:hep-ph/0503188].

[26] C. Hastings, jr., Approximations for Digital Computers, (Princeton University Press, Princeton/NJ, 1953);
M. Abramowitz and I.A. Stegun, Handbook of Mathematical Functions, (NBS, Washington, 1964);
L.A. Lyusternik, O.A. Chervonenkis, and A.R. Yanpol'skii, Handbook for Computing of Elementary Functions, Russian ed. (Fizmatgiz, Moscow, 1963); (Pergamon Press, New York,

1964);
C. Lanczos, J. Math. Phys. **17** (1938) 123;
F. Lösch and F. Schoblik, Die Fakultät und verwandte Funktionen, (Teubner, Stuttgart, 1951).

[27] S. I. Alekhin and J. Blümlein, Phys. Lett. B **594** (2004) 299 [arXiv:hep-ph/0404034].

[28] D. Graudenz, M. Hampel, A. Vogt and C. Berger, Z. Phys. C **70** (1996) 77 [arXiv:hep-ph/9506333].

[29] see also: A. V. Kotikov and V. N. Velizhanin, arXiv:hep-ph/0501274.

[30] C. Costermans, J.N. Enjalbert, Hoang Ngoc Minh, and M. Petitot, Proceedings ISSAC '05, Beijing, China, p. 100.

[31] J. Blümlein, KITP Workshop on Collider Physics Jan. 12- Apr. 2, 2004, Santa Barbara, CA, `http://online.kitp.ucsb.edu/online/collider04/bluemlein/`;
Transparencies presented at the Lille MZV2005 Workshop, June 2005.

[32] J. Blümlein, Nucl. Phys. Proc. Suppl. **135** (2004) 225 [arXiv:hep-ph/0407044].

[33] J. Blümlein and H. Kawamura, Nucl. Phys. B **708** (2005) 467 [arXiv:hep-ph/0409289].

[34] I. Bierenbaum, J. Blümlein and S. Klein, Phys. Lett. **B672** (2009) 401, [arXiv:0812.2427 [hep-ph]].

DEUTSCHES ELEKTRONEN–SYNCHROTRON, DESY,, PLATANENALLEE 6, D-15735 ZEUTHEN, GERMANY

E-mail address: `Johannes.Bluemlein@desy.de`

Clay Mathematics Proceedings
Volume 12, 2010

Hopf Subalgebras of Rooted Trees from Dyson-Schwinger Equations

Loïc Foissy

ABSTRACT. We consider the combinatorial Dyson-Schwinger equation $X = B^+(f(X))$ in the Connes-Kreimer Hopf algebra of rooted trees $\mathcal{H}$, where B^+ is the operator of grafting on a root, and f a formal series. The unique solution X of this equation generates a graded subalgebra $\mathcal{H}_f$ of $\mathcal{H}$. We characterize here all the formal series f such that $\mathcal{H}_f$ is a Hopf subalgebra. We obtain in this way a 2-parameter family of Hopf subalgebras of $\mathcal{H}$, organized into three isomorphism classes:

(1) A first (degenerate) one, restricted to a polynomial ring in one variable.
(2) A second one, restricted to the Hopf subalgebra of ladders, isomorphic to the Hopf algebra of symmetric functions.
(3) A last (infinite) one, which gives a family of isomorphic Hopf subalgebras of $\mathcal{H}$. These Hopf algebras can be seen as the coordinate ring of the group G of formal diffeomorphisms of the line tangent to the identity: in other terms, we obtain a family of embeddings of the Faà di Bruno Hopf algebra in $\mathcal{H}$.

In the second and the third cases, $\mathcal{H}_f$ is the graded dual of the enveloping algebra of a graded, connected Lie algebra $\mathfrak{g}$, such that the homogeneous components $\mathfrak{g}_n$ of $\mathfrak{g}$ are 1-dimensional when $n \geq 1$. Under a condition of commutativity, we prove that there exist three such Lie algebras:

(1) The Faà di Bruno Lie algebra, that is to say the Lie algebra of the group of formal diffeomorphisms G.
(2) The Lie algebra of corollas.
(3) A third one.

Embeddings in $\mathcal{H}$ of the dual of the enveloping algebra of the first case are given by the Dyson-Schwinger equations. For the second case, such an embedding is given by the subalgebra generated by corollas. We also describe an embedding in $\mathcal{H}$ for the third case.

CONTENTS

2010 *Mathematics Subject Classification.* Primary 16T05. Secondary 81T15, 81T18.

Introduction

The Connes-Kreimer algebra $\mathcal{H}$ of rooted trees was introduced in [**8**]. This graded Hopf algebra is commutative, non-cocommutative, and is given a linear basis by the set of rooted forests. A particularly important operator of $\mathcal{H}$ is the grafting on a root B^+, which satisfies the following equation:

$$\Delta \circ B^+(x) = B^+(x) \otimes 1 + (\mathrm{Id} \otimes B^+) \circ \Delta(x).$$

In other words, B^+ is a 1-cocycle for the Cartier-Quillen cohomology of coalgebras. Moreover, the couple $(\mathcal{H}, B^+)$ satisfies a universal property; see Theorem 3 of the present text.

We consider here a family of subalgebras of $\mathcal{H}$, associated to the combinatorial Dyson-Schwinger equation [**1**, **9**, **10**]:

$$X = B^+(f(X)),$$

where $f(h) = \sum p_n h^n$ is a formal series such that $p_0 = 1$, and X is an element of the completion of $\mathcal{H}$ for the topology given by the gradation of $\mathcal{H}$. This equation admits a unique solution $X = \sum x_n$, where x_n is, for all $n \geq 1$, a linear span of rooted trees of weight n, inductively given by

$$\left\{ \begin{array}{rcl} x_1 & = & p_0 \bullet, \\ x_{n+1} & = & \displaystyle\sum_{k=1}^{n} \sum_{a_1+\cdots+a_k=n} p_k B^+(x_{a_1} \cdots x_{a_k}). \end{array} \right.$$

We denote by $\mathcal{H}_f$ the subalgebra of $\mathcal{H}$ generated by the x_n's.

For the usual Dyson-Schwinger equation, $f(h) = (1-h)^{-1}$. It turns out that, in this case, $\mathcal{H}_f$ is a Hopf subalgebra. This is not the case in general; we characterise here the formal series $f(h)$ such that $\mathcal{H}_f$ is Hopf. Namely, $\mathcal{H}_f$ is a Hopf subalgebra of $\mathcal{H}$ if and only if there exists $(\alpha, \beta) \in K^2$, such that $f(h) = 1$ if $\alpha = 0$, or $f(h) = e^{\alpha h}$ if $\beta = 0$, or $f(h) = (1 - \alpha\beta h)^{-\frac{1}{\beta}}$ if $\alpha\beta \neq 0$. We obtain in this way a two-parameter family $\mathcal{H}_{\alpha,\beta}$ of Hopf subalgebras of $\mathcal{H}$ and we explicitly describe a system of generators of these algebras. In particular, if $\alpha = 0$, then $\mathcal{H}_{\alpha,\beta} = K[\bullet]$; if $\alpha \neq 0$, then $\mathcal{H}_{\alpha,\beta} = \mathcal{H}_{1,\beta}$.

The Hopf algebra $\mathcal{H}_{\alpha,\beta}$ is commutative, graded and connected. By the Milnor-Moore theorem [**11**], its dual is the enveloping algebra of a Lie algebra $\mathfrak{g}_{\alpha,\beta}$. Computing this Lie algebra, we find three isomorphism classes of $\mathcal{H}_{\alpha,\beta}$'s:

(1) $\mathcal{H}_{0,1}$, equal to $K[\bullet]$.
(2) $\mathcal{H}_{1,-1}$, the subalgebra of ladders, isomorphic to the Hopf algebra of symmetric functions.
(3) The $\mathcal{H}_{1,\beta}$'s, with $\beta \neq -1$, isomorphic to the Faà di Bruno Hopf algebra.

Note that non-commutative versions of these results are presented in [**6**].

In particular, if $\mathcal{H}_{\alpha,\beta}$ is non-cocommutative, it is isomorphic to the Faà di Bruno Hopf algebra. We try to explain this fact in the third section of this text. The dual Lie algebra $\mathfrak{g}_{\alpha,\beta}$ satisfies the following properties:

(1) $\mathfrak{g}_{\alpha,\beta}$ is graded and connected.

(2) The homogeneous component $\mathfrak{g}(n)$ of degree n of $\mathfrak{g}$ is 1-dimensional for all $n \geq 1$.

Moreover, if $\mathcal{H}_{\alpha,\beta}$ is non-cocommutative, then $[\mathfrak{g}(1), \mathfrak{g}(n)] \neq (0)$ if $n \geq 2$. Such a Lie algebra will be called a FdB Lie algebra. We prove here that there exist, up to isomorphism, only three FdB Lie algebras:

(1) The Faà di Bruno Lie algebra, which is the Lie algebra of the group of formal diffeomorphisms tangent to the identity at 0.
(2) The Lie algebra of corollas.
(3) A third Lie algebra.

In particular, with a stronger condition of non-commutativity, a FdB Lie algebra is isomorphic to the Faà di Bruno Lie algebra, and this result can be applied to all $\mathcal{H}_{1,\beta}$'s when $\beta \neq -1$. The dual of the enveloping algebras of the two other FdB Lie algebras can also be embedded in $\mathcal{H}$, using corollas for the second, giving in a certain way a limit of $\mathcal{H}_{1,\beta}$ when β goes to ∞, and the third one with a different construction.

Notation. We denote by K a commutative field of characteristic zero.

1. The Hopf algebra of rooted trees and Dyson-Schwinger equations

1.1. The Connes-Kreimer Hopf algebra. Let us first recall the construction of the Connes-Kreimer Hopf algebra of rooted trees.

DEFINITION 1. [**13, 14**]

(1) A *rooted tree* is a finite graph, connected and without loops, with a special vertex called the *root*.
(2) The *weight* of a rooted tree is the number of its vertices.
(3) The set of rooted trees will be denoted by $\mathcal{T}$.

Examples. The rooted trees of weight ≤ 5 are

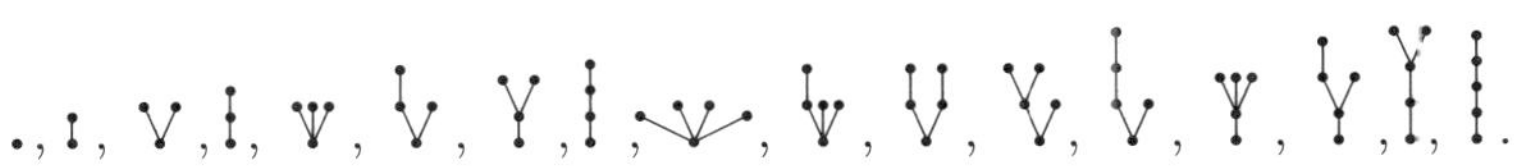

The Connes-Kreimer Hopf algebra of rooted trees $\mathcal{H}$ was introduced in [**2**]. As an algebra, $\mathcal{H}$ is the free associative, commutative, unitary algebra generated by the elements of $\mathcal{T}$. In other terms, a K-basis of $\mathcal{H}$ is given by rooted forests, that is to say not necessarily connected graphs F such that each connected component of F is a rooted tree. The set of rooted forests will be denoted by $\mathcal{F}$. The product of $\mathcal{H}$ is given by the concatenation of rooted forests, and the unit is the empty forest, denoted by 1.

Examples. The rooted forests of weight ≤ 4 are

In order to make $\mathcal{H}$ a bialgebra, we now introduce the notion of cut of a tree t. A *non-total cut* c of a tree t is a choice of edges of t. Deleting the chosen edges, the cut makes t into a forest, denoted by $W^c(t)$. The cut c is *admissible* if any oriented path[1] in the tree meets at most one cut edge. For such a cut, the tree of $W^c(t)$

[1]The edges of the tree are oriented from the root to the leaves.

which contains the root of t is denoted by $R^c(t)$ and the product of the other trees of $W^c(t)$ is denoted by $P^c(t)$. We also add the total cut, which is by convention an admissible cut such that $R^c(t) = 1$ and $P^c(t) = W^c(t) = t$. The set of admissible cuts of t is denoted by $\mathrm{Adm}_*(t)$. Note that the empty cut of t is admissible; we denote $\mathrm{Adm}(t) = \mathrm{Adm}_*(t) - \{\text{empty cut, total cut}\}$.

Example. Let us consider the rooted tree $t =$ . As it has 3 edges, it has 2^3 non-total cuts.

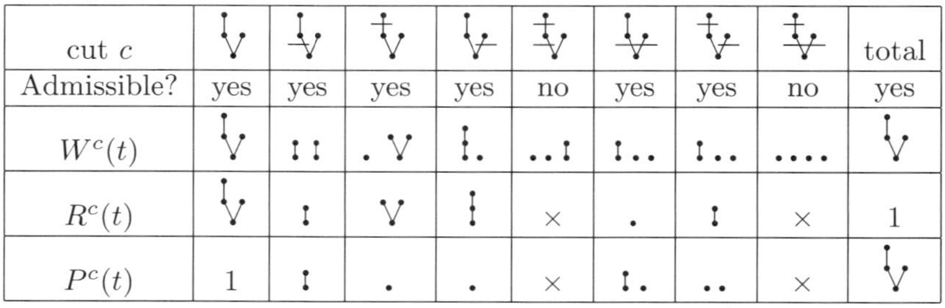

cut c									total
Admissible?	yes	yes	yes	yes	no	yes	yes	no	yes
$W^c(t)$									
$R^c(t)$					$\times$			$\times$	1
$P^c(t)$	1				$\times$			$\times$	

The coproduct of $\mathcal{H}$ is defined as the unique algebra morphism from $\mathcal{H}$ to $\mathcal{H}\otimes\mathcal{H}$ such that, for all rooted tree $t \in \mathcal{T}$,

$$\Delta(t) = \sum_{c\in \mathrm{Adm}_*(t)} P^c(t) \otimes R^c(t) = t \otimes 1 + 1 \otimes t + \sum_{c \in \mathrm{Adm}(t)} P^c(t) \otimes R^c(t).$$

As $\mathcal{H}$ is the polynomial algebra generated by $\mathcal{T}$, this makes sense.

Example.

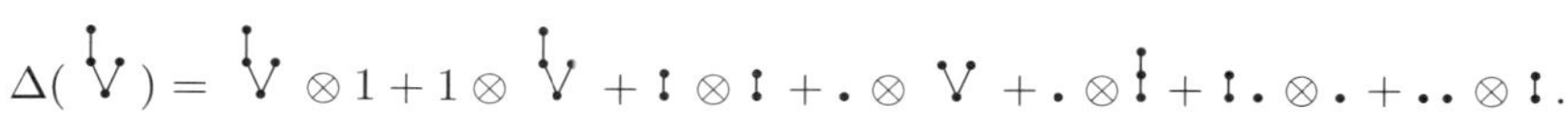

THEOREM 2. [2] *With this coproduct, $\mathcal{H}$ is a bialgebra. The counit of $\mathcal{H}$ is given by*

$$\varepsilon : \begin{cases} \mathcal{H} & \longrightarrow & K \\ F \in \mathcal{F} & \longrightarrow & \delta_{1,F}. \end{cases}$$

The antipode is the algebra endomorphism defined for all $t \in \mathcal{T}$ by

$$S(t) = -\sum_{c \text{ non-total cut of } t} (-1)^{n_c} W^c(t),$$

where n_c is the number of cut edges in c.

1.2. Gradation of $\mathcal{H}$ and completion. We grade $\mathcal{H}$ by declaring the forests of weight n homogeneous of degree n. We denote by $\mathcal{H}(n)$ the homogeneous component of $\mathcal{H}$ of degree n. Then $\mathcal{H}$ is a graded bialgebra, that is to say

(1) For all $i, j \in \mathbb{N}$, $\mathcal{H}(i)\mathcal{H}(j) \subseteq \mathcal{H}(i+j)$.
(2) For all $k \in \mathbb{N}$, $\Delta(\mathcal{H}(k)) \subseteq \sum_{i+j=k} \mathcal{H}(i) \otimes \mathcal{H}(j)$.

We define, for all $x, y \in \mathcal{H}$,

$$\left\{\begin{array}{lcl} \mathrm{val}(x) & = & \max\left\{n \in \mathbb{N} \mid x \in \bigoplus_{k \geq n} \mathcal{H}(k)\right\}, \\ d(x,y) & = & 2^{-\mathrm{val}(x-y)}, \end{array}\right.$$

with the convention $2^{-\infty} = 0$. Then d is a distance on $\mathcal{H}$. The metric space $(\mathcal{H}, d)$ is not complete; its completion will be denoted by $\widehat{\mathcal{H}}$. As a vector space,

$$\widehat{\mathcal{H}} = \prod_{n \in \mathbb{N}} \mathcal{H}(n).$$

The elements of $\widehat{\mathcal{H}}$ will be denoted $\sum x_n$, where $x_n \in \mathcal{H}(n)$ for all $n \in \mathbb{N}$. The product $m : \mathcal{H} \otimes \mathcal{H} \longrightarrow \mathcal{H}$ is homogeneous of degree 0, so is continuous. So it can be extended from $\widehat{\mathcal{H}} \otimes \widehat{\mathcal{H}}$ to $\widehat{\mathcal{H}}$, which is then an associative, commutative algebra. Similarly, the coproduct of $\mathcal{H}$ can be extended as a map

$$\Delta : \widehat{\mathcal{H}} \longrightarrow \mathcal{H} \widehat{\otimes} \mathcal{H} = \prod_{i,j \in \mathbb{N}} \mathcal{H}(i) \otimes \mathcal{H}(j).$$

Let $f(h) = \sum p_n h^n \in K[[h]]$ be any formal series, and let $X = \sum x_n \in \widehat{\mathcal{H}}$, such that $x_0 = 0$. The series of $\widehat{\mathcal{H}}$ of terms $p_n X^n$ is Cauchy, so converges. Its limit will be denoted by $f(X)$. In other words, $f(X) = \sum y_n$, with

$$y_n = \sum_{k=1}^{n} \sum_{a_1 + \cdots + a_k = n} p_k x_{a_1} \cdots x_{a_k}.$$

Remark. If $f(h) \in K[[h]]$, $g(h) \in K[[h]]$, without constant terms, and $X \in \widehat{\mathcal{H}}$, without constant terms, it is easy to show that $(f \circ g)(X) = f(g(X))$.

1.3. 1-cocycle of $\mathcal{H}$ and Dyson-Schwinger equations. We define the operator $B^+ : \mathcal{H} \longrightarrow \mathcal{H}$, sending a forest $t_1 \cdots t_n$ to the tree obtained by grafting $t_1, \cdots, t_n$ to a common root. For example, $B^+($$) =$ . This operator satisfies the following relation: for all $x \in \mathcal{H}$,

$$\Delta \circ B^+(x) = B^+(x) \otimes 1 + (\mathrm{Id} \otimes B^+) \circ \Delta(x). \tag{1}$$

This means that B^+ is a 1-cocycle for a certain cohomology, namely the Cartier-Quillen cohomology for coalgebras, the notion dual to the Hochschild cohomology [**2**]. Moreover, $(\mathcal{H}, B^+)$ satisfies the following universal property:

THEOREM 3 (Universal property). *Let A be a commutative algebra and let $L : A \longrightarrow A$ be a linear map.*

(1) *There exists a unique algebra morphism $\phi : \mathcal{H} \longrightarrow A$, such that $\phi \circ B^+ = L \circ \phi$.*
(2) *If moreover A is a Hopf algebra and L satisfies (1), then ϕ is a Hopf algebra morphism.*

The operator B^+ is homogeneous of degree 1, so is continuous. As a consequence, it can be extended as an operator $B^+ : \widehat{\mathcal{H}} \longrightarrow \widehat{\mathcal{H}}$. This operator still satisfies (1).

DEFINITION 4. **[1, 9, 10]** Let $f \in K[[h]]$. The *Dyson-Schwinger equation* associated to f is

$$X = B^+(f(X)), \tag{2}$$

where X is an element of $\widehat{\mathcal{H}}$, without constant term.

PROPOSITION 5. *The Dyson-Schwinger equation associated to the formal series $f(h) = \sum p_n h^n$ admits a unique solution $X = \sum x_n$, inductively defined by*

$$\begin{cases} x_0 &= 0, \\ x_1 &= p_0 \bullet, \\ x_{n+1} &= \displaystyle\sum_{k=1}^{n} \sum_{a_1+\cdots+a_k=n} p_k B^+(x_{a_1}\cdots x_{a_k}). \end{cases}$$

PROOF. It is enough to identify the homogeneous components of the two members of (2). $\square$

DEFINITION 6. The subalgebra of $\mathcal{H}$ generated by the homogeneous components x_n of the unique solution X of the Dyson-Schwinger equation (2) associated to f will be denoted by $\mathcal{H}_f$.

The aim of this text is to give a necessary and sufficient condition on f for $\mathcal{H}_f$ to be a Hopf subalgebra of $\mathcal{H}$.

Remarks.

(1) If $f(0) = 0$, the unique solution of (2) is 0. As a consequence, $\mathcal{H}_f = K$ is a Hopf subalgebra.
(2) For all $\alpha \in K$, if $X = \sum x_n$ is the solution of the Dyson-Schwinger equation associated to f, the unique solution of the Dyson-Schwinger equation associated to αf is $\sum \alpha^n x_n$. As a consequence, if $\alpha \neq 0$, $\mathcal{H}_f = \mathcal{H}_{\alpha f}$. We shall then suppose in the sequel that $p_0 = 1$. In this case, $x_1 = \bullet$.

Examples.

(1) We take $f(h) = 1 + h$. Then $x_1 = \bullet$, $x_2 =$, $x_3 =$, $x_4 =$. More generally, x_n is the ladder with n vertices, that is to say $(B^+)^n(1)$ (Definition 7). As a consequence, for all $n \geq 1$,

$$\Delta(x_n) = \sum_{i+j=n} x_i \otimes x_j.$$

So $\mathcal{H}_{1+h}$ is Hopf. Moreover, it is cocommutative.
(2) We take $f(h) = 1 + h + h^2 + 2h^3$. Then

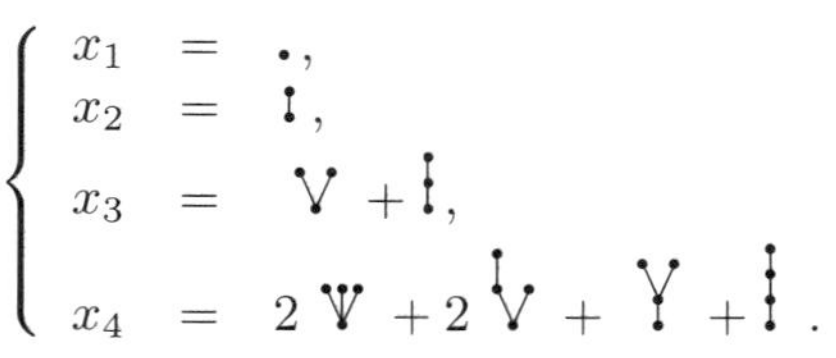

Hence

$$\begin{aligned}
\Delta(x_1) &= x_1 \otimes 1 + 1 \otimes x_1,\\
\Delta(x_2) &= x_2 \otimes 1 + 1 \otimes x_2 + x_1 \otimes x_1,\\
\Delta(x_3) &= x_3 \otimes 1 + 1 \otimes x_3 + x_1^2 \otimes x_1 + 3x_1 \otimes x_2 + x_2 \otimes x_1,\\
\Delta(x_4) &= x_4 \otimes 1 + 1 \otimes x_4 + 10x_1^2 \otimes x_2 + x_1^3 \otimes x_1 + 3x_2 \otimes x_2\\
&\quad +2x_1x_2 \otimes x_1 + x_3 \otimes x_1 + x_1 \otimes (8
\end{aligned}$$
 $+5$  $),$

so $\mathcal{H}_f$ is not Hopf.

We shall need later these two families of rooted trees:

DEFINITION 7. Let $n \geq 1$.

(1) The *ladder* l_n of weight n is the rooted tree $(B^+)^n(1)$. For example,

$l_1 = \bullet$, $l_2 =$ , $l_3 =$, $l_4 =$  .

(2) The *corolla* c_n of weight n is the rooted tree $B^+(\bullet^{n-1})$. For example,

$c_1 = \bullet$, $c_2 =$, $c_3 =$ , $c_4 =$  .

The following lemma is an immediate corollary of proposition 5:

LEMMA 8. *The coefficient of the ladder of weight n in x_n is p_1^{n-1}. The coefficient of the corolla of weight n in x_n is p_{n-1}.*

Using (1):

LEMMA 9. *For all $n \geq 1$,*

(1) $\Delta(l_n) = \sum_{i=0}^{n} l_i \otimes l_{n-i}$, *with the convention $l_0 = 1$.*

(2) $\Delta(c_n) = c_n \otimes 1 + \sum_{i=0}^{n-1} \binom{n-1}{i} \bullet^i \otimes c_{n-i}$.

2. Formal series giving Hopf subalgebras

2.1. Statement of the main theorem. The aim of this section is to prove the following result:

THEOREM 10. *Let $f(h) \in K[[h]]$, such that $f(0) = 1$. The following assertions are equivalent:*

1. *$\mathcal{H}_f$ is a Hopf subalgebra of $\mathcal{H}$.*
2. *There exists $(\alpha, \beta) \in K^2$ such that $(1 - \alpha\beta h)f'(h) = \alpha f(h)$.*
3. *There exists $(\alpha, \beta) \in K^2$ such that $f(h) = 1$ if $\alpha = 0$, or $f(h) = e^{\alpha h}$ if $\beta = 0$, or $f(h) = (1 - \alpha\beta h)^{-\frac{1}{\beta}}$ if $\alpha\beta \neq 0$.*

It is an easy exercise to prove that the second and the third statements are equivalent.

2.2. Proof of (1) $\Longrightarrow$ (2). We suppose that $\mathcal{H}_f$ is Hopf.

LEMMA 11. *Let us suppose that* $p_1 = 0$. *Then* $f(h) = 1$, *so* (2) *holds with* $\alpha = 0$.

PROOF. Let us suppose that $p_n \neq 0$ for a certain $n \geq 2$. Let us choose a minimal n. Then $x_1 = \bullet$, $x_2 = \cdots = x_n = 0$, and $x_{n+1} = p_n c_{n+1}$. So

$$\Delta(x_{n+1}) = x_{n+1} \otimes 1 + 1 \otimes x_{n+1} + \sum_{i=1}^{n} \binom{n}{i} p_n \bullet^i \otimes c_{n+1-i} \in \mathcal{H}_f \otimes \mathcal{H}_f.$$

In particular, for $i = n-1$, $c_2 = \mathfrak{l} \in \mathcal{H}_f$, so $x_2 \neq 0$: contradiction. □

We now assume that $p_1 \neq 0$. Let $Z_\bullet : \mathcal{H} \longrightarrow K$, defined by $Z_\bullet(F) = \delta_{\bullet,F}$ for all $F \in \mathcal{F}$. This map $Z_\bullet$ is homogeneous of degree -1, so is continuous and can be extended to a map $Z_\bullet : \widehat{\mathcal{H}} \longrightarrow K$. We put $(Z_\bullet \otimes \mathrm{Id}) \circ \Delta(X) = \sum y_n$, where X is the unique solution of (2). A direct computation shows that y_n can be computed by induction with

$$\begin{cases} y_0 = 1, \\ y_{n+1} = \displaystyle\sum_{k=1}^{n} \sum_{a_1+\cdots+a_k=n} (k+1) p_{k+1} B^+(x_{a_1} \cdots x_{a_k}) \\ \qquad\quad + \displaystyle\sum_{k=1}^{n} \sum_{a_1+\cdots+a_k=n} k p_k B^+(y_{a_1} x_{a_2} \cdots x_{a_k}). \end{cases}$$

As $\mathcal{H}_f$ is Hopf, $y_n \in \mathcal{H}_f$ for all $n \in \mathbb{N}$. Moreover, y_n is a linear span of rooted trees of weight n, so is a multiple of x_n; we put $y_n = \alpha_n x_n$.

Let us consider the coefficient of the ladder of weight n in y_n. By lemma 8, this is $\alpha_n p_1^{n-1}$. So, for all $n \geq 1$,

$$p_1^n \alpha_{n+1} = 2 p_1^{n-1} p_2 + p_1^n \alpha_n.$$

As $\alpha_1 = p_1$, for all $n \geq 1$, $\alpha_n = p_1 + 2\dfrac{p_2}{p_1}(n-1)$. Let us consider the coefficient of the corolla of weight n in y_n. By lemma 8, this is $\alpha_n p_n$. So, for all $n \geq 1$,

$$\alpha_n p_n = (n+1) p_{n+1} + n p_n p_1.$$

Summing all these relations, putting $\alpha = p_1$ and $\beta = 2\dfrac{p_2}{p_1} - 1$, we obtain the differential equation $(1 - \alpha\beta h) f'(h) = f(h)$, so (2) holds.

2.3. Proof of (2) $\Longrightarrow$ (1). Let us suppose (2) or, equivalently, (3). We now write $\mathcal{H}_{\alpha,\beta}$ instead of $\mathcal{H}_f$. We first give a description of the x_n's.

DEFINITION 12.

(1) Let $F \in \mathcal{F}$. The coefficient s_F is inductively computed by

$$\begin{cases} s_\bullet = 1, \\ s_{t_1^{a_1} \cdots t_k^{a_k}} = a_1! \cdots a_k! s_{t_1}^{a_1} \cdots s_{t_k}^{a_k}, \\ s_{B^+(t_1^{a_1} \cdots t_k^{a_k})} = a_1! \cdots a_k! s_{t_1}^{a_1} \cdots s_{t_k}^{a_k}, \end{cases}$$

where $t_1, \cdots, t_k$ are distinct elements of $\mathcal{T}$.

(2) Let $F \in \mathcal{F}$. The coefficient e_F is inductively computed by

$$\left\{\begin{array}{rcl} e_\bullet &=& 1,\\ e_{t_1^{a_1}\cdots t_k^{a_k}} &=& \dfrac{(a_1+\cdots+a_k)!}{a_1!\cdots a_k!}e_{t_1}^{a_1}\cdots e_{t_k}^{a_k},\\ e_{B^+(t_1^{a_1}\cdots t_k^{a_k})} &=& \dfrac{(a_1+\cdots+a_k)!}{a_1!\cdots a_k!}e_{t_1}^{a_1}\cdots e_{t_k}^{a_k}, \end{array}\right.$$

where $t_1, \cdots, t_k$ are distinct elements of $\mathcal{T}$.

Remarks.

(1) The coefficient s_F is the number of symmetries of F, that is to say the number of graph automorphisms of F respecting the roots.

(2) The coefficient e_F is the number of embeddings of F in the plane, that is to say the number of planar forests whose underlying rooted forest is F.

We now give β-equivalents of these coefficients. For all $k \in \mathbb{N}^*$, we put $[k]_\beta = 1+\beta(k-1)$ and $[k]_\beta! = [1]_\beta \cdots [k]_\beta$. We then inductively define $[s_F]_\beta$ and $[e_F]_\beta$ for all $F \in \mathcal{F}$ by

$$\left\{\begin{array}{rcl} [s_\bullet]_\beta &=& 1,\\ [s_{t_1^{a_1}\cdots t_k^{a_k}}]_\beta &=& [a_1]_\beta!\cdots[a_k]_\beta![s_{t_1}]_\beta^{a_1}\cdots[s_{t_k}]_\beta^{a_k},\\ [s_{B^+(t_1^{a_1}\cdots t_k^{a_k})}]_\beta &=& [a_1]_\beta!\cdots[a_k]_\beta![s_{t_1}]_\beta^{a_1}\cdots[s_{t_k}]_\beta^{a_k}, \end{array}\right.$$

$$\left\{\begin{array}{rcl} [e_\bullet] &=& 1,\\ [e_{t_1^{a_1}\cdots t_k^{a_k}}]_\beta &=& \dfrac{[a_1+\cdots+a_k]_\beta!}{[a_1]_\beta!\cdots[a_k]_\beta!}[e_{t_1}]_\beta^{a_1}\cdots[e_{t_k}]_\beta^{a_k},\\ [e_{B^+(t_1^{a_1}\cdots t_k^{a_k})}]_\beta &=& \dfrac{[a_1+\cdots+a_k]_\beta!}{[a_1]_\beta!\cdots[a_k]_\beta!}[e_{t_1}]_\beta^{a_1}\cdots[e_{t_k}]_\beta^{a_k}, \end{array}\right.$$

where $t_1, \cdots, t_k$ are distinct elements of $\mathcal{T}$. In particular, $[s_t]_1 = s_t$ and $[e_t]_1 = e_t$, whereras $[s_t]_0 = 1$ and $[e_t]_0 = 1$ all $t \in \mathcal{T}$.

Examples.

t	s_t	$[s_t]_\beta$	e_t	$[e_t]_\beta$
$\bullet$	1	1	1	1
	1	1	1	1
	2	$(1+\beta)$	1	1
	1	1	1	1
	6	$(1+\beta)(1+2\beta)$	1	1
	1	1	2	$(1+\beta)$
	2	$(1+\beta)$	1	1
	1	1	1	1

PROPOSITION 13. *For all $n \in \mathbb{N}^*$, in $\mathcal{H}_{\alpha,\beta}$,*

$$x_n = \alpha^{n-1} \sum_{t\in\mathcal{T},\ \mathrm{weight}(t)=n} \frac{[s_t]_\beta [e_t]_\beta}{s_t} t.$$

Examples.

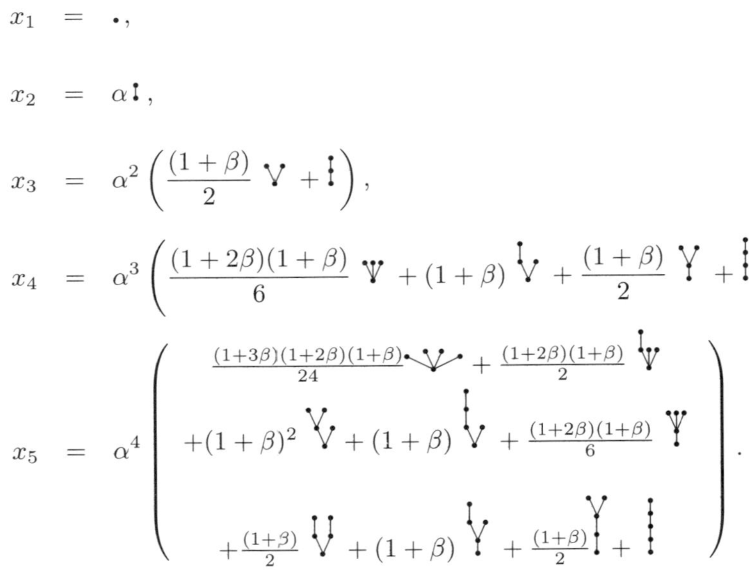

PROOF. For any $t \in \mathcal{T}$, we denote by b_t the coefficient of t in $x_{\mathrm{weight}(t)}$. Then $b_{\bullet} = 1$. The formal series $f(h)$ is given by

$$f(h) = \sum_{n=0}^{\infty} \alpha^n \frac{[n]_\beta!}{n!} h^n.$$

If $t = B^+(t_1^{a_1} \cdots t_k^{a_k})$, where $t_1, \cdots, t_k$ are distinct elements of $\mathcal{T}$, then

$$b_t = \alpha^{a_1+\cdots+a_k} \frac{[a_1+\cdots+a_k]_\beta!}{(a_1+\cdots+a_k)!} \frac{(a_1+\cdots+a_k)!}{a_1!\cdots a_k!} b_{t_1}^{a_1} \cdots b_{t_k}^{a_k}.$$

The result comes from an easy induction. $\square$

As a consequence, $\mathcal{H}_{0,\beta} = K[\bullet]$, so $\mathcal{H}_{0,\beta}$ is a Hopf subalgebra. Moreover, $\mathcal{H}_{\alpha,\beta} = \mathcal{H}_{1,\beta}$ if $\alpha \neq 0$. So we can restrict ourselves to the case $\alpha = 1$. In order to ease the notation, we put $n_t = s_t e_t$ and $[n_t]_\beta = [s_t]_\beta [e_t]_\beta$ for all $t \in \mathcal{T}$. Then

$$\left\{ \begin{array}{rcl} n_{\bullet} &=& 1, \\ n_{B^+(t_1\cdots t_k)} &=& k! n_{t_1} \cdots n_{t_k}, \end{array} \right.$$

$$\left\{ \begin{array}{rcl} [n_{\bullet}]_\beta &=& 1, \\ [n_{B^+(t_1\cdots t_k)}]_\beta &=& [k]_\beta! [n_{t_1}]_\beta \cdots [n_{t_k}]_\beta. \end{array} \right.$$

As a consequence, an easy induction proves that

$$n_t = \prod_{s \text{ vertex of } t} (\text{fertility of } s)!, \qquad [n_t]_\beta = \prod_{s \text{ vertex of } t} [\text{fertility of } s]_\beta!.$$

We shall use the following result, proved in [**5, 7**]:

LEMMA 14. *For all forests $F \in \mathcal{F}$, $G, H \in \mathcal{T}$, denote by $n(F,G;H)$ the coefficient of $F \otimes G$ in $\Delta(H)$, and by $n'(F,G;H)$ the number of graftings of the trees of F over G giving the tree H. Then $n'(F,G;H) s_H = n(F,G;H) s_F s_G$.*

LEMMA 15. *Let $k, n \in \mathbb{N}^*$. We put, in $K[X_1, \cdots, X_n]$, $S = X_1 + \cdots + X_n$. Then*

$$\sum_{\alpha_1+\cdots+\alpha_n=k} \prod_{i=1}^{n} \frac{X_i(X_i+1)\cdots(X_i+\alpha_i-1)}{\alpha_i} = \frac{S(S+1)\cdots(S+k-1)}{k!}.$$

PROOF. By induction on k, see [**6**]. □

PROPOSITION 16. *If $\alpha = 1$,*

$$\Delta(X) = X \otimes 1 + \sum_{n=1}^{\infty} (1-\beta X)^{-n(1/\beta+1)+1} \otimes x_n.$$

So $\mathcal{H}_{1,\beta}$ is a Hopf subalgebra.

PROOF. As for all $n \geq 1$, x_n is a linear span of trees, we can write

$$\Delta(X) = X \otimes 1 + \sum_{F \in \mathcal{F},\, t \in \mathcal{T}} a_{F,t} F \otimes t.$$

Then, if $F \in \mathcal{F}$, $G \in \mathcal{T}$,

$$a_{F,G} = \sum_{H \in \mathcal{T}} \frac{[n_H]_\beta}{s_H} n(F,G;H) = \sum_{H \in \mathcal{T}} \frac{[n_H]_\beta}{s_F s_G} n'(F,G;H).$$

We put $F = t_1 \cdots t_k$, and we denote by $s_1, \cdots, s_n$ the vertices of the tree G, of respective fertility $f_1, \cdots, f_n$. Let us consider a grafting of F over G, such that α_i trees of F are grafted on the vertex s_i. Then $\alpha_1 + \cdots + \alpha_n = k$. Denoting by H the result of this grafting,

$$[n_H]_\beta \quad = \quad [n_G]_\beta [n_{t_1}]_\beta \cdots [n_{t_k}]_\beta \frac{[f_1+\alpha_1]_\beta!}{[f_1]_\beta!} \cdots \frac{[f_n+\alpha_n]_\beta!}{[f_n]_\beta!}.$$

Moreover, the number of such graftings is $\dfrac{k!}{\alpha_1! \cdots \alpha_n!}$. So, with lemma 15, putting $x_i = f_i + 1/\beta$ and $s = x_1 + \cdots + x_n$,

$$\begin{aligned}
a_{F,G} &= \sum_{\alpha_1+\cdots+\alpha_n=k} \frac{k!}{\alpha_1!\cdots\alpha_k!} \frac{1}{s_F s_G} [n_G]_\beta \prod_{i=1}^{k} [n_{t_i}]_\beta \frac{[f_1+\alpha_i]_\beta!}{[f_i]_\beta!} \\
&= \frac{k![n_G]_\beta!}{s_G s_F} \left(\prod_{i=1}^{k} [n_{t_i}]_\beta \right) \sum_{\alpha_1+\cdots+\alpha_n=k} \prod_{i=1}^{n} \frac{(1+f_i\beta)\cdots(1+(f_i+\alpha_i-1)\beta)}{\alpha_i!} \\
&= \frac{k![n_G]_\beta!}{s_G s_F} \left(\prod_{i=1}^{k} [n_{t_i}]_\beta \right) \sum_{\alpha_1+\cdots+\alpha_n=k} \prod_{i=1}^{n} \beta^{\alpha_i} \frac{x_i(x_i+1)\cdots(x_i+\alpha_i+1)}{\alpha_i!} \\
&= \frac{k![n_G]_\beta!}{s_G s_F} \left(\prod_{i=1}^{k} [n_{t_i}]_\beta \right) \beta^k \sum_{\alpha_1+\cdots+\alpha_n=k} \prod_{i=1}^{n} \frac{x_i(x_i+1)\cdots(x_i+\alpha_i+1)}{\alpha_i!} \\
&= \frac{k![n_G]_\beta!}{s_G s_F} \left(\prod_{i=1}^{k} [n_{t_i}]_\beta \right) \beta^k \frac{s(s+1)\cdots(s+k-1)}{k!}.
\end{aligned}$$

Moreover, as G is a tree, $s = f_1 + \cdots + f_n + n/\beta = n - 1 + n/\beta = n(1+1/\beta) - 1$.

We now write $F = t_1 \cdots t_k = u_1^{a_1} \cdots u_l^{a_l}$, where $u_1, \cdots, u_l$ are distinct elements of $\mathcal{T}$. Then

$$s_F = s_{u_1}^{a_1} \cdots s_{u_l}^{a_l} a_1! \cdots a_l!,$$

so

$$\frac{k![n_{t_1}]_\beta \cdots [n_{t_k}]_\beta}{s_F} = \frac{(a_1 + \cdots + a_l)!}{a_1! \cdots a_l!} \left(\frac{[n_{t_1}]_\beta}{s_{t_1}}\right)^{a_1} \cdots \left(\frac{[n_{t_l}]_\beta}{s_{t_l}}\right)^{a_l}.$$

As a conclusion, putting $Q_k(S) = \dfrac{S(S+1)\cdots(S+k-1)}{k!}$,

$$\begin{aligned}
\Delta(X) &= \sum_{n \geq 1} \sum_{t_1^{a_1} \cdots t_l^{a_l} \in \mathcal{F}} \frac{(a_1 + \cdots + a_l)!}{a_1! \cdots a_l!} \beta^{a_1 + \cdots + a_l} Q_{a_1 + \cdots + a_l}(n(1 + 1/\beta) - 1) \\
&\quad \left(\frac{[n_{t_1}]_\beta}{s_{t_1}} t_1\right)^{a_1} \cdots \left(\frac{[n_{t_l}]_\beta}{s_{t_l}} t_l\right)^{a_l} \otimes \left(\sum_{\substack{G \in \mathcal{T} \\ \text{weight}(G) = n}} \frac{[n_G]_\beta!}{s_G} G \right) + X \otimes 1 \\
&= X \otimes 1 + \sum_{n=1}^{\infty} (1 - \beta X)^{-n(1/\beta + 1) + 1} \otimes x_n.
\end{aligned}$$

So $\Delta(X) \in \mathcal{H} \widehat{\otimes} \mathcal{H}$. Projecting on the homogeneous component of degree n, we obtain $\Delta(x) \in \mathcal{H} \otimes \mathcal{H}$, so $\mathcal{H}_{1,\beta}$ is a Hopf subalgebra. $\square$

Remarks.

(1) For $(\alpha, \beta) = (1, 0)$, $f(h) = e^h$ and for all $n \in \mathbb{N}$, $x_n = \displaystyle\sum_{\substack{t \in \mathcal{T} \\ \text{weight}(t) = n}} \frac{1}{s_t} t$.

(2) For $(\alpha, \beta) = (1, 1)$, $f(h) = (1-h)^{-1}$ and for all $n \in \mathbb{N}$, $x_n = \displaystyle\sum_{\substack{t \in \mathcal{T} \\ \text{weight}(t) = n}} e_t t$.

(3) For $(\alpha, \beta) = (1, -1)$, $f(h) = 1 + h$ and, as $[i]_{-1} = 0$ if $i \geq 2$, for all $n \in \mathbb{N}^*$, x_n is the ladder of weight n.

2.4. What is $\mathcal{H}_{\alpha,\beta}$? If $\alpha = 0$, then $\mathcal{H}_{0,\beta} = K[\bullet]$. If $\alpha \neq 0$, then obviously $\mathcal{H}_{\alpha,\beta} = \mathcal{H}_{1,\beta}$; let us suppose that $\alpha = 1$. The Hopf algebra $\mathcal{H}_{1,\beta}$ is graded, connected and commutative. Dually, its graded dual $\mathcal{H}_{1,\beta}^*$ is a graded, connected, cocommutative Hopf algebra. By the Milnor-Moore theorem [**11**], it is isomorphic to the enveloping algebra of the Lie algebra of its primitive elements. We now denote this Lie algebra by $\mathfrak{g}_{1,\beta}$. The dual of $\mathfrak{g}_{1,\beta}$ is identified with the quotient space

$$\text{coPrim}(\mathcal{H}_{1,\beta}) = \frac{\mathcal{H}_{1,\beta}}{(1) \oplus \text{Ker}(\varepsilon)^2},$$

and the transposition of the Lie bracket is the Lie cobracket δ induced by

$$(\varpi \otimes \varpi) \circ (\Delta - \Delta^{op}),$$

where ϖ is the canonical projection on $\mathrm{coPrim}(\mathcal{H}_{1,\beta})$. As $\mathcal{H}_{1,\beta}$ is the polynomial algebra generated by the x_n's, a basis of $\mathrm{coPrim}(\mathcal{H}_{1,\beta})$ is $(\varpi(x_n))_{n\in\mathbb{N}^*}$. By Proposition 16,

$$\begin{aligned}(\varpi\otimes\varpi)\circ\Delta(X) &= (\varpi\otimes\varpi)\left(\sum_{n=1}^{\infty}(1-\beta X)^{-n(1/\beta+1)+1}\otimes x_n\right)\\ &= \sum_{n\geq 1}(n(1+\beta)-\beta)\,\varpi(X)\otimes\varpi(x_n).\end{aligned}$$

Projecting on the homogeneous component of degree k,

$$(\varpi\otimes\varpi)\circ\Delta(x_k)=\sum_{i+j=k}^{k}(j(1+\beta)-\beta)\,\varpi(x_i)\otimes\varpi(x_j).$$

As a consequence,

$$\delta(\varpi(x_k))=\sum_{i+j=k}(1+\beta)(j-i)\varpi(x_i)\otimes\varpi(x_j).$$

Dually, the Lie algebra $\mathfrak{g}_{1,\beta}$ has the dual basis $(Z_n)_{n\geq 1}$, with bracket given by

$$[Z_i,Z_j]=(1+\beta)(j-i)Z_{i+j}.$$

So, if $\beta\neq -1$, this Lie algebra is isomorphic to the Faà di Bruno Lie algebra $\mathfrak{g}_{\mathrm{FdB}}$, which has a basis $(f_n)_{n\geq 1}$, and whose bracket defined by $[f_i,f_j]=(j-i)f_{i+j}$. So $\mathcal{H}_{1,\beta}$ is isomorphic to the Hopf algebra $\mathcal{U}(\mathfrak{g}_{\mathrm{FdB}})^*$, namely the Faà di Bruno Hopf algebra [**3**], coordinate ring of the group of formal diffeomorphisms of the line tangent to Id, that is to say

$$G_{\mathrm{FdB}}=\left(\left\{\sum a_nh^n\in K[[h]]\mid a_0=0,\ a_1=1\right\},\circ\right).$$

THEOREM 17. (1) *If $\alpha\neq 0$ and $\beta\neq -1$, $\mathcal{H}_{\alpha,\beta}$ is isomorphic to the Faà di Bruno Hopf algebra.*
(2) *If $\alpha\neq 0$ and $\beta=-1$, $\mathcal{H}_{\alpha,\beta}$ is isomorphic to the Hopf algebra of symmetric functions.*
(3) *If $\alpha=0$, $\mathcal{H}_{\alpha,\beta}=K[\bullet]$.*

Remark. If β and $\beta'\neq -1$, then $\mathcal{H}_{1,\beta}$ and $\mathcal{H}_{1,\beta'}$ are isomorphic but are not equal, as shown by considering x_3.

3. FdB Lie algebras

In the preceding section, we considered Hopf subalgebras of $\mathcal{H}$, generated in each degree by a linear span of trees. Their graded dual is then the enveloping algebra of a Lie algebra $\mathfrak{g}$, graded, with Poincaré-Hilbert formal series

$$\frac{h}{1-h}=\sum_{n=1}^{\infty}h^n.$$

Under a hypothesis of commutativity, we show that such a $\mathfrak{g}$ is isomorphic to the Faà di Bruno Lie algebra, so the considered Hopf subalgebra is isomorphic to the Faà di Bruno Hopf algebra.

Remark. The proofs of this section were completed using MuPAD pro 4. The notebook of the computations can be found at [**4**].

3.1. Definitions and first properties.

DEFINITION 18. Let $\mathfrak{g}$ be an $\mathbb{N}$-graded Lie algebra. For all $n \in \mathbb{N}$, we denote by $\mathfrak{g}(n)$ the homogeneous component of degree n of $\mathfrak{g}$. We shall say that $\mathfrak{g}$ is FdB if

(1) $\mathfrak{g}$ is connected, that is to say $\mathfrak{g}(0) = (0)$.
(2) For all $i \in \mathbb{N}^*$, $\mathfrak{g}(i)$ is one-dimensional.
(3) For all $n \geq 2$, $[\mathfrak{g}(1), \mathfrak{g}(n)] \neq (0)$.

Let $\mathfrak{g}$ be a FdB Lie algebra. For all $i \in \mathbb{N}^*$, we fix a non-zero element Z_i of $\mathfrak{g}(i)$. By conditions (1) and (2), $(Z_i)_{i\geq 1}$ is a basis of $\mathfrak{g}$. By homogeneity of the bracket of $\mathfrak{g}$, for all $i, j \geq 1$, there exists an element $\lambda_{i,j} \in K$, such that

$$[Z_i, Z_j] = \lambda_{i,j} Z_{i+j}.$$

The Jacobi relation gives, for all $i, j, k \geq 1$,

$$\lambda_{i,j}\lambda_{i+j,k} + \lambda_{j,k}\lambda_{j+k,i} + \lambda_{k,i}\lambda_{k+i,j} = 0. \tag{3}$$

Moreover, by antisymmetry, $\lambda_{j,i} = -\lambda_{i,j}$ for all $i, j \geq 1$. Condition (3) is expressed by $\lambda_{1,j} \neq 0$ for all $j \neq 1$.

LEMMA 19. *Up to a change of basis, we can suppose that $\lambda_{1,j} = 1$ for all $j \geq 2$ and that $\lambda_{2,3} \in \{0, 1\}$.*

PROOF. We define a family of scalars by

$$\left\{\begin{array}{lcl} \alpha_1 & = & 1, \\ \alpha_2 & \neq & 0, \\ \alpha_n & = & \lambda_{1,2} \cdots \lambda_{1,n-1}\alpha_2 \text{ if } n \geq 3. \end{array}\right.$$

By condition (3), all these scalars are non-zero. We put $Z'_i = \alpha_i Z_i$. Then, for all $j \geq 2$,

$$[Z'_1, Z'_j] = \alpha_j \lambda_{1,j} Z_{1+j} = \frac{\alpha_j \lambda_{1,j}}{\alpha_{j+1}} Z'_{1+j} = Z'_{1+j}.$$

So, replacing the Z_i's by the Z'_i's, we can suppose that $\lambda_{1,j} = 1$ if $j \geq 2$.

Let us suppose now that $\lambda_{2,3} \neq 0$. We then choose

$$\alpha_2 = \frac{\lambda_{1,3}\lambda_{1,4}}{\lambda_{2,3}}.$$

Then

$$[Z'_2, Z'_3] = \frac{\lambda_{2,3}\alpha_2\alpha_3}{\alpha_5} Z'_5 = \frac{\lambda_{2,3}\alpha_2\lambda_{1,2}\alpha_2}{\lambda_{1,2}\lambda_{1,3}\lambda_{1,4}\alpha_2} Z'_5 = Z'_5.$$

So, replacing the Z_i's by the Z'_i's, we can suppose that $\lambda_{2,3} = 1$. □

LEMMA 20. *If $i, j \geq 2$, $\lambda_{i,j} = \sum_{k=0}^{i-2} \binom{i-2}{k} (-1)^k \lambda_{2,j+k}$.*

PROOF. Let us write (3) with $i = 1$,

$$\lambda_{1,j}\lambda_{j+1,k} + \lambda_{j,k}\lambda_{j+k,1} + \lambda_{k,1}\lambda_{k+1,j} = 0.$$

If $j, k \geq 2$, then $\lambda_{1,j} = -\lambda_{k,1} = -\lambda_{j+k,1} = 1$, so

$$\lambda_{k+1,j} = \lambda_{k,j} - \lambda_{k,j+1}. \tag{4}$$

If $k = 2$, this gives the announced formula for $i = 3$.

Let us prove the result by induction on i. This is obvious for $i = 2$ and done for $i = 3$. Let us assume the result at rank $i - 1$. Then, by (4),

$$\begin{aligned}
\lambda_{i,j} &= \lambda_{i-1,j} - \lambda_{i-1,j+1} \\
&= \sum_{k=0}^{i-3} \binom{i-3}{k} (-1)^k \lambda_{2,j+k} - \sum_{k=0}^{i-3} \binom{i-3}{k} (-1)^k \lambda_{2,j+1+k} \\
&= \sum_{k=0}^{i-3} \binom{i-3}{k} (-1)^k \lambda_{2,j+k} + \sum_{k=1}^{i-2} \binom{i-3}{k-1} (-1)^k \lambda_{2,j+k} \\
&= \lambda_{2,j} + \sum_{k=1}^{i-3} \binom{i-2}{k} (-1)^k \lambda_{2,j+k} + (-1)^{i-2} \lambda_{2,j+i-2} \\
&= \sum_{k=0}^{i-2} \binom{i-2}{k} (-1)^k \lambda_{2,j+k}.
\end{aligned}$$

So the result is true for all $i \geq 2$. □

As a consequence, the $\lambda_{i,j}$'s are entirely determined by the $\lambda_{2,j}$'s. We can improve this result, using the following lemma:

LEMMA 21. *For all* $k \geq 2$, $\lambda_{2,2k} = \frac{1}{2k-3} \sum_{l=0}^{2k-4} \binom{2k-2}{l} (-1)^l \lambda_{2,l+3}$.

PROOF. Let us write the relation of Lemma 20 for $(i,j) = (3, 2k)$ and $(i,j) = (2k, 3)$,

$$\begin{aligned}
\lambda_{3,2k} &= \lambda_{2,2k} - \lambda_{2,2k+1}, \\
\lambda_{2k,3} &= \sum_{l=0}^{2k-2} \binom{2k-2}{l} (-1)^l \lambda_{2,3+l} \\
&= \lambda_{2,2k+1} - (2k-2)\lambda_{2,2k} + \sum_{l=0}^{2k-4} \binom{2k-2}{l} (-1)^l \lambda_{2,3+l}.
\end{aligned}$$

Summing these two relations,

$$-(2k-3)\lambda_{2,2k} + \sum_{l=0}^{2k-4} \binom{2k-2}{l} (-1)^l \lambda_{2,3+l} = 0.$$

This gives the announced result. □

As a consequence, the $\lambda_{i,j}$'s are entirely determined by the $\lambda_{2,j}$'s, with j odd. In order to ease the notation, we put $\mu_j = \lambda_{2,j}$ for all j odd. Then, for example,

$$\left\{ \begin{aligned}
\lambda_{2,4} &= \mu_3, \\
\lambda_{2,6} &= 2\mu_5 - \mu_3, \\
\lambda_{2,8} &= 3\mu_7 - 5\mu_5 + 3\mu_3, \\
\lambda_{2,10} &= 4\mu_9 - 14\mu_7 + 28\mu_5 - 17\mu_3, \\
\lambda_{2,12} &= 5\mu_{11} - 30\mu_9 + 126\mu_7 - 255\mu_5 + 155\mu_3.
\end{aligned} \right.$$

Moreover, we showed that we can assume $\mu_3 = 0$ or 1.

Remark. The coefficient $\lambda_{2,2k+4}$ is then a linear span of coefficients μ_{2i+3}, $0 \leq i \leq k$. We put, for all $k \in \mathbb{N}$,

$$\lambda_{2,2k+4} = \sum_{i=0}^{k} a_{k,i}\mu_{2i+3}.$$

We can prove inductively the following results:

(1) For all $k \in \mathbb{N}$, $a_{k,k} = k + 1$.
(2) For all $k \geq 1$, $a_{k,k-1} = -\frac{1}{4}\binom{2k+2}{3}$. Up to the sign, this is the sequence *A*000330 of [**12**] (pyramidal numbers).
(3) For all $k \geq 2$, $a_{k,k-2} = \frac{1}{2}\binom{2k+2}{5}$. This is the sequence *A*053132 of [**12**].
(4) The sequence $(-a_{k,0})$ is the sequence of signed Genocchi numbers, *A*001469 in [**12**].

It seems that for all $i \leq k$,

$$a_{k,k-i} = \frac{2^{2i+2} - 1}{i + 1} B_{2i+2} \binom{2k+2}{2i+1},$$

where the B_{2n}'s are the Bernoulli numbers (see sequence *A*002105 of [**12**]).

3.2. Case where $\mu_3 = 1$. In this case:

LEMMA 22. *Suppose that $\mu_3 = 1$. Then $\mu_5 = 1$ or $\frac{9}{10}$.*

PROOF. By relation (3) for $(i, j, k) = (2, 3, 4)$,

$$5\mu_5 - 3\mu_7 + \mu_5\mu_7 - 3 = 0.$$

If $\mu_5 = 3$, we obtain $12 = 0$, absurd. So $\mu_7 = -\frac{5\mu_5 - 3}{\mu_5 - 3}$. By relation (3) for $(i, j, k) = (2, 3, 6)$,

$$\frac{-2}{\mu_5 - 3}\left((2\mu_5 - 4)\mu_5\mu_9 + 3 - 7\mu_5 + \mu_5^2 - 5\mu_5^3\right) = 0.$$

If $\mu_5 = 0$, we obtain $2 = 0$, absurd. If $\mu_5 = 2$, we obtain $66 = 0$, absurd. So

$$\mu_9 = -\frac{3 - 7\mu_5 + \mu_5^2 - 5\mu_5^3}{(2\mu_5 - 4)\mu_5}.$$

Writing relation (3) for $(i, j, k) = (3, 4, 5)$,

$$-\frac{9(\mu_5 - 1)^5(10\mu_5 - 9)}{\mu_5(\mu_5 - 2)(\mu_5 - 3)^2} = 0.$$

So $\mu_5 = 1$ or $\mu_5 = \frac{9}{10}$. □

PROPOSITION 23. *Let us suppose that $\mu_3 = \mu_5 = 1$. Then*

$$\begin{cases} \lambda_{1,j} &= 1 \text{ if } j \geq 2, \\ \lambda_{2,j} &= 1 \text{ if } j \geq 3, \\ \lambda_{i,j} &= 0 \text{ if } i, j \geq 3. \end{cases}$$

PROOF. Let us first prove inductively on j that $\lambda_{2,j} = 1$ if $j \geq 3$. This is immediate if $j = 3$ or 5 and comes from $\lambda_{2,4} = \mu_3$ for $j = 4$. Let us suppose that $\lambda_{2,j} = 1$ for $3 \leq j < n$. If $n = 2k$ is even, then

$$\lambda_{2,2k} = \frac{1}{2k-3} \sum_{l=0}^{2k-4} \binom{2k-2}{l} (-1)^l = 1 + \frac{1}{2k-3} \sum_{l=0}^{2k-2} \binom{2k-2}{l} (-1)^l = 1.$$

If $n = 2k+1$ is odd, write relation (3) for $(i,j,k) = (2,3,2k-2)$,

$$\begin{aligned} \lambda_{2,3}\lambda_{5,2k-2} + \lambda_{3,2k-2}\lambda_{2k+1,2} + \lambda_{2k-2,2}\lambda_{2k,3} &= 0, \\ \sum_{l=0}^{3} \binom{3}{l} (-1)^l \lambda_{2,2k-2+l} - \lambda_{2,2k-2} + \lambda_{2,2k-1} + \lambda_{2,2k-2}(\lambda_{2,2k} - \lambda_{2,2k+1}) &= 0, \\ \lambda_{2,2k-2} - 3\lambda_{2,2k-1} + 3\lambda_{2,2k} - \lambda_{2,2k+1} + 1 - \lambda_{2,2k+1} &= 0, \\ 1 - 3 + 3 - 2\lambda_{2,2k+1} + 1 &= 0, \end{aligned}$$

so $\lambda_{2,2k+1} = 1$. Finally, if $i,j \geq 3$, $\lambda_{i,j} = \sum_{k=0}^{i-2} \binom{i-2}{k} (-1)^k = 0$. □

LEMMA 24. *For all $N \geq 2$,*

$$S_N = \sum_{l=0}^{N} \binom{N}{l} (-1)^l \frac{(l+1)}{(l+2)(l+3)} = \frac{N-1}{(N+3)(N+2)(N+1)}.$$

PROOF. Indeed,

$$\begin{aligned} S_N &= \sum_{l=0}^{N} \frac{N!}{(l+3)!(N-l)!} (-1)^l (l+1)^2 \\ &= \frac{1}{(N+3)(N+2)(N+1)} \sum_{j=3}^{N+3} \binom{N+3}{j} (-1)^j (j-2)^2 \\ &= \frac{1}{(N+3)(N+2)(N+1)} \sum_{j=0}^{N+3} \binom{N+3}{j} (-1)^j (j-2)^2 \\ &\quad - \frac{1}{(N+3)(N+2)(N+1)} (4 - (N+3)) \\ &= 0 + \frac{N-1}{(N+3)(N+2)(N+1)}. \end{aligned}$$

□

PROPOSITION 25. *Let us suppose that $\mu_3 = 1$ and $\mu_5 = \frac{9}{10}$. Then, for all $i, j \geq 1$,*

$$\lambda_{i,j} = \frac{6(i-j)(i-2)!(j-2)!}{(i+j-2)!}.$$

PROOF. We first prove that $\lambda_{2,n} = \frac{6(n-2)}{(n-1)n}$. This is immediate for $n = 1$, 2, 3, 4, 5. Let us assume the result for all $j < n$, with $n \geq 6$. If $n = 2k$ is even, using

Lemma 20,

$$\lambda_{2,2k} = \frac{6}{2k-3} \sum_{l=0}^{2k-4} \binom{2k-2}{l} (-1)^l \frac{l+1}{(l+2)(l+3)}.$$

Then Lemma 24 gives the result. If $n = 2k+3$ is odd, let us write the relation (3) with $(i,j,k) = (2,3,2k)$,

$$\lambda_{2,3}\lambda_{5,2k} + \lambda_{3,2k}\lambda_{2k+3,2} + \lambda_{2k,2}\lambda_{2k+2,3} = 0.$$

So, with relation (4)

$$\begin{array}{rcl} \lambda_{2,2k} - 3\lambda_{2,2k+1} + 3\lambda_{2,2k+2} - \lambda_{2,2k+3} & & \\ -\lambda_{2,2k+3}(\lambda_{2,2k} - \lambda_{2,2k+1}) + \lambda_{2,2k}(\lambda_{2,2k+2} - \lambda_{2,2k+3}) & = & 0, \\ \lambda_{2,2k+3}(-1 - 2\lambda_{2,2k} + \lambda_{2,2k+1}) + \lambda_{2,2k} - 3\lambda_{2,2k+1} & & \\ +3\lambda_{2,2k+2} + \lambda_{2,2k}\lambda_{2,2k+2} & = & 0, \\ -\lambda_{2,2k+3}\dfrac{(2k+3)(k-1)(2k+5)}{k(2k+1)(2k-1)} + \dfrac{3(2k+5)(k-1)}{k(k+1)(2k-1)} & = & 0, \end{array}$$

which implies the result.

Let us now prove the result by induction on i. This is immediate if $i = 1$, and the first part of this proof for $i = 2$. Let us assume the result at rank i. Then, by relation (4),

$$\begin{array}{rcl} \lambda_{i+1,j} & = & \lambda_{i,j} - \lambda_{i,j+1} \\ & = & 6\dfrac{(j-i)(i-2)!(j-2)!}{(i+j-2)!} - 6\dfrac{(j+1-i)(i-2)!(j-1)!}{(i+j-1)!} \\ & = & 6\dfrac{(i-1)!(j-2)!(j+1-i)}{(i+j-1)!}. \end{array}$$

So the result is true for all i,j. □

3.3. Case where $\mu_3 = 0$.

In this case:

PROPOSITION 26. *If $\mu_3 = 0$, then $\lambda_{i,j} = 0$ for all $i,j \geq 2$.*

PROOF. We first prove that $\mu_5 = 0$. If not, by (3) for $(i,j,k) = (2,3,4)$, $\mu_5\mu_7 = 0$, so $\mu_7 = 0$. By (3) with $(i,j,k) = (2,3,7)$, $-5\mu_5(28\mu_5 + 4\mu_9) = 0$, so $\mu_9 = -7\mu_5$. By (3) with $(i,j,k) = (3,4,5)$, $-36\mu_5^2 = 0$: contradiction. So $\mu_5 = 0$.

Let us then prove that all the μ_{2k+1}'s, $k \geq 1$, are zero. We assume that $\mu_3 = \mu_5 = \cdots = \mu_{2k-1} = 0$, and $\mu_{2k+1} \neq 0$, with $l \geq 3$. By lemma 21, $\lambda_{2,2} = \cdots = \lambda_{2,2k-1} = \lambda_{2,2k} = 0$ and $\lambda_{2,2k+1} \neq 0$. By relation (3) for $(i,j,k) = (2,3,n)$, combined with (4),

$$\begin{array}{rcl} \lambda_{2,3}\lambda_{5,n} + \lambda_{3,n}\lambda_{n+3,2} + \lambda_{n,2}\lambda_{n+2,3} & = & 0, \\ -(\lambda_{2,n} - \lambda_{2,n+1})\lambda_{2,n+3} + \lambda_{2,n}(\lambda_{2,n+2} - \lambda_{2,n+3}) & = & 0. \end{array}$$

For $n = 2k$, this gives $\lambda_{2,2k+1}\lambda_{2,2k+3} = 0$, so $\lambda_{2,2k+3} = 0$. For $n = 2k+2$,

$$\lambda_{2,2k+2}(\lambda_{2,2k+4} - 2\lambda_{2,2k+5}) = 0. \tag{5}$$

By Lemma 21,

$$\begin{aligned}\lambda_{2,2k+2} &= \frac{1}{2k-1}\binom{2k}{2k-2}\lambda_{2,2k+1}\\ &= k\lambda_{2,2k+1},\\ \lambda_{2,2k+4} &= \frac{1}{2k+1}\left(\binom{2k+2}{2k}\lambda_{2,2k+3}-\binom{2k+2}{2k-1}\lambda_{2,2k+2}+\binom{2k+2}{2k-2}\lambda_{2,2k+1}\right)\\ &= -\frac{k(k+1)(2k+1)}{6}\lambda_{2,2k+1}.\end{aligned}$$

With (5),

$$\lambda_{2,2k+5} = -\frac{k(k+1)(2k+1)}{12}\lambda_{2,2k+1}.$$

By relation (3) for $(i,j,k)=(3,4,2k)$,

$$\lambda_{3,4}\lambda_{7,2k}+\lambda_{4,2k}\lambda_{4+2k,3}+\lambda_{2k,3}\lambda_{2k+3,4}=0.$$

Moreover, using Lemma 20,

$$\left\{\begin{array}{rcl}\lambda_{3,4} &=& \lambda_{2,4}-\lambda_{2,5}=0,\\ \lambda_{4,2k} &=& \lambda_{2,2k}-2\lambda_{2,2k+1}+\lambda_{2,2k+2},\\ \lambda_{3,4+2k} &=& \lambda_{2,4+2k}-\lambda_{2,5+2k},\\ \lambda_{2k,3} &=& -\lambda_{2,2k}+\lambda_{2,2k+1},\\ \lambda_{3+2k,4} &=& -\lambda_{2,3+2k}+2\lambda_{2,4+2k}-\lambda_{2,5+2k}.\end{array}\right.$$

This gives

$$\frac{\lambda_{2,2k+1}^2 k(3k-11)(2k+1)(k+1)}{12}=0,$$

so $\lambda_{2,2k+1}=\mu_{2k+1}=0$: contradiction. So all the μ_{2k+1}, $k\geq 1$, are zero. By Lemma 21, the $\lambda_{2,i}$'s, $i\geq 2$, are zero. By Lemma 20, the $\lambda_{i,j}$'s, $i,j\geq 2$, are zero. □

THEOREM 27. *Up to isomorphism, there are three* FdB *Lie algebras:*

(1) *The Faà di Bruno Lie algebra $\mathfrak{g}_{\mathrm{FdB}}$, with basis $(e_i)_{i\geq 1}$, and the bracket given by $[e_i,e_j]=(j-i)e_{i+j}$ for all $i,j\geq 1$.*
(2) *The corolla Lie algebra $\mathfrak{g}_c$, with basis $(e_i)_{i\geq 1}$, and the bracket given by $[e_1,e_j]=e_{j+1}$ and $[e_i,e_j]=0$ for all $i,j\geq 2$.*
(3) *Another Lie algebra $\mathfrak{g}_3$, with basis $(e_i)_{i\geq 1}$, and the bracket given by $[e_1,e_i]=e_{i+1}$, $[e_2,e_j]=e_{j+2}$, and $[e_i,e_j]=0$ for all $i\geq 2$, $j\geq 3$.*

PROOF. We have first to prove that these are indeed Lie algebras: this is done by direct computations. Let $\mathfrak{g}$ be a FdB Lie algebra. We showed that three cases are possible:

(1) $\mu_3=1$ and $\mu_5=\frac{9}{10}$. By Proposition 25, putting $e_i=\frac{Z_i}{6(i-2)!}$ if $i\geq 2$ and $e_1=Z_1$, we obtain the Faà di Bruno Lie algebra.
(2) $\mu_3=\mu_5=1$. By Proposition 23, we obtain the third Lie algebra.
(3) $\mu_3=0$. By Proposition 26, we obtain the corolla Lie algebra.

□

COROLLARY 28. *Let $\mathfrak{g}$ be a* FdB *Lie algebra, such that if i and j are two distinct elements of $\mathbb{N}^*$, then $[\mathfrak{g}(i),\mathfrak{g}(j)]\neq(0)$. Then $\mathfrak{g}$ is isomorphic to the Faà di Bruno Lie algebra.*

4. Dual of enveloping algebras of FdB Lie algebras

We realized in the first section the Faà di Bruno Hopf algebra, the dual of the enveloping algebra of the Faà di Bruno Lie algebra, as a Hopf subalgebra of $\mathcal{H}$. We now give a similar result for the two other FdB Lie algebra.

4.1. The corolla Lie algebra.

DEFINITION 29. We denote by $\mathcal{H}_c$ the subalgebra of $\mathcal{H}$ generated by the corollas.

PROPOSITION 30. *$\mathcal{H}_c$ is a graded Hopf subalgebra of $\mathcal{H}$. Its dual is isomorphic to the enveloping algebra of the corolla Lie algebra.*

PROOF. The subalgebra $\mathcal{H}_c$, being generated by homogeneous elements, is graded. By Lemma 9, $\mathcal{H}_c$ is a Hopf subalgebra of $\mathcal{H}$. As it is commutative, its dual is the enveloping algebra of the Lie algebra $\mathrm{Prim}(\mathcal{H}_c^*)$. The dual of this Lie algebra is the Lie coalgebra $\mathrm{coPrim}(\mathcal{H}_c) = \dfrac{\mathcal{H}_c}{(1) \oplus \mathrm{Ker}(\varepsilon)^2}$, with cobracket δ induced by $(\varpi \otimes \varpi) \circ (\Delta - \Delta^{op})$. As $\mathcal{H}_c$ is generated by the corollas, a basis of $\mathrm{coPrim}(\mathcal{H}_c)$ is $(\varpi(c_n))_{n \geq 1}$. Moreover, if $n \geq 1$,

$$\begin{aligned}
(\varpi \otimes \varpi) \circ \Delta(c_n) &= \varpi(c_1) \otimes \varpi(c_{n-1}), \\
\delta(c_n) &= \varpi(c_1) \otimes \varpi(c_{n-1}) - \varpi(c_{n-1}) \otimes \varpi(c_1).
\end{aligned}$$

Let $(Z_n)_{n \geq 1}$ be the basis of $\mathrm{Prim}(\mathcal{H}_c^*)$, the dual of the basis $(\varpi(c_n))_{n \geq 1}$. By duality, for all $i, j \in \mathbb{N}^*$, such that $i \neq j$,

$$[Z_i, Z_j] = \begin{cases} Z_{1+j} \text{ if } i = 1, \\ -Z_{i+1} \text{ if } j = 1, \\ 0 \text{ otherwise.} \end{cases}$$

So $\mathrm{Prim}(\mathcal{H}_c^*)$ is isomorphic to the corolla Lie algebra, via the isomorphism

$$\begin{cases} \mathfrak{g}_c & \longrightarrow & \mathrm{Prim}(\mathcal{H}_c^*) \\ e_i & \longrightarrow & Z_i. \end{cases}$$

Dually, $\mathcal{H}_c$ is isomorphic to $\mathcal{U}(\mathfrak{g}_c)^*$. □

Remark. We work in $K[\mathcal{T}][\beta]$. The generators of $\mathcal{H}_{1,\beta}$ then satisfy

$$x_{n+1} = \frac{[n]_\beta!}{n!} c_{n+1} + \mathcal{O}(\beta^{n-2}).$$

Note that the degree of $[n]_\beta!$ in β is $n-1$. So

$$\lim_{\beta \to \infty} \frac{n!}{[n]_\beta!} x_{n+1} = c_{n+1}.$$

In this sense, the Hopf algebra $\mathcal{H}_c$ is the limit of $\mathcal{H}_{1,\beta}$ when β goes to infinity.

4.2. The third FdB Lie algebra. We consider the following element of $\widehat{\mathcal{H}}$:

$$Y = B^+ \left(\exp \left(\mathfrak{l} - \frac{1}{2} \bullet^2 + \bullet \right) \right) = \sum_{n \geq 1} y_n.$$

For example

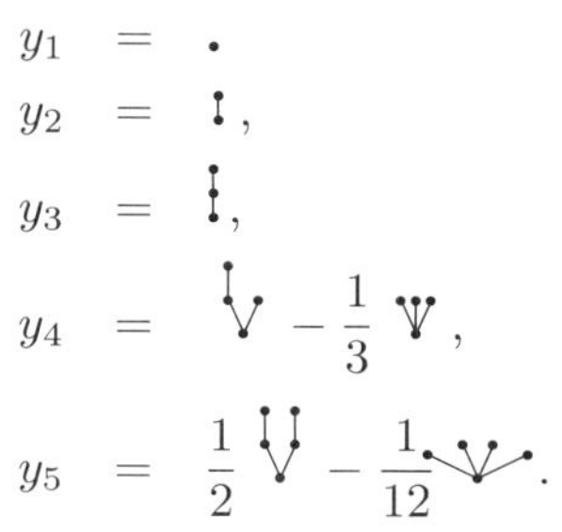

DEFINITION 31. We denote by $\mathcal{H}_3$ the subalgebra of $\mathcal{H}$ generated by the y_n's.

PROPOSITION 32. *$\mathcal{H}_3$ is a graded Hopf subalgebra of $\mathcal{H}$. Its dual is isomorphic to the enveloping algebra of the third* FdB *Lie algebra.*

PROOF. The subalgebra $\mathcal{H}_3$, being generated by homogeneous elements, is graded. An easy computation proves that $X =$ $- \frac{1}{2}\bullet^2 + \bullet$ is a primitive element of $\mathcal{H}$. As a consequence, in $\widehat{\mathcal{H}}$, by (1),

$$\begin{aligned}
\Delta(X) &= X \otimes 1 + 1 \otimes X, \\
\Delta(\exp(X)) &= \exp(X \otimes 1 + 1 \otimes X) \\
&= \exp(X \otimes 1)\exp(1 \otimes X) \\
&= (\exp(X) \otimes 1)(1 \otimes \exp(X)) \\
&= \exp(X) \otimes \exp(X), \\
\Delta(Y) &= \Delta \circ B^+(\exp(X)) \\
&= Y \otimes 1 + \exp(X) \otimes Y.
\end{aligned}$$

Moreover, $X = y_2 - \frac{1}{2}y_1^2 + y_1 \in \mathcal{H}_3$, so taking the homogeneous component of degree n of $\Delta(Y)$, we obtain

$$\Delta(y_n) = y_n \otimes 1 + \sum_{k=1}^{n} \sum_{l=1}^{n-k} \sum_{a_1+\cdots+a_l=n-k} \frac{1}{l!} x_{a_1} \cdots x_{a_l} \otimes y_k,$$

where $x_1 = \bullet = y_1$, $x_2 =$ $- \frac{1}{2}\bullet\bullet = y_2 - \frac{1}{2}y_1^2$ and $x_i = 0$ if $i \geq 3$, so $\Delta(y_n) \in \mathcal{H}_3 \otimes \mathcal{H}_3$ and $\mathcal{H}_3$ is a Hopf subalgebra of $\mathcal{H}$. As it is commutative, its dual is the enveloping algebra of the Lie algebra $\mathrm{Prim}(\mathcal{H}_3^*)$. The dual of this Lie algebra is the Lie coalgebra $\mathrm{coPrim}(\mathcal{H}_3) = \dfrac{\mathcal{H}_3}{(1) \oplus \mathrm{Ker}(\varepsilon)^2}$, with cobracket δ induced by $(\varpi \otimes \varpi) \circ (\Delta - \Delta^{op})$. As $\mathcal{H}_3$ is generated by the y_n's, a basis of $\mathrm{coPrim}(\mathcal{H}_3)$ is $(\varpi(y_n))_{n\geq 1}$. Moreover,

$$\begin{aligned}
(\varpi \otimes \varpi) \circ \Delta(Y) &= \varpi(\exp(X)) \otimes \varpi(Y) \\
&= \varpi(X) \otimes \varpi(Y) \\
&= (\varpi(y_2) + \varpi(y_1)) \otimes \varpi(Y).
\end{aligned}$$

Taking the homogeneous component of degree n, with the convention $y_{-1} = y_0 = 0$,

$$\begin{array}{rcl} (\varpi \otimes \varpi) \circ \Delta(y_n) & = & \varpi(y_2) \otimes \varpi(y_{n-2}) + \varpi(y_1) \otimes \varpi(y_{n-1}), \\ \delta(\varpi(y_n)) & = & \varpi(y_2) \otimes \varpi(y_{n-2}) + \varpi(y_1) \otimes \varpi(y_{n-1}) \\ & & -\varpi(y_{n-2}) \otimes \varpi(y_2) - \varpi(y_{n-1}) \otimes \varpi(y_1). \end{array}$$

Let $(Z_n)_{n\geq 1}$ be the basis of $\mathrm{Prim}(\mathcal{H}_3^*)$ dual to the basis $(\varpi(c_n))_{n\geq 1}$. By duality for all $i, j \in \mathbb{N}^*$, such that $i \geq 2$ and $j \geq 3$,

$$\left\{ \begin{array}{rcl} [Z_1, Z_i] & = & Z_{1+j}, \\ [Z_2, Z_j] & = & Z_{2+j}, \\ [Z_i, Z_j] & = & 0. \end{array} \right.$$

So $\mathrm{Prim}(\mathcal{H}_3^*)$ is isomorphic to third FdB Lie algebra, via the morphism

$$\left\{ \begin{array}{rcl} \mathfrak{g}_3 & \longrightarrow & \mathrm{Prim}(\mathcal{H}_c^*) \\ e_i & \longrightarrow & Z_i. \end{array} \right.$$

Dually, $\mathcal{H}_3$ is isomorphic to $\mathcal{U}(\mathfrak{g}_3)^*$. □

References

[1] Christoph Bergbauer and Dirk Kreimer, *Hopf algebras in renormalization theory: locality and Dyson-Schwinger equations from Hochschild cohomology*, IRMA Lect. Math. Theor. Phys., vol. 10, Eur. Math. Soc., Zürich, 2006, arXiv:hep-th/0506190.

[2] Alain Connes and Dirk Kreimer, *Hopf algebras, renormalization and noncommutative geometry*, Comm. Math. Phys **199** (1998), no. 1, 203–242, arXiv:hep-th/9808042.

[3] Héctor Figueroa and José M. Gracia-Bondia, *On the antipode of Kreimer's Hopf algebra*, Modern Phys. Lett. A **16** (2001), no. 22, 1427–1434, arXiv:hep-th/9912170.

[4] Loïc Foissy, *page personnelle*, http://loic.foissy.free.fr/pageperso/publications.html.

[5] ———, *Les algèbres de Hopf des arbres enracinés, II*, Bull. Sci. Math. **126** (2002), no. 4, 1249–288, arXiv:math.QA/0105212.

[6] ———, *Faà di Bruno subalgebras of the Hopf algebra of planar trees from combinatorial Dyson-Schwinger equations*, Advances in Mathematics **218** (2008), 136–162, ArXiv:0707.1204.

[7] Michael E. Hoffman, *Combinatorics of rooted trees and Hopf algebras*, Trans. Amer. Math. Soc. **355** (2003), no. 9, 3795–3811.

[8] Dirk Kreimer, *Combinatorics of (perturbative) Quantum Field Theory*, Phys. Rep. **4–6** (2002), 387–424, arXiv:hep-th/0010059.

[9] ———, *Dyson-Schwinger equations: from Hopf algebras to number theory*, Universality and renormalization, Fields Inst. Commun., no. 50, Amer. Math. Soc., Providence, RI, 2007, arXiv:hep-th/0609004.

[10] Dirk Kreimer and Karen Yeats, *An étude in non-linear Dyson-Schwinger equations*, Nuclear Phys. B Proc. Suppl. **160** (2006), 116–121, arXiv:hep-th/0605096.

[11] John W. Milnor and John C. Moore, *On the structure of Hopf algebras*, Ann. of Math. (2) **81** (1965), 211–264.

[12] N. J. A Sloane, *On-line encyclopedia of integer sequences*, http://www.research.att.com/ ~njas/ sequences/ Seis.html.

[13] Richard P. Stanley, *Enumerative combinatorics. Vol. 1.*, Cambridge Studies in Advanced Mathematics, no. 49, Cambridge University Press, Cambridge, 1997.

[14] ———, *Enumerative combinatorics. Vol. 2.*, Cambridge Studies in Advanced Mathematics, no. 62, Cambridge University Press, Cambridge, 1999.

Laboratoire de Mathématiques EA4535, Université de Reims, Moulin de la Housse - BP 1039, 51687 REIMS Cedex 2 France

E-mail address: loic.foissy@univ-reims.fr

Clay Mathematics Proceedings
Volume 12, 2010

From Gauge Anomalies to Gerbes and Gerbal Actions

Jouko Mickelsson

ABSTRACT. The purpose of this contribution is to point out connections between recent ideas about the gerbes and gerbal actions (as higher categorical extension of representation theory) and the old discussion in quantum field theory on commutator anomalies, gauge group extensions, and 3-cocycles. The unifying concept is the classical obstruction theory for group extensions as explained in the reference [M].

1. INTRODUCTION

It was first realized through perturbative analysis of gauge theories that gauge symmetry is broken in the presence of chiral fermions, [ABJ]. Later, it was found that this phenomenon is related to the index theory of (families) of Dirac operators. In particular, the effective action functional, defined as a regularized determinant of the Dirac operator, is not always gauge invariant and the lack of invariance can be formulated as the curvature of a complex line bundle, the determinant line bundle, over the moduli space of gauge connections, [AS].

In the Hamiltonian formulation of gauge theory the symmetry breaking manifests itself as a modification of the commutation relations of the Lie algebra of infinitesimal gauge transformations. The gauge algebra (in the case of a trivial vector bundle) is the Lie algebra of functions $M\mathfrak{g}$ from the physical space M to a finite-dimensional Lie algebra $\mathfrak{g}$. The commutation relations of the modified algebra can be written as

$$[(X,a),(Y,b)] = ([X,Y], \mathcal{L}_X b - \mathcal{L}_Y a + c(A;X,Y))$$

where $[X,Y]$ is the point wise commutator in $M\mathfrak{g}$ and a, b are complex-valued functions of the gauge potential A, $\mathcal{L}_X$ is the Lie derivative defined by an infinitesimal gauge transformation X, and c is a Lie algebra 2-cocycle determining an extension of $M\mathfrak{g}$. In the case when M is the unit circle S^1 it turns out that c is independent of A and we have a central extension defining (when $\mathfrak{g}$ is simple) an affine Kac-Moody algebra (here a, b are constant functions).

When $\dim M > 1$ the cocycle c depends explicitly on A. It is still an open question whether this algebra has interesting faithful Hilbert space representations,

2010 *Mathematics Subject Classification.* Primary 17B56, Secondary 17B65, 22E41, 22E67.

analogous to the highest weight representations of affine Lie algebras extensively used in string theory and constructions of quantum field theory models in 1+1 space-time dimensions. What is known at present that there are natural unitary *Hilbert bundle* actions of the extended gauge Lie algebras and groups. These come from quantizing chiral fermions in background gauge fields. For each gauge connection A there is a fermionic Fock space $\mathcal{F}_A$ where the quantized Dirac Hamiltonian $\hat{D}_A$ acts. This family of (essentially positive) Dirac operators transforms equivariantly with respect to the action of the extension of the group MG of functions from M to G (point wise multiplication of functions). The gauge transformations are defined as projective unitary operators between the fibers $\mathcal{F}_A$ and $\mathcal{F}_{A^g}$ of the Fock bundle, corresponding to a true action of the gauge group extension. As a consequence, the Fock bundle is defined only as a projective bundle over the moduli space $\mathcal{A}/MG$. Actually, in order that the moduli space be a smooth manifold, one has to restrict MG to the *based gauge transformations* which are functions on M taking the value $e \in G$ at a fixed base point $x_0 \in M$.

A projective bundle is completely determined, up to equivalence, by the Dixmier-Douady class, which is an element of $H^3(\mathcal{A}/MG, \mathbb{Z})$. This is the origin of gerbes in quantum field theory, [CMM]. Topologically a gerbe on a space X is just an equivalence class of $PU(H) = U(H)/S^1$ bundles over X. Here $U(H)$ is the (contractible) unitary group in a complex Hilbert space H. In terms of Čech cohomology subordinate to a good cover $\{U_\alpha\}$ of X, the gerbe is given as a $\mathbb{C}^\times$-valued cocycle $\{f_{\alpha\beta\gamma}\}$,

$$f_{\alpha\beta\gamma} f^{-1}_{\alpha\beta\delta} f_{\alpha\gamma\delta} f^{-1}_{\beta\gamma\delta} = 1$$

on intersections $U_\alpha \cap U_\beta \cap U_\gamma \cap U_\delta$. This cocycle arises from the lifting problem: A $PU(H)$ bundle is given in terms of transition functions $g_{\alpha\beta}$ with values in $PU(H)$. After lifting these to $U(H)$ one gets a family of functions $\hat{g}_{\alpha\beta}$ which satisfy the 1-cocycle condition up to a phase,

$$\hat{g}_{\alpha\beta}\hat{g}_{\beta\gamma}\hat{g}_{\gamma\alpha} = f_{\alpha\beta\gamma}\mathbf{1}.$$

The notion of gerbal action was introduced in the recent paper [FZ]. This is to be viewed as the next level after projective actions related to central extensions of groups and is given in terms of third group cohomology. In fact, the appearance of third cohomology in this context is not new and is related to group extensions as explained in [M]. In the simplest form, the problem is the following. Let F be an extension of G by the group N,

$$1 \to N \to F \to G \to 1$$

an exact sequence of groups. Suppose that $1 \to a \to \hat{N} \to N \to 1$ is a central extension by the abelian group a. Then one can ask whether the extension F of G by N can be prolonged to an extension of G by the group $\hat{N}$. The obstruction to this is an element in the group cohomology $H^3(G, a)$ with coefficients in a. In the case of Lie groups, there is a corresponding Lie algebra cocycle representing a class in $H^3(\mathfrak{g}, \mathfrak{a})$, where $\mathfrak{a}$ is the Lie algebra of a. We shall demonstrate this in detail for an example arising from the quantization of gauge theory. It is closely related to the idea in [Ca], further elaborated in [CGRS], which in turn was a response to a discussion in the 80's on breaking of the Jacobi identity for the field algebra in Yang-Mills theory [GJJ].

The paper is organized as follows. In Section 2 we explain the gauge group extensions arising from the action on bundles of fermionic Fock spaces over background gauge fields and the corresponding Lie algebra cocycles. Section 3 consists of a general discussion how gerbal action arises from the group of outer automorphisms of an associative algebra, how this leads to a 3-cocycle on the symmetry group, and finally we give an example coming from Yang-Mills theory in 1+1 space-time dimensions. Section 4 contains a generalization to Yang-Mills theory in higher space-time dimensions. Finally in Section 5 we explain an application to twisted K-theory on moduli space of gauge connections.

2. BUNDLES OF FOCK SPACES OVER GAUGE CONNECTIONS

A basic problem in quantum field theory in more than two space-time dimensions is that the representations of the canonical anticommutation relations algebra (CAR) are not equivalent in different background gauge fields, and this leads to various divergences in perturbation theory. However, in the case of the linear problem of quantizing fermions in a background gauge field one can construct the hamiltonian and the Hilbert space in a nonperturbative way. One can actually avoid the divergences by taking systematically into account the need of dealing with a family of nonequivalent CAR algebra representations.

The method introduced in [Mi93] and generalized in [LM] is based on the observation that, for each gauge connection A in the family $\mathcal{A}$ of all gauge connections on a vector bundle E over a compact spin manifold, one can choose a unitary operator T_A in the Hilbert space H of L^2 sections in the tensor product of the spin bundle and the vector bundle E such that the Dirac Hamiltonian D_A is conjugated to $\tilde{D}_A = T_A D_A T_A^{-1}$ such that the equivalent Hamiltonian $\tilde{D}_A$ can be quantized in the "free" Fock space, the Fock space for a fixed background connection A_0. In the case of a trivial bundle E one can take as A_0 the globally defined gauge connection represented by the 1-form equal to zero.

The action of the group $\mathcal{G}$ of smooth gauge transformations $A \mapsto A^g = g^{-1}Ag + g^{-1}dg$ on the family $\tilde{D}_A$ is then given by

$$\tilde{D}_A \mapsto \omega(A;g)^{-1}\tilde{D}_A\omega(A;g)$$

corresponding to $D_A \mapsto g^{-1}D_A g$ where, $\omega(A;g) = T_A g T_{A^g}^{-1}$ satisfies the 1-cocycle relation

$$\omega(A;gg') = \omega(A;g)\omega(A^g;g').$$

Furthermore, the cocycle satisfies the condition that $[\epsilon, \omega(A;g)]$ is Hilbert-Schmidt. Here ϵ is the sign $D_{A_0}/|D_{A_0}|$ of the free Dirac operator. This means that the operators $\omega(A;g)$ belong to the restricted unitary group $U_{\text{res}}(H_+ \oplus H_-)$ where $H = H_+ \oplus H_-$ is the polarization of H with respect to the sign operator ϵ, [PS].

The quantization of the operator $\tilde{D}_A$ is obtained in a fermionic Fock space $\mathcal{F}$ which carries an irreducible representation of the CAR algebra $\mathcal{B}$, which is a completion of the algebra defined by generators and relations according to

$$a^*(u)a(v) + a(v)a^*(u) = 2\langle v, u\rangle$$

and all other anticommutators equal to zero. Here $u, v \in H$ and $\langle \cdot, \cdot \rangle$ is the Hilbert space inner product (antilinear in the first argument). The representation is fixed (up to equivalence) by the requirement that there exists a vacuum vector $|0\rangle \in \mathcal{F}$ such that

$$a(u)|0\rangle = 0 = a^*(v)|0\rangle, \text{ for } u \in H_+, v \in H_-.$$

The group $U_{\rm res}(H)$ has a central extension by S^1 such that the Lie algebra central extension is given by the 2-cocycle $c(X,Y) = \frac{1}{4}\mathrm{tr}\, \epsilon[\epsilon, X][\epsilon, Y]$. The central extension $\hat{U}_{res}$ has a unitary representation $g \mapsto \hat{g}$ in $\mathcal{F}$ fixed by the requirement

$$\hat{g}a^*(u)\hat{g}^{-1} = a^*(gu)$$

for all $u \in H$.

If we choose a lift $\hat{\omega}(A;g)$ of the element $\omega(A,g)$ to unitaries in the Fock space $\mathcal{F}$ we can write

$$\hat{\omega}(A;gg') = \Phi(A;g,g')\hat{\omega}(A;g)\hat{\omega}(A^g;g')$$

where Φ takes values in S^1. It is a 2-cocycle by construction,

$$\Phi(A;g,g')\Phi(A;gg',g'') = \Phi(A;g,g'g'')\Phi(A^g;g',g''),$$

which is smooth in an open neighborhood of the neutral element in $\mathcal{G}$. This just reflects the associativity in the group multiplication in the central extension $\hat{U}_{\rm res}$.

Taking the second derivative

$$\frac{d^2}{dtds}|_{t=s=0}\Phi(A;e^{tX},e^{sY}) = \frac{1}{2}c(A;X,Y)$$

gives a 2-cocycle c for the Lie algebra of $\mathcal{G}$ with coefficients in the ring of complex functions of the variable A.

The cocycle depends on the lift $\omega \mapsto \hat{\omega}$ but two lifts are related by a multiplication by a circle-valued function $\psi(A;g)$ and the corresponding 2-cocycles are related by a coboundary,

$$\Phi'(A;g,g') = \Phi(A,g,g')\psi(A;gg')\psi(A;g)^{-1}\psi(A^g;g')^{-1}.$$

The Lie algebra cocycle c satisfies

$$c(A;X,[Y,Z]) + \mathcal{L}_X c(A;Y,Z) + \text{cyclic permutations} = 0,$$

where $\mathcal{L}_X$ is the Lie derivative acting on functions $f(A)$ through infinitesimal gauge transformations, $(\mathcal{L}_X f)(A) = Df(A)\cdot([A,X]+dX)$.

Explicit expressions for the cocycle c have been computed in the literature; for example if the physical space is a circle we get the central extension of a loop algebra (affine Kac-Moody algebra),

$$c(A;X,Y) = \frac{1}{2\pi}\int_{S^1} \mathrm{tr}\, XdY, \tag{2.1}$$

where the trace is evaluated in a finite dimensional representation of G. In this case c does not depend on A and the abelian extension reduces to a central extension. This reflects the fact that elements of LG act in the Hilbert space H through an embedding $LG \to U_{\rm res}$ and we can simply choose $T_A \equiv 1$ for all gauge connections A.

In three dimensions the simplest expression for the cocycle is, [Fa], [Mi85],

$$c(A;X,Y) = \frac{1}{24\pi^2}\int_M \mathrm{tr}\, A[dX,dY]. \tag{2.2}$$

3. GERBAL ACTIONS AND 3-COCYCLES

Let $\mathcal{B}$ be an associative algebra and G a group. Assume that we have a group homomorphism $s : G \to \mathrm{Out}(\mathcal{B})$ where $\mathrm{Out}(\mathcal{B})$ is the group of outer automorphisms of $\mathcal{B}$, that is, $\mathrm{Out}(\mathcal{B}) = \mathrm{Aut}(\mathcal{B})/\mathrm{In}(\mathcal{B})$, all automorphisms modulo the normal subgroup of inner automorphisms. If one chooses any lift $\tilde{s} : G \to \mathrm{Aut}(\mathcal{B})$ then we can write

$$\tilde{s}(g)\tilde{s}(g') = \sigma(g, g') \cdot \tilde{s}(gg')$$

for some $\sigma(g, g') \in \mathrm{In}(\mathcal{B})$. From the definition follows immediately the cocycle property

$$\sigma(g, g')\sigma(gg', g'') = [\tilde{s}(g)\sigma(g', g'')\tilde{s}(g)^{-1}]\sigma(g, g'g'') \text{ for all } g, g', g'' \in G. \tag{3.1}$$

Next, let H be any central extension of $\mathrm{In}(\mathcal{B})$ by an abelian group a. That is, we have an exact sequence of groups,

$$1 \to a \to H \to \mathrm{In}(\mathcal{B}) \to 1.$$

Let $\hat{\sigma}$ be a lift of the map $\sigma : G \times G \to \mathrm{In}(\mathcal{B})$ to a map $\hat{\sigma} : G \times G \to H$ (by a choice of section $\mathrm{In}(\mathcal{B}) \to H$). We have then

$$\hat{\sigma}(g, g')\hat{\sigma}(gg', g'') = [\tilde{s}(g)\hat{\sigma}(g', g'')\tilde{s}(g)^{-1}]\hat{\sigma}(g, g'g'') \cdot \alpha(g, g', g'') \text{ for all } g, g', g'' \in G$$

where $\alpha : G \times G \times G \to a$. Here the action of the outer automorphism $s(g)$ on $\hat{\sigma}(*)$ is defined by $s(g)\hat{\sigma}(*)s(g)^{-1}$ = the lift of $s(g)\sigma(*)s(g)^{-1} \in \mathrm{In}(\mathcal{B})$ to an element in H. One can show that α is a 3-cocycle [M, Lemma 8.4],

$$\alpha(g_2, g_3, g_4)\alpha(g_1g_2, g_3, g_4)^{-1}\alpha(g_1, g_2g_3, g_4)\alpha(g_1, g_2, g_3g_4)^{-1}\alpha(g_1, g_2, g_3) = 1.$$

Remark If we work in the category of topological groups (or Lie groups) the lifts above are in general discontinuous; normally, we can require continuity (or smoothness) only in an open neighborhood of the unit element.

Next we construct an example from quantum field theory. Let G be a compact simply connected Lie group and P the space of smooth paths $f : [0, 1] \to G$ with initial point $f(0) = e$, the neutral element, and quasiperiodicity condition that $f^{-1}df$ should be a smooth function.

P is a group under point wise multiplication but it is also a principal ΩG bundle over G. Here $\Omega G \subset P$ is the loop group with $f(0) = f(1) = e$ and $\pi : P \to G$ is the projection to the end point $f(1)$. Fix a unitary representation ρ of G in $\mathbb{C}^N$ and denote $H = L^2(S^1, \mathbb{C}^N)$.

For each polarization $H = H_- \oplus H_+$ we have a vacuum representation of the CAR algebra $\mathcal{B}(H)$ in a Hilbert space $\mathcal{F}(H_+)$. Denote by $\mathcal{C}$ the category of these representations. Denote by $a(v), a^*(v)$ the generators of $\mathcal{B}(H)$ corresponding to a vector $v \in H$,

$$a^*(u)a(v) + a(v)a^*(u) = 2\langle v, u\rangle$$

and all the other anticommutators equal to zero.

Any element $f \in P$ defines a unique automorphism of $\mathcal{B}(H)$ with $\phi_f(a^*(v)) = a^*(f \cdot v)$, where $f \cdot v$ is the function on the circle defined by $\rho(f(x))v(x)$. These automorphisms are in general not inner except when f is periodic. We have now a map $s : G \to \mathrm{Aut}(\mathcal{B})/\mathrm{In}(\mathcal{B})$ given by $g \mapsto F(g)$ where $F(g)$ is an arbitrary smooth quasiperiodic function on $[0, 1]$ such that $F(g)(1) = g$. Any two such functions $F(g), F'(g)$ differ by an element σ of ΩG, $F(g)(x) = F'(g)(x)\sigma(x)$. Now σ is an inner automorphism through a projective representation of the loop group ΩG in $\mathcal{F}(H_+)$.

In an open neighborhood U of the neutral element e in G we can fix in a smooth way for any $g \in U$ a path $F(g)$ with $F(g)(0) = e$ and $F(g)(1) = g$. Of course, for a connected group G we can make this choice globally on G but then the dependence of the path $F(g)$ would not be a continuous function of the end point. For a pair $g_1, g_2 \in G$ we have $\sigma(g_1, g_2)F(g_1 g_2) = F(g_1)F(g_2)$ with $\sigma(g_1, g_1) \in \Omega G$.

For a triple of elements g_1, g_2, g_3 we have now

$$F(g_1)F(g_2)F(g_3) = \sigma(g_1, g_2)F(g_1 g_2)F(g_3) = \sigma(g_1, g_2)\sigma(g_1 g_2, g_3)F(g_1 g_2 g_3).$$

In the same way,

$$\begin{aligned} F(g_1)F(g_2)F(g_3) &= F(g_1)\sigma(g_2, g_3)F(g_2 g_3) = [g_1 \sigma(g_2, g_3) g_1^{-1}]F(g_1)F(g_2 g_3) \\ &= [g_1 \sigma(g_2, g_3) g_1^{-1}]\sigma(g_1, g_2 g_3)F(g_1 g_2 g_3) \end{aligned}$$

which proves the cocycle relation (3.1) .

Lifting the loop group elements σ to inner automorphisms $\hat{\sigma}$ through a projective representation of ΩG we can write

$$\hat{\sigma}(g_1, g_2)\hat{\sigma}(g_1 g_2, g_3) = \mathrm{Aut}(g_1)[\hat{\sigma}(g_2, g_3)]\hat{\sigma}(g_1, g_2 g_3)\alpha(g_1, g_2, g_3),$$

where $\alpha : G \times G \times G \to S^1$ is some phase function arising from the fact that the projective lift is not necessarily a group homomorphism. Since (in the case of a Lie group) the function $F(\cdot)$ is smooth only in a neighborhood of the neutral element, the same is true also for σ and finally for the cocycle α.

An equivalent point of view to the construction of the 3-cocycle α is this: We are trying to construct a central extension $\hat{P}$ of the group P of paths in G (with initial point $e \in G$) as an extension of the central extension over the subgroup ΩG. The failure of this central extension is measured by the cocycle α, as an obstruction to associativity of $\hat{P}$. On the Lie algebra level, we have a corresponding cocycle $c_3 = d\alpha$ which is easily computed. The cocycle c of $\Omega\mathfrak{g}$ extends to the path Lie algebra $P\mathfrak{g}$ as

$$c(X, Y) = \frac{1}{4\pi i} \int_{[0,2\pi]} \mathrm{tr}\,(XdY - YdX).$$

This is an antisymmetric bilinear form on $P\mathfrak{g}$ but it fails to be a Lie algebra 2-cocycle. The coboundary is given by

$$\begin{aligned} (\delta c)(X, Y, Z) &= c(X, [Y, Z]) + c(Y, [Z, X]) + c(Z, [X, Y]) \\ &= -\frac{1}{4\pi i} \mathrm{tr}\, X[Y, Z]|_{2\pi} = d\alpha(X, Y, Z). \end{aligned}$$

Thus δc reduces to a 3-cocycle of the Lie algebra $\mathfrak{g}$ of G on the boundary $x = 2\pi$. This cocycle defines by left translations on G the left-invariant de Rham form $-\frac{1}{12\pi i}\mathrm{tr}\,(g^{-1}dg)^3$; this is normalized as $2\pi i$ times an integral 3-form on G.

Let $f_1, f_2 \in P$ and $f_{12} \in P$ with the property $f_1(2\pi)f_2(2\pi) = f_{12}(2\pi)$. Then we have a fiber S^1 over the loop $\phi_{12} = f_1(t)f_2(t)f_{12}(t)^{-1}$ coming from the central extension $\widehat{\Omega G} \to \Omega G$. Concretely, this fiber can be realized geometrically as a pair (f, λ) where $f : D \to G$, D is the unit disk with boundary S^1 such that the restriction of f to S^1 is the loop above, and $\lambda \in S^1$. Two pairs $(f, \lambda), (f', \lambda')$ are equivalent if the restrictions of f, f' to the boundary are equal and $\lambda' = \lambda e^{W(f,f')}$, where

$$W(f,f') = W(g) = \frac{1}{12\pi i}\int_B \mathrm{tr}\,(g^{-1}dg)^3,$$

where B is a unit ball with boundary $S^2 = S^2_+ \cup S^2_-$ and g is any extension to B of the map $f \cup f'$ on S^2 obtained by joining f, f' on the boundary circle S^1 of the unit disk D. The value of $W(g)$ depends on the extension g of $f \cup f'$ only modulo $2\pi i$ times an integer and therefore $\exp(W(f,f'))$ is well-defined. The product in the central extension of the full loop group LG is then defined as

$$[(f,\lambda)]\cdot[(f',\lambda')] = [(ff', \lambda\lambda' e^{\gamma(f,f')})],$$

where

$$\gamma(f,f') = \frac{1}{4\pi i}\int_D \mathrm{tr}\, f^{-1}df \wedge df' f'^{-1},$$

see [Mi87] for details; here the square brackets mean equivalence classes of pairs, subject to the equivalence defined above.

The 3-cocycle α can now be written in terms of the local data as

$$\alpha = \exp[\gamma(\phi_{12},\phi_{12,3}) - \gamma(\mathrm{Aut}_{f_1}\phi_{23},\phi_{1,23}) + W(h)]$$

where $\phi_{12,3}$ is an extension to the disk of the loop composed from the paths f_1f_2 and f_3 and the path $f_{12,3}$ joining the identity e to $g_1g_2g_3$. h is the function on D, equal to the neutral element on the boundary, such that $\phi_{12}\phi_{12,33} = \phi_{12}\phi_{12,3}h$. The value of α now depends, besides on the paths f_i, on the extensions ϕ to the disk D of the boundary loops determined by $f_i, f_{12}, f_{23}, f_{12,3}, f_{1,23}$. However, different choices of extensions are related by phase factors which can be obtained from the equivalence relation

$$(\phi,\lambda) \equiv (\phi h, \lambda e^{\gamma(\phi,h)+W(h)})$$

defining the central extension of the loop group.

4. THE CASE OF HIGHER DIMENSIONS

The construction of the gerbal action of G has a generalization which comes from a study of gauge anomalies in higher dimensions. Let M be the boundary of a compact contractible manifold N. Fix again a compact Lie group G and denote now by MG the group of smooth maps from a compact manifold M to G, which can be extended to N. This is an infinite-dimensional Lie group under point wise multiplication of maps. Denote by NG the smooth maps from N to G such that the normal derivatives at the boundary M vanish to all orders. Finally, let $\mathcal{G}$ be the normal subgroup of NG consisting of maps equal to the constant $e \in G$ on the boundary. This will play the role of ΩG in the previous section. Now $NG/\mathcal{G} = MG$.

We also assume that M is the boundary of another copy N' of N and denote by $\overline{N}$ the manifold obtained from N, N' by gluing along the common boundary. We also assume a spin structure and Riemannian metric given on $\overline{N}$.

We may view elements $f \in NG$ as $\mathfrak{g}$-valued vector potentials on the space $\overline{N}$. This correspondence is given by $f \mapsto A = f^{-1}df$ on N and A' is fixed from the boundary values $A|_M$ and from a contraction of N' to one point. This construction gives a map from the group MG to the moduli space of gauge connections in a trivial vector bundle over $\overline{N}$.

Example Let $M = S^n$, viewed as the equator in the sphere S^{n+1}. Fix a path α from the South Pole of S^{n+1} to the North Pole. Then for any great circle joining the North Pole to the South Pole we can take the union with α, giving a loop starting from the South Pole and traveling via the North Pole. For a given (globally defined) vector potential A let g_A be the holonomy around this loop. The great circles are parametrized by points on the equator S^n and thus we obtain a map $S^n \to G$. The group $\mathcal{G}$ of *based* gauge transformations, those which are equal to the identity on the South Pole, does not affect the holonomy, thus we obtain a map from the moduli space $\mathcal{A}/\mathcal{G}$ of gauge connections (in a trivial vector bundle) to S^nG, the group of contractible maps from S^n to G. The same construction can be made for any n-sphere S^n around the South Pole, and we may view S^{n+1} as the $(n+1)$-dimensional solid ball N with boundary $M = S^n$ contracted to one point, the South Pole of S^{n+1}. Here $\mathcal{G}$ is viewed as the group G-valued maps on N equal to the constant e on the boundary. In this case one can actually show that the gauge moduli space $\mathcal{A}/\mathcal{G}$ is homotopy equivalent to MG.

The above example can be extended to the case of gauge connections in a nontrivial vector bundle, but then the moduli space is disconnected and we should allow S^nG to consist of all smooth maps, not only the contractible ones. In the case $n = 2$ there is no difference, since for any finite-dimensional Lie group $\pi_2 G = 0$.

The main difference as compared to the case of loop groups is that the transformations $f \in \mathcal{G}$ are not in general implementable Bogoliubov automorphisms. However, as explained in Section 2, they are well-defined automorphism of a Hilbert bundle $\mathcal{F}$ over $\mathcal{A}$, the space of $\mathfrak{g}$-valued connections on $\overline{N}$. The fibers of this bundle are fermionic Fock spaces and each of them carries an (inequivalent) representation, in the category $\mathcal{C}$, of the canonical anticommutation relations, with a Dirac vacuum which depends on the background field $A \in \mathcal{A}$. The group $\mathcal{G}$ acts on this bundle through an abelian extension $\widehat{\mathcal{G}}$.

The 3-cocycle is constructed in essentially the same way as in Section 3. So for an element $g \in MG$ select $f \in NG$ such that the restriction of f to the boundary is equal to g. For any pair $g_1, g_2 \in MG$ we have then $f_1 f_2 f_{12}^{-1} \in \mathcal{G}$ where again $f_{12} \in NG$ such that $f_{12}|_M = g_1 g_2$. For a triple $g_1, g_2, g_3 \in MG$ we then construct $\alpha(A; g_1, g_3, g_3) \in S^1$ as before, but now it depends on the connection A since the operators $\hat{\sigma}$ now all depend on A.

$$\hat{\sigma}_{A'}(g_1, g_2)\hat{\sigma}_A(g_1 g_2, g_3) = \mathrm{Aut}(g_1)[\hat{\sigma}_{A''}(g_2, g_3)]\hat{\sigma}_A(g_1, g_2 g_3)\alpha(A; g_1, g_2, g_3),$$

where $A' = A^{\sigma(g_1 g_2, g_3)}$ is the gauge transform of A by $\sigma(g_1 g_2, g_3) \in \mathcal{G}$ and similarly A'' is the gauge transform of A by $\sigma(g_1, g_2 g_3)$.

Example Again, passing to the Lie algebra cocycles one gets reasonably simple expressions. For example, in the case $\dim M = 2$ the Lie algebra extension of Lie($\mathcal{G}$) is given by the 2-cocycle (2.2) and for a manifold N with boundary M this formula is not a cocycle but its coboundary is the Lie algebra 3-cocycle

$$d\alpha(X, Y, Z) = -\frac{1}{8\pi^2}\int_M \mathrm{tr}\, X[dY, dZ]. \tag{4.1}$$

In this case the cocycle does not depend on the variable A but when $\dim M > 2$ it does.

5. TWISTED K-THEORY ON MODULI SPACES OF GAUGE CONNECTIONS

Let P be a principal bundle over a space X with model fiber equal to the projective unitary group $PU(H) = U(H)/S^1$ of a complex Hilbert space. It is known that equivalence classes of such bundles are classified by elements in $H^3(X, \mathbb{Z})$, the Dixmier-Douady class of the bundle, [DD].

K-theory of X twisted by P, denoted by $K^*(X, P)$, is defined as the abelian group of homotopy classes of sections of a bundle Q, defined as an associated bundle with fiber equal to the (Z_2 graded) space of Fredholm operators in H with $PU(H)$ action given by the conjugation $T \mapsto gTg^{-1}$. The grading is as in ordinary complex K-theory: The even sector is defined by the space of *all* Fredholm operators whereas the odd sector is defined by self-adjoint operators with both positive and negative essential spectrum. As a model, one can use either bounded Fredholm operators, or unbounded operators, for example with the graph topology, [AtSe].

If $X = G$ is a compact Lie group one can construct elements of $K^*(G, P)$ in terms of highest weight representations of the central extension $\widehat{LG}$, [Mi04]. Actually, these come as G-equivariant classes, under the conjugation action of G on itself. In the equivariant case the construction of $K^*(G, P)$ is related to the Verlinde algebra in conformal field theory, [FHT]. Although for simple compact Lie groups there exist classification theorems [Do], [Br] also in the nonequivariant case it is still an open problem how to give explicit constructions for all classes in the nonequivariant case, in terms of families of Fredholm operators, using representation theory, even for unitary groups $SU(n)$ when $n > 3$.

Let $\omega : \mathcal{A} \times \mathcal{G} \to U_{\text{res}}(H_+ \oplus H_-)$ be the 1-cocycle constructed in Section 2. Let Y be a family of Fredholm operators in $\mathcal{F}$ which is is mapped onto itself under a projective representation $g \mapsto \hat{g}$ of U_{res} in $\mathcal{F}$, $T \mapsto \hat{g}T\hat{g}^{-1} \in Y$ for any $T \in Y$.

Now we have an action of a central extension of the groupoid $(\mathcal{A}, \mathcal{G})$ on Y by

$$(A, g) : Y \to Y, T \mapsto \hat{\omega}(A; g)T\hat{\omega}(A; g)^{-1}.$$

We can also view this as a central extension of the transformation groupoid defined by the action of the gauge group $\mathcal{G}$ on the space $\mathcal{A} \times Y$. If $\mathcal{G}$ is the group of based gauge transformations then it acts freely on $\mathcal{A}$ and therefore also freely on $\mathcal{A} \times Y$. If furthermore Y is contractible then $(\mathcal{A} \times Y)/\mathcal{G} \simeq \mathcal{A}/\mathcal{G}$ is the gauge moduli space.

A system of Fredholm operators transforming covariantly under U_{res} can be constructed from a Dirac operator on the infinite-dimensional Grassmann manifold $Gr_{\text{res}} = U_{\text{res}}(H_+ \oplus H_-)/(U(H_+) \times U(H_-))$, [Tä]. The members of the family are parametrized by a gauge connection on a complex line bundle L over Gr_{res}. The line bundle L is used as twisting of the spin bundle over Gr_{res}, and can be viewed as defining a Spin^c structure on Gr_{res}.

In the case when $\mathcal{A}$ is the space of gauge connections on the circle and $\mathcal{G} = LG$ is a loop group the cocycle $\omega(A; g)$ does not depend on A and it gives a unitary representation of LG in the Hilbert space H and $g \mapsto \hat{g}$ is given by a representation of a central extension $\widehat{LG}$ in $\mathcal{F}$. As Y we can take the family Q_A of supercharges in [Mi04] parametrized by points in $\mathcal{A}$. This means that in the notation above, we can identify $\mathcal{A}$ as the diagonal in $\mathcal{A} \times Y$ and we have a natural identification of the groupoid moduli space with the moduli space of gauge connections on the circle.

Here the groupoid action defines an element in the twisted G-equivariant K-theory on G. In fact, in the case of the circle, one can directly work with highest weight representations of the loop group without using the embedding $LG \subset U_{\text{res}}$.

References

[ABJ] S.L. Adler: Axial-vector vertex in spinor electrodynamics. Phys. Rev, **177,** p. 2426 (1969). W.A. Bardeen: Anomalous Ward identities in spinor field theories. Phys. Rev. **184,** p.1848 (1969) J. Bell and R. Jackiw: The PCAC puzzle: $\pi^0 \to \gamma\gamma$ in the σ-model. Nuevo Cimento **A60,** p. 47 (1969).

[AS] M.F. Atiyah and I.M. Singer: Dirac operators coupled to vector potentials. Proc. Natl. Acad. Sci. USA, **81,** pp. 2597-2600 (1984)

[AtSe] M. Atiyah and G. Segal: Twisted K-theory. Ukr. Mat. Visn. **1,** , no. 3, 287–330 (2004); translation in Ukr. Math. Bull. **1,** no. 3, pp. 291–334 (2004)

[Br] V. Braun: Twisted K-theory of Lie groups. J. High Energy Phys. no. 3, 029, 15 pp (2004) (electronic)

[Ca] A. L. Carey: The origin of three-cocycles in quantum field theory. Phys. Lett. B **194,** pp. 267-270 (1987)

[CGRS] A.L. Carey, H. Grundling, I. Raeburn, and C. Sutherland: Group actions on C^*-algebras, 3-cocycles and quantum field theory. Commun. Math. Phys. **168,** pp. 389-416 (1995)

[DD] J. Dixmier and A. Douady: Champs continus d'espaces hilbertiens et de C^*-algebres. Bull. Soc. Math. France **91,** pp. 227–284 (1963)

[Do] C. Douglas: On the twisted K-homology of simple Lie groups. Topology **45,** pp. 955–988 (2006)

[Fa] L.D. Faddeev: Operator anomaly for the Gauss law. Phys. Lett. B **145,** p.81 (1984)

[FHT] D. Freed, M. Hopkins, and C. Teleman: Twisted K-theory and loop group representations, arXiv:math/0312155. Twisted equivariant K-theory with complex coefficients. J. Topol. **1** (2008), no. 1, 16–44.

[FZ] E. Frenkel and Xinwen Zhu: Gerbal representations of double loop groups. arXiv.math/0810.1487

[GJJ] B. Grossman. The meaning of the third cocycle in the group cohomology of nonabelian gauge theories. Phys. Lett. B **160,** pp. 94–100 (1985). R. Jackiw, R: Three-cocycle in mathematics and physics. Phys. Rev. Lett. **54,** pp.159–162 (1985). S.G. Jo: Commutator of gauge generators in nonabelian chiral theory. Nuclear Phys. B **259,** pp. 616–636 (1985)

[LM] E. Langmann and J. Mickelsson: Scattering matrix in external field problems. J. Math. Phys. **37,** pp. 3933–3953 (1996)

[Mi85] J. Mickelsson: Chiral anomalies in even and odd dimensions. Comm. Math. Phys. **97 (1985),** pp. 361–370 (1985)

[Mi87] J. Mickelsson: Kac-Moody groups, topology of the Dirac determinant bundle, and fermionization. Comm. Math. Phys. **110,** pp. 173–183 (1987)

[Mi90] J. Mickelsson: Commutator anomalies and the Fock bundle. Comm. Math. Phys. **127,** pp. 285–294 (1990)

[Mi93] J. Mickelsson: Hilbert space cocycles as representations of (3+1)-D current algebras. Lett. Math. Phys. **28,** pp.97–106 (1993). Wodzicki residue and anomalies of current algebras. Integrable models and strings (Espoo, 1993), pp. 123–135, Lecture Notes in Phys., **436,** Springer, Berlin (1994)

[Mi04] J. Mickelsson: Gerbes, (twisted) K-theory, and the supersymmetric WZW model. Infinite dimensional groups and manifolds, 93–107, IRMA Lect. Math. Theor. Phys., **5,** de Gruyter, Berlin (2004)

[M] Saunders Mac Lane: *Homology.* Die Grundlehren der Mathematischen Wissenschaften, Band 114. Springer Verlag (1963)

[PS] A. Pressley and G. Segal: *Loop Groups.* Clarendon Press, Oxford (1986)

[Tä] V. Tähtinen: The Dirac operator on the restricted Grassmannian, in progress

Department of Mathematics and Statistics, University of Helsinki, and Department of Theoretical Physics, Royal Institute of Technology, Stockholm

Clay Mathematics Proceedings
Volume 12, 2010

A Microlocal Approach to Fefferman's Program in Conformal and CR Geometry

Raphaël Ponge

Motivated by the analysis of the singularity of the Bergman kernel of a strictly pseudoconvex domain $D \subset \mathbb{C}^{n+1}$, Fefferman [**Fe2**] launched the program of determining *all* local invariants of a strictly pseudoconvex CR structure. This program was subsequently extended to deal with local invariants of other parabolic geometries, including conformal geometry (see [**FG1**]). Since Fefferman's seminal paper further progress has been made, especially recently (see, e.g., [**Al2**], [**BEG**], [**GH**], [**Hi1**], [**Hi2**]). In addition, there is a very recent upsurge of new conformally invariant Riemannian differential operators (see [**Al2**], [**Ju**]).

In this article we present the results of [**Po4**] on the logarithmic singularities of the Schwartz kernels and Green kernels of *general* invariant pseudodifferential operators in conformal and CR geometry. This connects nicely with results of Hirachi ([**Hi1**], [**Hi2**]) on the logarithmic singularities of the Bergman and Szegö kernels on boundaries of strictly pseudoconvex domains.

The main result in the conformal case (Theorem 3) asserts that in odd dimension, as well as in even dimension below the critical weight (i.e. half of the dimension), the logarithmic singularities of Schwartz kernels and Green kernels of conformally invariant Riemannian ΨDOs are linear combinations of Weyl conformal invariants, that is, of local conformal invariants arising from complete tensorial contractions of covariant derivatives of the ambient Lorentz metric of Fefferman-Graham ([**FG1**], [**FG2**]). Above the critical weight the description in even dimension involves the ambiguity-independent Weyl conformal invariants recently defined by Graham-Hirachi [**GH**], as well as the exceptional local conformal invariants of Bailey-Gover [**BG**]. In particular, by specializing this result to the GJMS operators of [**GJMS**], including the Yamabe and Paneitz operators, we obtain invariant expressions for the logarithmic singularities of the Green kernels of these operators (see Theorem 4).

In the CR setting the relevant class of pseudodifferential operators is the class of Ψ_HDOs introduced by Beals-Greiner [**BGr**] and Taylor [**Tay**]. In this context the main result (Theorem 6) asserts that the logarithmic singularities of Schwartz kernels and Green kernels of CR invariant Ψ_HDOs are local CR invariants, and below the critical weight are linear combinations of complete tensorial contractions

Research partially supported by NSERC Discovery Grant 341328-07 (Canada) and by a New Staff Matching Grant of the Connaught Fund of the University of Toronto (Canada).

of covariant derivatives of the curvature of the ambient Kähler-Lorentz metric of Fefferman [**Fe2**]. As a consequence this allows us to get invariant expressions for the logarithmic singularities of the Green kernels of the CR GJMS operators of [**GG**] (see Theorem 7).

The setup of the the first part of the paper, Sections 1–5, is conformal geometry. The main results on the logarithmic singularities of conformally invariant ΨDOs are presented in Section 5. In the previous sections, we review the main definitions and examples concerning local conformal invariants and conformally invariant differential operators (Section 1), conformally invariant ΨDOs (Section 2), the logarithmic singularity of the Schwartz kernel of a ΨDO (Section 3) and the program of Fefferman in conformal geometry (Section 4).

The setup of the second part, Sections 6–10, is pseudo-Hermitian and CR geometry. In Section 6, we review the motivating example of the Bergman kernel of a strictly pseudoconvex domain. Section 7 is an overview of the main facts about the Heisenberg calculus. In Section 8, we present important definitions and properties concerning pseudo-Hermitian geometry, local pseudo-Hermitian invariants and pseudo-Hermitian invariant Ψ_HDOs. In Section 9, we review the main facts about local CR invariants, CR invariant operators and the program of Fefferman in CR geometry. In Section 10, we present the main results concerning the logarithmic singularities of CR invariant operators.

This proceeding is a survey of the main results of [**Po4**].

1. Conformal Invariants

Up to Section 5, we denote by M a Riemannian manifold of dimension n and we denote by g_{ij} and R_{ijkl} its metric and curvature tensors. As usual we shall use the metric and its inverse g^{ij} to lower and raise indices. For instance, the Ricci tensor is $\rho_{jk} = R_{ijk}{}^{i} = g^{il}R_{ijkl}$ and the scalar curvature is $\kappa_g = \rho_j^{\ j} = g^{ji}\rho_{ij}$.

1.1. Local conformal invariants. In the sequel we denote by $M_n(\mathbb{R})_+$ the open subset of $M_n(\mathbb{R})$ consisting of positive definite matrices.

DEFINITION 1. *A local Riemannian invariant of weight w is the datum on each Riemannian manifold (M^n, g) of a function $\mathcal{I}_g \in C^\infty(M)$ such that:*

(i) *There exist finitely many functions $a_{\alpha\beta}$ in $C^\infty(M_n(\mathbb{R})_+)$ such that, in any local coordinates,*

$$\mathcal{I}_g(x) = \sum a_{\alpha\beta}(g(x))(\partial^\alpha g(x))^\beta. \tag{1}$$

(ii) *For all $t > 0$,*

$$\mathcal{I}_{tg}(x) = t^{-w}\mathcal{I}_g(x). \tag{2}$$

Using Weyl's invariant theory for $O(n)$ (see, e.g, [**Gi**]) we obtain the following determination of local Riemannian invariants.

THEOREM 1 (Weyl, Cartan). *Any local Riemannian invariant is a linear combination of Weyl Riemannian invariants, that is, of complete contractions of the curvature tensor and its covariant derivatives.*

For instance, the only Weyl Riemannian invariant of weight 1 is the scalar curvature κ_g. In weight 2 the Weyl Riemannian invariants are

$$|\kappa_g|^2, \qquad |\rho|^2 := \rho^{ij}\rho_{ij}, \qquad |R|^2 := R^{ijkl}R_{ijkl}, \qquad \Delta_g\kappa_g, \tag{3}$$

where Δ_g denotes the Laplace operator of (M, g). In weight 3 there are 17 Weyl invariants (see [**Gi**]).

DEFINITION 2. *A local conformal invariant of weight w is a local Riemannian invariant $\mathcal{I}_g$ such that*

$$\mathcal{I}_{e^f g}(x) = e^{-wf(x)}\mathcal{I}_g(x) \qquad \forall f \in C^\infty(M, \mathbb{R}). \tag{4}$$

The most fundamental conformally invariant tensor is the Weyl curvature tensor W_{ijkl}. Using complete contractions of k-fold tensor powers of W we get local conformal invariants of various weights. For instance, the following are local conformal invariants

$$|W|^2 := W^{ijkl}W_{ijkl}, \tag{5}$$

$$W_{ij}{}^{kl}W_{lk}{}^{pq}W_{pq}{}^{ij}, \qquad W_i{}^{jk}{}_l W^i{}_{pk}{}^q W_j{}^{pl}{}_q. \tag{6}$$

Here $|W|^2$ has weight 2, while the other two invariants have weight 3.

Other local conformal invariants can be obtained in terms of the ambient metric of Fefferman-Graham ([**FG1**], [**FG2**]; see Section 4 below).

1.2. Conformally invariant operators.

DEFINITION 3. *A Riemannian invariant differential operator of weight w is the datum on each Riemannian manifold (M^n, g) of a differential operator P_g on M such that:*

(i) *There exist finitely many functions $a_{\alpha\beta\gamma}$ in $C^\infty(M_n(\mathbb{R})_+)$ such that, in any local coordinates,*

$$P_g = \sum a_{\alpha\beta\gamma}(g(x))(\partial^\alpha g(x))^\beta D_x^\gamma. \tag{7}$$

(ii) *We have*

$$P_{tg} = t^{-w}P_g \qquad \forall t > 0. \tag{8}$$

DEFINITION 4. *A conformally invariant differential operator of biweight (w, w') is a Riemannian invariant differential operator P_g such that*

$$P_{e^f g} = e^{w'f}P_g e^{-wf} \qquad \forall f \in C^\infty(M, \mathbb{R}). \tag{9}$$

An important example of a conformally invariant differential operator is the Yamabe operator,

$$\Box_g := \Delta_g + \frac{n-2}{4(n-1)}\kappa_g, \tag{10}$$

where Δ_g denotes the Laplace operator. In particular,

$$\Box_{e^{2f}g} = e^{-(\frac{n}{2}+1)f}\Box_g e^{(\frac{n}{2}-1)f} \qquad \forall f \in C^\infty(M, \mathbb{R}). \tag{11}$$

A generalization of the Yamabe operator is provided by the GJMS operators of Graham-Jenne-Mason-Sparling [**GJMS**]. For $k = 1, \ldots, \frac{n}{2}$ when n is even, and for all non-negative integers k when n is odd, the GJMS operator of order k is a differential operator $\Box_g^{(k)}$ such that

$$\Box_g^{(k)} = \Delta_g^{(k)} + \text{ lower order terms}, \tag{12}$$

and which satisfies

$$\Box_{e^{2f}g}^{(k)} = e^{-(\frac{n}{2}+k)f}\Box_g^{(k)}e^{(\frac{n}{2}-k)f} \qquad \forall f \in C^\infty(M, \mathbb{R}). \tag{13}$$

For $k = 1$ this operator agrees with the Yamabe operator, while for $k = 2$ we recover the Paneitz operator.

Recently, Alexakis ([**Al2**], [**Al1**]) and Juhl [**Ju**] constructed new families of conformally invariant operators. Furthermore, Alexakis proved that, under some restrictions, his family of operators exhausts *all* conformally invariant differential operators.

2. Conformally Invariant ΨDOs

Let U be an open subset of $\mathbb{R}^n$. The (classical) symbols on $U \times \mathbb{R}^n$ are defined as follows.

DEFINITION 5. *1) $S_m(U \times \mathbb{R}^n)$, $m \in \mathbb{C}$, is the space of functions $p(x,\xi)$ contained in $C^\infty(U \times (\mathbb{R}^n \setminus \{0\}))$ such that $p(x,t\xi) = t^m p(x,\xi)$ for any $t > 0$.*

2) $S^m(U \times \mathbb{R}^n)$, $m \in \mathbb{C}$, consists of functions $p \in C^\infty(U \times \mathbb{R}^n)$ with an asymptotic expansion $p \sim \sum_{j \geq 0} p_{m-j}$, $p_k \in S_k(U \times \mathbb{R}^n)$, in the sense that, for any integer N, any compact $K \subset U$ and any multi-orders α, β, there exists a constant $C_{NK\alpha\beta} > 0$ such that, for any $x \in K$ and any $\xi \in \mathbb{R}^{d+1}$ so that $|\xi| \geq 1$, we have

$$|\partial_x^\alpha \partial_\xi^\beta (p - \sum_{j<N} p_{m-j})(x,\xi)| \leq C_{NK\alpha\beta} |\xi|^{\Re m - |\beta| - N}. \tag{14}$$

Given a symbol $p \in S^m(U \times \mathbb{R}^n)$ we let $p(x,D)$ be the continuous linear operator from $C_c^\infty(U)$ to $C^\infty(U)$ such that

$$p(x,D)u(x) = (2\pi)^{-n} \int e^{ix\cdot\xi} p(x,\xi)\hat{u}(\xi) d\xi \qquad \forall u \in C_c^\infty(U). \tag{15}$$

We define ΨDOs on the manifold M^n as follows.

DEFINITION 6. *$\Psi^m(M)$, $m \in \mathbb{C}$, consists of continuous operators P from $C_c^\infty(M)$ to $C^\infty(M)$ such that:*

(i) The Schwartz kernel of P is smooth off the diagonal;

(ii) In any local coordinates the operator P can be written as

$$P = p(x,D) + R, \tag{16}$$

where p is a symbol of order m and R is a smoothing operator.

Recall that the principal symbol of a ΨDO makes sense intrinsically as a function $p_m(x,\xi) \in C^\infty(T^*M \setminus \{0\})$ such that

$$p_m(x,\lambda\xi) = \lambda^m p_m(x,\xi) \qquad \forall (x,\xi) \in T^*M \setminus \{0\} \quad \forall \lambda > 0. \tag{17}$$

Recall also that P is said to be *elliptic* if $p_m(x,\xi) \neq 0$ for all $(x,\xi) \in T^*M \setminus \{0\}$. This is equivalent to the existence of a parametrix in $\Psi^{-m}(M)$, i.e., an inverse modulo smoothing operators.

This said, in order to define Riemannian and conformally invariant ΨDOs, we need to consider the following class of symbols.

DEFINITION 7. *$S_m(M_n(\mathbb{R})_+ \times \mathbb{R}^n)$, $m \in \mathbb{C}$, consists of functions $a(g,\xi)$ in $C^\infty(M_n(\mathbb{R})_+ \times (\mathbb{R}^n \setminus \{0\}))$ such that $a(g,t\xi) = t^m a(g,\xi)$ $\forall t > 0$.*

In the sequel we let $\Psi^{-\infty}(M)$ denote the space of smoothing operators on M.

DEFINITION 8. *A Riemannian invariant ΨDO of order m and weight w is the datum for every Riemannian manifold (M^n, g) of an operator $P_g \in \Psi^m(M)$ in such a way that:*

(i) *For $j = 0, 1, \ldots$ there are finitely many $a_{j\alpha\beta} \in S_{m-j}(M_n(\mathbb{R})_+ \times \mathbb{R}^n)$ such that, in any local coordinates, P_g has symbol*

$$\sigma(P_g)(x,\xi) \sim \sum_{j\geq 0} \sum_{\alpha,\beta} (\partial^\alpha g(x))^\beta a_{j\alpha\beta}(g(x),\xi). \tag{18}$$

(ii) *For all $t > 0$ we have*

$$P_{tg} = t^{-w} P_g \qquad \bmod \Psi^{-\infty}(M). \tag{19}$$

REMARK. For differential operators this definition is equivalent to Definition 3, because two differential operators differing by a smoothing operator must agree.

DEFINITION 9. *A conformally invariant ΨDO of order m and biweight (w, w') is a Riemannian invariant ΨDO of order m such that, for all $f \in C^\infty(M, \mathbb{R})$,*

$$P_{e^f g} = e^{w'f} P_g e^{-wf} \qquad \bmod \Psi^{-\infty}(M). \tag{20}$$

In the sequel we say that a Riemannian invariant is *admissible* if its principal symbol does not depend on the derivatives of the metric (i.e. in (18) we can take $a_{0\alpha\beta} = 0$ for $(\alpha, \beta) \neq 0$).

PROPOSITION 1. *Let P_g be a conformally invariant ΨDO of order m and biweight (w, w').*

(1) *Let Q_g be a conformally invariant ΨDO of order m' and biweight (w, w''), and assume that P_g or Q_g is properly supported. Then $Q_g P_g$ is a conformally invariant ΨDO of order $m + m'$ and biweight (w, w'').*

(2) *Assume that P_g is elliptic and admissible. Then the datum on every Riemannian manifold (M^n, g) of a parametrix $Q_g \in \Psi^{-m}(M)$ for P_g gives rise to a conformally invariant ΨDO of biweight (w', w).*

For instance, if $Q_g^{(k)}$ is a parametrix for the kth order GJMS operator $\Box_g^{(k)}$, then $Q_g^{(k)}$ is a conformally invariant ΨDO of biweight $(\frac{n+2k}{4}, \frac{n-2k}{4})$. By multiplying these operators with the operators of Alexakis and Juhl we obtain various examples of conformally invariant ΨDOs that are not differential operators or parametrices of elliptic differential operators

3. The Logarithmic Singularity of a ΨDO

We can give a precise description of the singularity of the Schwartz kernel of a ΨDO near the diagonal and, in fact, the general form of these singularities can be used to characterize ΨDOs (see, e.g., [**Hö**], [**Me**], [**BGr**]). In particular, if $P : C_c^\infty(M) \to C^\infty(M)$ if a ΨDO of integer order $m \geq -n$, then in local coordinates its Schwartz kernel $k_P(x, y)$ has a behavior near the diagonal $y = x$ of the form

$$k_P(x,y) = \sum_{-(m+n)\leq j\leq -1} a_j(x, x-y) - c_P(x)\log|x-y| + O(1), \tag{21}$$

where $a_j(x, y) \in C^\infty(U \times (\mathbb{R}^n \setminus \{0\}))$ is homogeneous of degree j in y and we have

$$c_P(x) = \frac{1}{(2\pi)^n} \int_{S^{n-1}} p_{-n}(x,\xi) d\sigma(\xi), \tag{22}$$

where $p_{-n}(x,\xi)$ is the symbol of degree $-n$ of P and we have denoted by $d\sigma(\xi)$ is the surface measure of S^{n-1}.

It seems to have been first observed by Connes-Moscovici [**CMo**] (see also [**GVF**]) that the coefficient $c_P(x)$ makes sense globally on M as a 1-density.

In the sequel we refer to the density $c_P(x)$ as the *logarithmic singularity* of the Schwartz kernel of P.

If P is elliptic, then we shall call a *Green kernel for P* the Schwartz kernel of any parametrix $Q \in \Psi^{-m}(M, \mathcal{E})$ for P. Such a parametrix is uniquely defined only modulo smoothing operators, but the singularity near the diagonal of the Schwartz kernel of Q, including the logarithmic singularity $c_Q(x)$, does not depend on the choice of Q.

DEFINITION 10. *If $P \in \Psi^m(M)$, $m \in \mathbb{Z}$, is elliptic, then the Green kernel logarithmic singularity of P is the density*

$$\gamma_P(x) := c_Q(x), \tag{23}$$

where $Q \in \Psi^{-m}(M)$ is any given parametrix for P.

Next, because of (22) the density $c_P(x)$ is related to the noncommutative residue trace of Wodzicki ([**Wo1**], [**Wo3**]) and Guillemin [**Gu1**] as follows.

Let $\Psi^{<-n}(M) = \bigcup_{\Re m<-n} \Psi^m(M)$ denote the class of ΨDOs whose symbols are integrable with respect to the ξ-variable. If P is a ΨDO in this class then the restriction to the diagonal of its Schwartz kernel $k_P(x,y)$ defines a smooth density $k_P(x,x)$. Therefore, if M is compact then P is trace-class on $L^2(M)$ and we have

$$\operatorname{Trace} P = \int_M k_P(x,x). \tag{24}$$

In fact, the map $P \to k_P(x,x)$ admits an analytic continuation $P \to t_P(x)$ to the class $\Psi^{\mathbb{C}\setminus\mathbb{Z}}(M)$ of non-integer ΨDOs, where analyticity is meant with respect to holomorphic families of ΨDOs as in [**Gu2**] and [**KV**]. Furthermore, if $P \in \Psi^{\mathbb{Z}}(M)$ and if $(P(z))_{z\in\mathbb{C}}$ is a holomorphic family of ΨDOs such that $\operatorname{ord} P(z) = \operatorname{ord} P + z$ and $P(0) = P$. Then, at $z = 0$, the map $z \to t_{P(z)}(x)$ has at worst a simple pole singularity with residue given by

$$\operatorname{Res}_{z=0} t_{P(z)}(x) = -c_P(x). \tag{25}$$

Suppose now that M is compact. Then the *noncommutative residue* is the linear functional on $\Psi^{\mathbb{Z}}(M)$ defined by

$$\operatorname{Res} P := \int_M c_P(x) \qquad \forall P \in \Psi^{\mathbb{C}\setminus\mathbb{Z}}(M). \tag{26}$$

Thanks to (22) this definition agrees with the usual definition of the noncommutative residue. Moreover, by using (25) we see that if $(P(z))_{z\in\mathbb{C}}$ is a holomorphic family of ΨDOs such that $\operatorname{ord} P(z) = \operatorname{ord} P + z$ and $P(0) = P$, then the map $z \to \operatorname{Trace} P(z)$ has an analytic extension to $\mathbb{C} \setminus \mathbb{Z}$ and, at $z = 0$, it has at worst a simple pole singularity with residue given by

$$\operatorname{Res} P = -\operatorname{Res}_{z=0} \operatorname{Trace} P(z). \tag{27}$$

Using this it is not difficult to see that the noncommutative residue is a trace on $\Psi^{\mathbb{Z}}(M)$. Wodzicki [**Wo2**] even proved that his is the unique trace up to constant multiple when M is connected and has dimension ≥ 2.

Finally, let $P : C^\infty(M) \to C^\infty(M)$ be a ΨDO of integer order $m \geq 0$ with a positive principal symbol. For $t > 0$ let $k_t(x,y)$ denote the Schwartz kernel of e^{-tP}. Then $k_t(x,y)$ is a smooth kernel and, as $t \to 0^+$,

$$k_t(x,x) \sim t^{-\frac{n}{m}} \sum_{j\geq 0} t^{\frac{j}{m}} a_j(P)(x) + \log t \sum_{j\geq 0} t^j b_j(P)(x), \tag{28}$$

where we further have $a_{2j+1}(P)(x) = b_j(P)(x) = 0$ for all $j \in \mathbb{N}_0$ when P is a differential operator (see, e.g., [**Gi**], [**Gr**]).

Using the Mellin Formula, we can explicitly relate the coefficients of the above heat kernel asymptotics to the singularities of the local zeta function $t_{P^{-s}}(x)$ (see, e.g., [**Wo3**, 3.23]). In particular, if for $j = 0, \ldots, n-1$ we set $\sigma_j = \frac{n-j}{m}$, then

$$mc_{P^{-\sigma_j}}(x) = \Gamma(\sigma_j)^{-1} a_j(P)(x). \tag{29}$$

The above equalities provide us with an immediate connection between the Green kernel logarithmic singularity of P and the heat kernel asymptotics (28). Indeed, as the partial inverse P^{-1} is a parametrix for P in $\Psi^{-m}(M)$, setting $j = n - m$ in (29) gives

$$a_{n-m}(P)(x) = mc_{P^{-1}}(x) = m\gamma_P(x). \tag{30}$$

4. Fefferman's Program in Conformal Geometry

In the sequel by *Green kernel* of an elliptic ΨDO we shall mean the Schwartz kernel of a parametrix, and by *null kernel* of a selfadjoint ΨDO we shall mean the Schwartz kernel of the orthogonal projection onto its null space.

The program of Fefferman in conformal geometry can be described as follows.

FEFFERMAN'S PROGRAM (Analytic Aspect). *Give a precise geometric description of the singularities of the Schwartz, Green and null kernels of conformally invariant operators in terms of local conformal invariants.*

As stated by Theorem 1, any local Riemannian invariant is a linear combination of Weyl Riemannian invariants. Is there a similar description for local conformal invariants? Establishing such a description is the aim of the geometric aspect of Fefferman's program:

FEFFERMAN'S PROGRAM (Geometric Aspect). *Determine all local invariants of a conformal structure.*

4.1. Ambient metric and Weyl conformal invariants. The analogues in conformal geometry of the Weyl Riemannian invariants are obtained via the ambient metric construction of Fefferman-Graham ([**FG1**], [**FG2**]).

In this section we denote by (M^n, g) a general Riemannian manifold of dimension n. Consider the metric ray-bundle,

$$\mathcal{G} := \{t^2 g(x); x \in M,\ t > 0\} \subset S^2T^*M \xrightarrow{\pi} M. \tag{31}$$

It carries the family of dilations,

$$\delta_s(x, \bar{g}) := s^2 \bar{g} \qquad \forall x \in M \quad \forall \bar{g} \in \mathcal{G}_x \quad \forall s > 0, \tag{32}$$

It also carries the (degenerate) tautological metric,

$$g_0(x, \bar{g}) := (d\pi(x))^* \, \bar{g} \qquad \forall (x, \bar{g}) \in \mathcal{G}. \tag{33}$$

Thus, if $\{x^j\}$ are local coordinates with respect to which $g(x) = g_{ij}dx^i \otimes dx^j$ and if we denote by t the fiber coordinate on $\mathcal{G}$ defined by the metric g, then in the local coordinates $\{x^j, t\}$ we have

$$g_0(x,t) = t^2 g_{ij} dx^i \otimes dx^j. \tag{34}$$

The *ambient space* is defined to be

$$\tilde{\mathcal{G}} := \mathcal{G} \times (-1,1). \tag{35}$$

In the sequel we shall use the letter ρ to denote the variable with values in $(-1,1)$. Then $\mathcal{G}$ can be identified with the hypersurface $\mathcal{G}_0 := \{\rho = 0\} \subset \tilde{\mathcal{G}}$.

THEOREM 2 ([**FG1**], [**FG2**]). *There exists a unique Lorentzian metric $\tilde{g}$ on $\tilde{\mathcal{G}}$ defined formally near $\rho = 0$ such that:*

$$\delta_s^* \tilde{g} = s^2 g_0, \qquad \tilde{g}_{|\rho=0} = g_0, \tag{36}$$

$$\operatorname{Ric}(\tilde{g}) = \begin{cases} O(\rho^\infty) & \text{if } n \text{ is odd,} \\ O(\rho^{\frac{n}{2}}) & \text{if } n \text{ is even.} \end{cases} \tag{37}$$

The ambient metric depends only on the conformal class of g, so any local Riemannian invariant of $(\tilde{\mathcal{G}}, \tilde{g})$ gives rise to a local conformal invariant of (M^n, g).

DEFINITION 11. *The Weyl conformal invariants are the local conformal invariants arising from the Weyl Riemannian invariants of $(\tilde{\mathcal{G}}, \tilde{g})$.*

For instance, the Weyl tensor is obtained by pushing down to M the curvature tensor $\tilde{R}$ of $\tilde{\mathcal{G}}$. Therefore, the invariants in (5)–(6) are Weyl conformal invariants.

In fact, if we use the Ricci-flatness of the ambient metric, then we see that there is no Weyl conformal invariant of weight 1 and the only of these invariants in weight 2 is $|W|^2$. In addition, in weight 3 we only have the invariants in (6) together with the invariant arising from $|\tilde{\nabla}\tilde{R}|^2$, namely, the Fefferman-Graham invariant,

$$\Phi_g := |V|^2 + 16\langle W, U\rangle + 16|C|^2, \tag{38}$$

where $C_{jkl} = \nabla_l A_{jk} - \nabla_k A_{jl}$ is the Cotton tensor and V and U are the tensors

$$V_{sijkl} = \nabla_s W_{ijkl} - g_{is}C_{jkl} + g_{js}C_{ikl} - g_{ks}C_{lij} + g_{ls}C_{kij}, \tag{39}$$

$$U_{sjkl} = \nabla_s C_{jkl} + g^{pq} A_{sp} W_{qjkl}. \tag{40}$$

Next, a very important result is the following.

PROPOSITION 2 ([**BEG**]).

(1) *If n is odd, then any local conformal invariant is a linear combination of Weyl conformal invariants.*
(2) *If n is even, the same holds in weight $\leq \frac{n}{2}$.*

In even dimension a description of the scalar local conformal invariants of weight $w \geq \frac{n}{2} + 1$ was recently presented by Graham-Hirachi [**GH**]. More precisely, they modified the construction of the ambient metric in such way as to obtain a metric on the ambient space $\tilde{\mathcal{G}}$ which is smooth of any order near $\mathcal{G}_0$. There is an ambiguity on the choice of a smooth ambient metric, but such a metric agrees with the ambient metric of Fefferman-Graham up to order $< \frac{n}{2}$ near $\mathcal{G}_0$.

Using a smooth ambient metric we can construct Weyl conformal invariants in the same way as we do by using the ambient metric of Fefferman-Graham. If such an invariant does not depend on the choice of the smooth ambient metric we then say that it is an *ambiguity-independent* Weyl conformal invariant. Not every conformal invariant arises this way, since in dimension $n = 4m$ this construction does not encapsulate the exceptional local conformal invariants of [**BG**].

PROPOSITION 3 (Graham-Hirachi [**GH**]). *Let w be an integer $\geq \frac{n}{2}$.*

1) If $n \equiv 2 \mod 4$, or if $n \equiv 0 \mod 4$ and w is even, then every scalar local conformal invariant of weight w is a linear combination of ambiguity-independent Weyl conformal invariants.

2) If $n \equiv 0 \mod 4$ and w is odd, then every scalar local conformal invariant of weight w is a linear combination of ambiguity-independent Weyl conformal invariants and of exceptional conformal invariants.

5. Logarithmic Singularities of Conformally Invariant Operators

One aim of this paper is to look at the logarithmic singularities (as defined in (21)–(22)) of conformally invariant ΨDOs.

In the sequel we denote by $|v_g(x)|$ the volume density of (M^n, g), i.e., in local coordinates $|v_g(x)| = \sqrt{g(x)}|dx|$, where $|dx|$ is the Lebesgue density. We also denote by $[g]$ the conformal class of g.

PROPOSITION 4. *Consider a family $(P_{\hat{g}})_{\hat{g}\in[g]} \subset \Psi^m(M)$ for which there are real numbers w and w' such that, for all $f \in C^\infty(M,\mathbb{R})$, we have*

$$P_{e^f g} \equiv e^{w'f} P_g e^{-wf} \mod \Psi^{-\infty}(M). \tag{41}$$

Then, at the level of the logarithmic singularities,

$$c_{P_{e^f g}}(x) = e^{(w'-w)f(x)} c_{P_g}(x) \qquad \forall f \in C^\infty(M,\mathbb{R}). \tag{42}$$

This result generalizes a well-known result of Parker-Rosenberg [**PR**] about the logarithmic singularity of the Green kernel of the Yamabe operator. Moreover, using (22) and (30) this also allows us to recover and extend results of Gilkey [**Gi**] and Paycha-Rosenberg [**PRo**] on the noncommutative residue densities of elliptic ΨDOs satisfying (41). In particular, all the assumptions on the compactness of M or on the invertibility and the values of the principal symbol of P_g can be removed from those statements.

PROPOSITION 5 (see [**Po4**]). *Let P_g be a Riemannian invariant ΨDO of weight w and integer order. Then*

$$c_{P_g}(x) = \mathcal{I}_{P_g}(x)|v_g(x)| \tag{43}$$

where $\mathcal{I}_{P_g}(x)$ is a local Riemannian invariant of weight w.

Combining this with Proposition 4 allows us to prove the following.

THEOREM 3 ([**Po4**]). *Let P_g be a conformally invariant Riemannian ΨDO of integer order and biweight (w, w'). In odd dimension, as well as in even dimension when $w' > w$, the logarithmic singularity $c_{P_g}(x)$ is of the form*

$$c_{P_g}(x) = \mathcal{I}_{P_g}(x)|v_g(x)|, \tag{44}$$

where $\mathcal{I}_{P_g}(x)$ is a universal linear combination of Weyl conformal invariants of weight $\frac{n}{2}+w-w'$. If n is even and we have $w' \leq w$, then $c_{P_g}(x)$ still is of a similar form, but in this case $\mathcal{I}_{P_g}(x)$ is a local conformal invariant of weight $\frac{n}{2}+w-w'$ of the type described in Proposition 3.

As an application of this result we can obtain a precise description of the logarithmic singularities of the Green kernels of the GJMS operators.

THEOREM 4. *1) In odd dimension the Green kernel logarithmic singularity $\gamma_{\Box_g^{(k)}}(x)$ is always zero.*

2) In even dimension and for $k = 1, \ldots, \frac{n}{2}$ we have

$$\gamma_{\Box_g^{(k)}}(x) = c_g^{(k)}(x) d\nu_g(x), \tag{45}$$

where $c_g^{(k)}(x)$ is a linear combination of Weyl conformal invariants of weight $\frac{n}{2} - k$. In particular, we have

$$c_g^{(\frac{n}{2})}(x) = (4\pi)^{-\frac{n}{2}} \frac{n}{(n/2)!}, \qquad c_g^{(\frac{n}{2}-1)}(x) = 0, \qquad c_g^{(\frac{n}{2}-2)}(x) = \alpha_n |W(x)|_g^2, \tag{46}$$

$$c_g^{(\frac{n}{2}-3)}(x) = \beta_n W_{ij}{}^{ki} W_{lk}{}^{pq} W_{pq}{}^{ij} + \gamma_n W_i{}^{jk}{}_l W^i{}_{pk}{}^q W_j{}^{pl}{}_q + \delta_n \Phi_g, \tag{47}$$

where W is the Weyl curvature tensor, Φ_g is the Fefferman-Graham invariant (38) and α_n, β_n, γ_n and δ_n are universal constants depending only on n.

Finally, we can get an explicit expression for $c_g^{(1)}(x)$ in dimensions 6 and 8 by making use of the computations by Parker-Rosenberg [**PR**] of the coefficient $a_{n-2}(\Box_g)(x)$ of t^{-1} in the heat kernel asymptotics (28) for the Yamabe operator. Indeed, as by (30) we have $2\gamma_{\Box_g}(x) = a_{n-2}(\Box_g)(x)$, using [**PR**, Prop. 4.2] we see that, in dimension 6,

$$c_g^{(1)}(x) = \frac{1}{360} |W(x)|^2, \tag{48}$$

and, in dimension 8,

$$c_g^{(1)}(x) = \frac{1}{90720} (81\Phi_g + 352 W_{ij}{}^{kl} W_{lk}{}^{pq} W_{pq}{}^{ij} + 64 W_i{}^{jk}{}_l W^i{}_{pk}{}^q W_j{}^{pl}{}_q). \tag{49}$$

In order to use the results of [**PR**] the manifold M has to be compact. However, as $c_g^{(1)}(x)$ is a local Riemannian invariant which makes sense independently of whether M is compact or not, the above formulas for $c_g^{(1)}(x)$ remain valid when M is non-compact.

REMARK. The logarithmic singularities of the Green kernels of the GJMS operator have been computed explicitly after this volume was finalized. The details will appear in a forthcoming paper.

6. The Bergman Kernel of a Strictly Pseudoconvex Domain

Let $D = \{r(z) < 0\} \subset \mathbb{C}^{n+1}$ be a strictly pseudoconvex domain with boundary $\partial D = \{r(z) = 0\}$. The fact that D is strictly pseudoconvex means that the defining function $r(z)$ can be chosen so that $\bar{\partial}\partial r$ defines a positive definite Hermitian form on the holomorphic tangent space $T^{1,0}D$.

Let $\mathcal{O}(D)$ denote the space of holomorphic functions on D. The *Bergman projection*,

$$B : L^2(D) \longrightarrow \mathcal{O}(D) \cap L^2(D), \tag{50}$$

is the orthogonal projection of $L^2(D)$ onto the space of holomorphic L^2-functions on D. The *Bergman kernel*, denoted $B(z, w)$, is the Schwartz kernel of B defined so that

$$Bu(z) = \int B(z,w)u(w)dw \qquad \forall u \in L^2(D). \tag{51}$$

Equivalently, $B(z,w)$ is the reproducing kernel of the Hilbert space $\mathcal{O}(D) \cap L^2(D)$.

In the analysis of the Bergman kernel an important result is the following.

THEOREM 5 (Fefferman [**Fe1**]). *Near ∂D we have*

$$B(z,z) = \varphi(z)r(z)^{-(n+1)} - \psi(z)\log r(z), \tag{52}$$

where $\varphi(z)$ and $\psi(z)$ are smooth up to the boundary.

The original motivation for the program of Fefferman [**Fe2**] was to give a precise description of the singularity of the Bergman kernel near ∂D in terms of local geometric invariants of ∂D. In this case the complex structure of D induces on ∂D a *CR structure* and, as D is strictly pseudoconvex, the CR structure of ∂D is *strictly pseudoconvex*. Thus, the original goals of Fefferman were the following:

(i) Express the singularity in terms of local invariants of the strictly pseudoconvex CR structure of ∂D.

(ii) Determine *all* local invariants of a strictly pseudoconvex CR structure.

We refer to Section 9 for the precise definition of a local invariant of a strictly pseudoconvex CR structure. For now let us recall that, in general, a *CR structure* on an oriented manifold M^{2n+1} is given by the datum of a hyperplane bundle $H \subset TM$ equipped with an (integrable) complex structure J_H. For instance, the CR structure on the boundary ∂D above is given by the complex hyperplane bundle

$$H := T(\partial D) \cap iT(\partial D) \subset T(\partial D). \tag{53}$$

Let (M,H,J) be a CR manifold. Set $T_{1,0} = \ker(J-i) \subset T_{\mathbb{C}}M$ and $T_{0,1} = \ker(J+i)$, so that $H \otimes \mathbb{C} = T_{1,0} \otimes T_{0,1}$. Since M is orientable there is a non-vanishing 1-form θ on M annihilating H. The *Levi form* is the Hermitian form L_θ on $T_{1,0}$ defined by

$$L_\theta(Z,W) = -id\theta(Z,\overline{W}) \qquad \forall Z,W \in C^\infty(M,T_{1,0}). \tag{54}$$

When we can choose θ so that L_θ is positive definite we say that M is *strictly pseudoconvex*. Notice that this implies that θ is a contact form.

Examples of CR manifolds include:

- Boundaries of complex domains in $\mathbb{C}^{n+1}$, e.g., the sphere $S^{2n+1} \subset \mathbb{C}^{n+1}$ and the hyperquadric $Q^{2n+1} := \{z \in \mathbb{C}^{n+1}; \Im z_{n+1} = |z_1|^2 + \cdots + |z_n|^2\}$.
- The Heisenberg group $\mathbb{H}^{2n+1}$ and its quotients $\Gamma\backslash\mathbb{H}^{2n+1}$ by discrete cocompact subgroups.
- Circle bundles over complex manifolds.

Recall that the Heisenberg group $\mathbb{H}^{2n+1}$ can be realized as $\mathbb{R}^{n+1}$ equipped with the group law and dilations,

$$x.y = (x_0 + y_0 + \sum_{j=1}^{n}(x_{n+j}y_j - x_jy_{n+j}), x_1 + y_1, \ldots, x_{2n} + y_{2n}), \tag{55}$$

$$t.(x_0, \ldots, x_{2n}) = (t^2x_0, tx_1, \ldots, tx_{2n}), \qquad t \in \mathbb{R}. \tag{56}$$

Notice that the group-law (55) is homogeneous with respect to the anisotropic dilations (56).

The Lie algebra $\mathfrak{h}^{2n+1}$ of $\mathbb{H}^{2n+1}$ is spanned by the left-invariant vector fields,

$$X_0 = \frac{\partial}{\partial x_0}, \qquad X_j = \frac{\partial}{\partial x_j} + x_{n+j}\frac{\partial}{\partial x_0}, \qquad X_{n+j} = \frac{\partial}{\partial x_{n+j}} - x_j\frac{\partial}{\partial x_0}, \tag{57}$$

where j ranges over $1, \ldots, n$. Notice that, with respect to the dilations (56), the vector field X_0 is homogeneous of degree -2, while $X_1, \ldots . X_{2n}$ are homogeneous of degree -1. Moreover, for $j, k = 1, \ldots, n$, we have the Heisenberg relations,

$$[X_j, X_{n+k}] = -2\delta_{jk}X_0, \qquad [X_j, X_k] = [X_{n+j}, X_{n+k}] = 0. \tag{58}$$

The CR structure of $\mathbb{H}^{2n+1}$ is defined by the hyperplane bundle

$$H = \operatorname{Span}\{X_1, \ldots, X_{2n}\} \tag{59}$$

equipped with the complex structure J defined by

$$JX_j = X_{n+j}, \qquad JX_{n+j} = -X_j, \qquad j = 1, \ldots, n. \tag{60}$$

The hyperplane H is the annihilator of the 1-form,

$$\theta_0 := dx_0 - \sum_{j=1}^{n}(x_{n+j}dx_j - x_j dx_{n+j}). \tag{61}$$

One can check that the associated Levi form is positive definite, so $\mathbb{H}^{2n+1}$ is a strictly pseudoconvex CR manifold. This is in fact the local model of such a manifold.

7. Heisenberg Calculus

In this section, we briefly recall the main facts about the Heisenberg calculus. This calculus was introduced by Beals-Greiner [**BGr**] and Taylor [**Tay**] (see also [**EM**], [**Po3**]). This is the most relevant calculus to study the main geometric operators on CR manifolds.

7.1. Overview of the Heisenberg calculus. The Heisenberg calculus holds in full generality for *Heisenberg manifolds*, that is, manifolds M^{d+1} together with a distinguished hyperplane bundle $H \subset TM$. This terminology stems from the fact that, for a Heisenberg manifold, the relevant notion of tangent bundle is that of a bundle of 2-step nilpotent Lie groups whose fibers are isomorphic to $\mathbb{H}^{2n+1} \times \mathbb{R}^d$ for some k and n such that $2n + k = d$ (see, e.g., [**BGr**], [**Po1**]). This tangent Lie group bundle can be described as follows.

First, there is an intrinsic Levi form obtained as the 2-form $\mathcal{L} : H \times H \to TM/H$ such that, for any point $a \in M$ and any sections X and Y of H near a, we have

$$\mathcal{L}_a(X(a), Y(a)) = [X, Y](a) \qquad \text{mod } H_a. \tag{62}$$

In other words the class of $[X, Y](a)$ modulo H_a depends only on $X(a)$ and $Y(a)$, not on the germs of X and Y near a (see [**Po1**]).

We define the tangent Lie algebra bundle $\mathfrak{g}M$ as the graded Lie algebra bundle consisting of $(TM/H) \oplus H$ together with the fields of Lie bracket and dilations such that, for sections X_0, Y_0 of TM/H and X', Y' of H and for $t \in \mathbb{R}$, we have

$$[X_0 + X', Y_0 + Y'] = \mathcal{L}(X', Y'), \qquad t.(X_0 + X') = t^2 X_0 + tX'. \tag{63}$$

Each fiber $\mathfrak{g}_a M$ is a two-step nilpotent Lie algebra so, by requiring the exponential map to be the identity, the associated tangent Lie group bundle GM appears as $(TM/H) \oplus H$ together with the grading above and the product law such that, for sections X_0, Y_0 of TM/H and X', Y' of H, we have

$$(X_0 + X').(Y_0 + Y') = X_0 + Y_0 + \frac{1}{2}\mathcal{L}(X', Y') + X' + Y'. \tag{64}$$

A motivating example for the Heisenberg calculus is the *horizontal sub-Laplacian* Δ_b on a Heisenberg manifold (M^{d+1}, H) equipped with a Riemannian metric. This is the operator $\Delta_b : C^\infty(M) \to C^\infty(M)$ defined by

$$\Delta_b = d_b^* d_b, \qquad d_b = \pi \circ d, \tag{65}$$

where π is the orthogonal projection onto H^* (identified with a subbundle of T^*M using the Riemannian metric).

An H-frame of TM is a frame $X_0, X_1, \ldots, X_d$ of TM such that $X_1, \ldots, X_d$ span H. Locally, we always can find an H-frame $X_0, X_1, \ldots, X_d$ such that Δ_b takes the form

$$\Delta_b = -(X_1^2 + \cdots + X_{2n}^2) + \sum_{j=1}^{d} a_j(x) X_j. \tag{66}$$

As the differentiation along X_0 is missing we see that Δ_b is not elliptic. However, whenever the Levi form (62) is everywhere non-zero, a celebrated theorem of Hörmander [**Hö2**] ensures us that Δ_b is hypoelliptic with gain of one derivative (i.e., if $\Delta_b u$ is in L^2_{loc} then u must be in the Sobolev space $W^{2,1}_{\text{loc}}$).

In the case of the Heisenberg group, we can explicitly construct a fundamental solution for Δ_b (see [**BGr**], [**FS1**]). This fundamental solution comes from a symbol of type $(\frac{1}{2}, \frac{1}{2})$ in the sense of Hörmander [**Hö**]. As the usual symbolic calculus does not hold anymore for ΨDOs of type $(\frac{1}{2}, \frac{1}{2})$, the full strength of the classical pseudodifferential calculus cannot be used to study natural operators on Heisenberg manifolds such as the horizontal sub-Laplacian Δ_b.

The relevant substitute for the classical pseudodifferential calculus is precisely provided by the Heisenberg calculus. The idea is to construct a class of pseudodifferential operators, called Ψ_HDOs, which near each point $a \in M$ are approximated (in a suitable sense) by left-invariant convolution operators on G_aM. This allows us to get a pseudodifferential calculus with a full symbolic calculus with inverses and which is invariant under changes of charts preserving the hyperplane bundle H.

The symbols that we consider in the Heisenberg calculus are the following.

DEFINITION 12. *1) $S_m(U \times \mathbb{R}^{d+1})$, $m \in \mathbb{C}$, is the space of functions $p(x,\xi)$ in $C^\infty(U \times (\mathbb{R}^{d+1} \setminus \{0\}))$ such that $p(x, t.\xi) = t^m p(x,\xi)$ for any $t > 0$.*

2) $S^m(U \times \mathbb{R}^{d+1})$, $m \in \mathbb{C}$, consists of functions $p \in C^\infty(U \times \mathbb{R}^{d+1})$ with an asymptotic expansion $p \sim \sum_{j \geq 0} p_{m-j}$, $p_k \in S_k(U \times \mathbb{R}^{d+1})$, in the sense that, for any integer N, any compact $K \subset U$ and any multi-orders α, β, there exists a constant $C_{NK\alpha\beta} > 0$ such that, for any $x \in K$ and any $\xi \in \mathbb{R}^{d+1}$ so that $\|\xi\| \geq 1$, we have

$$|\partial_x^\alpha \partial_\xi^\beta (p - \sum_{j<N} p_{m-j})(x,\xi)| \leq C_{NK\alpha\beta} \|\xi\|^{\Re m - \langle\beta\rangle - N}, \tag{67}$$

where we have set $\langle\beta\rangle = 2\beta_0 + \beta_1 + \ldots + \beta_d$.

Next, for $j = 0, \ldots, d$ let $\sigma_j(x,\xi)$ denote the symbol (in the classical sense) of the vector field $\frac{1}{i}X_j$ and set $\sigma = (\sigma_0, \ldots, \sigma_d)$. Then for $p \in S^m(U \times \mathbb{R}^{d+1})$ we let

$p(x,-iX)$ be the continuous linear operator from $C_c^\infty(U)$ to $C^\infty(U)$ such that

$$p(x,-iX)u(x) = (2\pi)^{-(d+1)} \int e^{ix.\xi} p(x,\sigma(x,\xi))\hat{u}(\xi)d\xi \qquad \forall u \in C_c^\infty(U). \tag{68}$$

Let (M^{d+1},H) be a Heisenberg manifold. We define the Ψ_HDOs on M as follows.

DEFINITION 13. *$\Psi_H^m(M)$, $m \in \mathbb{C}$, consists of continuous operators P from $C_c^\infty(M)$ to $C^\infty(M)$ such that:*

(i) The Schwartz kernel of P is smooth off the diagonal;

(ii) In any local coordinates equipped with an H-frame $X_0,\ldots,X_d$ the operator P can be written as

$$P = p(x,-iX) + R, \tag{69}$$

where $p(x,\xi)$ is a Heisenberg symbol of order m and R is a smoothing operator.

For any $a \in M$ the convolution on G_aM gives rise under the (linear) Fourier transform to a bilinear product for homogeneous symbols,

$$*^a : S_{m_1}(\mathfrak{g}_a^*M) \times S_{m_2}(\mathfrak{g}_a^*) \longrightarrow S_{m_1+m_2}(\mathfrak{g}_a^*M). \tag{70}$$

This product depends smoothly on a, so it gives rise to the product,

$$* : S_{m_1}(\mathfrak{g}^*M) \times S_{m_2}(\mathfrak{g}^*M) \longrightarrow S_{m_1+m_2}(\mathfrak{g}^*M), \tag{71}$$

$$p_1 * p_2(a,\xi) = [p_1(a,\cdot) *^a p_2(a,\cdot)](\xi). \tag{72}$$

This provides us with the right composition for principal symbols, since for any operators $P_1 \in \Psi_H^{m_1}(M)$ and $P_2 \in \Psi_H^{m_2}(M)$ such that P_1 or P_2 is properly supported we have

$$\sigma_{m_1+m_2}(P_1P_2) = \sigma_{m_1}(P_1) * \sigma_{m_2}(P_2). \tag{73}$$

Notice that when G_aM is not commutative, i.e., when $\mathcal{L}_a \neq 0$, the product $*^a$ is no longer the pointwise product of symbols and, in particular, it is not commutative. As a consequence, unless H is integrable, the product for Heisenberg symbols, while local, is not microlocal (see [**BGr**]).

When the principal symbol of $P \in \Psi_H^m(M)$ is invertible with respect to the product $*$, the symbolic calculus of [**BGr**] allows us to construct a parametrix for P in $\Psi_H^{-m}(M)$. In particular, although not elliptic, P is hypoelliptic with a controlled loss/gain of derivatives (see [**BGr**]).

In general, it may be difficult to determine whether the principal symbol of a given operator $P \in \Psi_H^m(M)$ is invertible with respect to the product $*$, but this can be completely determined in terms of a representation theoretic criterion on each tangent group G_aM, the so-called Rockland condition (see [**Po3**], Thm. 3.3.19). In particular, if $\sigma_m(P)(a,\cdot)$ is *pointwise* invertible with respect to the product $*^a$ for all $a \in M$.

7.2. The logarithmic singularity of a Ψ_HDO. It is possible to characterize the Ψ_HDOs in terms of their Schwartz kernels (see [**BGr**]). As a consequence we get the following description of the singularity near the diagonal of the Schwartz kernel of a Ψ_HDO.

In the sequel, given an open subset of local coordinates $U \subset \mathbb{R}^{d+1}$ equipped with an H-frame $X_0,\ldots,X_d$ of TU, for any $a \in U$ we let ψ_a denote the unique affine change of variables such that $\psi_a(a) = 0$ and $(\psi_{a*}X_j)(0) = \frac{\partial}{\partial x_j}$ for $j = 0,1,\ldots,d+1$.

DEFINITION 14. *The local coordinates provided by ψ_a are called privileged coordinates centered at a.*

Throughout the rest of the paper the notion of homogeneity refers to homogeneity with respect to the anisotropic dilations (63).

PROPOSITION 6 ([**Po2**, Prop. 3.11]). *Let $\Psi_H^m(M)$, $m \in \mathbb{Z}$.*

1) In local coordinates equipped with an H-frame the kernel $k_P(x,y)$ has a behavior near the diagonal $y = x$ of the form

$$k_P(x,y) = \sum_{-(m+d+2)\leq j\leq -1} a_j(x, -\psi_x(y)) - c_P(x)\log\|\psi_x(y)\| + O(1), \tag{74}$$

where $a_j(x,y) \in C^\infty(U \times (\mathbb{R}^n \setminus \{0\}))$ is homogeneous of degree j in y, and we have

$$c_P(x) = (2\pi)^{-(d+1)} \int_{\|\xi\|=1} p_{-(d+2)}(x,\xi)\iota_E d\xi, \tag{75}$$

where $p_{-(d+2)}(x,\xi)$ is the symbol of degree $-(d+2)$ of P and E denotes the anisotropic radial vector $2x^0\partial_{x^0} + x^1\partial_{x^1} + \cdots + x^d\partial_{x^d}$.

2) The coefficient $c_P(x)$ makes sense globally on M as a 1-density.

Let $P \in \Psi_H^m(M)$ be such that its principal symbol is invertible in the Heisenberg calculus sense and let $Q \in \Psi_H^{-m}(M)$ be a parametrix for P. Then Q is uniquely defined modulo smoothing operators, so the logarithmic singularity $c_Q(x)$ does not depend on the particular choice of Q.

DEFINITION 15. *If $P \in \Psi_H^m(M)$, $m \in \mathbb{Z}$, has an invertible principal symbol, then its Green kernel logarithmic singularity is the density*

$$\gamma_P(x) := c_Q(x), \tag{76}$$

where $Q \in \Psi_H^{-m}(M)$ is any given parametrix for P.

In the same way as for classical ΨDOs, the logarithmic singularity densities are related to the construction of the noncommutative residue trace for the Heisenberg calculus (see [**Po2**]).

8. Pseudo-Hermitian Invariants

8.1. Pseudo-Hermitian geometry. Let (M^{2n+1}, H, J) be a strictly pseudoconvex CR manifold. In the terminology of [**We**] a *pseudo-Hermitian structure* on M is given by the datum of real 1-form on M such that θ annihilates H and the associated Levi form (62) is positive definite. Notice that θ is uniquely determined up to a conformal factor. Conversely, the conformal class of θ is uniquely determined by the strictly pseudoconvex CR structure of M.

Since θ is a contact form there exists a unique vector field X_0 on M, called the *Reeb field*, such that $\iota_{X_0}\theta = 1$ and $\iota_{X_0}d\theta = 0$. Let $\mathcal{N} \subset T_{\mathbb{C}}M$ be the complex line bundle spanned by X_0. We then have the splitting

$$T_{\mathbb{C}}M = \mathcal{N} \oplus T_{1,0} \oplus T_{0,1}. \tag{77}$$

The Levi metric h_θ is the unique Hermitian metric on $T_{\mathbb{C}}M$ such that:

- The splitting (77) is orthogonal with respect to h_θ;
- h_θ commutes with complex conjugation;
- We have $h(X_0, X_0) = 1$ and h_θ agrees with L_θ on $T_{1,0}$.

Notice that the volume form of h_θ is $\frac{1}{n!}(d\theta)^n \wedge \theta$.

As proved by Tanaka [**Ta**] and Webster [**We**], the datum of the pseudo-Hermitian contact form θ uniquely defines a connection, the *Tanaka-Webster connection*, which preserves the pseudo-Hermitian structure of M, i.e., such that $\nabla\theta = 0$ and $\nabla J = 0$. It can be defined as follows.

Let $\{Z_j\}$ be a frame of $T_{1,0}$. We set $Z_{\bar{j}} = \overline{Z_j}$. Then $\{X_0, Z_j, Z_{\bar{j}}\}$ forms a frame of $T_{\mathbb{C}}M$. In the sequel such a frame will be called an *admissible frame* of $T_{\mathbb{C}}M$. Let $\{\theta, \theta^j, \theta^{\bar{j}}\}$ be the coframe of $T^*_{\mathbb{C}}M$ dual to $\{X_0, Z_j, Z_{\bar{j}}\}$. With respect to this coframe we can write $d\theta = ih_{j\bar{k}}\theta^j \wedge \theta^{\bar{k}}$.

Using the matrix $(h_{j\bar{k}})$ and its inverse $(h^{j\bar{k}})$ to lower and raise indices, the connection 1-form $\omega = (\omega_j{}^k)$ and the torsion form $\tau_j = A_{jk}\theta^k$ of the Tanaka-Webster connection are uniquely determined by the relations

$$d\theta^k = \theta^j \wedge \omega_j{}^k + \theta \wedge \tau^k, \qquad \omega_{j\bar{k}} + \omega_{\bar{k}j} = dh_{j\bar{k}}, \qquad A_{jk} = A_{kj}. \tag{78}$$

In addition, we have the structure equations

$$d\omega_j{}^k - \omega_j{}^l \wedge \omega_l{}^k = R_j{}^k{}_{l\bar{m}}\theta^l \wedge \theta^{\bar{m}} + W_{j\bar{k}l}\theta^l \wedge \theta - W_{\bar{k}j\bar{l}}\theta^{\bar{l}} \wedge \theta + i\theta_j \wedge \tau_{\bar{k}} - i\tau_j \wedge \theta_{\bar{k}}. \tag{79}$$

The *pseudo-Hermitian curvature tensor* of the Tanaka-Webster connection is the tensor with components $R_{j\bar{k}l\bar{m}}$, its *Ricci tensor* is $\rho_{j\bar{k}} := R_l{}^l{}_{j\bar{k}}$ and its *scalar curvature* is $\kappa_\theta := \rho_j{}^j$.

8.2. Local pseudo-Hermitian invariants. Let us now define local pseudo-Hermitian invariants. The definition is more involved than that of local Riemannian invariants, because:

- The components of the Tanaka-Webster connection and its curvature and torsion tensors are defined with respect to the datum of a local frame $Z_1, \ldots, Z_n$ which never is a frame $\frac{\partial}{\partial z^1}, \ldots \frac{\partial}{\partial z^n}$ associated to local coordinates $z^1, \ldots, z^n$;

- In order to get local pseudo-Hermitian invariants from pseudo-Hermitian invariant Ψ_HDOs it is important to take into account the tangent group bundle of a CR manifold, in which the Heisenberg group comes into play.

Before defining local pseudo-Hermitian invariants, some notation needs to be introduced.

Let $U \subset \mathbb{R}^n$ be an open subset of local coordinates equipped with a frame $Z_1, \ldots, Z_n$ of $T_{1,0}$. Write $Z_j = X_j - iX_{n+j}$, where X_j and X_{n+j} are real vector fields. Then $X_0, \ldots, X_{2n}$ is a local H-frame of TM. We shall call this frame the *H-frame associated to $Z_1, \ldots, Z_n$*.

Let $\eta^0, \ldots, \eta^{2n}$ be the coframe of T^*M dual to $X_0, \ldots, X_{2n}$ (so that $\eta^0 = \theta$). We write $X_j = X_j{}^k\partial_{x^k}$ and $\eta^j = \eta^j{}_k dx^k$. We also write $Z_j = Z_j{}^k\partial_{x^k}$. It will be convenient to identify $X_0(x)$ with the vector $(X_0{}^k(x)) \in \mathbb{R}^{2n+1}$ and $Z(x) := (Z_1(x), \ldots, Z_n(x))$ with the matrix $(Z_j{}^k(x))$ in $M_{n,2n+1}(\mathbb{C})^\times$, where the latter denotes the open subset of $M_{n,2n+1}(\mathbb{C})$ consisting of regular matrices.

For $j, \bar{k} = 1, \ldots, n$ we set $h_{j\bar{k}} = h_\theta(Z_j, Z_k) = i\theta([Z_j, Z_{\bar{k}}])$, and for $j, k = 1, \ldots, 2n$ we set $L_{jk} = \theta([X_j, X_k])$. Let $M_n(\mathbb{C})_+$ denote the open cone of positive definite Hermitian $n \times n$ matrices. In the sequel it will also be convenient to identify h_θ with the matrix $h_\theta(x) := (h_{j\bar{k}}(x)) \in M_n(\mathbb{C})_+$.

Thanks to the integrability of $T_{1,0}$ we have $\theta([Z_j,Z_k]) = 0$. As we have $[Z_j,Z_k]=[X_j,X_k]-[X_{n+j},X_{n+k}]-i([X_{n+j},X_k]+[X_j,X_{n+k}])$ we see that

$$L_{n+j,n+k}=L_{j,k} \quad \text{and} \quad L_{j,n+k}=-L_{n+j,k}. \tag{80}$$

Since $[Z_j,Z_{\bar{k}}]=[X_j,X_k]+[X_{n+j},X_{n+k}]+i([X_{n+j},X_k]-[X_j,X_{n+k}])$ we get

$$h_{j\bar{k}}=i\theta([Z_j,Z_{\bar{k}}])=2iL_{j,k}+2L_{n+j,k}. \tag{81}$$

In other words, we have

$$(L_{jk})=\frac{1}{2}\begin{pmatrix} \Im h & -\Re h \\ \Re h & \Im h \end{pmatrix}. \tag{82}$$

For any $a\in U$ we let ψ_a be the affine change of variables to the privileged coordinates centered at a (cf. Definition 14). One checks that $\psi_a(x)^j=\eta^j{}_k(x^k-a^k)$, so we have

$$\psi_{a*}X_j=X_j{}^k(\psi_a(x))\eta^l{}_k(a)\partial_l. \tag{83}$$

Given a vector field X defined near $x=0$ let us denote by $X(0)_l$ the vector field obtained as the part in the Taylor expansion at $x=0$ of X which is homogeneous of degree l with respect to the Heisenberg dilations (63). Then the Taylor expansions at $x=0$ of the vector fields $\psi_{a*}X_0,\ldots,\psi_{a*}X_{2n}$ take the form

$$X_0=X_0^{(a)}+X_0(0)_{(-1)}+\cdots, \tag{84}$$

$$X_j=X_j^{(a)}+X_j(0)_{(0)}+\cdots, \quad 1\leq j\leq 2n, \tag{85}$$

with

$$X_0^{(a)}=\partial_{x^0}, \qquad X_j^{(a)}=\partial_{x^j}+b_{jk}(a)x^k\partial_{x^0}, \quad 1\leq j\leq 2n, \tag{86}$$

where we have set $b_{jk}(a):=\partial_k[X_j{}^l(\psi_a(x))]_{|x=0}\eta^0{}_l(a)$. Notice that $X_0^{(a)}$ is homogeneous of degree -2, while $X_1^{(a)},\ldots,X_{2n}^{(a)}$ are homogeneous of degree -1.

The linear span of the vector fields $X_0^{(a)},\ldots,X_{2n}^{(a)}$ is a 2-step nilpotent Lie algebra under the Lie bracket of vector fields. Therefore, this is the Lie algebra of left-invariant vector fields on a 2-step nilpotent Lie group $G^{(a)}$. The latter can be realized as $\mathbb{R}^{2n+1}$ equipped with the product

$$x.y=(x^0+y^0+b_{kj}(a)x^jy^k,x^1+y^1,\ldots,x^{2n}+y^{2n}). \tag{87}$$

Notice that $[X_j^{(a)},X_k^{(a)}]=(b_{kj}(a)-b_{jk}(a))X_0^{(a)}$. In addition, we can check that $[\psi_{a*}X_j,\psi_{a*}X_k](0)=(b_{kj}(a)-b_{jk}(a))\partial_{x^0}$ mod H_0. Thus,

$$\begin{aligned} L_{jk}(a)=\theta(X_j,X_k)(a) &= (\psi_{a*}\theta)([\psi_{a*}X_j,\psi_{a*}X_k])(0) \\ &= \langle dx^0,[\psi_{a*}X_j,\psi_{a*}X_k](0)\rangle = b_{kj}(a)-b_{jk}(a). \end{aligned} \tag{88}$$

This shows that $G^{(a)}$ has the same structure constants as the tangent group G_aM, hence is isomorphic to it (see [**Po1**]). This also implies that $(-\frac{1}{2}L_{jk}(a))$ is the skew-symmetric part of $(b_{jk}(a))$. For $j,k=1,\ldots,2n$ set $\mu_{jk}(a)=b_{jk}(a)+\frac{1}{2}L_{jk}(a)$. The matrix $(\mu_{jk}(a))$ is the symmetric part of $(b_{jk}(a))$, so it belongs to the space $S_{2n}(\mathbb{R})$ of symmetric $2n\times 2n$ matrices with real coefficients.

In the sequel we set

$$\Omega=M_n(\mathbb{C})_+\times\mathbb{R}^{2n+1}\times M_{n,2n+1}(\mathbb{C})^\times\times S_{2n}(\mathbb{R}). \tag{89}$$

This is a manifold, and for any $x\in U$ the quadruple $(h(x),X_0(x),Z(x),\mu(x))$ is an element of Ω depending smoothly on x.

In addition, we let $\mathcal{P}$ be the set of monomials in the indeterminate variables $\partial^\alpha {X_0}^k$, $\partial^\alpha {Z_j}^k$ and $\partial^\alpha \overline{{Z_j}^k}$, where the integer j ranges over $\{1,\ldots,n\}$, the integer k ranges over $\{0,\ldots,2n\}$, and α ranges over all multi-orders in $\mathbb{N}_0^{2n}$. Given the Reeb field X_0 and a local frame $Z_0,\ldots,Z_n$ of $T_{1,0}$ by plugging $\partial_x^\alpha {X_0}^k(x)$, $\partial_x^\alpha {Z_j}^k(x)$ and $\partial^\alpha \overline{{Z_j}^k}(x)$ into a monomial $\mathfrak{p}\in\mathcal{P}$ we get a function which we shall denote by $\mathfrak{p}(X_0,Z,\overline{Z})(x)$.

Bearing all this mind we define local pseudo-Hermitian invariants as follows.

DEFINITION 16. *A local pseudo-Hermitian invariant of weight w is the datum on each pseudo-Hermitian manifold (M^{2n+1},θ) of a function $\mathcal{I}_\theta \in C^\infty(M)$ such that:*

(i) There exists a finite family $(a_{\mathfrak{p}})_{\mathfrak{p}\in\mathcal{P}}\subset C^\infty(\Omega)$ such that, in any local coordinates equipped with a frame $Z_1,\ldots,Z_n$ of $T_{1,0}$, we have

$$\mathcal{I}_\theta(x) = \sum_{\mathfrak{p}\in\mathcal{P}} a_{\mathfrak{p}}(h(x),X_0(x),Z(x),\mu(x))\mathfrak{p}(X_0,Z,\overline{Z})(x). \tag{90}$$

(ii) We have $\mathcal{I}_{t\theta}(x)=t^{-w}\mathcal{I}_\theta(x)$ for any $t>0$.

Any local Riemannian invariant of h_θ is a local pseudo-Hermitian invariant. However, the above notion of weight for pseudo-Hermitian invariant is anisotropic with respect to h_θ. For instance if we replace θ by $t\theta$ then h_θ is rescaled by t on $T_{1,0}\oplus T_{0,1}$ and by t^2 on the vertical line bundle $\mathcal{N}\otimes\mathbb{C}$.

On the other hand, as shown in [**JL2**, Prop. 2.3], by means of parallel translation along parabolic geodesics any orthonormal frame $Z_1(a),\ldots,Z_n(a)$ of $T_{1,0}$ at a point $a\in M$ can be extended to a local frame $Z_1,\ldots,Z_n$ of $T_{1,0}$ near a. Such a frame is called a *special orthonormal frame.*

Furthermore, as also shown in [**JL2**, Prop. 2.3] any special orthonormal frame $Z_1,\ldots,Z_n$ near a allows us to construct *pseudo-Hermitian normal coordinates* $x_0, z^1=x^1+ix^{n+1},\ldots,z^n=x^n+ix_{2n}$ centered at a in such way that in the notation of (84)–(85) we have

$$X_0(0)_{(-2)}=\partial_{x^0},\qquad Z_j(0)_{(-1)}=\partial_{z^j}+\frac{i}{2}\bar{z}^j\partial_{x^0},\qquad \omega_{j\bar{k}}(0)=0. \tag{91}$$

Write $Z_j=X_j-iX_{n+j}$, where X_j and X_{n+j} are real vector fields. Then we have $X_j(0)_{(-1)}=\partial_{x^j}-\frac{1}{2}x^{n+j}\partial_{x^0}$ and $X_{n+j}(0)_{(-1)}=\partial_{x^{n+j}}+\frac{1}{2}x^j\partial_{x^0}$. In particular, we have $X_j(0)=\partial_{x^j}$ for $j=0,\ldots,2n$. This implies that the affine change of variables ψ_0 to the privileged coordinates at 0 is just the identity. Moreover, in the notation of (86) for $j=1,\ldots,n$ we have

$$X_j^{(0)}=\partial_{x^j}-\frac{1}{2}x^{n+j}\partial_{x^0},\qquad X_{n+j}^{(0)}=\partial_{x^{n+j}}+\frac{1}{2}x^j\partial_{x^0}. \tag{92}$$

Incidentally, this shows that the matrix $(b_{jk}(0))$ is skew-symmetric, so its symmetric part vanishes, i.e., $\mu(0)=0$.

PROPOSITION 7 ([**Po4**]). *Assume each pseudo-Hermitian manifold (M^{2n+1},θ) gifted with a function $\mathcal{I}_\theta\in C^\infty(M)$ in such a way that $\mathcal{I}_{t\theta}(x)=t^{-w}\mathcal{I}_\theta(x)$ for any $t>0$. Then the following are equivalent:*

(i) $\mathcal{I}_\theta(x)$ is a local pseudo-Hermitian invariant;

(ii) There exists a finite family $(a_{\mathfrak{p}})_{\mathfrak{p}\in\mathcal{P}} \subset \mathbb{C}$ such that, for any pseudo-Hermitian manifold (M^{2n+1},θ) and any point $a \in M$, in any pseudo-Hermitian normal coordinates centered at a associated to any given special orthonormal frame $Z_1,\ldots,Z_n$ of $T_{1,0}$ near a, we have

$$\mathcal{I}_\theta(a) = \sum_{\mathfrak{p}\in\mathcal{P}} a_{\mathfrak{p}}\mathfrak{p}(X_0, Z, \overline{Z})(x)_{|x=0}. \tag{93}$$

(iii) $\mathcal{I}_\theta(x)$ is a universal linear combination of complete tensorial contractions of covariant derivatives of the pseudo-Hermitian curvature tensor and of the torsion tensor of the Tanaka-Webster connection.

8.3. Pseudo-Hermitian invariant Ψ_HDOs. We define homogeneous symbols on $\Omega \times \mathbb{R}^{2n+1}$ as follows.

DEFINITION 17. *$S_m(\Omega\times\mathbb{R}^{2n+1})$, $m \in \mathbb{C}$, consists of functions $a(h, X_0, Z, \xi)$ in $C^\infty(\Omega \times (\mathbb{R}^{2n+1}\setminus\{0\}))$ such that $a(\theta, Z, t\xi) = t^m a(\theta, Z, \xi)$ $\forall t > 0$.*

In addition, recall that if $Z_1,\ldots,Z_n$ is a local frame of $T_{1,0}$ then its associated H-frame is the frame $X_0,\ldots,X_{2n}$ of TM such that $Z_j = X_j - iX_{n+j}$ for $j = 1,\ldots,n$.

DEFINITION 18. *A pseudo-Hermitian invariant Ψ_HDO of order m and weight w is the datum on each pseudo-Hermitian manifold (M^{2n+1},θ) of an operator P_θ in $\Psi_H^m(M)$ such that:*

(i) For $j = 0,1,\ldots$ there exists a finite family $(a_{j\mathfrak{p}})_{\mathfrak{p}\in\mathcal{P}} \subset S_{m-j}(\Omega\times\mathbb{R}^{2n+1})$ such that, in any local coordinates equipped with the H-frame associated to a frame $Z_1,\ldots,Z_n$ of $T_{1,0}$, the operator P_θ has symbol $p_\theta \sim \sum p_{\theta,m-j}$ with

$$p_{\theta,m-j}(x,\xi) = \sum_{\mathfrak{p}\in\mathcal{P}} \mathfrak{p}(X_0, Z, \overline{Z})(x) a_{j\mathfrak{p}}(h(x), X_0(x), Z(x), \mu(x), \xi). \tag{94}$$

(ii) For all $t > 0$ we have $P_{t\theta} = t^{-w}P_\theta$ modulo $\Psi^{-\infty}(M)$.

In addition, we will say that P_θ is admissible *if in (94) we can take $a_{0\mathfrak{p}}(h, X_0, Z, \mu, \xi)$ to be zero for $\mathfrak{p} \neq 1$.*

For instance, the horizontal sub-Laplacian Δ_b is an admissible pseudo-Hermitian invariant differential operator of weight 1.

We gather the main properties of pseudo-Hermitian invariant Ψ_HDOs in the following.

PROPOSITION 8 ([**Po4**]). *Let P_θ be a pseudo-Hermitian invariant Ψ_HDO of order m and weight w.*

1. *Let Q_θ be a pseudo-Hermitian invariant Ψ_HDO of order m' and weight w', and assume that P_θ or Q_θ is uniformly properly supported. Then $P_\theta Q_\theta$ is a pseudo-Hermitian invariant Ψ_HDO of order $m+m'$ and weight $w+w'$.*
2. *Assume that P_θ is admissible and its principal symbol is invertible in the Heisenberg calculus sense. Then the datum on every pseudo-Hermitian manifold (M^{2n+1},θ) of a parametrix $Q_\theta \in \Psi^{-m}(M)$ gives rise to a pseudo-Hermitian invariant Ψ_HDO of order $-m$ and weight $-w$.*

Finally, concerning the logarithmic singularities of pseudo-Hermitian invariant Ψ_HDOs the following holds.

PROPOSITION 9 ([**Po4**]). *Let P_θ be a pseudo-Hermitian invariant $\Psi_H DO$ of order m and weight w. Then the logarithmic singularity $c_{P_\theta}(x)$ takes the form*

$$c_{P_\theta}(x) = \mathcal{I}_{P_\theta}(x)|(d\theta)^n \wedge \theta|, \tag{95}$$

where $\mathcal{I}_\theta(x)$ is a local pseudo-Hermitian invariant of weight $n+1+w$.

9. CR Invariants and Fefferman's Program

9.1. Local CR Invariants. The local CR invariants can be defined as follows.

DEFINITION 19. *A local scalar CR invariant of weight w is a local scalar pseudo-Hermitian invariant $\mathcal{I}_\theta(x)$ such that*

$$\mathcal{I}_{e^f\theta}(x) = e^{-wf(x)}\mathcal{I}_\theta(x) \qquad \forall f \in C^\infty(M, \mathbb{R}). \tag{96}$$

When M is a real hypersurface the above definition of a local CR invariant agrees with the definition in [**Fe2**] in terms of Chern-Moser invariants (with our convention about weight a local CR invariant that has weight w in the sense of (96) has weight $2w$ in [**Fe2**]).

The analogue of the Weyl curvature in CR geometry is the Chern-Moser tensor ([**CM**], [**We**]). Its components with respect to any local frame $Z_1, \ldots, Z_n$ of $T_{1,0}$ are

$$S_{j\bar{k}l\bar{m}} = R_{j\bar{k}l\bar{m}} - (P_{j\bar{k}}h_{l\bar{m}} + P_{l\bar{k}}h_{j\bar{m}} + P_{l\bar{m}}h_{j\bar{k}} + P_{j\bar{m}}h_{l\bar{k}}), \tag{97}$$

where $P_{j\bar{k}} = \frac{1}{n+2}(\rho_{j\bar{k}} - \frac{\kappa}{2(n+1)}h_{j\bar{k}})$ is the CR Schouten tensor. The Chern-Moser tensor is a CR invariant tensor of weight 1, so we get scalar local CR invariants by taking complete tensorial contractions. For instance, as a scalar invariant of weight 2 we have

$$|S|_\theta^2 = S^{\bar{j}k\bar{l}m}S_{j\bar{k}l\bar{m}}, \tag{98}$$

and as scalar invariants of weight 3 we get

$$S_{i\bar{j}}{}^{\bar{k}l}S_{k\bar{l}}{}^{\bar{p}q}S_{p\bar{q}}{}^{\bar{i}j} \quad \text{and} \quad S_{i}{}^{j\bar{k}}{}_{\bar{l}}S^{\bar{i}}{}_{\bar{j}p}{}^{q}S^{\bar{p}}{}_{\bar{q}k}{}^{l}. \tag{99}$$

More generally, the Weyl CR invariants are obtained as follows. Let $\mathcal{K}$ be the canonical line bundle of M, i.e., the annihilator of $T_{1,0} \wedge \Lambda^n T^*_{\mathbb{C}}M$ in $\Lambda^{n+1}T^*_{\mathbb{C}}M$. The Fefferman bundle is the total space of the circle bundle,

$$\mathcal{F} := (\mathcal{K} \setminus \{0\})/\mathbb{R}^*_+. \tag{100}$$

It carries a natural S^1-invariant Lorentzian metric g_θ whose conformal class depends only the CR structure of M, for we have $g_{e^f\theta} = e^f g_\theta$ for any $f \in C^\infty(M, \mathbb{R})$ (see [**Fe1**], [**Le**]). Notice also that the Levi metric defines a Hermitian metric h_θ^* on $\mathcal{K}$, so we have a natural isomorphism of circle bundles $\iota_\theta : \mathcal{F} \to \Sigma_\theta$, where $\Sigma_\theta \subset \mathcal{K}$ denotes the unit sphere bundle of $\mathcal{K}$.

LEMMA 1 ([**Fe2**]; see also [**Po4**]). *Any local scalar conformal invariant $\mathcal{I}_g(x)$ of weight w uniquely defines a local scalar CR invariant of weight w.*

9.2. CR invariant operators.

DEFINITION 20. *A CR invariant $\Psi_H DO$ of order m and biweight (w, w') is a pseudo-Hermitian invariant $\Psi_H DO$ P_θ such that*

$$P_{e^f\theta} = e^{w'f} P_\theta e^{-wf} \qquad \forall f \in C^\infty(M, \mathbb{R}). \tag{101}$$

We summarize the algebraic properties of CR invariant Ψ_HDOs in the following.

PROPOSITION 10 ([**Po4**]). *Let P_θ be a CR invariant $\Psi_H DO$ of order m and biweight (w, w').*

(1) *Let Q_θ be a CR invariant $\Psi_H DO$ of order m' and biweight (w'', w), and assume that P_θ or Q_θ is uniformly properly supported. Then $P_\theta Q_\theta$ is a CR invariant $\Psi_H DO$ of order $m + m'$ and biweight (w'', w').*

(2) *Assume that P_θ is admissible and its principal symbol is invertible in the Heisenberg calculus sense. Then the datum on every pseudo-Hermitian manifold (M^{2n+1}, θ) of a parametrix $Q_\theta \in \Psi^{-m}(M)$ gives rise to a CR invariant $\Psi_H DO$ of order $-m$ and biweight (w', w).*

Next, we have plenty of CR invariant operators thanks to the following result.

PROPOSITION 11 ([**JL1**], [**GG**]; see also [**Po4**]). *Any conformally invariant Riemannian differential operator L_g of weight w uniquely defines a CR invariant differential operator L_θ of the same weight.*

When L_g is the Yamabe operator the corresponding CR invariant operator is the CR Yamabe operator introduced by Jerison-Lee [**JL1**] in their solution of the Yamabe problem on CR manifolds. Namely,

$$\boxdot_\theta = \Delta_b + \frac{n}{n+2}\kappa_\theta, \tag{102}$$

where κ_θ is the Tanaka-Webster scalar curvature. This is a CR invariant differential operator of biweight $(\frac{-n}{2}, -\frac{n+2}{2})$.

More generally, Gover-Graham [**GG**] proved that for $k = 1, \ldots, n+1$ the GJMS operator $\Box_g^{(k)}$ on the Fefferman bundle gives rise to a selfadjoint differential operator,

$$\boxdot_\theta^{(k)} : C^\infty(M) \longrightarrow C^\infty(M). \tag{103}$$

This is a CR invariant operator of biweight $(\frac{k-(n+1)}{2}, -\frac{k+n+1}{2})$ and it has the same principal symbol as

$$(\Delta_b + i(k-1)X_0)(\Delta_b + i(k-3)X_0) \cdots (\Delta_b - i(k-1)X_0). \tag{104}$$

In particular, except for the critical value $k = n+1$, the principal symbol of $\boxdot_\theta^{(k)}$ is invertible in the Heisenberg calculus sense (see [**Po3**, Prop. 3.5.7]). The operator $\boxdot_\theta^{(k)}$ is called the CR GJMS operator of order k. For $k = 1$ we recover the CR Yamabe operator. Notice that by making use of the CR tractor calculus we also can define CR GJMS operators of order $k \geq n+2$ (see [**GG**]). These operators can also be obtained by means of geometric scattering theory (see [**HPT**]).

9.3. Fefferman's program. In the same way as in conformal geometry, in the setting of CR geometry the program of Fefferman has two main aspects:

FEFFERMAN'S PROGRAM (Analytic Aspect). *Give a precise geometric description of the singularities of the Schwartz, Green and null kernels of CR invariant operators in terms of local conformal invariants.*

FEFFERMAN'S PROGRAM (Geometric Aspect). *Determine all local invariants of a strictly pseudoconvex CR structure.*

Concerning the latter aspect, the analogues of the Weyl conformal invariants are the *Weyl CR invariants* which are the local CR invariants arising from the Weyl conformal invariants of the Fefferman as described by Lemma 1. Notice that, for the Fefferman bundle, the ambient metric was constructed by Fefferman [**Fe2**] as a Kähler-Lorentz metric. Therefore, the Weyl CR invariants are the local CR invariants that arise from complete tensorial contractions of covariant derivatives of the curvature tensor of Fefferman's ambient Kähler-Lorentz metric.
Bearing this in mind the CR analogue of Proposition 2 is given by the following.

PROPOSITION 12 ([**Fe2**, Thm. 2], [**BEG**, Thm. 10.1]). *Every local CR invariant of weight $\leq n+1$ is a linear combination of local Weyl CR invariants.*

In particular, we recover the fact that there is no local CR invariant of weight 1. Furthermore, we see that every local CR invariant of weight 2 is a constant multiple of $|S|_\theta$. Similarly, the local CR invariants of weight 3 are linear combinations of the invariants (99) and of the invariant Φ_θ that arises from the Fefferman-Graham invariant Φ_{g_θ} of the Fefferman Lorentzian space $\mathcal{F}$.

10. Logarithmic singularities of CR invariant Ψ_HDOs

Let us now look at the logarithmic singularities of CR invariant Ψ_HDOs. To this end let us denote by $[\theta]$ the conformal class of θ.

PROPOSITION 13. *Consider a family $(P_{\hat{\theta}})_{\hat{\theta}\in[\theta]} \subset \Psi^m(M)$ such that*

$$P_{e^f\theta} = e^{w'f}P_\theta e^{-wf} \qquad \forall f \in C^\infty(M,\mathbb{R}). \tag{105}$$

Then

$$c_{P_{e^f\theta}}(x) = e^{(w'-w)f(x)}c_{P_\theta}(x) \qquad \forall f \in C^\infty(M,\mathbb{R}). \tag{106}$$

This result generalizes a previous result of N.K. Stanton [**St**]. Combining it with Proposition 9, and using Proposition 12 , we obtain the following.

THEOREM 6 ([**Po4**]). *Let P_θ be a CR invariant $\Psi_H DO$ of order m and biweight (w,w'). Then the logarithmic singularity $c_{P_\theta}(x)$ takes the form*

$$c_{P_\theta}(x) = \mathcal{I}_{P_\theta}(x)|(d\theta)^n \wedge \theta|, \tag{107}$$

where $\mathcal{I}_\theta(x)$ is a scalar local CR invariant of weight $n+1+w-w'$. If we further have $w \leq w'$, then $\mathcal{I}_\theta(x)$ is a linear combination of Weyl CR invariants of weight $n+1+w-w'$.

We can make use of this result to study the logarithmic singularities of the Green kernels of the CR GJMS operators.

THEOREM 7. *For $k = 1, \ldots, n$ we have*

$$\gamma_{\square_\theta^{(k)}}(x) = c_\theta^k(x)|d\theta^n \wedge \theta|, \tag{108}$$

where $c_\theta^k(x)$ is a linear combination of scalar Weyl CR invariants of weight $n+1-k$. In particular,

$$c_\theta^{(n)}(x) = 0, \qquad c_\theta^{(n-1)}(x) = \alpha_n |S|_\theta^2, \tag{109}$$

$$c_\theta^{(n-2)}(x) = \beta_n S_{i\bar{j}}{}^{\bar{k}l} S_{k\bar{l}}{}^{\bar{p}q} S_{p\bar{q}}{}^{\bar{i}j} + \gamma_n S_i{}^{j\bar{k}}{}_{\bar{l}} S^{\bar{i}}{}_{\bar{j}p}{}^q S^{\bar{p}}{}_{\bar{q}k}{}^l + \delta_n \Phi_\theta, \tag{110}$$

where S is the Chern-Moser curvature tensor, Φ_θ is the CR Fefferman-Graham invariant, and the constants α_n, β_n, γ_n and δ_n depend only on n.

References

[Al1] Alexakis, S.: *On conformally invariant differential operators in odd dimension.* Proc. Nat. Acad. Sci. USA **100** (2003), no. 2, 4409–4410.

[Al2] Alexakis, S.: *On conformally invariant differential operators.* E-print, arXiv, August 2006, 50 pages.

[BEG] Bailey, T.N.; Eastwood, M.G.; Graham, C.R.: *Invariant theory for conformal and CR geometry.* Ann. Math. **139** (1994) 491–552.

[BG] Bailey, T.N.; Gover, A.R.: *Exceptional invariants in the parabolic invariant theory of conformal geometry* Proc. A.M.S. **123** (1995), 2535–2543.

[BGr] Beals, R.; Greiner, P.C.: *Calculus on Heisenberg manifolds.* Annals of Mathematics Studies, vol. 119. Princeton University Press, Princeton, NJ, 1988.

[CM] Chern, S. S.; Moser, J. K. *Real hypersurfaces in complex manifolds.* Acta Math. **133** (1974), 219–271.

[CMo] Connes, A.; Moscovici, H.: *The local index formula in noncommutative geometry.* Geom. Funct. Anal. **5** (1995), no. 2, 174–243.

[EM] Epstein, C.L.; Melrose, R.B.: *The Heisenberg algebra, index theory and homology.* Preprint, 2000. Available online at `http://www-math.mit.edu/~rbm/book.html`.

[Fe1] Fefferman, C.: *Monge-Ampère equations, the Bergman kernel, and geometry of pseudo-convex domains.* Ann. of Math. (2) **103** (1976), no. 2, 395–416.

[Fe2] Fefferman, C.: *Parabolic invariant theory in complex analysis.* Adv. in Math. **31** (1979), no. 2, 131–262.

[FG1] Fefferman, C.; Graham, C.R.: *Conformal invariants.* Élie Cartan et les Mathématiques d'Aujourd'hui, Astérisque, hors série, (1985), 95–116.

[FG2] Fefferman, C.; Graham, C.R.: *The ambient metric.* E-print, arXiv, Oct. 07.

[FS1] Folland, G.; Stein, E.: *Estimates for the $\bar\partial_b$ complex and analysis on the Heisenberg group.* Comm. Pure Appl. Math. **27** (1974) 429–522.

[Gi] Gilkey, P.B.: *Invariance theory, the heat equation, and the Atiyah-Singer index theorem.* 2nd edition. Studies in Advanced Mathematics. CRC Press, Boca Raton, FL, 1995.

[GG] Gover, A.R.; Graham, C.R.: *CR invariant powers of the sub-Laplacian.* J. Reine Angew. Math. **583** (2005), 1–27.

[GVF] Gracia-Bondía, J.M.; Várilly, J.C.; Figueroa, H.: *Elements of noncommutative geometry.* Birkhäuser Boston, Boston, MA, 2001.

[GJMS] Graham, C.R.; Jenne, R.; Mason, L.J.; Sparling, G.A.: *Conformally invariant powers of the Laplacian. I. Existence.* J. London Math. Soc. (2) **46** (1992), no. 3, 557–565

[GH] Graham, C.R.; Hirachi, M.: *Inhomogeneous ambient metrics. Symmetries and overdetermined systems of partial differential equations*, 403–420, IMA Vol. Math. Appl., 144, Springer, New York, 2008.

[Gr] Greiner, P.: *An asymptotic expansion for the heat equation.* Arch. Rational Mech. Anal. **41** (1971) 163–218.

[Gu1] Guillemin, V.W.: *A new proof of Weyl's formula on the asymptotic distribution of eigenvalues.* Adv. in Math. **55** (1985), no. 2, 131–160.

[Gu2] Guillemin, V.W.: *Gauged Lagrangian distributions.* Adv. Math. **102** (1993), no. 2, 184–201.

[Hi1] Hirachi, K.: *Construction of boundary invariants and the logarithmic singularity of the Bergman kernel.* Ann. of Math. **151** (2000) 151–191.

[Hi2] Hirachi, K.: *Ambient metric construction of CR invariant differential operators. Symmetries and overdetermined systems of partial differential equations*, 61–75, IMA Vol. Math. Appl., 144, Springer, New York, 2008.

[HPT] Hislop, P.D.; Perry, P.A.: Tang, S.-H.: *CR-invariants and the scattering operator for complex manifolds with boundary.* Anal. PDE **1** (2008), no. 2, 197–227.

[Hö2] Hörmander, L.: *Hypoelliptic second order differential equations.* Acta Math. **119** (1967), 147–171.

[Hö] Hörmander, L.: *The analysis of linear partial differential operators. III. Pseudo-differential operators.* Grundlehren der Mathematischen Wissenschaften, 274. Springer-Verlag, Berlin, 1994.

[JL1] Jerison, D.; Lee, J.M.: *The Yamabe problem on CR manifolds.* J. Differential Geom. **25** (1987), no. 2, 167–197.

[JL2] Jerison, D.; Lee, J.M.: *Intrinsic CR normal coordinates and the CR Yamabe problem.* J. Differential Geom. **29** (1989), no. 2, 303–343.

[Ju] Juhl, A.: *Families of conformally covariant differential operators, Q-curvature and holography.* Progress in Mathematics, Birkhäuser, Vol. 275, 2009, 500 pages.

[KV] Kontsevich, M.; Vishik, S.: *Geometry of determinants of elliptic operators. Functional analysis on the eve of the 21st century*, Vol. 1 (New Brunswick, NJ, 1993), 173–197, Progr. Math., 131, Birkhäuser Boston, Boston, MA, 1995.

[Le] Lee, J.M.: *The Fefferman metric and pseudo-Hermitian invariants.* Trans. Amer. Math. Soc. **296** (1986), no. 1, 411–429.

[Me] Melrose, R.: *The Atiyah-Patodi-Singer index theorem.* A.K. Peters, Boston, 1993.

[PR] Parker, T; Rosenberg, S.: *Invariants of conformal Laplacians.* J. Differential Geom. **25** (1987), no. 2, 199–222.

[PRo] Paycha, S.; Rosenberg, S.: *Conformal anomalies via canonical traces. Analysis, geometry and topology of elliptic operators*, World Scientific, 2006.

[Po1] Ponge, R.: *The tangent groupoid of a Heisenberg manifold.* Pacific Math. J. **227** (2006) 151–175.

[Po2] Ponge, R.: *Noncommutative residue for Heisenberg manifolds and applications in CR and contact geometry.* J. Funct. Anal. **252** (2007), 399–463.

[Po3] Ponge, R.: *Heisenberg calculus and spectral theory of hypoelliptic operators on Heisenberg manifolds.* Mem. Amer. Math. Soc. **194** (2008), no. 906, viii+ 134 pp.

[Po4] Ponge, R.: *Logarithmic singularities of Schwartz kernels and local invariants of conformal and CR structures.* E-print, arXiv, October 2007, 44 pages.

[Ro] Rockland, C.: *Intrinsic nilpotent approximation.* Acta Appl. Math. **8** (1987), no. 3, 213–270.

[St] Stanton, N.K.: *Spectral invariants of CR manifolds.* Michigan Math. J. **36** (1989), no. 2, 267–288.

[Ta] Tanaka, N.: *A differential geometric study on strongly pseudo-convex manifolds.* Lectures in Mathematics, Department of Mathematics, Kyoto University, No. 9. Kinokuniya Book-Store Co., Ltd., Tokyo, 1975.

[Tay] Taylor, M.E.: *Noncommutative microlocal analysis. I.* Mem. Amer. Math. Soc. 52 (1984), no. 313,

[We] Webster, S.: *Pseudo-Hermitian structures on a real hypersurface.* J. Differential Geom. **13** (1978), no. 1, 25–41.

[Wo1] Wodzicki, M.: *Local invariants of spectral asymmetry.* Invent. Math. **75** (1984), no. 1, 143–177.

[Wo2] Wodzicki, M.: *Spectral asymmetry and noncommutative residue* (in Russian), Habilitation Thesis, Steklov Institute, (former) Soviet Academy of Sciences, Moscow, 1984.

[Wo3] Wodzicki, M.: *Noncommutative residue. I. Fundamentals. K-theory, arithmetic and geometry* (Moscow, 1984–1986), 320–399, Lecture Notes in Math., 1289, Springer, Berlin-New York, 1987.

Department of Mathematics, University of Toronto, Canada.

Current address: Graduate School of Mathematical Sciences, University of Tokyo, Japan

E-mail address: `ponge@ms.u-tokyo.ac.jp`

Clay Mathematics Proceedings
Volume 12, 2010

Mirror Symmetry for Elliptic Curves: The A-Model (Fermionic) Counting

Mike Roth and Noriko Yui

Abstract. We will make an attempt to understand the results about mirror symmetry and elliptic curves in the paper of Dijkgraaf [3], more precisely on the generating functions counting simply ramified curves of genus $g \geq 1$ over a fixed elliptic curve with $2g-2$ marked points and their relations to the integrals on trivalent Feynman graphs. The result which appeals to us most is the "modularity" of the generating functions. It asserts that the generating function counting simply ramified covers of an elliptic curve with $2g-2$ marked points is a quasimodular form of weight $6g-6$ on $\Gamma = SL_2(\mathbb{Z})$. Two different counting methods, the fermionic and bosonic countings, are discussed in Dijkgraaf [3].

In this note, we will confine ourselves to the A-model – fermionic counting. In this case, one computes the generating function based on the classical theory of Riemann surfaces, monodromy representations, representation theory of symmetric groups, etc. The method of arriving at the generating function is not due to us, and our exposition on this part is based on the article of Dijkgraaf [3]. To establish the quasimodularity of the generating function, we will follow the exposition of Kaneko and Zagier [9]. The fermionic counting method is purely mathematical and we can present an algorithm to compute the generating functions explicitly in terms of traces (or eigenvalues) of certain matrices.

Contents

2010 *Mathematics Subject Classification.* Primary: 14N10, Secondary: 05E10, 11F23, 14H52.

Key words and phrases. elliptic curves, counting simply ramified covers, generating functions, fermionic counting, quasimodular forms, a generalized Jacobi theta function, mirror symmetry.

Mike Roth and Noriko Yui were partially supported by Natural Sciences and Engineering Research Council of Canada (NSERC) Discovery Grants, Canada.

1. Introduction

1.1. In this note we will study the formula in Dijkgraaf [**3**] on the generating functions counting simply ramified curves of genus $g \geq 1$ over a fixed elliptic curve with $2g-2$ marked points. The striking result stated in therein is the *Fermion Theorem* which asserts:

For each $g \geq 2$, the generating function, $F_g(q)$ (with $q = e^{2\pi i\tau}$, $\tau \in \mathfrak{H}$), is a quasimodular form of weight $6g-6$ on $\Gamma = SL_2(\mathbb{Z})$.

Consequently it is expressed as a polynomial in $\mathbb{Q}[E_2, E_4, E_6]$ of weight $6g-6$. Here E_{2i}, $i = 1,2,3$ are Eisenstein series of weight $2i$, $i = 1,2,3$ on Γ.

Several natural questions arise: For a fixed $g \geq 1$,

(a) how to calculate the number of simply ramified covers?

(b) characterize the generating function,

(c) why do quasimodular forms appear?

Dijkgraaf [**3**] and Douglas [**5**] arrived at this result using reasoning from physics (e.g., Gromov–Witten Theory, mirror symmetry, quantum field theory, Feynman diagrams and integrals). Therefore it is imperative to provide purely mathematical proofs, and that is exactly the purpose of this paper, and along the way we try to give answers to the above questions. The method of arriving at the generating function is not due to us. It is straight from the article of Dijkgraaf [**3**], and our exposition is based on that article.

As the modular form theoretic proof for the quasimodularity of the generating function was supplied by Kaneko and Zagier [**9**], one of our purposes of this note is to study their proof. The quasimodularity of the generating function can be established with mathematical rigour. However, the proofs do not reveal conceptually why it is quasimodular! For this we need to appeal to the B-model (bosonic) theory.

One reason why physicists are interested in the generating functions of simply ramified covers of an elliptic curve is that it can be connected to the partition function of two-dimensional Yang–Mills theory on an elliptic curve [**2**].

1.2. We should mention that a main motivation in Dijkgraaf's paper [**3**] was to discuss mirror symmetry on Calabi–Yau manifolds of dimension 1, namely on elliptic curves. Here is a brief description of mirror symmetry for elliptic curves. It is a well known fact that the moduli space of elliptic curves over $\mathbb{C}$ is the space $\mathfrak{H}/\Gamma$, where $\mathfrak{H} := \{\,\tau \in \mathbb{C} \,|\, \mathrm{Im}(\tau) > 0\,\}$ denotes the upper-half complex plane. This means that any elliptic curve E over $\mathbb{C}$ is determined, up to isomorphism, by a complex number $\tau \in \mathfrak{H}$, in fact, E is isomorphic to a complex torus:

$$E \simeq \mathbb{C}/\mathbb{Z} + \tau\mathbb{Z}.$$

Write $\tau = \tau_1 + i\tau_2$ with $\tau_2 > 0$. The complex modulus $\tau \in \mathfrak{H}$ is called a *complex structure modulus* of E, and we denote the elliptic curve by E_τ.

There is another way of defining an elliptic curve E over $\mathbb{C}$. As an elliptic curve is a Calabi–Yau manifold of dimension 1, there is a unique complexified Kähler class $[\omega] \in H^2(E, \mathbb{C})$ associated to it. We will describe how to associate another modulus to an elliptic curve. With τ_2 as above, we parameterize ω with t as

$$t = \frac{1}{2\pi i}\int_E \omega, \quad \omega = -\frac{-\pi t}{\tau_2} dz \wedge d\bar{z}$$

Again, write $t = -t_1 + it_2$ with $t_2 > 0$. (The condition $t_2 > 0$ corresponds to the fact that ω lies in the positive Kähler cone). The complex number $t \in \mathfrak{H}$ is called a *Kähler modulus* or *(complexified) symplectic modulus* and determines an elliptic curve, E_t.

Thus an elliptic curve E is endowed with two moduli, τ (a complex structure modulus) and t (a Kähler modulus), and we may write $E = E_{\tau,t}$. The dependence on τ is up to the lattice generated by 1 and τ, so that

$$E_{\tau,t} = E_{\tau+1,t} = E_{-\frac{1}{\tau},t}.$$

1.3. The mirror symmetry conjecture (prediction) for elliptic curves has been formulated in Dijkgraaf, Verlinde and Verlinde [**4**].

Mirror symmetry for elliptic curves is simply the interchange of the two parameters τ (the complex structure modulus) and t (the Kähler modulus) :

$$E = E_{\tau,t} \Leftrightarrow \tilde{E} = E_{t,\tau}.$$

That is, the Kähler modulus of the original elliptic curve is the complex structure modulus of the mirror elliptic curve, and vice versa. So the mirror map sends τ to t.

The mirror symmetry conjecture suggests that there are two ways of computing the generating functions of simply ramified covers. One is the fermionic count (the A-model with a Kähler modulus t), and the other is the bosonic count (the B-model with a complex structure modulus τ). The fermionic counting is done on the original elliptic curve and involves Hurwitz numbers. This fermionic side of the story is a purely mathematical theory, and is accessible to mathematicians. In fact, the classical Hurwitz theory, and the more modern Gromov–Witten theory, are the main ingredients in the fermionic enumerations of not necessarily simply ramified covers but more general types of covers of curves (e.g., $\mathbf{P}^1$, an elliptic curve with marked points). The reader is referred to a series of papers by Okounkov and Pandharipande, e.g., [**11**].

On the other hand, the bosonic count is based on physical theories, e.g., quantum field theory, involving path integrals on trivalent Feynman diagrams. This arises from the interpretation of the generating functions $F_g(q)$ in terms of the geometry of the mirror partner. The B-model enumerates "constant maps" from simply ramified covers to points in the mirror elliptic curve. The partition function is defined by a path integral

$$Z^* = \int_{\mathcal{F}} \exp(-S(\phi))\mathcal{D}\phi$$

over the space of fields $\mathcal{F}$ for some formal measure $\mathcal{D}\phi$ on $\mathcal{F}$, where $S : \mathcal{F} \to \mathbb{R}$ is the action functional determined by the Lagrangian $L(\phi)$. Then by the standard quantum field theory argument, this partition function may be computed using Feynman diagrams. In particular, for the model that we are using here (e.g., Chern–Simons theory), the action functional S is given by a cubic potential,

$$S(\phi) = \int \frac{1}{2}\partial\phi\bar{\partial}\phi + \frac{\lambda}{6}(\partial\phi)^3.$$

Then Wick's theorem tells us how to represent these integrals in terms of Feynman diagrams of trivalent graphs. The upshot of these techniques is that $F_g(q)$ (with $q = e^{2\pi i\tau}$) should be computed by evaluating the integrals associated to trivalent Feynman diagrams.

However, at this moment, we are not able to present mathematically rigorous discussions on the bosonic counting. We plan to come back to the B-model–bosonic counting in subsequent articles. One of our future projects is to "understand" the bosonic counting for elliptic curves by carrying out some computations of Feynman integrals, first for small genera. Then we would like to illustrate that fermionic counting coincides with bosonic counting (at least for small genera), thereby establishing the mirror symmetry conjecture for elliptic curves for the cases of small genera. For higher genera, we ought to study inductive structures on Feynman diagrams, and structures of the Hopf algebras associated to Feynman diagrams.

Also, through the bosonic counting, and mirror symmetry, we hope that it becomes conceptually clear why quasimodular forms enter the scene in the description of the generating function $F_g(q)$.

1.4. In this note, we will consider only the A-model (fermionic) counting for the generating function $F_g(q)$ where we set $q = e^{2\pi i t}$ with a Kähler parameter $t \in \mathfrak{H}$. Since the discussions below involve only one parameter, we will use the notation τ to mean a Kähler parameter t throughout.

2. Statement of the problem

Fix an elliptic curve E, and a set $S = \{b_1, \ldots, b_{2g-2}\}$ of $2g-2$ distinct points of E. We define a *degree d, genus g, cover of E* to be an irreducible smooth curve C of genus g along with a finite degree d map $p : C \longrightarrow E$ simply branched over the points $b_1, \ldots, b_{2g-2}$ in S.

We consider two such covers $p_1 : C_1 \longrightarrow E$ and $p_2 : C_2 \longrightarrow E$ to be equivalent if they are isomorphic as schemes over E, that is, if there is an isomorphism $\phi : C_1 \tilde{\to} C_2$ commuting with the structure maps p_i to E.

For any cover $p : C \longrightarrow E$ we define the *automorphism group of the cover*, $\mathrm{Aut}_p(C)$ to be the automorphism group of C as a scheme over E, that is, the group

$$\mathrm{Aut}_p(C) := \{\phi : C \tilde{\to} C \mid p \circ \phi = p\}.$$

$\mathrm{Aut}_p(C)$ is always a finite group. We will usually abuse notation and write $\mathrm{Aut}(C)$ for this group, with the understanding that it depends on the structure map p to E.

2.1. Let $\mathrm{Cov}(E,S)^{\circ}_{g,d}$ be the set of degree d, genus g covers of E, up to equivalence. We will refer to an element of $\mathrm{Cov}(E,S)^{\circ}_{g,d}$ by a representative (C,p) of the equivalence class.

If (C,p) is a cover, and (C',p') any cover equivalent to it, then $|\mathrm{Aut}_p(C)| = |\mathrm{Aut}_{p'}(C')|$. For any element of $\mathrm{Cov}(E,S)^{\circ}_{g,d}$ we assign it the weight $1/|\mathrm{Aut}_p(C)|$, where (C,p) is any representative of the class. This is well defined by the previous remark.

2.2. Let $N_{g,d}$ be the weighted count of the elements of $\mathrm{Cov}(E,S)^{\circ}_{g,d}$, each equivalence class being weighted as above.

The number $N_{g,d}$ is purely topological, and does not depend on the particular elliptic curve E chosen, nor on the set S of $2g-2$ distinct points, and we therefore omit them from the notation.

The first goal of this note is to explain how to calculate $N_{g,d}$ for all g and d; the second to link the generating functions to quasimodular forms, which we now describe.

2.3. Here we recall some facts about modular and quasimodular forms from Kaneko and Zagier [**9**], which are relevant to our discussions. For detailed accounts on modular and quasimodular forms, the reader is referred to the excellent recent lecture note *The 1-2-3 of Modular Forms* [**1**].

For any even integer $k \geq 2$ define the *Eisenstein series of weight* k to be the series

$$E_k := 1 - \frac{2k}{B_k} \sum_{n \geq 1} \sigma_{k-1}(n) q^n,$$

where B_k is the k-th Bernoulli number, and $\sigma_{k-1}(n)$ denotes the sum $\sigma_{k-1}(n) := \sum_{m|n} m^{(k-1)}$.

The first three Eisenstein series are

$$\begin{aligned} E_2 &= 1 - 24q - 72q^2 - 96q^3 - 168q^4 - 144q^4 - \cdots, \\ E_4 &= 1 + 240q + 2160q^2 + 6720q^3 + 17520q^4 + \cdots, \quad \text{and} \\ E_6 &= 1 - 504q - 16632q^2 - 122976q^3 - 532728q^4 - \cdots. \end{aligned}$$

Let $\Gamma = \mathrm{SL}_2(\mathbb{Z})$ be the full modular group. If we set $q = \exp(2\pi i \tau)$ with $\tau \in \mathfrak{H}$, each E_k becomes a holomorphic function on the upper half plane, which is also, by virtue of its defining q-expansion, holomorphic at infinity. For $k \geq 4$ each E_k is a modular form of weight k.

2.4. The series E_2 is not modular. For any $\gamma \in \Gamma$, $\gamma = \begin{pmatrix} a & b \\ c & d \end{pmatrix}$,

$$E_2(\gamma \cdot \tau) = (c\tau + d)^2 E_2(\tau) + \frac{6c(c\tau + d)}{\pi i}. \tag{2.4.1}$$

This can be easily computed using the identity

$$E_2(q) = \frac{1}{2\pi i} \frac{d}{d\tau} \log(\Delta(q)),$$

where $\Delta(q)$ is the weight 12 cusp form

$$\Delta(q) := q \prod_{n \geq 1} (1 - q^n)^{24}.$$

2.5. Let $\Im(\tau)$ be the imaginary part of τ and $Y(\tau)$ the function $Y(\tau) = 4\pi\Im(\tau)$.

Kaneko and Zagier ([**9**], p. 166) define an *almost holomorphic modular form of weight* k to be a function $F(\tau)$ on the upper half plane of the form

$$F(\tau) = \sum_{m=0}^{M} f_m(\tau) Y^{-m}$$

where each $f_m(\tau)$ is holomorphic and grows at most polynomially in $1/Y$ as $Y \to 0$. In addition, $F(\tau)$ must satisfy the usual weight k modular transformation rule

$$F(\gamma \cdot \tau) = (c\tau + d)^k F(\tau).$$

They define a *quasimodular form of weight* k to be any holomorphic function $f_0(\tau)$ appearing as the "constant term with respect to $1/Y$" in such an expansion.

Since

$$\frac{1}{Y(\gamma \cdot \tau)} = \frac{(c\tau + d)^2}{Y(\tau)} + \frac{c(c\tau + d)}{2\pi i},$$

equation 2.4.1 shows that

$$E_2^* := E_2 - \frac{12}{Y}$$

is an almost holomorphic modular form of weight 2, and therefore that E_2 is a quasimodular form of the same weight.

2.6. Kaneko and Zagier ([**9**], p. 166) further define $\widetilde{M}_k(\Gamma)$ to be the vector space of quasimodular forms of weight k, and $\widetilde{M}(\Gamma) = \oplus_k \widetilde{M}_k(\Gamma)$ to be the graded ring of quasimodular forms.

They prove ([**9**], p. 167, Proposition 1) that $\widetilde{M}(\Gamma) = \mathbb{C}[E_2, E_4, E_6]$.

The monomials of weight k in E_2, E_4 and E_6 are linearly independent over $\mathbb{C}$ as functions on the upper half plane. One way to see this is to use the fact that monomials in E_4 and E_6 are modular forms, and linearly independent, and then apply equation 2.4.1 to any supposed linear relation among monomials involving E_2.

It follows that

$$\dim_{\mathbb{C}}(\widetilde{M}_k(\Gamma)) = \begin{cases} \left\lfloor \frac{(k+6)^2}{48} \right\rfloor & \text{if } k \not\equiv 0 \mod 12 \\ \left\lfloor \frac{(k+6)^2}{48} \right\rfloor + 1 & \text{if } k \equiv 0 \mod 12 \end{cases}$$

or just $\dim_{\mathbb{C}}(\widetilde{M}_k(\Gamma)) = \left\lfloor \dfrac{(k+6)^2+12}{48} \right\rfloor$ for any even k.

See also the article of Kaneko and Koike [**8**] for the dimension formula for the vector space $\widetilde{M}_k(\Gamma)$ of quasimodular forms of even weight k for Γ.

2.7. We now quote the important result on quasimodular forms from [**1**] (Proposition 20, on page 59), which is relevant to our subsequent discussions. We consider derivatives of modular and quasimodular forms. For $\tau \in \mathfrak{H}$, and $q = e^{2\pi i \tau}$, let

$$D := \frac{1}{2\pi i}\frac{d}{d\tau} = q\frac{d}{dq}$$

be a differential operator.

Proposition (Zagier). —

(a) *The space of quasimodular forms on Γ is closed under differentiation. Let $D := \frac{1}{2\pi i}\frac{d}{d\tau}$ be a differential operator. If f is a modular form of weight k for $\Gamma = SL_2(\mathbb{Z})$, then Df is a quasimodular form of weight $k+2$ for Γ. The m-th derivative of a quasimodular form is again a quasimodular form of weight $k+2m$.*

(b) *Every quasimodular form for Γ can be written uniquely as a linear combination of derivatives of modular forms and of E_2.*

Note that the above proposition is valid for certain subgroups of Γ as well, for instance, Γ_2 appearing in section 9.

Examples. Here are the derivatives of E_2, E_4 and E_6:

$$
\begin{aligned}
D(E_2) &= \frac{1}{12}(E_2^2 - E_4) = -24\sum_{n=1}^{\infty}\frac{n^2q^n}{(1-q^n)^2} = -24\sum_{n=1}^{\infty} n\,\sigma_1(n)q^n \\
D(E_4) &= \frac{1}{3}(E_2E_4 - E_6) = 240\sum_{n=1}^{\infty}\frac{n^4q^n}{(1-q^n)^2} = 540\sum_{n=1}^{\infty} n\,\sigma_3(n)q^n \\
D(E_6) &= \frac{1}{2}(E_2E_6 - E_4^2) = \sum_{n=1}^{\infty}\frac{n^6q^n}{(1-q^n)^2} = -504\sum_{n=1}^{\infty} n\,\sigma_5(n)q^n \\
D^2(E_2) &= \frac{1}{36}(E_6 - E_2^3 - 18E_2D(E_2)) = \frac{1}{36}(E_6 - \frac{5}{2}E_2^3 + \frac{3}{2}E_2E_4) \\
D^3(E_2) &= \frac{3}{2}D(E_2)^2 - E_2D^2(E_2)
\end{aligned}
$$

2.8. For any $g \geq 2$, let

$$F_g(q) := \sum_{d\geq 1} N_{g,d}\, q^d$$

be the generating series counting covers of genus g.

(Note that for $g = 1$, the definition of $F_1(q)$ should be modified by adding the term $-\frac{1}{24}\log q$. See, for instance, [**10**].)

The interest in counting covers is because of the following surprising result due to Dijkgraaf [**3**] and Douglas [**5**].

Theorem (Dijkgraaf–Douglas). — For $g \geq 2$, $F_g(q)$ is a quasimodular form of weight $6g-6$ for Γ.

3. From connected to disconnected covers

3.1. A general feature of combinatorial problems is that it is often easier to count something if we remove the restriction that the objects in question are connected.

A standard example is to count the number of graphs with n labelled vertices. If $n = 3$ then there are four connected graphs with 3 labelled vertices (Figure 1):

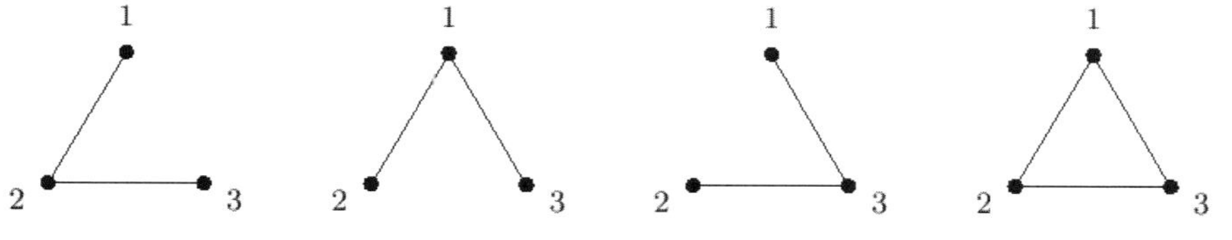

FIGURE 1

If we allow (possibly) disconnected graphs, then there are a total of eight such graphs; the four above, and an additional four labelled graphs (Figure 2):

FIGURE 2

Let c_n be the number of connected graphs on n vertices, and d_n the (possibly) disconnected ones. It may not be clear how to calculate c_n, but since each graph on n vertices has a total of $\binom{n}{2}$ possible edges which we can either draw or omit, we have $d_n = 2^{\binom{n}{2}}$.

3.2. Suppose that we did know the number of connected graphs on n vertices. We would then have a second way to compute the number of possibly disconnected graphs. For instance, knowing that $c_1 = 1$, $c_2 = 2$, and $c_3 = 4$ we can compute that

$$d_3 = \binom{3}{3} c_3 + \binom{3}{2,1} c_2 c_1 + \frac{1}{3!}\binom{3}{1,1,1} c_1^3 = 8.$$

In this calculation, the sum is over the number of ways to break a graph with $n = 3$ vertices into connected pieces. In each term of the sum the binomial symbol (and factorial term) keeps track of the number of ways to split the vertices into connected pieces of the appropriate size. Finally the monomial in the c's calculates the number of such graphs given that particular division of the vertex set.

3.3. If we set

$$C(\lambda) := \sum_{n\geq 1} \frac{c_n}{n!}\lambda^n,$$

and

$$D(\lambda) := \sum_{n\geq 1} \frac{d_n}{n!}\lambda^n,$$

then our method of calculating the d_n's from the c_n's shows that

$$D(\lambda) = \exp(C(\lambda)) - 1.$$

This follows since the coefficient of λ^n in $\exp(C(\lambda))$ is clearly a sum of monomials in the c's, and the $n!$ terms in the definition of C and D, as well as the factorials appearing in the exponential ensure that these monomials appear with the correct coefficients. Working out the coefficient of λ^3 in $\exp(C(\lambda))$ is a good way to see how this happens.

Thanks to this identity, it is a formal matter to compute the d_n's if we know the c_n's, or conversely, by taking the logarithm, we can compute the c_n's from the d_n's. Since we do know a formula for the d_n's, this gives us a method to compute c_n for all n.

3.4. A typical strategy for attacking a combinatorial problem is the one used above: first solve the disconnected version of the problem, and then use that to compute the numbers for the connected case.

The process of passing from the connected to the possibly disconnected version of a combinatorial problem usually involves taking the exponential of the generating function. Describing a disconnected version generally requires partitioning some of the data in the problem (like the vertices in our example); the factorials and exponential take into account the combinatorics of making this division.

In order to apply this strategy to our counting problem, we first need to define the disconnected version of a cover.

3.5. Returning to the setup of sections 2.1 and 2.2, let E be an elliptic curve and $S = \{b_1, \ldots, b_{2g-2}\}$ a set of $2g-2$ distinct points of E. We define a *degree d, genus g, disconnected cover of E* to be a union $C = \cup_i C_i$ of smooth irreducible curves, along with a finite degree d map $p : C \longrightarrow E$ simply branched over the points $b_1, \ldots, b_{2g-2}$ in S.

The condition on branching means that each ramification point of p is a simple ramification point, and that there are a total of $2g-2$ ramification points, mapped bijectively under p to $b_1, \ldots, b_{2g-2}$.

Note that by "disconnected cover" we mean disconnected in the weak sense: we allow the possibility that C is connected, so that this includes the case of degree d genus g covers from sections 2.1 and 2.2. A more accurate but also more clumsy name would be "possibly disconnected cover".

The restriction $p_i := p|_{C_i}$ of p to any component C_i of C is a finite map of some degree d_i. If C_i of C is of genus g_i, then by the Riemann-Hurwitz formula the map p_i will have $2g_i - 2$ ramification points on C_i. We therefore have the relations

$$\sum_i d_i = d, \text{ and } \sum_i (2g_i - 2) = 2g - 2.$$

3.6. As before, we define two covers to be equivalent if they are isomorphic over E. We define the automorphism group of a cover $\mathrm{Aut}_p(C)$ to be the automorphism group of C as a scheme over E:

$$\mathrm{Aut}_p(C) := \{\phi : C \tilde{\to} C \mid p \circ \phi = p\}.$$

If C has no components of genus $g_i = 1$, then the automorphism group of the cover is the direct product of the automorphism groups of the components.

$$\mathrm{Aut}_p(C) = \prod_i \mathrm{Aut}_{p_i}(C_i).$$

This follows from the fact that in this case no two components can be isomorphic as curves over E since they have different branch points over E.

If C has components of genus $g_i = 1$, then the map p_i has no ramification points and it is possible that some genus 1 components are isomorphic over E. In this case the automorphism group of the cover is the semi-direct product of $\prod_i \mathrm{Aut}_{p_i}(C_i)$ and the permutations of the genus 1 components isomorphic over E.

3.7. Let $\mathrm{Cov}(E,S)_{g,d}$ be the set of genus g, degree d, disconnected covers of E up to equivalence. We will refer to an element of $\mathrm{Cov}(E,S)_{g,d}$ by a representative (C,p) of the equivalence class.

For any element (C,p) of $\mathrm{Cov}(E,S)_{g,d}$, we give it the weight $1/|\mathrm{Aut}_p(C)|$. As in section 2.1, this is well defined.

3.8. Let $\widehat{N}_{g,d}$ be the number of genus g, degree d, disconnected covers counted with the weighting above. As before, the number $\widehat{N}_{g,d}$ is purely topological and does not depend on the curve E or the particular distinct branch points $b_1,\dots,b_{2g-2}$ chosen.

Let $Z(q,\lambda)$ be the generating function for the $N_{g,d}$'s:

$$\begin{aligned} Z(q,\lambda) &:= \sum_{g\geq 1}\sum_{d\geq 1} \frac{N_{g,d}}{(2g-2)!}\, q^d \lambda^{2g-2} \\ &= \sum_{g\geq 1} \frac{F_g(q)}{(2g-2)!}\, \lambda^{2g-2}, \end{aligned}$$

and $\widehat{Z}(q,\lambda)$ the corresponding generating function for the $\widehat{N}_{g,d}$'s:

$$\widehat{Z}(q,\lambda) := \sum_{g\geq 1}\sum_{d\geq 1} \frac{\widehat{N}_{g,d}}{(2g-2)!}\, q^d \lambda^{2g-2}.$$

3.9. *Lemma.* — *The generating functions are related by*

$$\widehat{Z}(q,\lambda) = \exp(Z(q,\lambda)) - 1.$$

Taking logarithms,

$$Z(q,\lambda) = \log(\widehat{Z}(q,\lambda) + 1).$$

Proof. Let us organize the data of a disconnected cover by keeping track of the genus of each component, the degree of the map restricted to that component, and the number of times that each such pair occurs. We will call such data the *combinatorial type* of the cover.

Suppose that in a particular genus g, degree d, disconnected cover there are k_1 components of type (g_1,d_1), k_2 components of type (g_2,d_2), ..., and k_r components of type (g_r,d_r).

The numbers $\{k_j\}$, $\{g_j\}$, $\{d_j\}$, g and d satisfy the relations

$$\sum_{j=1}^{r} k_j d_j = d, \text{ and } \sum_{j=1}^{r} k_j(2g_j - 2) = 2g - 2.$$

We want to compute the weighted count of all disconnected genus g degree d covers of this combinatorial type. Clearly this number is some multiple of

$$N_{g_1,d_1}^{k_1} N_{g_2,d_2}^{k_2} \cdots N_{g_r,d_r}^{k_r},$$

the multiple depending on some combinatorial choices of branch points and some accounting for the automorphisms among the genus 1 components. We want to compute this multiple and see that it is the same as the coefficient of $N_{g_1,d_1}^{k_1} \cdots N_{g_r,d_r}^{k_r}$ appearing in $(2g-2)!\exp(Z(q,\lambda))$.

Since $\widehat{N}_{g,d}$ is the sum over the weighted counts of such topological types, and since the factor $(2g-2)!$ is included in the definition of $\widehat{Z}(q,\lambda)$, it will follow that $\widehat{Z}(q,\lambda) = \exp(Z(q,\lambda)) - 1$.

We first need to split up the $2g-2$ branch points into k_1 sets of $2g_1 - 2$, k_2 sets of $2g_2 - 2$, ..., and k_r sets of $2g_r - 2$ branch points. There are

$$\binom{2g-2}{(2g_1-2),(2g_1-2),\ldots,(2g_r-2)} \prod_{k_j \text{ such that } g_j > 1} \frac{1}{k_j!}$$

ways to make such a choice. Here, in the binomial symbol, each $2g_j - 2$ appears k_j times.

The binomial symbol alone counts the number of ways to split the $2g-2$ points into a set of $2g_1 - 2$ points, a second set of $2g_1 - 2$ points, ..., a k_r-th set of $2g_r - 2$ points. However there is no natural order among sets of the same size which correspond to maps of the same degree (which is the "first" component of genus g_j and degree d_j?). Therefore we need to multiply by $1/k_j!$ for each k_j with $g_j > 1$, i.e., those k_j for which there are branch points, to account for this symmetrization.

From the description of the automorphism group in section 3.6 we see that the monomial $N_{g_1,d_1}^{k_1} \cdots N_{g_r,d_r}^{k_r}$ already takes care of most of the weighting coming from the automorphism group of the reducible cover. What is left is to account for the automorphisms coming from permuting genus 1 components isomorphic over E, as well as taking care of some overcounting in the monomial involving genus 1 components.

Suppose that $g_j = 1$ for some j, and that (C,p) is a particular genus g, degree d cover of the combinatorial type we are currently analyzing. Suppose that in our particular cover C, all of the components of type (g_j, d_j) are isomorphic over E. Then the automorphism group $\mathrm{Aut}_p(C)$ includes permutations of these components, automorphisms which are not taken care of by the monomial. We should therefore multiply by $1/k_j!$ to include this automorphism factor.

At the other extreme, suppose that in our particular cover C none of the components of type (g_j, d_j) are isomorphic over E. Then the claim is that the monomial $N_{g_1,d_1}^{k_1} \cdots N_{g_r,d_r}^{k_r}$ overcounts this cover by a factor of $k_j!$. Indeed, the product

$N_{g_j,d_j} \cdot N_{g_j,d_j} \cdots N_{g_j,d_j} = N^{k_j}_{g_j,d_j}$ represents the act of choosing a genus g_j degree d_j cover k_j times. Since the resulting disjoint union of these components (as a scheme over E) will not depend on the order in which they are chosen, we have overcounted by a factor of $k_j!$ if the components are all distinct over E. Therefore in this case we should again multiply by $1/k_j!$ to compensate.

In the general case we should break the components of type (g_j, d_j) (still with $g_j = 1$) into subsets of components which are mutually isomorphic over E. The total contribution from permutation automorphisms and from correcting for overcounting is always $1/k_j!$.

Putting all this together, we see that the total contribution coming from covers of our particular combinatorial type is

$$\binom{2g-2}{(2g_1-2),(2g_1-2),\ldots,(2g_r-2)}\left(\prod_{j=1}^{r}\frac{1}{k_j!}\right)N^{k_1}_{g_1,d_1}\cdots N^{k_r}_{g_r,d_r}$$

$$=(2g-2)!\prod_{j=1}^{r}\left(\frac{N_{g_j,d_j}}{(2g_j-2)!}\right)^{k_j}\frac{1}{k_j!}.$$

Since this is exactly the expression appearing in $(2g-2)!\exp(Z(q,\lambda))$, the lemma is proved. □

3.10. Thanks to the above lemma, we are now reduced to counting disconnected genus g, degree d covers. This problem is naturally equivalent to a counting problem in the symmetric group, by a method due to Hurwitz.

4. The monodromy map

4.1. Returning to the notation of section 3.5 let E be the elliptic curve and $S = \{b_1, \ldots, b_{2g-2}\}$ the set of distinct branch points. Pick any point b_0 of E, $b_0 \notin S$.

We define a *marked disconnected cover* (or *marked cover* for short) of genus g and degree d to be a disconnected cover (C, p) of topological Euler characteristic $2-2g$ and degree d, along with a labelling from 1 to d of the d distinct points $p^{-1}(b_0)$.

We consider two marked covers (C, p), (C', p') to be equivalent if there is an isomorphism over E preserving the markings. Marked covers have no nontrivial automorphisms.

Let $\widetilde{\mathrm{Cov}}(E,S)_{g,d}$ be the set of marked covers up to equivalence. We will denote an element of $\widetilde{\mathrm{Cov}}(E,S)_{g,d}$ by $(\widetilde{C}, p)$.

4.2. Thanks to the marking, we have a natural monodromy map

$$\widetilde{\mathrm{Cov}}(E,S)_{g,d} \xrightarrow{\text{mon}} \mathrm{Hom}\left(\pi_1(E \setminus S, b_0), S_d\right)$$

where Hom is homomorphism of groups and S_d is the symmetric group on the set $\{1,\ldots,d\}$. We wish to describe the image of this map.

Let $\pi_1 := \pi_1(E \setminus S, b_0)$. Cutting open the torus into a square with b_0 in the bottom left corner, the generators for the group π_1 are the loops α_1, α_2, and γ_i, $i = 1,\ldots 2g-2$, where each γ_i is a small loop passing around b_i and returning to b_0 (Figure 3).

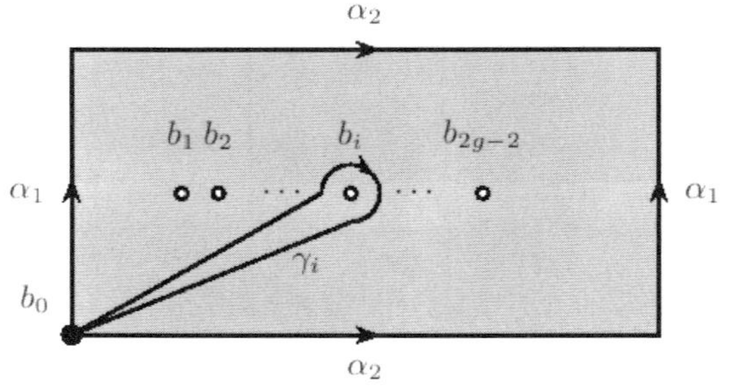

FIGURE 3

The generators satisfy the single relation $\gamma_1\gamma_2\cdots\gamma_{2g-2} = \alpha_1\alpha_2\alpha_1^{-1}\alpha_2^{-1}$.

Giving a homomorphism from π_1 to any group H is equivalent to giving elements $h_1,\ldots, h_{2g-2}$, h, h' of H satisfying the relation $h_1\cdots h_{2g-2} = hh'h^{-1}(h')^{-1}$.

Because of our conditions on the branching over b_i, $i = 1,\ldots,2g-2$, for any marked cover $(\widetilde{C}, p)$ the monodromy homomorphism sends γ_i to a simple transposition τ_i in S_d.

Let

$$\textbf{(4.2.1)}\quad \widehat{T}_{g,d} := \left\{ (\tau_1,\tau_2,\ldots,\tau_{2g-2},\sigma_1,\sigma_2) \;\middle|\; \begin{array}{l} \tau_1,\ldots,\tau_{2g-2},\sigma_1,\sigma_2 \in S_d, \\ \text{each } \tau_i \text{ is a simple transposition,} \\ \tau_1\tau_2\cdots\tau_{2g-2} = \sigma_1\sigma_2\sigma_1^{-1}\sigma_2^{-1} \end{array} \right\}$$

The set $\widehat{T}_{g,d}$ classifies the homomorphisms $\mathrm{Hom}(\pi_1, S_d)$ of the type that can arise from covers in $\widetilde{\mathrm{Cov}}(E,S)_{g,d}$, and so we get a classifying map

$$\rho : \widetilde{\mathrm{Cov}}(E,S)_{g,d} \longrightarrow \widehat{T}_{g,d}$$

induced by monodromy.

4.3. *Lemma.* — *The map* $\rho : \widetilde{\mathrm{Cov}}(E,S)_{g,d} \longrightarrow \widehat{T}_{g,d}$ *is surjective.*

Proof. Suppose that t is an element of $\widehat{T}_{g,d}$ and $\psi_t : \pi_1 \longrightarrow S_d$ the corresponding homomorphism from π_1 to S_d. From the theory of covering spaces, this homomorphism gives rise to a finite étale map $p'' : C'' \longrightarrow (E\setminus S)^{\mathrm{an}}$ along with a labelling of $(p')^{-1}(b_0)$ with monodromy equal to ψ_t. Here $(E\setminus S)^{\mathrm{an}}$ is the analytic space associated to $E \setminus S$.

The theory of covering spaces only guarantees the existence of C'' as a complex manifold. However, by Grothendieck's version of the Riemann existence theorem ([**7**], Exposé XII, Théorème 5.1) we may assume that $C'' = (C')^{\mathrm{an}}$ and $p'' = (p')^{\mathrm{an}}$ where C' is algebraic and $p' : C' \longrightarrow E \setminus S$ is a finite étale map.

Let C_i' be any component of C' and p_i' the restriction of p' to C_i'. Let C_i be the integral closure of E in the function field of C_i', and $p_i : C_i \longrightarrow E$ the associated map. Finally we set $\widetilde{C} = \cup_i C_i$ and define the map $p : \widetilde{C} \longrightarrow E$ to be the one given by p_i on each C_i.

Then $(\widetilde{C}, p)$ is an element of $\widetilde{\mathrm{Cov}}(E,S)_{g,d}$ and $\rho((\widetilde{C}, p)) = t$. □

4.4. The symmetric group S_d acts naturally on $\widetilde{\mathrm{Cov}}(E,S)_{g,d}$. The natural map

$$\eta : \widetilde{\mathrm{Cov}}(E,S)_{g,d} \longrightarrow \mathrm{Cov}(E,S)_{g,d}$$

forgetting the marking makes $\widetilde{\mathrm{Cov}}(E,S)_{g,d}$ a principal S_d set over $\mathrm{Cov}(E,S)_{g,d}$. The orbits of S_d on $\widetilde{\mathrm{Cov}}(E,S)_{g,d}$ are therefore naturally in one to one correspondence with the elements of $\mathrm{Cov}(E,S)_{g,d}$.

The symmetric group S_d also acts on $\widehat{T}_{g,d}$ by conjugation; the map ρ is S_d-equivariant.

We will prove in Lemma 4.6 that ρ is injective as a map from the set of S_d orbits of $\widetilde{\mathrm{Cov}}(E,S)_{g,d}$ to the set of S_d orbits of $\widehat{T}_{g,d}$. Using this fact in conjunction with Lemma 4.3, we will be able to conclude that the orbits of S_d on $\widehat{T}_{g,d}$ are also naturally in one to one correspondence with the elements of $\mathrm{Cov}(E,S)_{g,d}$.

For any element t of $\widehat{T}_{g,d}$ let $\mathrm{Stab}(t)$ be the stabilizer subgroup of t under the S_d action.

4.5. *Proposition.* — *Let $(\widetilde{C}, p)$ be any element of $\widetilde{\mathrm{Cov}}(E,S)_{g,d}$, $t = \rho((\widetilde{C}, p))$ the corresponding element of the classifying set $\widehat{T}_{g,d}$, and $C = \eta(\widetilde{C})$ the curve $\widetilde{C}$ with the markings forgotten. Then $\mathrm{Aut}_p(C) \cong \mathrm{Stab}(t)$.*

Proof. Let $\mathrm{Fet}(E \setminus S)$ be the category of finite étale covers of $E \setminus S$. Each element of $\mathrm{Cov}(E,S)_{g,d}$ gives an element of $\mathrm{Fet}(E \setminus S)$, and the automorphisms of (C, p) as a cover over E are the same as the automorphisms of the corresponding object in $\mathrm{Fet}(E \setminus S)$.

By Grothendieck's theory of the algebraic fundamental group the fibre functor ("fibre over b_0") gives an equivalence of categories between $\mathrm{Fet}(E \setminus S)$ and the category of finite π_1 sets, i.e., the category of finite sets with π_1 action ([**7**], Exposé V, Théorème 4.1).

More accurately, the theorem gives an equivalence of categories between $\mathrm{Fet}(E \setminus S)$ and the category of finite $\widehat{\pi}_1$ sets, where $\widehat{\pi}_1$ is the profinite completion of π_1, and the action of $\widehat{\pi}_1$ on finite sets is continuous. By the definition of profinite completion, this is the same as the category of finite π_1 sets.

Since we have an equivalence of categories, the automorphism group of an element of $\mathrm{Fet}(E \setminus S)$ is the same as the automorphism group of the associated π_1 set.

If D is a finite set with π_1 action, then an automorphism of D in the category of π_1 sets is a permutation $\sigma' \in S_D$ commuting with the π_1 action, where S_D is the permutation group of D. The action of π_1 on D is given by a homomorphism $\pi_1 \longrightarrow S_D$, and so the automorphism group of D as a π_1 set is just the group of elements in S_D commuting with the image of π_1. We may check if $\sigma' \in S_D$ commutes with the entire image of π_1 by checking if it commutes with the images of the generators.

Since $t \in \widehat{T}_{g,d}$ is the list of images of the generators of π_1 on the fibre $D := p^{-1}(b_0) \cong \{1, \ldots, d\}$ of C, and since for σ' to commute with a generator is the same as saying that the action by conjugation of σ' on the generator is trivial, the proposition follows. □

The equivalence of categories lets us clear up one point left over from section 4.4.

4.6. *Lemma.* — *The map ρ is injective as a map from the set of S_d orbits on $\widetilde{\mathrm{Cov}}(E,S)_{g,d}$ to the set of S_d orbits on $\widehat{T}_{g,d}$.*

Proof. Suppose that $(\widetilde{C}, p)$ and $(\widetilde{C}', p')$ are two elements of $\widetilde{\mathrm{Cov}}(E,S)_{g,d}$ whose images under ρ are in the same S_d orbit. By using the S_d action, we may in fact assume that they have the same image $t \in \widehat{T}_{g,d}$ under ρ. We want to show that $(\widetilde{C}, p)$ and $(\widetilde{C}', p')$ are in the same S_d orbit in $\widetilde{\mathrm{Cov}}(E,S)_{g,d}$.

But, since the π_1 set associated to both $\widetilde{C}$ and $\widetilde{C}'$ is the same, it follows from the fact that the fibre functor defines an equivalence of categories that $\eta(\widetilde{C})$ and $\eta(\widetilde{C}')$ are isomorphic as objects of $\mathrm{Fet}(E \setminus S)$, and therefore that $\eta(\widetilde{C})$ and $\eta(\widetilde{C}')$ are isomorphic over E, and so (by the definition of $\mathrm{Cov}(E,S)_{g,d}$) that $\eta(\widetilde{C})$ and $\eta(\widetilde{C}')$ are the same element of $\mathrm{Cov}(E,S)_{g,d}$.

Since $\widetilde{\mathrm{Cov}}(E,S)_{g,d}$ is a a principal S_d set over $\mathrm{Cov}(E,S)_{g,d}$ via η, this implies that $(\widetilde{C}, p)$ and $(\widetilde{C}', p')$ are in the same S_d orbit, and therefore that ρ is an injective map between the sets of orbits. □

4.7. Let G be a finite group and X a finite set with G action. For any $x \in X$ let G_x be the stabilizer subgroup of x. If x and x' are in the same G orbit, then $G_x \cong G_{x'}$ and so $|G_x| = |G_{x'}|$.

We wish to compute a weighted sum of the orbits in X, with the following weighting: if $\mathcal{O}$ is any G orbit in X, assign it the weighting $1/|G_x|$ where x is any element $x \in \mathcal{O}$. This is well defined by the previous remark.

4.8. *Lemma.* — *Let G be a finite group acting on a finite set X. Then the weighted count of G orbits of X is $|X|/|G|$. In particular, it is independent of the G action.*

Proof. This follows immediately from the orbit-stabilizer theorem. □

By Lemma 4.6 and the discussion in section 4.4 the elements of $\mathrm{Cov}(E, S)_{g,d}$ are naturally in one to one correspondence with the S_d orbits on $\widehat{T}_{g,d}$. By proposition 4.5 the group $\mathrm{Aut}_p(C)$ is isomorphic to $\mathrm{Stab}(t)$ for any t in the orbit corresponding to (C, p). Applying Lemma 4.8 and recalling the definition of $\widehat{N}_{g,d}$ from section 3.8 we have proved the following:

4.9. *Reduction Step I.* — $\widehat{N}_{g,d} = |\widehat{T}_{g,d}|/d!$.

Our counting problem has now been reduced to computing the size of $\widehat{T}_{g,d}$.

5. A calculation in the symmetric group

5.1. In order to compute the size of $\widehat{T}_{g,d}$ let us fix a σ_2 in S_d, and ask how many elements t of $\widehat{T}_{g,d}$ end with σ_2. This is easier to analyze if we rewrite the defining condition in 4.2.1 as

(5.1.1)
$$(\tau_1\tau_2\cdots\tau_{2g-2})\sigma_2 = \sigma_1\sigma_2\sigma_1^{-1},$$

from which we see that the main obstacle is that $(\tau_1\cdots\tau_{2g-2})\sigma_2$ should be conjugate to σ_2.

For any $\sigma_2 \in S_d$, let

$$P_{g,d}(\sigma_2) := \left\{ (\tau_1, \tau_2, \ldots, \tau_{2g-2}) \;\middle|\; \begin{array}{l} \text{each } \tau_i \in S_d \text{ is a simple transposition,} \\ \text{and } (\tau_1\tau_2\cdots\tau_{2g-2})\sigma_2 \text{ is conjugate to } \sigma_2 \end{array} \right\}$$

By the definition of $P_{g,d}(\sigma_2)$, for any $(\tau_1, \ldots, \tau_{2g-2})$ in $P_{g,d}(\sigma_2)$ there exists at least one $\sigma_1 \in S_d$ with $\tau_1\ldots\tau_{2g-2}\sigma_2 = \sigma_1\sigma_2\sigma_1^{-1}$.

Once we have found one such σ_1, the others differ from it by an element commuting with σ_2. In other words, once we fix σ_2, and $(\tau_1, \ldots, \tau_{2g-2}) \in P_{g,d}(\sigma_2)$, there are exactly as many possibilities for σ_1 satisfying 5.1.1 as there are elements which commute with σ_2.

If $c(\sigma_2)$ is the conjugacy class of σ_2, and $|c(\sigma_2)|$ the size of $c(\sigma_2)$, then the number of elements commuting with σ_2 is $|S_d|/|c(\sigma_2)|$, or $d!/|c(\sigma_2)|$, and so we get the formula

$$|\widehat{T}_{g,d}| = \sum_{\sigma_2\in S_d} \frac{d!}{|c(\sigma_2)|}|P_{g,d}(\sigma_2)|.$$

5.2. In general, if c is any conjugacy class of S_d, let $|c|$ denote the size of c. If $f : S_d \longrightarrow \mathbb{C}$ is any function from S_d to $\mathbb{C}$ which is constant on conjugacy classes, then let $f(c)$ denote the value of f on any element $\sigma \in c$ of that conjugacy class. Finally, let c^{-1} denote the conjugacy class made up of inverses of elements in c.

For any $\sigma, \sigma_2 \in S_d$ we have $P_{g,d}(\sigma\sigma_2\sigma^{-1}) = \sigma P_{g,d}(\sigma_2)\sigma^{-1}$ where the conjugation of the set $P_{g,d}(\sigma_2)$ means to conjugate the entries of all elements in $P_{g,d}(\sigma_2)$. This calculation shows that the function $|P_{g,d}(\,\cdot\,)| : S_d \longrightarrow \mathbb{N}$ is constant on conjugacy classes.

Using our notational conventions, we can now write the previous formula as

$$|\widehat{T}_{g,d}| = \sum_{c}\sum_{\sigma_2\in c}\frac{d!}{|c|}|P_{g,d}(c)| = \sum_{c}|c|\frac{d!}{|c|}|P_{g,d}(c)| = \sum_{c} d!|P_{g,d}(c)|$$

where the sum $\sum_c$ means to sum over the conjugacy classes of S_d.

On account of reduction 4.9 we then have

$$\widehat{N}_{g,d} = \sum_{c}|P_{g,d}(c)|. \tag{5.2.1}$$

5.3. The problem of computing $|P_{g,d}(c)|$ looks very much like the problem of computing the number of cycles in a graph.

Imagine a graph where the vertices are indexed by the conjugacy classes c of S_d, and the edges are the conjugacy classes which can be connected by multiplying by a transposition. For any conjugacy class c, the number $|P_{g,d}(c)|$ then looks like an enumeration of the paths of length $2g-2$ starting and ending at vertex c.

This picture is not quite correct since there is fundamental asymmetry in the definition of $P_{g,d}$ – we are really only allowed to pick a particular representative σ_2 of the conjugacy class c, and ask for "paths" (i.e. sequences of transpositions) which join σ_2 to c.

If we take this asymmetry into account we can define a version of the adjacency matrix which will allow us to compute $|P_{g,d}(c)|$ just as in the case of graphs.

5.4. For any $d \geq 1$ let M_d be the square matrix whose rows and columns are indexed by the conjugacy classes of S_d.

In the column indexed by the conjugacy class c and the row indexed by the conjugacy class c' we define the entry $(M_d)_{c',c}$ as follows: Pick any representative σ_2 of c, and let $M_{c',c}$ be the number of transpositions $\tau \in S_d$ such that $\tau\sigma_2 \in c'$. This number is independent of the representative $\sigma_2 \in c$ picked.

As an example, here are the matrices for $d = 3$ and $d = 4$, with the conjugacy classes represented by the partitioning of the dots (Figure 4).

5.5. *Lemma.* — *For any $k \geq 1$, the entry in column c and row c' of M_d^k is given by:*

FIGURE 4

$$(M_d^k)_{c',c} = \left| \left\{ (\tau_1, \ldots, \tau_k) \;\middle|\; \begin{array}{l} \textit{Each } \tau_i \in S_d \textit{ is a transposition,} \\ \textit{and } (\tau_1 \cdots \tau_k)\sigma_2 \in c' \end{array} \right\} \right|$$

where σ_2 *is any element* $\sigma_2 \in c$. *The calculation of this number does not depend on the representative* σ_2 *chosen.*

Proof. Straightforward induction argument. The case $k = 1$ is the definition of M_d. □

Lemma 5.5 has the following useful corollary.

5.5.1. *Corollary.* —

(a) *If* k *is odd, then* $(M_d^k)_{c,c} = 0$ *for all conjugacy classes* c.

(b) *If* $k = 2g - 2$ *is even, then* $(M_d^{2g-2})_{c,c} = |P_{g,d}(c)|$ *for all conjugacy classes* c.

Proof. Part (a) follows from the lemma and parity considerations – an odd number of transpositions can never take an element of a conjugacy class back to the same class (or to any other class of the same parity).

Part (b) follows immediately from the lemma and the definition of $|P_{g,d}(c)|$. □

5.6. Applying part (b) of the corollary and equation 5.2.1 we now have

$$\widehat{N}_{g,d} = \mathrm{Tr}(M_d^{2g-2}),$$

where Tr is the trace of a matrix.

Along with part (a) of the corollary and the definition (section 3.8) of $\widehat{Z}(q, \lambda)$ this gives

(5.6.1) $$\widehat{Z}(q,\lambda) = \sum_{g\geq 1}\sum_{d\geq 1} \frac{\mathrm{Tr}(M_d^{2g-2})}{(2g-2)!}\, q^d\lambda^{2g-2} = \sum_{d\geq 1} \mathrm{Tr}(\exp(M_d\cdot\lambda))\, q^d.$$

5.7. For any integer $d \geq 1$, let $\mathrm{part}(d)$ be the number of partitions of the integer d. This is exactly the number of conjugacy classes in S_d, so that each M_d is a $\mathrm{part}(d)\times\mathrm{part}(d)$ matrix.

For each d, let $\{\mu_{i,d}\}$, $i = 1,\ldots,\mathrm{part}(d)$, be the eigenvalues of M_d. Since

$$\mathrm{Tr}(M_d^k) = \sum_{i=1}^{\mathrm{part}(d)} \mu_{i,d}^k,$$

we can also write equation 5.6.1 as

5.8. *Reduction Step II.* — $\widehat{Z}(q,\lambda) = \sum_{d\geq 1}\sum_{i=1}^{\mathrm{part}(d)} \exp(\mu_{i,d}\lambda)q^d$.

Therefore we have reduced our counting problem yet again – this time to computing the eigenvalues of the matrices M_d.

6. A calculation in the group algebra

6.1. Let $\mathbb{C}[S_d]$ be the group algebra of S_d, and $\mathcal{H}_d$ its center. For any conjugacy class c of S_d, let

$$z_c := \sum_{\sigma\in c}\sigma$$

be the sum in $\mathbb{C}[S_d]$ of the group elements in the class c. It is well known that the $\{z_c\}$ form a basis for the center $\mathcal{H}_d$ and therefore that $\mathcal{H}_d$ is a $\mathrm{part}(d)$ dimensional $\mathbb{C}$-algebra.

Since $\mathcal{H}_d$ is an algebra, multiplication by any element $z\in\mathcal{H}_d$ is a $\mathbb{C}$-linear map, and can be represented by a $\mathrm{part}(d)\times\mathrm{part}(d)$ matrix.

Let τ stand for the conjugacy class of a transposition, and z_τ the corresponding basis element of $\mathcal{H}_d$. In the $\{z_c\}$ basis, the matrices for multiplication by z_τ in $\mathcal{H}_3$ and $\mathcal{H}_4$ are (Figure 5)

6.2. *Proposition.* — *In the $\{z_c\}$ basis, the matrix for multiplication by z_τ in $\mathcal{H}_d$ is the transpose of M_d.*
Proof. The entry in column c' and row c of the multiplication matrix is the coefficient of z_c in the expansion of $z_\tau\cdot z_{c'}$ into basis vectors.

To compute this, pick any $\sigma_2\in c$, and ask how many times σ_2 appears in the product expansion

$$z_\tau\cdot z_{c'} = \left(\sum_{\tau_i\in\tau}\tau_j\right)\cdot\left(\sum_{\sigma'_j\in c'}\sigma'_j\right).$$

	•••	•(••)	(•••)
•••	0	3	0
•(••)	1	0	2
(•••)	0	3	0

in $\mathcal{H}_3$

and

	••••	••(••)	•(•••)	(••)(••)	(••••)
••••	0	6	0	0	0
••(••)	1	0	4	1	0
•(•••)	0	3	0	0	3
(••)(••)	0	2	0	0	4
(••••)	0	0	4	2	0

in $\mathcal{H}_4$.

FIGURE 5

This is clearly the number of times that there exists a $\tau_i \in \tau$ and $\sigma'_j \in c'$ with $\tau_i \sigma'_j = \sigma_2$, which is perhaps more easily phrased as the number of times that there is a transposition $\tau_i \in \tau$ with $\tau_i^{-1}\sigma_2 \in c'$.

Since the inverse of a transposition is a transposition, we see that this is the same as the entry $(M_d)_{c',c}$ in column c and row c' of M_d, and therefore that the multiplication matrix is the transpose of M_d. $\square$

It is therefore enough to understand the eigenvalues for z_τ acting on $\mathcal{H}_d$ by multiplication.

6.3. Since $\mathcal{H}_d$ is a commutative algebra, for any action of $\mathcal{H}_d$ on a finite dimensional vector space, we might hope to diagonalize the action by finding a basis of simultaneous eigenvectors. In particular, we might hope to find such a basis for $\mathcal{H}_d$ acting on itself.

6.4. Let χ be an irreducible character of S_d, and define

$$w_\chi := \frac{\dim(\chi)}{d!} \sum_c \chi(c^{-1})\, z_c = \frac{\dim(\chi)}{d!} \sum_{\sigma \in S_d} \chi(\sigma^{-1})\, \sigma,$$

where $\dim(\chi) = \chi(1)$ is the dimension of the irreducible representation associated to χ. Each w_χ is by definition in $\mathcal{H}_d$, and we will see below that the $\{w_\chi\}$ form a basis for $\mathcal{H}_d$.

There are two well known orthogonality formulas involving the characters. For the first, pick any σ_1 in S_d, then

$$\textbf{(6.4.1)} \qquad \sum_{\sigma \in S_d} \chi(\sigma)\chi'(\sigma^{-1}\sigma_1) = \begin{cases} 0 & \text{if } \chi \neq \chi' \\ \frac{d!}{\dim(\chi)}\chi(\sigma_1) & \text{if } \chi = \chi' \end{cases}.$$

The second is an orthogonality between the conjugacy classes. For any conjugacy classes c and c',

(6.4.2)
$$\sum_{\chi}\chi(c)\chi(c') = \begin{cases} 0 & \text{if } c \neq c' \\ \frac{d!}{|c|} & \text{if } c = c' \end{cases},$$

where the sum is over the characters χ of S_d.

Formula 6.4.1 is equivalent to the multiplication formula

(6.4.3)
$$w_{\chi}\cdot w_{\chi'} = \begin{cases} 0 & \text{if } \chi \neq \chi' \\ w_{\chi} & \text{if } \chi = \chi' \end{cases},$$

while formula 6.4.2 is equivalent to the following expression for z_c in terms of the w_{χ},

(6.4.4)
$$z_c = \sum_{\chi}\left(\frac{|c^{-1}|\chi(c^{-1})}{\dim(\chi)}\right) w_{\chi}.$$

To see this last equation, expand the term on the right using the definition of w_{χ},

$$\sum_{\chi}\left(\frac{|c^{-1}|\chi(c^{-1})}{\dim(\chi)}\right) w_{\chi} = \sum_{\sigma\in S_d}\left(\sum_{\chi}\frac{|c^{-1}|}{d!}\chi(c^{-1})\chi(\sigma^{-1})\right)\sigma = \sum_{\sigma\in c}\sigma = z_c;$$

the middle equality comes from applying 6.4.2.

Equation 6.4.4 shows that the $\{w_{\chi}\}$ span $\mathcal{H}_d$. Either formula 6.4.3 or the fact that the number of characters is the same as the number of conjugacy classes shows that the $\{w_{\chi}\}$ are linearly independent. The $\{w_{\chi}\}$ therefore form a basis for $\mathcal{H}_d$.

6.5. Formula 6.4.3 is the most important; it shows that the basis $\{w_{\chi}\}$ diagonalizes the action of $\mathcal{H}_d$ on itself by multiplication. In particular, if

$$z = \sum_{\chi} a_{\chi}w_{\chi}$$

is any element in $\mathcal{H}_d$, then the eigenvalues of the matrix for multiplication by z are precisely the coefficients a_{χ} of z expressed in the $\{w_{\chi}\}$ basis.

The coefficients of z_{τ} in the $\{w_{\chi}\}$ basis are given by equation 6.4.4 with $c = \tau$. Since $\tau^{-1} = \tau$ (the inverse of a transpose is a transpose), and since by proposition 6.2 the eigenvalues for multiplication by z_{τ} are the same as the eigenvalues of M_d, we have proved

6.6. *Reduction Step III.* — The eigenvalues $\{\mu_{i,d}\}$ of M_d are given by

$$\frac{\binom{d}{2}\chi(\tau)}{\dim(\chi)}$$

as χ runs through the irreducible characters of S_d.

7. A formula of Frobenius

7.1. In order to use reduction 6.6 to calculate $\widehat{Z}(q, \lambda)$ we need to have a method to compute $\chi(\tau)/\dim(\chi)$ for the characters χ of S_d.

The irreducible representations of S_d are in one to one correspondence with the partitions of d. Any such partition can be represented by a *Young diagram* (Figure 6):

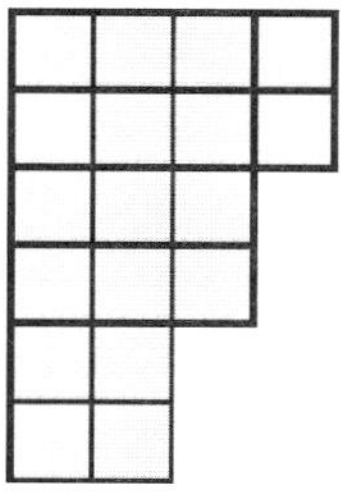

FIGURE 6

The picture represents the partition $4 + 4 + 3 + 3 + 2 + 2$ of $d = 18$.

There are many formulas relating the combinatorics of the Young diagram to the data of the irreducible representation associated to it. For example, the hook-length formula computes the dimension $\dim(\chi)$ of the irreducible representation, while the Murnaghan-Nakayama formula computes the value of the character on any conjugacy class.

More convenient for us is the following somewhat startling formula of Frobenius.

7.2. Given a Young diagram, split it diagonally into two pieces (Figure 7)

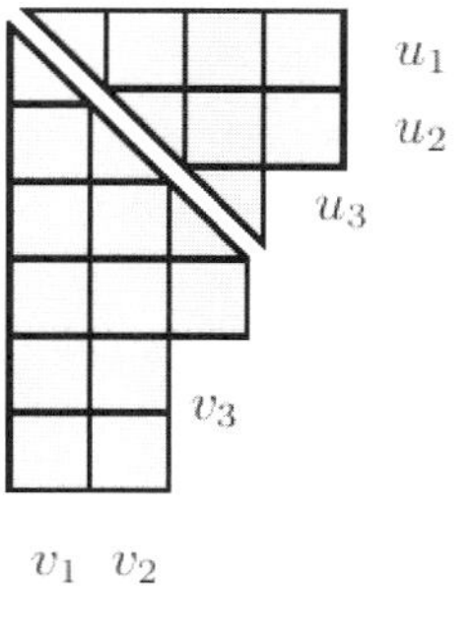

FIGURE 7

because the splitting is diagonal, there are as many rows in the top piece as columns in the bottom piece.

Suppose that there are r rows in the top and r columns in the bottom. Let u_i, $i = 1, \ldots, r$ be the number of boxes in the i-th row of the top piece, and v_i, $i = 1, \ldots, r$ the number of boxes in the i-th column of the bottom piece. The numbers u_i, v_i are *half integers*. In the example $r = 3$ and the numbers are $u_1 = 3\frac{1}{2}$, $u_2 = 2\frac{1}{2}$, $u_3 = \frac{1}{2}$ and $v_1 = 5\frac{1}{2}$, $v_2 = 4\frac{1}{2}$, $v_3 = 1\frac{1}{2}$.

If χ is the irreducible character associated to the partition, and τ the conjugacy class of transpositions in S_d, then Frobenius's formula is

$$\frac{\binom{d}{2}\chi(\tau)}{\dim(\chi)} = \frac{1}{2}\left(\sum_{i=1}^{r} u_i^2 - \sum_{i=1}^{r} v_i^2\right).$$

In our example this is

$$\frac{1}{2}\left((3\frac{1}{2})^2 + (2\frac{1}{2})^2 + (\frac{1}{2})^2 - (5\frac{1}{2})^2 - (4\frac{1}{2})^2 - (1\frac{1}{2})^2\right) = -17.$$

The formula produces exactly the eigenvalues we are looking for.

7.3. Let $\mathbb{Z}_{\geq 0+\frac{1}{2}}$ be the set

$$\mathbb{Z}_{\geq 0+\frac{1}{2}} := \left\{\frac{1}{2},\ \frac{3}{2},\ \frac{5}{2},\ \frac{7}{2},\ \cdots\cdots\right\}$$

of positive half integers.

The process of cutting Young diagram diagonally gives a one to one correspondence between partitions and subsets U and V of $\mathbb{Z}_{\geq 0+\frac{1}{2}}$ with $|U| = |V|$.

To reverse it, given any two such subsets with $|U| = |V| = r$, we organize the elements $u_1, \ldots, u_r$ so that $u_1 > u_2 > \cdots > u_r$, and similarly with the v_i's. We then recover the Young diagram by gluing together the appropriate row with u_i boxes to the column with v_i boxes, $i = 1, \ldots, r$.

The resulting Young diagram is a partition of $d = \sum_{u \in U} u + \sum_{v \in V} v$.

Since the data of the subsets U and V are sufficient to recover the degree d, and the eigenvalue associated to the corresponding irreducible representation, we see that all of the combinatorial information we are interested in is contained in these subsets.

7.4. Consider the infinite product

$$\prod_{u \in \mathbb{Z}_{\geq 0+\frac{1}{2}}} \left(1 + z\, q^u\, e^{\frac{u^2}{2}\lambda}\right) \prod_{v \in \mathbb{Z}_{\geq 0+\frac{1}{2}}} \left(1 + z^{-1}\, q^v\, e^{\frac{-v^2}{2}\lambda}\right).$$

Computing a term in the expansion of this product involves choosing finite subsets $U \subset \mathbb{Z}_{\geq 0+\frac{1}{2}}$ and $V \subset \mathbb{Z}_{\geq 0+\frac{1}{2}}$.

In the term corresponding to a pair of subsets U and V,

- The power of z appearing in the term is $|U| - |V|$,
- the power of q appearing is $\sum_{u\in U} u + \sum_{v\in V} v$, and
- the exponential term is $\exp\left(\frac{1}{2}(\sum_{u\in U} u^2 - \sum_{v\in V} v^2)\lambda\right)$.

The infinite product is a Laurent series in z with coefficients formal power series in q and λ. Combining the above discussion, the formula of Frobenius, and reduction 5.8 we have

7.5. *Reduction Step IV.* —

$$\widehat{Z}(q,\lambda) = \text{coeff of } z^0 \text{ in } \left(\prod_{u\in\mathbb{Z}_{\geq 0+\frac{1}{2}}} \left(1 + z\, q^u\, e^{\frac{u^2}{2}\lambda}\right) \prod_{v\in\mathbb{Z}_{\geq 0+\frac{1}{2}}} \left(1 + z^{-1}\, q^v\, e^{\frac{-v^2}{2}\lambda}\right)\right) - 1.$$

The "-1" appears because $\widehat{Z}(q,\lambda)$ doesn't have a constant term, or alternately, because we should ignore the term where both U and V are the empty set.

The fact that $F_g(q)$ is a quasimodular form of weight $6g-6$ follows from this formula and the work of Kaneko and Zagier.

8. The work of Kaneko and Zagier

8.1. Kaneko and Zagier **[9]** start with the series

$$\Theta(q,\lambda,z) := \prod_{n\geq 1}(1-q^n) \prod_{u\in\mathbb{Z}_{\geq 0+\frac{1}{2}}} \left(1 + z\, q^u\, e^{\frac{u^2}{2}\lambda}\right) \prod_{v\in\mathbb{Z}_{\geq 0+\frac{1}{2}}} \left(1 + z^{-1}\, q^v\, e^{\frac{-v^2}{2}\lambda}\right)$$

considered as a Laurent series in z with coefficients formal power series in q and λ.

Now instead of taking the infinite products over half integers u, v, it is equivalent to take $n/2$ where $n \geq 1$ and $2 \nmid n$. Then $\Theta(q,\lambda,z)$ is equivalently formulated as the following infinite product:

$$\Theta(q,\lambda,z) = \prod_{n\geq 1}(1-q^n) \prod_{n\geq 1, 2\nmid n} (1 + e^{n^2\lambda/8} q^{n/2}\, z)(1 + e^{-n^2\lambda/8} q^{n/2} z^{-1}).$$

Since we are interested in the coefficient of z^0 in this series, we may replace z by $-z$, and by abuse of notation we also call this new series $\Theta(q,\lambda,z)$.

Let $\Theta_0(q,\lambda)$ be the coefficient of z^0 in the triple product

$$\Theta(q,\lambda,z) = \prod_{n\geq 1}(1-q^n) \prod_{n\geq 1, 2\nmid n} (1 - e^{n^2\lambda/8} q^{n/2} z)(1 - e^{-n^2\lambda/8} q^{n/2} z^{-1}).$$

Since

$$\Theta_0(q,\lambda) = \Theta_0(q,-\lambda)$$

$\Theta_0(q,\lambda)$ has only even powers of λ in its Taylor expansion. Therefore, we can write

$$\Theta_0(q,\lambda)=\sum_{n=0}^{\infty}A_n(q)\lambda^{2n}.$$

They prove ([**9**], Theorem 1) that $A_n(q)$ is a quasimodular form of weight $6n$.

By reduction 7.5,

$$\Theta_0(q,\lambda)=(\prod_{n\geq 1}(1-q^n))(\widehat{Z}(q,\lambda)+1)$$

and so taking the logarithm and using Lemma 3.9 we get

$$\begin{aligned}\log(\Theta_0(q,\lambda)) &= \log\left(\prod_{n\geq 1}(1-q^n)\right)+\log(\widehat{Z}(q,\lambda)+1)\\ &= \sum_{n\geq 1}\log(1-q^n)+Z(q,\lambda)\end{aligned}$$

Using Kaneko and Zagier's theorem, the coefficient of λ^{2n} in $\log(\Theta_0(q,\lambda))$ is also a quasimodular form of weight $6n$. In particular, the coefficient of λ^{2g-2} is a quasimodular form of weight $6g-6$.

Since the $\log(1-q^n)$ terms contain no power of λ, as long as $g\geq 2$ the coefficient of λ^{2g-2} is $F_g(q)/(2g-2)!$, and therefore $F_g(q)$ is a quasimodular form of weight $6g-6$.

As for the λ^0 term, $F_1(q)$ is exactly equal to $-\sum_{n\geq 1}\log(1-q^n)$, and so this is cancelled out in the expression for $\log(\Theta_0(q,\lambda))$.

9. Proof of the quasimodularity

We follow the exposition of Kaneko and Zagier [**9**]. Their idea is to identify the generating function as the coefficient of a generalized theta series. Various kinds of theta series and functions have appeared in counting arguments in mathematics and physics, e.g., Eskin and Okounkov [**6**].

Now we change the variable λ to X in $\Theta_0(q,\lambda)$. We have a Taylor series expansion for $\Theta_0(X,q)$ given as

$$\Theta_0(X,q)=\sum_{n=0}^{\infty}A_n(q)X^{2n}$$

where $A_n(q)\in Q[[q]]$. The main result of Kaneko and Zagier is the assertion on $A_n(q)$ that it is a quasimodular form of weight $6n$ for Γ. For this, we will study the expansion of $\Theta(X,q,z)$ with respect to z, and in particular, the constant term in that expansion. We state the result that we are to prove in this section once again.

9.1. *Theorem* —. *$A_n(q)$ is a quasimodular form of weight $6n$ for $\Gamma=SL(2,\mathbb{Z})$.*

9.2. Recall the definition of $\Theta(X,q,z)$:

$$\Theta(X,q,z)=\prod_{n>0}(1-q^n)\prod_{n>0,2\nmid n}(1-e^{n^2X/8}q^{n/2}z)(1-e^{-n^2X/8}q^{n/2}z^{-1}).$$

Now introduce two new variables w and Z by putting

$$w=e^X \quad \text{and} \quad z=e^Z$$

Noting that $q=e^{2\pi i\tau}$, we may write $\Theta(X,q,z)$ as $\Theta(X,\tau,Z)$. Further, we need three more functions: let

$$\eta(q)=\eta(\tau)=q^{1/24}\prod_{n=1}^{\infty}(1-q^n)$$

be the eta-function, and let

$$\theta(\tau)=\sum_{r=1}^{\infty}(-1)^r q^{r^2/2}$$

be the theta-series of weight 1/2 on the subgroup $\Gamma_2=\{\begin{pmatrix}a & b\\ c & d\end{pmatrix}\mid b\equiv 0 \pmod 2\}$. (The usual notation for Γ_2 is $\Gamma^0(2)$, which is an index 3 subgroup of Γ.)

$\theta(\tau)$ is related to $\eta(\tau)$ by the following identity:

$$\theta(\tau)=\eta(\tau/2)^2\eta(\tau)^{-1}.$$

Finally, we introduce the identity similar to the Jacobi triple product:

$$H(w,q,z)=q^{-1/24}\prod_{n>0,2\nmid n}(1-w^{n^2/8}q^{n/2}z)(1-w^{-n^2/8}q^{n/2}z^{-1})$$

and $H_0(w,q)$ denotes the coefficient of $H(w,q,z)$ in its Laurent expansion with respect to z. We have the following relation:

$$H(X,\tau,Z)=\frac{1}{\eta(\tau)}\Theta(X,\tau,Z).$$

Reading off this identity at the term Z^0 and comparing the coefficients of the two sides, we obtain

$$H_0(Z,\tau)=\frac{1}{\eta(\tau)}\Theta_0(X,\tau)=\sum_{n=0}^{\infty}\frac{A_n(q)}{\eta(q)}X^{2n}.$$

9.3. *Proposition —. The function $\Theta(X,\tau,Z)$ has the expansion*

$$\Theta(X,\tau,Z)=\theta(\tau)\sum_{i,j\geq 0}H_{j,\ell}(\tau)\frac{X^j}{j!}\frac{Z^\ell}{\ell!}$$

where $H_{0,0}(\tau)=1$ and each $H_{j,\ell}(\tau)$ is a quasimodular form of weight $3j+\ell$ on Γ_2.

Proof. We consider the quotient $\frac{\Theta(X,\tau,Z)}{\theta(\tau)}$. Note that

$$\theta(\tau)=\eta(\tau/2)^2/\eta(\tau)=\prod_{n>0}(1-q^{n/2})^2(1-q^n)^{-1}.$$

We have the following infinite product expression:

$$\frac{\Theta(X,\tau,Z)}{\theta(\tau)} = \prod_{n>0}\frac{(1-q^n)^2}{(1-q^{n/2})^2}\prod_{n>0,2\nmid n}(1-e^{n^2X/8}q^{n/2}z)(1-e^{-n^2X/8}q^{n/2}z^{-1})$$
$$= \prod_{n>0,2\nmid n}(1-q^{n/2})^{-2}(1-e^{n^2X/8}q^{n/2}z)(1-e^{-n^2X/8}q^{n/2}z^{-1}).$$

Now take the logarithm of both sides. Using the formal power series expansion $\log(1-t) = -(\sum_{r=1}^{\infty}\frac{x^r}{r})$, we get

$$\log\left(\frac{\Theta(X,\tau,Z)}{\theta(\tau)}\right) = \sum_{n>0,2\nmid n}\sum_{r=1}^{\infty}\left(2\frac{q^{nr/2}}{r} - \frac{e^{n^2Xr/8}q^{nr/2}z^r}{r} - \frac{e^{-n^2Xr/8}q^{nr/2}z^{-r}}{r}\right)$$
$$= -\sum_{n,r>0,2\nmid n}\frac{1}{r}\left(e^{n^2Xr/8}z^r - 2 + e^{-n^2Xr/8}z^{-r}\right)q^{nr/2}.$$

Now recall that we put $z = e^Z$. Using the formal power series expansion for $e^t = \sum_{r=0}^{\infty}\frac{t^r}{r!}$, we have

$$e^{n^2Xr/8}z^r = \sum_{j=0}^{\infty}\frac{(n^2Xr/8)^j}{j!}\sum_{\ell=0}^{\infty}\frac{(rZ)^\ell}{\ell!} = \sum_{j,\ell\geq 0} n^{2j}r^{j+\ell}\frac{(X/8)^j}{j!}\frac{Z^\ell}{\ell!}.$$

Similarly, we have the expansion for $e^{-n^2Xr/8}z^{-r}$. Then the sum yields the following identity:

$$e^{n^2Xr/8}z^r + e^{-n^2Xr/8}z^{-r} = 2\sum_{j,\ell\geq 0,\, 2|j+\ell} n^{2j}r^{j+\ell}\frac{(X/8)^j}{j!}\frac{Z^\ell}{\ell!}$$

where the term with $j=\ell=0$ is 2.

Thus we obtain

$$\log\left(\frac{\Theta(X,\tau,Z)}{\theta(\tau)}\right) = -2\sum_{\substack{n,r>0\\ 2\nmid n}}\frac{1}{r}\left(\sum_{\substack{j,\ell\geq 0\\ j+\ell>0\\ 2|j+\ell}} n^{2j}r^{j+\ell}\frac{(X/8)^j}{j!}\frac{Z^\ell}{\ell!}\right)q^{nr/2}$$
$$= -2\sum_{\substack{j,\ell\geq 0\\ j+\ell>0\\ 2|j+\ell}}\left(\sum_{\substack{n,r>0\\ 2\nmid n}} n^{2j}r^{j+\ell-1}q^{nr/2}\right)\frac{(X/8)^j}{j!}\frac{Z^\ell}{\ell!}.$$

Now we put

$$\phi_{j,\ell}(\tau) = \sum_{\substack{n,r>0\\ 2\nmid n}} r^{j+\ell-1}n^{2j}q^{nr/2}.$$

Then we can express ϕ in terms of derivatives of quasimodular forms for Γ_2.

9.3.1. Proof of the quasimodularity of $\phi_{j,\ell}(\tau)$ for Γ_2. — For even $k\geq 2$, let

$$G_k^{(1)}(\tau) := E_k(\tau/2) - E_k(\tau)$$

and

$$G_k^{(2)}(\tau) = E_k(\tau/2) - 2^{k-1}E_k(\tau).$$

Then we see that $G_k^{(i)}$ for $i = 1, 2$ have the q-expansions as follows:

$$G_k^{(1)}(q) = \sum_{n=1}^{\infty}\Big(\sum_{d|n, 2\nmid d}(\frac{n}{d})^{k-1}\Big)q^{n/2}$$

and

$$\begin{aligned} G_k^{(2)}(q) &= (2^{k-1}-1)\frac{B_k}{k} + \sum_{n=1}^{\infty}\Big(\sum_{d|n} d^{k-1}q^{n/2} - 2^{k-1}d^{k-1}q^n\Big) \\ &= \sum_{n=1}^{\infty}\Big(\sum_{d|n,2\nmid d} d^{k-1}\Big)q^{n/2}. \end{aligned}$$

Then

$$\phi_{j,\ell}(\tau) = \begin{cases} 2^{2j}D^{2j}G_{\ell-j}^{(1)}(\tau) & \text{if } \ell > j, \\ 2^{j+\ell-1}D^{j+\ell-1}G_{j-\ell+2}^{(2)}(\tau) & \text{if } j \geq \ell \end{cases}$$

where D is the differential operator $D := \frac{1}{2\pi i}\frac{d}{d\tau}$.

Assuming that $G_k^{(i)}$ $(i = 1, 2)$ are quasimodular forms for Γ_2, (whose proof will be given in Lemma 9.4 below) then, in view of Proposition 2.7 applied to Γ and Γ_2, $\phi_{j,\ell}$ are quasimodular forms of weight $\ell - j + 4j = 3j + \ell$ if $\ell > j$ and $j - \ell + 2 + 2j + 2\ell - 2 = 3j + \ell$ if $j \geq \ell$ for Γ_2. □

To complete our proof of Proposition 9.3, we take the exponential of both sides.

$$\Theta(X,\tau,Z) = \theta(\tau)\exp\Big(-2\sum_{j,\ell\geq 0}\phi_{j,\ell}(\tau)\frac{(X/8)^j}{j!}\frac{Z^\ell}{\ell!}\Big)$$

where the inner sum runs over $j, \ell \geq 0$ subject to the conditions $j + \ell > 0$, odd. We write

$$\Theta(X,\tau,Z) = \theta(\tau)\sum_{j,\ell\geq 0}H_{j,\ell}(\tau)\frac{X^j}{j!}\frac{Z^\ell}{\ell!}.$$

Then it might be possible to express $H_{j,\ell}(\tau)$ as a polynomial in $\phi_{j,\ell}(\tau)$ by expanding out the exponential, but it is too complicated to do that.

9.3.2. Claim . — *$H_{j,\ell}(\tau)$ is a quasimodular form of weight $3j + \ell$ on Γ_2.*

Proof. This claim can be established as follows. Note that the coefficient of X^jZ^ℓ is a quasimodular form of weight $3j + \ell$ on Γ_2. When we expand the exponential, we need to multiply the terms $X^{j_1}Z^{\ell_1}$ and $X^{j_2}Z^{\ell_2}$. Then we see that the coefficient of $X^{j_1+j_2}Z^{\ell_1+\ell_2}$ has weight $3(j_1+j_2)+(\ell_1+\ell_2)$, as quasimodular forms form a graded ring. So the coefficient of X^jZ^ℓ, which is nothing but $H_{j,\ell}(\tau)$, is a quasimodular form of weight $3j + \ell$ on Γ_2. □

This completes our proof of Proposition 9.3 modulo the fact that $G_k(i)$ $(i = 1, 2)$ are quasimodular forms of weight k for Γ_2.

9.4. *Lemma .—* $G_k^{(i)}$ $(i = 1,2)$ *are quasimodular forms of weight* k *for* $\Gamma_2 = \{\begin{pmatrix} a & b \\ c & d \end{pmatrix} \in \mathrm{SL}_2(\mathbb{Z}) \,|\, b \equiv 0 \pmod 2 \}$.

Proof. If $\begin{pmatrix} a & b \\ c & d \end{pmatrix} \in \Gamma_2$, then note that $\begin{pmatrix} a & b \\ c & d \end{pmatrix} \in \Gamma$ and $\begin{pmatrix} a & b/2 \\ 2c & d \end{pmatrix} \in \Gamma$. Then for $k \geq 4$, we claim that $G_k^{(i)}$ for $i = 1,2$ are indeed *modular* forms of weight k for Γ_2. Since $E_k(\tau)$ is a modular form of weight k for Γ, we have

$$G_k^{(1)}\left(\frac{a\tau+b}{c\tau+d}\right) = E_k\left(\frac{1}{2}\frac{a\tau+b}{c\tau+d}\right) - E_k\left(\frac{a\tau+b}{c\tau+d}\right)$$

$$= E_k\left(\frac{a\frac{\tau}{2}+\frac{b}{2}}{2c\frac{\tau}{2}+d}\right) - E_k\left(\frac{a\tau+b}{c\tau+d}\right) = (2c\frac{\tau}{2}+d)^k E_k(\frac{\tau}{2}) - (c\tau+d)^k E_k(\tau)$$

$$= (c\tau+d)^k G_k^{(1)}(\tau).$$

Similarly, we have

$$G_k^{(2)}\left(\frac{a\tau+b}{c\tau+d}\right) = E_k\left(\frac{1}{2}\frac{a\tau+b}{c\tau+d}\right) - 2^{k-1}E_k\left(\frac{a\tau+b}{c\tau+d}\right)$$

$$= E_k\left(\frac{a\frac{\tau}{2}+\frac{b}{2}}{2c\frac{\tau}{2}+d}\right) - 2^{k-1}E_k\left(\frac{a\tau+b}{c\tau+d}\right) = (2c\frac{\tau}{2}+d)^k E_k(\frac{\tau}{2}) - 2^{k-1}(c\tau+d)^k E_k(\tau)$$

$$= (c\tau+d)^k (E_k(\frac{\tau}{2}) - 2^{k-1}E_k(\tau)) = (c\tau+d)^k G_k^{(2)}(\tau).$$

This proves the modularity (not just the quasimodularity) of $G_k^{(i)}$ for $i = 1,2$ for Γ_2.

For $k = 2$, if we show that $G_2^{(2)}(\tau)$ is a quasimodular form of weight 2 for Γ_2; then it follows that

$$G_2^{(1)}(\tau) = G_2^{(2)}(\tau) + E_2(\tau)$$

is quasimodular for Γ_2, as the sum of two quasimodular forms is again quasimodular. Therefore, it remains to establish the quasimodularity of $G_2^{(2)}$. For $\begin{pmatrix} a & b \\ c & d \end{pmatrix} \in \Gamma_2$, we have

$$G_2^{(2)}\left(\frac{a\tau+b}{c\tau+d}\right) = E_2\left(\frac{a\frac{\tau}{2}+\frac{b}{2}}{2c\frac{\tau}{2}+d}\right) - 2E_2\left(\frac{a\tau+b}{c\tau+d}\right)$$

$$= (c\tau+d)^2 E_2(\frac{\tau}{2}) - \frac{c(c\tau+d)}{4\pi i} - 2\left((c\tau+d)^2 E_2(\tau) - \frac{c(c\tau+d)}{4\pi i}\right)$$

$$= (c\tau+d)^2 G_2^{(2)}(\tau)$$

and this proves the quasimodularity of $G_2^{(2)}(\tau)$. □

Now we consider the functions $H(w,q,z)$ and $H_0(w,q,0)$ introduced in section 9.2.

9.5. *Proposition* —. *The function $H(w,q,z)$ has the expansion*

$$H(w,q,z)=\sum_{r\in\mathbb{Z}}(-1)^r H_0(w,w^r q)w^{r^3/6}q^{r^2/2}z^r.$$

Proof. Here we introduce the function $H(w,q,z)$ (similar to the Jacobi triple product formula)

$$H(w,q,z)=q^{-1/24}\prod_{n>0,2\nmid n}(1-w^{n^2/8}q^{n/2}z)(1-w^{-n^2/8}q^{n/2}z^{-1}),$$

as it is related to the $\Theta(X,q,z)$ by the identity

$$H(X,\tau,Z)=\frac{1}{\eta(\tau)}\Theta(X,\tau,Z).$$

Expanding out the first few terms, we see the pattern of this function:

$$\begin{aligned}H(w,q,z)&=q^{-1/24}(1-w^{1/8}q^{1/2}z)(1-w^{-1/8}q^{1/2}z^{-1})\\ &\times(1-w^{3^2/8}q^{3/2}z)(1-w^{-3^2/8}q^{3/2}z^{-1})(1-w^{5^2/8}q^{5/2}z)(1-w^{-5^2/8}q^{5/2}z^{-1})\cdots\end{aligned}$$

Let $H_0(w,q)$ denote the coefficient of z^0 in this expansion. We now consider the function $H(w,wq,w^{1/2}qz)$. From the product expansion of H, we obtain

$$\begin{aligned}&H(w,wq,w^{1/2}qz)\\ &=(wq)^{-1/24}\prod_{2\nmid n}(1-w^{n^2/8}(wq)^{n/2}w^{1/2}qz)(1-w^{-n^2/8}(wq)^{n/2}w^{-1/2}q^{-1}z^{-1})\\ &=(wq)^{-1/24}\prod_{2\nmid n}(1-w^{(n+2)^2/8}q^{(n+2)/2}z)(1-w^{-(n-2)^2/8}q^{(n-2)/2}z^{-1})\\ &=w^{-1/24}\frac{1-w^{-1/8}q^{-1}z^{-1}}{1-w^{1/8}q^{1/2}z}H(w,q,z)=-w^{-1/6}q^{-1/2}z^{-1}H(w,q,z).\end{aligned}$$

This means that

$$H(w,q,z)=-w^{1/6}q^{1/2}z\,H(w,wq,w^{1/2}qz).$$

Continuing this process, we can write the Laurent expansion for H with respect to z in the form

$$H(w,q,z)=\sum_{r\in\mathbb{Z}}(-1)^r H_r(w,q)w^{r^3/6}q^{r^2/2}z^r.$$

Then

$$H_{r+1}(w,q)=H_r(w,wq)=H_{r-1}(w,w^2q)=\cdots=H_0(w,w^r q)$$

for any $r\in\mathbb{Z}$. This gives rise to the expansion of Proposition 9.3.

□

Using the identities in the two propositions, the proof that $A_n(q)$ is quasimodular of weight $6n$ for Γ proceeds as follows.

9.6. *Proof of the quasimodularity of $A_n(q)$* —. We have two kinds of Laurent series expansions for $H(X,\tau,Z)$:

$$H(X,\tau,Z)=\frac{\theta(\tau)}{\eta(\tau)}\sum_{j,\ell\geq 0}H_{j,\ell}(\tau)\frac{X^j}{j!}\frac{Z^\ell}{\ell!}$$

$$= \sum_{r\in\mathbb{Z}} (-1)^r e^{r^3X/6+rZ} H_0(X, \tau + \frac{rX}{2\pi i}) q^{r^2/2}.$$

Then, comparing the coefficients of X^jZ^ℓ of both sides noticing that

$$e^{r^3X/6+rZ} = \sum_{p,\ell\geq 0} \frac{r^{3p+\ell}}{6^p p!\ell!} X^p Z^\ell,$$

and

$$\frac{A_n}{\eta}(\tau + \frac{rZ}{2\pi i}) = \sum_{m\geq 0} \frac{r^m}{m!} D^m(\frac{A_n}{\eta})(\tau))X^m,$$

we obtain

$$\frac{\theta(\tau)}{\eta(\tau)} H_{j,\ell}(\tau) = \sum \frac{j!}{6^p p! m!} D^m(\frac{A_n(\tau)}{\eta(\tau)}) \sum_{r\in\mathbb{Z}} (-1)^r r^{3p+\ell+m} q^{r^2/2}$$

where the first sum runs over $m, n, p \geq 0$ such that $p + 2n + m = j$. This is further reformatted as

$$\sum \frac{2^s(2n)!}{6^{j-2n-m}} \binom{2n+m}{m} \binom{j}{2n+m} D^m(\frac{A_n(\tau)}{\eta(\tau)}) D^s\theta(\tau)$$

where the sum runs over $m, n, s \geq 0$ such that $2m + 2s + 6n = 3j + \ell$.

Here is the punch line! The functions θ and η are modular forms for Γ and $H_{j,\ell}$ is a quasimodular form of weight $3j + \ell$ on Γ_2. The space of quasimodular forms is closed under the operator D, and indeed, weights increase by 2 under D. Therefore, the above identities show that A_n is a quasimodular form of weight $6n$ for Γ_2. But Γ is generated by Γ_2 and $\begin{pmatrix} 1 & 1 \\ 0 & 1 \end{pmatrix}$, so a modular or a quasimodular form on Γ_2 whose Fourier expansion involves only integral powers of q is a modular or a quasimodular form for Γ. Therefore, $A_n(\tau)$ is a quasimodular form of weight $6n$ for Γ.

9.7. *Corollary .— For $g \geq 2$, $F_g(q)$ is a quasimodular form of weight $6g - 6$.*

Proof. We have

$$\log(\Theta_0(X, q)) = \log(1 + \sum_{n\geq 1} A_n(q)X^{2n}) = \sum_{r\geq 1} \frac{(-1)^{r+1}}{r} \left(\sum_{n\geq 1} A_n(q)X^{2n}\right)^r.$$

The coefficient of X^{2g-2} in this expression is a quasimodular form of weight $6g - 6$ for Γ. But by the identity we established in section 8.1, this coefficient coincides with $F_g(q)/(2g - 2)!$. Consequently $F_g(q)$ is a quasimodular form of weight $6g - 6$ for Γ. □

The reader is also referred to the article of Zagier [**12**] for another proof of the quasimodularity of $F_g(q)$.

9.8. *Remark.* We have presented a mathematically rigorous proof of the theorem due to Dijkgraaf and Douglas asserting that the generating function $F_g(q)$ $(g \geq 2)$ of simply ramified covers of an elliptic curve with $2g - 2$ marked points is a quasi-modular form of weight $6g - 6$ for the full modular group $\Gamma = SL_2(\mathbb{Z})$. As the reader might have realized, our proofs involve a sequence of reduction steps, and the final proof consists of somewhat complicated analytic manoeuvres. Though everything is mathematically rigorous, this line of argument, however, does not reveal conceptually why quasimodular forms enter the scene in the A-model (fermionic) counting. We have to wait for the B-model (bosonic) counting for a conceptual explanation.

10. Tables

10.1. We tabulate the numbers of genus g, degree d, simply ramified covers of an elliptic curve, where g ranges from 2 to 10 and d from 2 to 18.

$g=2$	
d	#
2	2
3	16
4	60
5	160
6	360
7	672
8	1240
9	1920
10	3180
11	4400
12	6832
13	8736
14	12880
15	15840
16	22320
17	26112
18	36666

$g=3$	
d	#
2	2
3	160
4	2448
5	18304
6	90552
7	341568
8	1068928
9	2877696
10	7014204
11	15423200
12	32107456
13	61663104
14	115156144
15	200764608
16	346235904
17	561158400
18	911313450

$g=4$	
d	#
2	2
3	1456
4	91920
5	1931200
6	21639720
7	160786272
8	893985280
9	4001984640
10	15166797900
11	50211875600
12	149342289472
13	404551482816
14	1017967450960
15	2389725895200
16	5320611901440
17	11218821981312
18	22749778149786

$g = 5$

d	#
2	2
3	13120
4	3346368
5	197304064
6	5001497112
7	73102904448
8	724280109568
9	5371101006336
10	31830391591644
11	157705369657280
12	675306861112576
13	2559854615265024
14	8759525149882864
15	27434575456211328
16	79665883221602304
17	216263732895828480
18	553988245305680010

$g = 6$

d	#
2	2
3	118096
4	120815280
5	19896619840
6	1139754451080
7	32740753325472
8	577763760958720
9	7092667383039360
10	65742150901548780
11	487018342594703600
12	3004685799388645312
13	15919220244209484096
14	74163755181310506640
15	309440248335185814240
16	1173700610446435061760
17	4094919386905893808512
18	13274627847658663840506

$g = 7$

d	#
2	2
3	1062880
4	4352505888
5	1996102225024
6	258031607185272
7	14560223135464128
8	457472951327051008
9	9293626316677061376
10	134707212077147740284
11	1491667016717716134560
12	13258722309534523444096
13	98155445716515005756544
14	622608528358993525294384
15	3459690237503699953309248
16	17143154981017805400827904
17	76843825646212716425276160
18	315316749952804309551553770

$g = 8$	
d	#
2	2
3	9565936
4	156718778640
5	199854951398080
6	58230316414059240
7	6451030954702152672
8	360793945093731688960
9	12127449147074861834880
10	274847057797905019237260
11	4548825193274613857646800
12	58246387837051777276658752
13	602459738298815084682461376
14	5202820358556329365805383120
15	38499747480350614341732504480
16	249216184092355960780119674880
17	1435109816316883240864058627712
18	7453948595460163331625275982426

$g = 9$	
d	#
2	2
3	86093440
4	5642133787008
5	19994654452125184
6	13120458818999011032
7	2852277353239208548608
8	283889181859169785013248
9	15786934495235533394850816
10	559374323532926110389380124
11	13836013152938852920073095040
12	255210544832216420532477846016
13	3687933280407403025259141041664
14	43359873830374370211757798766704
15	427254495213241208775759565210368
16	3612921587646224114145820619464704
17	26726550587804552791079214953149440
18	175711339161472053202073003447846730

$g = 10$	
d	#
2	2
3	774840976
4	203119138758000
5	1999804372817081920
6	2954080786719122704200
7	1259649848110685616355872
8	2230624655322958757890224000
9	2051916951738606884143485120 0
10	11366305910063743293599690153 40
11	420155769332891431085733127052 00
12	1116355464862438151830378349593792
13	2253724594144986720212269471665 4656
14	3607365818826794857651226665190 88400
15	4733248189193492784514748822817 229920
16	5228500935459162214957636395465 7996800
17	4968548168200368239419142717183 61942912
18	4134625570069614451109511415147 720431546

10.2. Here we tabulate the generating functions $F_g(q)$ computed for $g \leq 7$.

$$F_2(q) = \frac{1}{2^6 3^4 5}\left(5\,{E_2}^3 - 3\,E_2\,E_4 - 2\,E_6\right)$$

$$\begin{aligned} F_3(q) &= \frac{1}{2^{11}3^6}\Big(- 6{E_2}^6 + 15\,{E_2}^4 E_4 + 4\,{E_2}^3 E_6 \\ &- 12\,{E_2}^2 {E_4}^2 - 12\,E_2\,E_4\,E_6 + 7\,{E_4}^3 + 4\,{E_6}^2 \Big) \end{aligned}$$

$$\begin{aligned} F_4(q) &= \frac{1}{2^{15}3^9}\Big(355{E_2}^9 - 1395\,{E_2}^7 E_4 - 600\,{E_2}^6 E_6 \\ &+ 1737\,{E_2}^5 {E_4}^2 + 4410\,{E_2}^4 E_4\,E_6 \\ &- 2145\,{E_2}^3 {E_4}^3 - 1860\,{E_2}^3 {E_6}^2 - 6300\,{E_2}^2 {E_4}^2 E_6 + 3600\,E_2\,{E_4}^4 \\ &+ 4860\,E_2\,E_4\,{E_6}^2 - 2238\,{E_4}^3 E_6 - 424\,{E_6}^3 \Big) \end{aligned}$$

$$\begin{aligned} F_5(q) &= \frac{1}{2^{28}3^{11}}\Big(- 44310{E_2}^{12} + 186900\,{E_2}^{10} E_4 \\ &+ 211120\,{E_2}^9 E_6 - 116067\,{E_2}^8 {E_4}^2 \\ &- 1854216\,{E_2}^7 E_4\,E_6 - 247940\,{E_2}^6 {E_4}^3 + 436688\,{E_2}^6 {E_6}^2 \\ &+ 5699400\,{E_2}^5 {E_4}^2 E_6 - 464520\,{E_2}^4 {E_4}^4 - 1758120\,{E_2}^4 E_4\,{E_6}^2 \\ &- 9725912\,{E_2}^3 {E_4}^3 E_6 - 1169056\,{E_2}^3 {E_6}^3 + 4277448\,{E_2}^2 {E_4}^5 \\ &+ 11020128\,{E_2}^2 {E_4}^2 {E_6}^2 - 5480664\,E_2\,{E_4}^4 E_6 \\ &- 2497824\,E_2\,E_4\,{E_6}^3 + 255897\,{E_4}^6 + 1165336\,{E_4}^3 {E_6}^2 + 105712\,{E_6}^4 \Big) \end{aligned}$$

$$\begin{aligned} F_6(q) &= \frac{1}{2^{23}3^{12}}\Big(90560820{E_2}^{10} E_4\,E_6 \\ &+ 225798000\,E_2\,E_4\,{E_6}^4 + 1010594160\,E_2\,{E_4}^4 {E_6}^2 \\ &- 1927375200\,{E_2}^2 {E_4}^2 {E_6}^3 - 2228443380\,{E_2}^2 {E_4}^5 E_6 \\ &+ 4832320680\,{E_2}^3 {E_4}^3 {E_6}^2 - 1249004400\,{E_2}^4 E_4\,{E_6}^3 \\ &- 3414090330\,{E_2}^4 {E_4}^4 E_6 + 148712760\,{E_2}^5 {E_4}^2 {E_6}^2 \\ &+ 1346850960\,{E_2}^6 {E_4}^3 E_6 - 229831560\,{E_2}^7 E_4\,{E_6}^2 \\ &- 362752020\,{E_2}^8 {E_4}^2 E_6 + 1057210\,{E_2}^{15} \\ &- 5687776\,{E_6}^5 - 111148848\,{E_4}^3 {E_6}^3 \\ &- 74719416\,{E_4}^6 E_6 + 152826525\,E_2\,{E_4}^7 \\ &+ 354656240\,{E_2}^3 {E_6}^4 + 1171877925\,{E_2}^3 {E_4}^6 \\ &+ 225055040\,{E_2}^6 {E_6}^3 + 240597000\,{E_2}^5 {E_4}^5 \\ &- 280860660\,{E_2}^7 {E_4}^4 + 98613165\,{E_2}^9 {E_4}^3 \\ &- 13749435\,{E_2}^{11} {E_4}^2 - 10397450\,{E_2}^{12} E_6 \\ &+ 11720440\,{E_2}^9 {E_6}^2 - 3180450\,{E_2}^{13} E_4 \Big) \end{aligned}$$

$$
\begin{aligned}
F_7(q) &= \frac{1}{2^{28}3^{14}}\Big(6485729456448 E_2{}^2 E_4{}^2 E_6{}^4 - 3652838628384\, E_2\, E_4{}^4 E_6{}^3 \\
&+ 343299239380\, E_4{}^6 E_6{}^2 + 248654161008\, E_4{}^3 E_6{}^4 \\
&+ 1708407276048\, E_2{}^2 E_4{}^8 - 1323241487040\, E_2{}^3 E_6{}^5 \\
&+ 8721747087735\, E_2{}^4 E_4{}^7 + 3683336152210\, E_2{}^6 E_4{}^6 \\
&- 2296344667155\, E_2{}^8 E_4{}^5 + 265041770400\, E_2{}^6 E_6{}^4 \\
&+ 40754175\, E_2{}^{16} E_4 - 176046929135\, E_2{}^{12} E_4{}^3 \\
&- 58261668080\, E_2{}^{12} E_6{}^2 + 716455989480\, E_2{}^{10} E_4{}^4 \\
&- 274557704960\, E_2{}^9 E_6{}^3 + 27304216170\, E_2{}^{14} E_4{}^2 \\
&+ 9096818500\, E_2{}^{15} E_6 + 7938163648\, E_6{}^6 \\
&+ 27967464684\, E_4{}^9 - 579294100\, E_2{}^{18} \\
&+ 860386259040\, E_2{}^{10} E_4\, E_6{}^2 - 474327657408\, E_2\, E_4\, E_6{}^5 \\
&- 30048009145440\, E_2{}^3 E_4{}^3 E_6{}^3 + 15276854478864\, E_2{}^2 E_4{}^5 E_6{}^2 \\
&- 1693078150368\, E_2\, E_4{}^7 E_6 - 21791387720960\, E_2{}^3 E_4{}^6 E_6 \\
&- 31633500031980\, E_2{}^5 E_4{}^5 E_6 - 23283960305760\, E_2{}^5 E_4{}^2 E_6{}^3 \\
&+ 52829353179240\, E_2{}^4 E_4{}^4 E_6{}^2 + 10411655662320\, E_2{}^4 E_4\, E_6{}^4 \\
&+ 3323307667680\, E_2{}^7 E_4\, E_6{}^3 + 10959473005360\, E_2{}^6 E_4{}^3 E_6{}^2 \\
&- 478499593880\, E_2{}^9 E_4{}^3 E_6 - 5009357600940\, E_2{}^8 E_4{}^2 E_6{}^2 \\
&+ 6224536233300\, E_2{}^7 E_4{}^4 E_6 - 59770219740\, E_2{}^{13} E_4\, E_6 \\
&+ 123175769640\, E_2{}^{11} E_4{}^2 E_6 \Big)
\end{aligned}
$$

Acknowledgements

We would like to thank a number of mathematicians and mathematical physicists for their help and for answering questions on the subject matter dealt with herein. This includes (in alphabetical order) V. Bouchard, P. Cartier, R. Dijkgraaf, S. Hosono, M. Kaneko, M. Kontsevich, B. Mazur, and D. Zagier.

During the preparation of this paper, N. Yui held a research (visiting) professorship at various institutions. This includes Fields Institute Toronto, Tsuda College Tokyo, Max-Planck-Institut für Mathematik Bonn, IHES Paris, and Department of Pure Mathematics University of Cambridge. She thanks these institutions for their hospitality and support.

At last but not the least, we thank the referee for carefully reading through the manuscript, and for suggesting future problems. Indeed, the referee has pointed out that there are a few papers that applied counting covers of elliptic curves with various ramification types to important questions in algebraic geometry. A. Eskin and A. Okounkov *Asymptotics of numbers of branched coverings of torus and volumes of moduli spaces of holomorphic differentials*, arXiv:math/0006171, studied the asymptotic behavior of the number of curves with prescribed ramification type, which has a meaning as the volume of the corresponding moduli space of holomorphic differentials. Dawei Chen *Simply branched covers of an elliptic curve and the moduli space of curves*, arXiv:0806.0674, studied covers of elliptic curves with a unique branch point, aiming to bounding the effective cone of the moduli space of curves. The referee feels that there is no reason to satisfy ourselves only in counting simply ramified covers, and has suggested that we try to improve our methods to counting covers with other ramification data.

References

[1] Bruinier, J.H., van der Geer, G., and Zagier, D., *The 1-2-3 of Modular Forms*, Lectures at a summer school in Nordfjordeid, Norway June 2004. Universitext, Springer 2008.

[2] Caporaso, N., Cirafici, M., Griguolo, L., Pasquetti, S., Seminara, D., and Szabo, R., *Topological strings, two-dimensional Yang–Mills theory and Chern–Simons theory on torus bundles*, Adv. Theor. Math. Phys. **12** (2008), 981–1058.

[3] Dijkgraaf, R., *Mirror symmetry and elliptic curves*, in *The moduli space of curves (Texel Island, 1994)*, 149–163. Progr. Math. **129**, Birkhäuser Boston, Boston MA, 1995.

[4] Dijkgraaf, R., Verline, E., and Verlinde, H., *On moduli spaces of conformal field theories with* $c \geq 1$, in Perspectives in String Theory, P.Di Vecchia and J.L. Petersen (Eds.), World Scientific 1988.

[5] Douglas, M. R., *Conformal field theory techniques in large N Yang–Mills theory*, in "Quantum Field Theory and String Theory (Cargèse, 1993), NATO Adv. Sci. Inst. Ser. B Phys. **328** (1995), 119–135. [arXiv:hep-th/9311130].

[6] Eskin, A., and Okounkov, A., *Pillowcases and quasimodular forms*, Prog. in Math. **253**, (2006), pp. 1–25.

[7] Grothendieck, A., *Revêtements Étales et Groupe Fondamental*, Springer Lecture Notes 224 (SGA1).

[8] Kaneko, M., and Koike, M., *On extremal quasimodular forms*, Kyushu J. Math. **60** (2006), no.2, 457–470.

[9] Kaneko, M., and Zagier, D., *A generalized Jacobi theta function and quasimodular forms*, in *The moduli space of curves (Texel Island, 1994)*, 165–172. Progr. Math. **129**, Birkhäuser Boston, Boston MA, 1995.

[10] Monni, S., Song, J.S., and Song, Y.S., *The Hurwitz enumeration problem of brached covers and Hodge integrals*, J. Geometry and Physics **50** (2004), 223–256.

[11] Okounkov, A., and Pandharipande, R., *Gromov–Witten theory, Hurwitz theory, and completed cycles*, Ann. of Math. (2) **163** (2006), no.2, pp. 517–560. [mathAG/0204305].

[12] Zagier, D., *Applications of the representation theory of finite groups*, Appendix to *Graphs and Surfaces and Their Applications* by Lando, S., and Zvonkin, A. Enclyclopedia Math. Sci. **141**, Low-Dimensional Topology II, Springer-Verlag, Berlin, 2004.

Department of Mathematics and Statistics, Queen's University, Kingston, Ontario Canada K7L 3N6

E-mail address: `mikeroth@mast.queensu.ca`

E-mail address: `yui@mast.queensu.ca`

Clay Mathematics Proceedings
Volume 12, 2010

A Symbolic Summation Approach to Find Optimal Nested Sum Representations

Carsten Schneider

ABSTRACT. We consider the following problem: Given a nested sum expression, find a sum representation such that the nested depth is minimal. We obtain a symbolic summation framework that solves this problem for sums defined, e.g., over hypergeometric, q–hypergeometric or mixed hypergeometric expressions. Recently, our methods have found applications in quantum field theory.

1. Introduction

Karr's algorithm (Kar81; Kar85) based on his difference field theory provides a general framework for symbolic summation. For example, his algorithm, or a simplified version presented in (Sch05c), covers summation over hypergeometric terms (Gos78; Zei91), q–hypergeometric terms (PR97) or mixed hypergeometric terms (BP99). More generally, indefinite nested product-sum expressions can be represented in his $\Pi\Sigma$-difference fields which cover as special cases, e.g., harmonic sums (BK99; Ver99) or generalized nested harmonic sums (MUW02).

In this article much emphasis is put on the problem of how these indefinite nested product-sum expressions can be simplified in a $\Pi\Sigma^*$-field. For example, with our algorithms we shall compute for the sum expression

$$A = \sum_{r=1}^{n} \frac{\sum_{l=1}^{r} \frac{H_l^2 + H_l^{(2)}}{l} + \sum_{l=1}^{r} \frac{H_l}{l}}{r} \tag{1.1}$$

the alternative representation

$$B = \tfrac{1}{12}\Big(H_n^4 + 2H_n^3 + 6(H_n+1)H_n^{(2)}H_n + 3\big(H_n^{(2)}\big)^2 + (8H_n+4)H_n^{(3)} + 6H_n^{(4)}\Big) \tag{1.2}$$

where $A(n) = B(n)$ for all $n \geq 0$ and where the nested depth of B is minimal; $H_n = \sum_{k=1}^{n} \frac{1}{k}$ denotes the nth harmonic numbers and $H_n^{(o)} = \sum_{k=1}^{n} \frac{1}{k^o}$ are the generalized harmonic numbers of order $o \geq 1$.

2010 *Mathematics Subject Classification.* Primary 33F10; Secondary 68W30, 81T18.
Supported by the SFB-grant F1305 and the grant P20347-N18 of the Austrian FWF.

In order to accomplish this task, we exploit a new difference field theory for symbolic summation (Sch08) that refines Karr's $\Pi\Sigma$-fields to the so-called depth-optimal $\Pi\Sigma^*$-fields. In particular, we construct explicitly a difference ring monomorphism (Sch09) which links elements from such a difference field to elements in the ring of sequences. Using this algorithmic machinery, we will derive for a given nested product-sum expression A a nested product-sum expression B with the following property: There is an explicit $\lambda \in \mathbb{N} = \{0, 1, 2, \dots\}$ such that

$$A(k) = B(k) \quad \forall k \geq \lambda \tag{1.3}$$

and among all such alternative representations for A the depth of B is minimal.

From the point of view of applications our algorithms are able to produce d'Alembertian solutions (Nör24; AP94; Sch01), a subclass of Liouvillian solutions (HS99), of a given recurrence with minimal nested depth; for applications arising from particle physics see, e.g., (BBKS07; BBKS08; BKKS09a; BKKS09b). The presented algorithms are implemented in the summation package Sigma (Sch07), that can be executed in the computer algebra system Mathematica.

The general structure of this article is as follows. In Section 2 we introduce the problem to find optimal sum representations which we supplement by concrete examples. In Section 3 we define depth-optimal $\Pi\Sigma^*$-extensions and show how indefinite summation can be handled accordingly in such fields. After showing how generalized d'Alembertian extensions can be embedded in the ring of sequences in Section 4, we are ready to prove that our algorithms produce sum representations with optimal nested depth in Section 5. Applications are presented in Section 6.

2. The problem description for indefinite nested sum expressions

Inspired by (BL82; NP97) one can consider the following general simplification problem. Let X be a set of expressions (i.e., terms of certain types), let $\mathbb{K}$ be a field[1], and let $\text{ev} : X \times \mathbb{N} \to \mathbb{K}$ with $(x, n) \mapsto x(n)$ be a function. Here one considers the so-called *evaluation function* ev as a procedure that computes $x(n)$ for a given $x \in X$ and $n \in \mathbb{N}$ in a finite number of steps. In addition, we suppose that we are given a function $\mathfrak{d} : X \to \mathbb{N}$ which measures the simplicity of the expressions in X; subsequently, we call such a triple $(X, \text{ev}, \mathfrak{d})$ also a *(measured) sequence domain*; cf. (NP97).
In this setting the following problem can be stated: *Given $A \in X$; find $B \in X$* and $\lambda \in \mathbb{N}$ such that (1.3) and such that among all such possible solutions $\mathfrak{d}(B)$ is minimal.

In this article, the expressions X are given in terms of indefinite nested sums and products and the measurement of simplicity is given by the nested depth of the occurring sum- and product-quantifiers. Subsequently, we shall make this more precise. Let $(X, \text{ev}, \mathfrak{d})$ be a measured sequence domain with $\text{ev} : X \times \mathbb{N} \to \mathbb{K}$ and $\mathfrak{d} : X \to \mathbb{N}$.

EXAMPLE 2.1. Let $X = \mathbb{K}(x)$ be a rational function field and define for $f = \frac{p}{q} \in \mathbb{K}(x)$ with $p, q \in \mathbb{K}[x]$ where $q \neq 0$ and p, q being coprime the evaluation function

$$\text{ev}(f, k) = \begin{cases} 0 & \text{if } q(k) = 0 \\ \frac{p(k)}{q(k)} & \text{if } q(k) \neq 0; \end{cases} \tag{2.1}$$

[1]Subsequently, all fields and rings are commutative and contain the rational numbers $\mathbb{Q}$.

here $p(k), q(k)$ is the usual evaluation of polynomials at $k \in \mathbb{N}$. In particular, we define $\mathfrak{d}(f) = 1$ if $f \in \mathbb{K}(x) \setminus \mathbb{K}$ and $\mathfrak{d}(f) = 0$ if $f \in \mathbb{K}$. In the following $(\mathbb{K}(x), \mathrm{ev}, \mathfrak{d})$ is called the *rational sequence domain*.

EXAMPLE 2.2. Suppose that $\mathbb{K} = \mathbb{K}'(q_1, \ldots, q_m)$ is a rational function field extension over $\mathbb{K}'$ and consider the rational function field $X := \mathbb{K}(x, x_1, \ldots, x_m)$ over $\mathbb{K}$. Then for $f = \frac{p}{q} \in X$ with $p, q \in \mathbb{K}[x, x_1, \ldots, x_m]$ where $q \neq 0$ and p, q being coprime we define

$$\mathrm{ev}(f, k) = \begin{cases} 0 & \text{if } q(k, q_1^k, \ldots, q_m^k) = 0 \\ \frac{p(k, q_1^k, \ldots, q_m^k)}{q(k, q_1^k, \ldots, q_m^k)} & \text{if } q(k, q_1^k, \ldots, q_m^k) \neq 0. \end{cases} \tag{2.2}$$

Note that there is a $\delta \in \mathbb{N}$ such that $q(k, q_1^k, \ldots, q_m^k) \neq 0$ for all $k \in \mathbb{N}$ with $k \geq \delta$; for an algorithm that determines δ see (BP99, Sec. 3.7). In particular, we define $\mathfrak{d}(f) = 0$ if $f \in \mathbb{K}$, and $\mathfrak{d}(f) = 1$ if $f \notin \mathbb{K}$. In the following $(\mathbb{K}(x, x_1, \ldots, x_m), \mathrm{ev}, \mathfrak{d})$ is called a *q–mixed sequence domain*. Note: If $m = 0$, we are back to the rational sequence domain. If we restrict to the setting $\mathbb{K}(x_1, \ldots, x_m)$ which is free of x, it is called a *q–rational sequence domain*.

More generally, X can contain hypergeometric, q–hypergeometric or mixed hypergeometric terms; for instance see Example 4.10. Over such a set X we consider the set of *(indefinite nested) product-sum expressions*, denoted by $\mathrm{ProdSum}(X)$, which is defined as follows. Let $\oplus$, $\otimes$, Sum, Prod be operations with the signatures

$$\begin{array}{lll} \oplus: & \mathrm{ProdSum}(X) \times \mathrm{ProdSum}(X) & \to \mathrm{ProdSum}(X) \\ \otimes: & \mathrm{ProdSum}(X) \times \mathrm{ProdSum}(X) & \to \mathrm{ProdSum}(X) \\ \mathrm{Sum}: & \mathbb{N} \times \mathrm{ProdSum}(X) & \to \mathrm{ProdSum}(X) \\ \mathrm{Prod}: & \mathbb{N} \times \mathrm{ProdSum}(X) & \to \mathrm{ProdSum}(X). \end{array}$$

Then $\mathrm{ProdSum}(X) \supseteq X$ is the smallest set that satisfies the following rules:

(1) For any $f, g \in \mathrm{ProdSum}(X)$, $f \oplus g \in \mathrm{ProdSum}(X)$ and $f \otimes g \in \mathrm{ProdSum}(X)$.
(2) For any $f \in \mathrm{ProdSum}(X)$ and any $r \in \mathbb{N}$, $\mathrm{Sum}(r, f) \in \mathrm{ProdSum}(X)$ and $\mathrm{Prod}(r, f) \in \mathrm{ProdSum}(X)$.

The set of all expressions in $\mathrm{ProdSum}(X)$ which are free of Prod is denoted by $\mathrm{Sum}(X)$. $\mathrm{Sum}(X)$ is called the set of *(indefinite nested) sum expressions over X*.

EXAMPLE 2.3. Given $(X, \mathrm{ev}, \mathfrak{d})$ from Example 2.1 with $\mathbb{K} = \mathbb{Q}$ and $X = \mathbb{Q}(x)$ the following indefinite nested sum expressions are in $\mathrm{Sum}(\mathbb{Q}(x))$:

$$E_1 = \frac{1}{x}, \quad E_2 = \mathrm{Sum}(1, \frac{1}{x}), \text{ and}$$
$$A = \mathrm{Sum}(1, \frac{1}{x}\Big(\mathrm{Sum}(1, \frac{1}{x}(\mathrm{Sum}(1, \frac{1}{x})^2 \oplus \mathrm{Sum}(1, \frac{1}{x^2}))) \oplus \mathrm{Sum}(1, \mathrm{Sum}(1, \frac{1}{x}))\Big)).$$

Finally, ev and $\mathfrak{d}$ are extended from X to $\mathrm{ev}' : \mathrm{ProdSum}(X) \times \mathbb{N} \to \mathrm{ProdSum}(X)$ with $(x, n) \mapsto x(n)$ and $\mathfrak{d}' : \mathrm{ProdSum}(X) \to \mathbb{N}$ as follows.

(1) For $f \in X$ we set $\mathfrak{d}'(f) := \mathfrak{d}(f)$ and $\mathrm{ev}'(f, k) := \mathrm{ev}(f, k)$.
(2) For $f, g \in \mathrm{ProdSum}(X)$ we set $\mathfrak{d}'(f \oplus g) = \mathfrak{d}'(f \otimes g) := \max(\mathfrak{d}'(f), \mathfrak{d}'(g)$

$$\mathrm{ev}'(f \oplus g, k) := \mathrm{ev}'(f, k) + \mathrm{ev}'(g, k) \quad \text{and} \quad \mathrm{ev}'(f \otimes g, k) := \mathrm{ev}'(f, k)\,\mathrm{ev}'(g, k);$$

here the operations on the right hand side are from the field $\mathbb{K}$.

(3) For $r \in \mathbb{N}$, $f \in \text{ProdSum}(X)$ define $\mathfrak{d}'(\text{Sum}(r,f)) = \mathfrak{d}'(\text{Prod}(r,f)) := \mathfrak{d}'(f)+1$,

$$\text{ev}'(\text{Sum}(r,f),k) = \sum_{i=r}^{k} \text{ev}'(f,i) \quad \text{and}^2 \quad \text{ev}'(\text{Prod}(r,f),k) = \prod_{i=r}^{k} \text{ev}'(f,i).$$

Since ev′ and ev, resp. $\mathfrak{d}$ and $\mathfrak{d}'$, agree on X, we do not distinguish them any longer. Subsequently, $(\text{ProdSum}(X), \text{ev}, \mathfrak{d})$ (resp. $(\text{Sum}(X), \text{ev}, \mathfrak{d})$) is called the *product-sum sequence domain over X* (resp. *sum sequence domain over X*).

EXAMPLE 2.4. The expressions from Example 2.3 are evaluated as

$$\text{ev}(E_1,k) = E_1(k) = \frac{1}{k}, \quad \text{ev}(E_2,k) = E_2(k) = \sum_{i=1}^{k} \text{ev}(\frac{1}{x},i) = \sum_{i=1}^{k} \frac{1}{i}, \quad \text{and}$$

$$\text{ev}(A,k) = A(k) = \sum_{r=1}^{k} \frac{\sum_{l=1}^{r} \frac{\left(\sum_{i=1}^{l} \frac{1}{i}\right)^2 + \sum_{i=1}^{l} \frac{1}{i^2}}{l} + \sum_{l=1}^{r} \frac{\sum_{i=1}^{l} \frac{1}{i}}{l}}{r}.$$

We have $\mathfrak{d}(E_1) = 1$, $\mathfrak{d}(E_2) = 2$ and $\mathfrak{d}(A) = 4$.

Usually, we stick to the following more convenient and frequently used notation.

- We write, e.g., $E = a \oplus \text{Sum}(1, c\,\text{Sum}(2,b)) \in \text{Sum}(X)$ with $a,b,c \in X$ in the form
$$E' = \text{ev}(a,n) \oplus \sum_{i=1}^{n} \text{ev}(c,i) \sum_{j=2}^{i} \text{ev}(b,j)$$
for a symbolic variable n. Clearly, fixing the variable n, the two encodings E and E' can be transformed into each other; if we want to emphasize the dependence on n, we also write $E' \in \text{Sum}_n(X)$.
- Even more, by abuse of notation, we use instead of $\oplus$ and $\otimes$ the usual field operations in $\mathbb{K}$. This "sloppy" notation immediately produces the evaluation mechanism: $\text{ev}(E,k) = E(k)$ for a concrete integer $k \in \mathbb{N}$ is produced by substitution in E' the variable n with the concrete value $k \in \mathbb{N}$.
- Finally, whenever possible, the evaluation $\text{ev}(a,n)$ for some $a \in X$ is expressed by well known functions, like, e.g., $\text{ev}(1/(x+1),n) = \frac{1}{(n+1)}$ (Ex. 2.1) or $\text{ev}(x_i,n) = q_i^n$ for $1 \le i \le e$ (Ex. 2.2).

EXAMPLE 2.5. We write $E_1, E_2, A \in \Sigma_n(\mathbb{Q}(x))$ from Example 2.4 in the more convenient notation $E_1 = \frac{1}{n}$, $E_2 = H_n$ and (1.1); note that for E_1 we must require that n is only evaluated for $n \ge 1$.

Let $(X, \text{ev}, \mathfrak{d})$ be a measured sequence domain and consider the sum sequence domain $(\text{Sum}(X), \text{ev}, \mathfrak{d})$ over X. We define the $\text{Sum}(X)$-*optimal depth* of $A \in \text{Sum}(X)$ as

$$\min\{\mathfrak{d}(B) | B \in \text{Sum}(X) \text{ such that (1.3) for some } \lambda \in \mathbb{N}\}.$$

Then we are interested in the following problem.

DOS: Depth Optimal Simplification. *Given* $A \in \text{Sum}(X)$; *find* $B \in \text{Sum}(X)$ and $\lambda \in \mathbb{N}$ such that (1.3) and such that $\mathfrak{d}(B)$ is the $\text{Sum}(X)$-optimal depth of A.

[1]Note that $\text{ev}'(\text{Prod}(r,f),k)$ might be 0 if r is too small. Later, products will be used only as described in Ex. 4.10 or in the general case 4.11; there we will take care of the bound r by (4.9).

EXAMPLE 2.6. Consider, e.g., $A \in \mathrm{Sum}(\mathbb{Q}(x))$ from Example 2.5. Then with our algorithms, see Example 3.13, we find $B \in \mathrm{Sum}(\mathbb{K}(x))$ with (1.2) such that $A(n) = B(n)$ for all $n \in \mathbb{N}$. At this point it is easy to see that B cannot be expressed with depth≤ 1, and thus B is a solution of DOS. In Section 5 we will show that this fact is an immediate consequence of our algebraic construction. Summarizing, 2 is the $\mathrm{Sum}(\mathbb{Q}(x))$-optimal depth of A.

We shall solve problem DOS algorithmically, if X is, e.g., the rational sequence domain (Ex. 2.1) or the q–mixed sequence domain (Ex. 2.2). More generally, X might be a sequence domain in which objects from $\mathrm{ProdSum}(X')$ can be represented; in this setting X' might be the rational, q–rational or q–mixed sequence domain. Note that the general case 4.11 (page 297) includes most of the (q–)hypergeometric or q–mixed hypergeometric terms (see Ex. 4.10).

3. Step I: Reducing the problem to difference fields by telescoping

Let $(X, \mathrm{ev}, \mathfrak{d})$ be a measured sequence domain. Then for $f \in \mathrm{ProdSum}(X)$ and $r \in \mathbb{N}$ the sum $S = \sum_{k=r}^{n} f(k) \in \mathrm{ProdSum}_n(X)$ satisfies the recurrence relation

$$S(n+1) = S(n) + f(n+1) \quad \forall n \geq r, \tag{3.1}$$

and the product $P = \sum_{k=r}^{n} f(k) \in \mathrm{ProdSum}_n(X)$ satisfies the recurrence relation

$$P(n+1) = f(n+1)P(n) \quad \forall n \geq r. \tag{3.2}$$

As a consequence, we can define a *shift operator* acting on the expressions $S(n)$ and $P(n)$. Subsequently, we shall restrict to sequence domains X such that the sums and products $S(n)$ and $P(N)$ can be modeled in difference rings.

In general, a *difference ring* (resp. *difference field*) $(\mathbb{A}, \sigma)$ is defined as a ring $\mathbb{A}$ (resp. field) with a ring automorphism (resp. field automorphism) $\sigma : \mathbb{A} \to \mathbb{A}$. The set of constants $\mathrm{const}_\sigma \mathbb{A} = \{k \in \mathbb{A} \mid \sigma(k) = k\}$ forms a subring[3] (resp. subfield) of $\mathbb{A}$. We call $\mathrm{const}_\sigma \mathbb{A}$ the *constant field* of $(\mathbb{A}, \sigma)$. A difference ring (resp. difference field) $(\mathbb{E}, \sigma)$ is a *difference ring extension* (resp. *difference field extension*) of a difference ring (resp. difference field) $(\mathbb{A}, \sigma')$ if $\mathbb{A}$ is a subring (resp. subfield) of $\mathbb{E}$ and $\sigma'(f) = \sigma(f)$ for all $f \in \mathbb{A}$; we call $(\mathbb{A}, \sigma')$ also a sub-difference ring (resp. field) of $(\mathbb{E}, \sigma)$. Since σ and σ' agree on $\mathbb{A}$, we do not distinguish them any longer.

EXAMPLE 3.1. For the rational function field $\mathbb{K}(x)$ we can define uniquely the automorphism $\sigma : \mathbb{K}(x) \to \mathbb{K}(x)$ such that $\sigma(x) = x + 1$ and such that $\sigma(c) = c$ for all $c \in \mathbb{K}$; $(\mathbb{K}(x), \sigma)$ is called the *rational difference field over* $\mathbb{K}$.

EXAMPLE 3.2. For the rational function field $\mathbb{F} := \mathbb{K}(x, x_1, \dots, x_m)$ from Ex. 2.2 we can define uniquely the field automorphism $\sigma : \mathbb{F} \to \mathbb{F}$ such that $\sigma(x) = x + 1$ and $\sigma(x_i) = q_i\, x_i$ for all $1 \leq i \leq m$ and such that $\sigma(c) = c$ for all $c \in \mathbb{K}$. The difference field $(\mathbb{F}, \sigma)$ is also called the *q–mixed difference field over* $\mathbb{K}$.

Then any expression in $\mathrm{ProdSum}(\mathbb{K}(x))$ (resp. in $\mathrm{ProdSum}(\mathbb{K}(x, x_1, \dots, x_m))$) with its shift behavior can be modeled by defining a tower of difference field extensions over $(\mathbb{K}(x), \sigma)$ (resp. of $(\mathbb{K}(x, x_1, \dots, x_m), \sigma)$). Subsequently, we restrict to those extensions in which the constants remain unchanged. We confine ourselves to $\Pi\Sigma^*$-extensions (Sch01) being slightly less general but covering all sums and products treated explicitly in Karr's $\Pi\Sigma$-extensions (Kar85).

[3] Subsequently, we assume that $\mathrm{const}_\sigma \mathbb{A}$ is always a field, which we usually denote by $\mathbb{K}$. Note that this implies that $\mathbb{Q}$ is a subfield of $\mathbb{K}$.

DEFINITION 3.3. A difference field extension $(\mathbb{F}(t),\sigma)$ of $(\mathbb{F},\sigma)$ is called a $\Pi\Sigma^*$-*extension* if both difference fields share the same field of constants, t is transcendental over $\mathbb{F}$, and $\sigma(t) = t + a$ for some $a \in \mathbb{F}^*$ or $\sigma(t) = a\,t$ for some $a \in \mathbb{F}^*$. If $\sigma(t)/t \in \mathbb{F}$ (resp. $\sigma(t) - t \in \mathbb{F}$), we call the extension also a Π-*extension* (resp. Σ^*-*extension*). In short, we say that $(\mathbb{F}(t_1)\cdots(t_e),\sigma)$ is a $\Pi\Sigma^*$-extension (resp. Π-extension, Σ^*-extension) of $(\mathbb{F},\sigma)$ if the extension is given by a tower of $\Pi\Sigma^*$-extensions (resp. Π-extensions, Σ^*-extensions). We call a $\Pi\Sigma^*$-extension $(\mathbb{F}(t_1)\cdots(t_e),\sigma)$ of $(\mathbb{F},\sigma)$ with $\sigma(t_i) = \alpha_i\,t_i + \beta_i$ *generalized d'Alembertian*, or in short *polynomial*, if $\alpha_i \in \mathbb{F}^*$ and $\beta_i \in \mathbb{F}[t_1,\dots,t_{i-1}]$ for all $1 \le i \le e$. A $\Pi\Sigma^*$-*field* $(\mathbb{K}(t_1)\cdots(t_e),\sigma)$ over $\mathbb{K}$ is a $\Pi\Sigma^*$-extension of $(\mathbb{K},\sigma)$ with constant field $\mathbb{K}$.

REMARK 3.4. If $(\mathbb{F}(t_1)\cdots(t_e),\sigma)$ is a polynomial $\Pi\Sigma^*$-extension of $(\mathbb{F},\sigma)$ then it follows that $(\mathbb{F}[t_1]\cdots[t_e],\sigma)$ is a difference ring extension of $(\mathbb{F},\sigma)$.

Karr's approach. The following result from (Kar81) tells us how one can design a $\Pi\Sigma^*$-field for a given product-sum expression.

THEOREM 3.5. *Let $(\mathbb{F}(t),\sigma)$ be a difference field extension of $(\mathbb{F},\sigma)$ with $\sigma(t) = a\,t + f$ where $a \in \mathbb{F}^*$ and $f \in \mathbb{F}$. Then the following holds.*

(1) $(\mathbb{F}(t),\sigma)$ is a Σ^*-extension of $(\mathbb{F},\sigma)$ iff $a = 1$ and there is no $g \in \mathbb{F}$ such that

$$\sigma(g) = g + f. \tag{3.3}$$

(2) $(\mathbb{F}(t),\sigma)$ is a Π-extension of $(\mathbb{F},\sigma)$ iff $t \neq 0$, $f = 0$ and there are no $g \in \mathbb{F}^*$ and $m > 0$ such that $\sigma(g) = a^m g$.

E.g., with Theorem 3.5 it is easy to see that the difference fields from Examples 3.1 and 3.2 are $\Pi\Sigma^*$-fields over $\mathbb{K}$.

From the algorithmic point of view we emphasize the following: For a given $\Pi\Sigma^*$-field $(\mathbb{F},\sigma)$ and $f \in \mathbb{F}$, Karr's summation algorithm (Kar81) can compute a solution $g \in \mathbb{F}$ for the telescoping equation (3.3), or it outputs that such a solution in $\mathbb{F}$ does not exist; for a simplified version see (Sch05c). In this case, we can adjoin a new Σ^*-extension which produces by construction a solution for (3.3).
Summarizing, Karr's algorithm in combination with Theorem 3.5 enables one to construct algorithmically a $\Pi\Sigma^*$-field that encodes the shift behavior of a given indefinite nested sum expression.

EXAMPLE 3.6. We start with the $\Pi\Sigma^*$-field $(\mathbb{Q}(x),\sigma)$ over $\mathbb{Q}$ with $\sigma(x) = x+1$. Now we consider the sum expressions of A in (1.1), say in the order

$$\overset{(1)}{\to} H_n = \sum_{i=1}^{n}\frac{1}{i} \overset{(2)}{\to} S := \sum_{i=1}^{n}\frac{H_i}{i} \overset{(3)}{\to} H_n^{(2)} = \sum_{i=1}^{n}\frac{1}{i^2} \overset{(4)}{\to} T := \sum_{i=1}^{n}\frac{H_i^2 + H_i^{(2)}}{i} \overset{(5)}{\to} A, \tag{3.4}$$

and represent them in terms of Σ^*-extensions following Theorem 3.5.1.

(1) Using, e.g., Gosper's algorithm (Gos78), Karr's algorithm (Kar81) or a simplified version of it presented in (Sch05c), we check that there is no $g \in \mathbb{Q}(x)$ with $\sigma(g) = g + \frac{1}{x+1}$. Hence, by Theorem 3.5.1 we adjoin H_n in the form of the Σ^*-extension $(\mathbb{Q}(x)(h),\sigma)$ of $(\mathbb{Q}(x),\sigma)$ with $\sigma(h) = h + \frac{1}{x+1}$; note that the shift behavior $H_{n+1} = H_n + \frac{1}{n+1}$ is reflected by the automorphism σ.
(2) With the algorithms from (Kar81) or (Sch05c) we show that there is no $g \in \mathbb{Q}(x)(h)$ with $\sigma(g) = g + \frac{\sigma(h)}{x+1}$. Thus we take the Σ^*-extension $(\mathbb{Q}(x)(h)(s),\sigma)$ of $(\mathbb{Q}(x)(h),\sigma)$ with $\sigma(s) = s + \frac{\sigma(h)}{x+1}$ and express S by s.

(3) With the algorithms from above, we find $g = 2s - h^2 \in \mathbb{Q}(x)(h)(s)$ with $\sigma(g) = g + \frac{1}{(x+1)^2}$, and represent[4] $H_n^{(2)}$ by g.
(4) There is no $g \in \mathbb{Q}(x)(h)(s)$ with $\sigma(g) = g + 2\frac{\sigma(s)}{x+1}$; thus we rephrase T as t in the Σ^*-extension $(\mathbb{Q}(x)(h)(s)(t), \sigma)$ of $(\mathbb{Q}(x)(h)(s), \sigma)$ with $\sigma(t) = t + 2\frac{\sigma(s)}{x+1}$.
(5) There is no $g \in \mathbb{Q}(x)(h)(s)(t)$ s.t. $\sigma(g) = g + \frac{\sigma(s+t)}{x+1}$; thus we represent A with a in the Σ^*-ext. $(\mathbb{Q}(x)(h)(s)(t)(a), \sigma)$ of $(\mathbb{Q}(x)(h)(s)(t), \sigma)$ with $\sigma(a) = a + \frac{\sigma(s+t)}{x+1}$.

Reformulating a as a sum expression (for more details see Section 4) yields

$$W = \sum_{r=1}^{n} \frac{\displaystyle\sum_{l=1}^{r} \frac{2\displaystyle\sum_{i=1}^{l} \frac{H_i}{i}}{l} + \sum_{l=1}^{r} \frac{H_l}{l}}{r} \tag{3.5}$$

with $A(n) = W(n)$ for all $n \in \mathbb{N}$.

We remark that the sums occurring in W pop up only in the numerator. Here the following result plays an important role.

THEOREM 3.7 ((Sch09),Thm. 2.7). *Let $(\mathbb{F}(t_1)\cdots(t_e), \sigma)$ be a polynomial $\Pi\Sigma^*$-extension of $(\mathbb{F}, \sigma)$; let $\mathbb{A} = \mathbb{F}[t_1, \dots, t_e]$. For all $g \in \mathbb{A}$, $\sigma(g) - g \in \mathbb{A}$ iff $g \in \mathbb{A}$.*

Namely, if, e.g., A consists only of sums that occur in the numerator, then by solving iteratively the telescoping problem, it is guaranteed that also the telescoping solutions will have only sums that occur in the numerators.

REMARK 3.8. Similar to the sum case, there exist algorithms (Kar81) which can handle the product case; for details and technical problems we refer to (Sch05b). Note that Π-extensions will occur later only in the framework of the general case 4.11. At this point one has explicit control over how the sequence domain $(X, \text{ev}, \mathfrak{d})$ for $\text{Sum}(X)$ is defined.

A depth-refined approach. The depth of W in (3.5) is reflected by the nested depth of the underlying difference field constructed in Example 3.6.

DEFINITION 3.9. Let $(\mathbb{F}, \sigma)$ be a $\Pi\Sigma^*$-field over $\mathbb{K}$ with $\mathbb{F} := \mathbb{K}(t_1)\cdots(t_e)$ where $\sigma(t_i) = a_i\, t_i$ or $\sigma(t_i) = t_i + a_i$ for $1 \leq i \leq e$. The *depth function for elements of* $\mathbb{F}$, $\delta_{\mathbb{K}} : \mathbb{F} \to \mathbb{N}$, is defined as follows.
(1) For any $g \in \mathbb{K}$, $\delta_{\mathbb{K}}(g) := 0$.
(2) If $\delta_{\mathbb{K}}$ is defined for $(\mathbb{K}(t_1)\cdots(t_{i-1}), \sigma)$ with $i > 1$, we define $\delta_{\mathbb{K}}(t_i) := \delta_{\mathbb{K}}(a_i) + 1$; for $g = \frac{g_1}{g_2} \in \mathbb{K}(t_1)\cdots(t_i)$, with $g_1, g_2 \in \mathbb{K}[t_1, \dots, t_i]$ coprime, we define

$$\delta_{\mathbb{K}}(g) := \max(\{\delta_{\mathbb{K}}(t_j) | 1 \leq j \leq i \text{ and } t_j \text{ occurs in } g_1 \text{ or } g_2\} \cup \{0\}).$$

The *extension depth* of a $\Pi\Sigma^*$-extension $(\mathbb{F}(x_1)\cdots(x_r), \sigma)$ of $(\mathbb{F}, \sigma)$ is defined by $\max(\delta_{\mathbb{K}}(x_1), \dots, \delta_{\mathbb{K}}(x_r), 0)$.

EXAMPLE 3.10. In Example 3.6 we have $\delta_{\mathbb{Q}}(x) = 1$, $\delta_{\mathbb{Q}}(h) = 2$, $\delta_{\mathbb{Q}}(s) = 3$, $\delta_{\mathbb{Q}}(t) = 4$, and $\delta_{\mathbb{Q}}(a) = 5$.

---

[4]Note that there is no way to adjoin a Σ^*-extension h_2 of the desired type $\sigma(h_2) = h_2 + 1/(x+1)^2$, since otherwise $\sigma(g - h_2) = (g - h_2)$, i.e., $\text{const}_\sigma \mathbb{Q}(x)(h)(s)(h_2) \neq \mathbb{Q}$.

With the approach sketched in Example 3.6 we obtain an alternative sum representation $W(n)$ for $A(n)$ with larger depth. Motivated by such problematic situations, Karr's $\Pi\Sigma^*$-fields have been refined in the following way; see (Sch05a; Sch08).

DEFINITION 3.11. Let $(\mathbb{F},\sigma)$ be a $\Pi\Sigma^*$-field over $\mathbb{K}$. A difference field extension $(\mathbb{F}(s),\sigma)$ of $(\mathbb{F},\sigma)$ with $\sigma(s)=s+f$ is called a *depth-optimal* Σ^*-extension, in short Σ^δ-extension, if there is no Σ^*-extension $(\mathbb{E},\sigma)$ of $(\mathbb{F},\sigma)$ with extension depth $\leq \delta_{\mathbb{K}}(f)$ such that there is a $g\in\mathbb{E}$ as in (3.3). A $\Pi\Sigma^*$-extension $(\mathbb{F}(t_1)\cdots(t_e),\sigma)$ of $(\mathbb{F},\sigma)$ is depth-optimal, in short a $\Pi\Sigma^\delta$-extension, if all Σ^*-extensions are depth-optimal. A $\Pi\Sigma^\delta$-field consists of Π- and Σ^δ-extensions.

Note that a Σ^δ-extension is a Σ^*-extension by Theorem 3.5.1. Moreover, a $\Pi\Sigma^*$-field $(\mathbb{F},\sigma)$ with depth ≤ 2 and $x\in\mathbb{F}$ such that $\sigma(x)=x+1$ is always depth-optimal; see (Sch08, Prop. 19). *In particular, the rational and the q–mixed difference fields from the Examples 3.1 and 3.2 are $\Pi\Sigma^\delta$-fields over $\mathbb{K}$.*
Given any $\Pi\Sigma^\delta$-field, we obtain the following crucial property which will be essential to solve problem DOS.

THEOREM 3.12 ((Sch08),Result 3). *Let $(\mathbb{F},\sigma)$ be a $\Pi\Sigma^\delta$-field over $\mathbb{K}$. Then for any $f,g\in\mathbb{F}$ such that (3.3) we have*

$$\delta_{\mathbb{K}}(f)\leq\delta_{\mathbb{K}}(g)\leq\delta_{\mathbb{K}}(f)+1. \tag{3.6}$$

In other words, in a given $\Pi\Sigma^\delta$-field we can guarantee that the depth of a telescoping solution is not bigger than the depth of the sum itself.

EXAMPLE 3.13. We consider again the sum expressions in (3.4), but this time we use the refined algorithm presented in (Sch08).
(1) As in Example 3.13 we compute the $\Pi\Sigma^\delta$-field $(\mathbb{Q}(x)(h),\sigma)$ and represent H_n with h. From this point on, our new algorithm works differently.
(2) Given $(\mathbb{Q}(x)(h),\sigma)$, we find the Σ^δ-extension $(\mathbb{Q}(x)(h)(h_2),\sigma)$ of $(\mathbb{Q}(x)(h),\sigma)$ with $\sigma(h_2)=h_2+\frac{1}{(x+1)^2}$ in which we find $s'=\frac{1}{2}(h^2+h_2)$ such that $\sigma(s')-s'=\frac{\sigma(h)}{x+1}$. Hence we represent S by s'.
(3) $H_n^{(2)}$ can be represented by h_2 in the already constructed $\Pi\Sigma^\delta$-field.
(4) Our algorithm finds the Σ^δ-extension $(\mathbb{Q}(x)(h)(h_2)(h_3),\sigma)$ of $(\mathbb{Q}(x)(h)(h_2),\sigma)$ with $\sigma(h_3)=h_3+\frac{1}{(x+1)^3}$ together with $t'=\frac{1}{3}(h^3+3hh_2+2h_3)$ such that $\sigma(t')-t'=\frac{\sigma(h^2+h_2)}{x+1}$; hence we rephrase T as t'.
(5) Finally, we find the Σ^δ-extension $(\mathbb{Q}(x)(h)(h_2)(h_3)(h_4),\sigma)$ of the difference field $(\mathbb{Q}(x)(h)(h_2)(h_3),\sigma)$ with $\sigma(h_4)=h_4+\frac{1}{(x+1)^4}$ and we represent A by $a'=\frac{1}{12}(h^4+2h^3+6(h+1)h_2h+3h_2^2+(8h+4)h_3+6h_4)$ with $\sigma(a')-a'=\frac{\sigma(t'+s')}{x+1}$.
Reinterpreting a' as a sum expression gives B in (1.2); see also Example 2.6.

To sum up, we can compute step by step a $\Pi\Sigma^\delta$-field in which we can represent nested sum expressions. To be more precise, we will exploit the following

THEOREM 3.14 ((Sch08),Result 1). *Let $(\mathbb{F},\sigma)$ be a $\Pi\Sigma^\delta$-field over $\mathbb{K}$ and $f\in\mathbb{F}$.*
(1) There is a Σ^δ-extension $(\mathbb{E},\sigma)$ of $(\mathbb{F},\sigma)$ with $g\in\mathbb{E}$ such that (3.3) holds; $(\mathbb{E},\sigma)$ and g can be given explicitly if $\mathbb{K}$ has the form as stated in Remark 3.15.
(2) Suppose that $(\mathbb{F},\sigma)$ with $\mathbb{F}=\mathbb{G}(y_1,\dots,y_r)$ is a polynomial $\Pi\Sigma^\delta$-extension of $(\mathbb{G},\sigma)$. If $f\in\mathbb{G}[y_1\dots,y_r]$, then $(\mathbb{E},\sigma)$ from part (1) can be given as a polynomial $\Pi\Sigma^\delta$-extension of $(\mathbb{G},\sigma)$; if $\mathbb{E}=\mathbb{F}(t_1,\dots,t_e)$, then $g\in\mathbb{G}[y_1,\dots,y_r][t_1,\dots,t_e]$.

REMARK 3.15. From the point of view of computation certain operations must be carried out in the constant field. For instance, Theorem 3.14 is completely constructive, if $\mathbb{K}$ is of the following from: $\mathbb{K} = \mathbb{A}(q_1, \dots, q_m)$ is a rational function field with variables $q_1, \dots, q_m$ over an algebraic number field $\mathbb{A}$. Due to the limitations of the computer algebra system Mathematica, the implementation in Sigma (Sch07) works only optimal if $\mathbb{A} = \mathbb{Q}$, i.e., $\mathbb{K} = \mathbb{Q}(q_1, \dots, q_m)$.

4. Step II: Reinterpretation as product-sum expressions

Let $(X, \text{ev}, \mathfrak{d})$, e.g., be the q–mixed sequence domain from Example 2.2 with $X = \mathbb{K}(x, x_1, \dots, x_m)$ and let $(\mathbb{F}, \sigma)$ with $X = \mathbb{F}$ be the q–mixed difference field from Example 3.2. Moreover, take $A \in \text{Sum}(X)$.

Then in the previous section we have demonstrated how one can compute a polynomial Σ^{δ}-extension $(\mathbb{F}(t_1) \cdots (t_e), \sigma)$ of $(\mathbb{F}, \sigma)$ in which one can model the shift-behavior of A by an element $a \in \mathbb{F}[t_1, \dots, t_e]$. Then, as illustrated in Example 3.13, we were able to reinterpret a as an element from $B \in \text{Sum}(X)$ such that $\delta_{\mathbb{K}}(a) = \mathfrak{d}(B)$ and such that (1.3) where $\lambda \in \mathbb{N}$ could be given explicitly.

In order to accomplish this task algorithmically, we will supplement the construction of the difference field $(\mathbb{F}(t_1) \cdots (t_e), \sigma)$ by defining in addition an explicitly given difference ring monomorphism. Namely, following (Sch09) we will embed the difference ring $(\mathbb{F}[t_1, \dots, t_e], \sigma)$ into the ring of sequences by a so-called $\mathbb{K}$-monomorphism. It turns out that any element $h \in \mathbb{F}[t_1, \dots, t_e]$ can be mapped injectively to $\langle \text{ev}(H, k) \rangle_{k \geq 0}$ for some properly chosen expression $H \in \text{ProdSum}(X)$.

Subsequently, we define the ring of sequences and $\mathbb{K}$-monomorphisms. Let $\mathbb{K}$ be a field and consider the set of sequences $\mathbb{K}^{\mathbb{N}}$ with elements $\langle a_n \rangle_{n \geq 0} = \langle a_0, a_1, a_2, \dots \rangle$, $a_i \in \mathbb{K}$. With componentwise addition and multiplication we obtain a commutative ring; the field $\mathbb{K}$ can be naturally embedded by identifying $k \in \mathbb{K}$ with the sequence $\langle k, k, k, \dots \rangle$; we write $\mathbf{0} = \langle 0, 0, 0, \dots \rangle$.
We follow the construction from (PWZ96, Sec. 8.2) in order to turn the shift

$$\mathcal{S} : \langle a_0, a_1, a_2, \dots \rangle \mapsto \langle a_1, a_2, a_3, \dots \rangle \tag{4.1}$$

into an automorphism: We define an equivalence relation $\sim$ on $\mathbb{K}^{\mathbb{N}}$ by $\langle a_n \rangle_{n \geq 0} \sim \langle b_n \rangle_{n \geq 0}$ if there exists a $d \geq 0$ such that $a_k = b_k$ for all $k \geq d$. The equivalence classes form a ring which is denoted by $S(\mathbb{K})$; the elements of $S(\mathbb{K})$ (also called germs) will be denoted, as above, by sequence notation. Now it is immediate that $\mathcal{S} : S(\mathbb{K}) \to S(\mathbb{K})$ with (4.1) forms a ring automorphism. The difference ring $(S(\mathbb{K}), \mathcal{S})$ is called the *ring of sequences (over $\mathbb{K}$)*.

A *difference ring homomorphism* $\tau : \mathbb{A}_1 \to \mathbb{A}_2$ between difference rings $(\mathbb{A}_1, \sigma_1)$ and $(\mathbb{A}_2, \sigma_2)$ is a ring homomorphism such that $\tau(\sigma_1(f)) = \sigma_2(\tau(f))$ for all $f \in \mathbb{A}_1$. If τ is injective, we call τ a *difference ring monomorphism.*

Let $(\mathbb{A}, \sigma)$ be a difference ring with constant field $\mathbb{K}$. Then a difference ring homomorphism (resp. difference ring monomorphism) $\tau : \mathbb{A} \to S(\mathbb{K})$ is called a $\mathbb{K}$-*homomorphism* (resp. $\mathbb{K}$-*monomorphism* or $\mathbb{K}$-*embedding*) if for all $c \in \mathbb{K}$ we have that $\tau(c) = \langle c, c, \dots \rangle$.

As mentioned already above, our final goal is to construct a $\mathbb{K}$-monomorphism $\tau : \mathbb{F}[t_1, \dots, t_e] \to S(\mathbb{K})$. For this task we exploit the following property.

If $\tau : \mathbb{A} \to S(\mathbb{K})$ is a $\mathbb{K}$-homomorphism, there is a map $\text{ev} : \mathbb{A} \times \mathbb{N} \to \mathbb{K}$ with

$$\tau(f) = \langle \text{ev}(f, 0), \text{ev}(f, 1), \dots \rangle \tag{4.2}$$

for all $f \in \mathbb{A}$ which has the following properties: For all $c \in \mathbb{K}$ there is a $\delta \geq 0$ with

$$\forall i \geq \delta : \operatorname{ev}(c, i) = c; \tag{4.3}$$

for all $f, g \in \mathbb{A}$ there is a $\delta \geq 0$ with

$$\forall i \geq \delta : \operatorname{ev}(f\,g, i) = \operatorname{ev}(f, i)\operatorname{ev}(g, i), \tag{4.4}$$

$$\forall i \geq \delta : \operatorname{ev}(f + g, i) = \operatorname{ev}(f, i) + \operatorname{ev}(g, i); \tag{4.5}$$

and for all $f \in \mathbb{A}$ and $j \in \mathbb{Z}$ there is a $\delta \geq 0$ with

$$\forall i \geq \delta : \operatorname{ev}(\sigma^j(f), i) = \operatorname{ev}(f, i + j). \tag{4.6}$$

Conversely, if there is a function $\operatorname{ev} : \mathbb{A} \times \mathbb{N} \to \mathbb{K}$ with (4.3), (4.4), (4.5) and (4.6), then the function $\tau : \mathbb{A} \to S(\mathbb{K})$ defined by (4.2) forms a $\mathbb{K}$-homomorphism.

Subsequently, we assume that a $\mathbb{K}$-homomorphism/$\mathbb{K}$-monomorphism is always defined by such a function ev; ev is also called a *defining function of* τ. To take into account the constructive aspects, we introduce the following functions for ev.

DEFINITION 4.1. Let $(\mathbb{A}, \sigma)$ be a difference ring and let $\tau : \mathbb{A} \to S(\mathbb{K})$ be a $\mathbb{K}$-homomorphism given by the defining function ev as in (4.2). ev is called *operation-bounded* by $L : \mathbb{A} \to \mathbb{N}$ if for all $f \in \mathbb{A}$ and $j \in \mathbb{Z}$ with $\delta = \delta(f, j) := L(f) + \max(0, -j)$ we have (4.6) and for all $f, g \in \mathbb{A}$ with $\delta = \delta(f, g) := \max(L(f), L(g))$ we have (4.4) and (4.5); moreover, we require that for all $f \in \mathbb{A}$ and all $j \in \mathbb{Z}$ we have $L(\sigma^j(f)) \leq L(f) + \max(0, -j)$. Such a function is also called an *o-function for* ev. ev is called *zero-bounded* by $Z : \mathbb{A} \to \mathbb{N}$ if for all $f \in \mathbb{A}^*$ and all $i \geq Z(f)$ we have $\operatorname{ev}(f, i) \neq 0$; such a function is also called *z-function for* ev.

EXAMPLE 4.2. Given the $\Pi\Sigma^\delta$-field $(\mathbb{K}(x), \sigma)$ over $\mathbb{K}$ with $\sigma(x) = x + 1$, we obtain a $\mathbb{K}$-homomorphism $\tau : \mathbb{K}(x) \to S(\mathbb{K})$ by taking the defining function (2.1); here we assume that $f = \frac{p}{q} \in \mathbb{K}(x)$ with $p \in \mathbb{K}[x]$ and $q \in \mathbb{K}[x]^*$ are coprime. For the o-function $L(f)$ we take the minimal non-negative integer l such that $q(k+l) \neq 0$ for all $k \in \mathbb{N}$, and as z-function we take $Z(f) = L(p\,q)$. Note: Since $p(x)$ and $q(x)$ have only finitely many roots, $\tau(\frac{p}{q}) = \mathbf{0}$ iff $\frac{p}{q} = 0$. Hence τ is injective.

Summarizing, the $\Pi\Sigma^\delta$-field $(\mathbb{K}(x), \sigma)$ with $\sigma(x) = x + 1$ can be embedded into $(S(\mathbb{K}), \mathcal{S})$. More generally, if $(\mathbb{F}, \sigma)$ is the q–mixed difference field, $\tau : \mathbb{F} \to S(\mathbb{K})$ with the defining function ev given in (2.2) is a $\mathbb{K}$-monomorphism. In addition, there are a computable o-function L and a computable z-function Z for ev; for details we refer to (Sch09, Cor. 4.10) which relies on (BP99).

EXAMPLE 4.3. Take the rational difference field $(\mathbb{K}(x), \sigma)$ and the $\mathbb{K}$-monomorphism τ with defining function ev and the o-function L from Example 4.2 and consider the Σ^δ-extension $(\mathbb{K}(x)(h), \sigma)$ of $(\mathbb{K}(x), \sigma)$ with $\sigma(h) = h + \frac{1}{x+1}$. We get a $\mathbb{K}$-homomorphism $\tau' : \mathbb{K}(x)[h] \to S(\mathbb{K})$ where the defining function ev' is given by $\operatorname{ev}'(h, k) = H_k$ and

$$\operatorname{ev}'(\sum_{i=0}^{d} f_i h^i, k) = \sum_{i=0}^{d} \operatorname{ev}(f_i, k)\operatorname{ev}'(h, k)^i.$$

As o-function we can take $L'(\sum_{i=0}^{d} f_i h^i) = \max(L(f_i) | 0 \leq i \leq d)$. Now suppose that τ' is not injective. Then we can take $f = \sum_{i=0}^{d} f_i h^i \in \mathbb{K}(x)[h] \setminus \{0\}$ with

$\deg(f) = d$ minimal such that $\tau'(f) = \mathbf{0}$. Since τ is injective, $f \notin \mathbb{K}(x)$ Define $g := \sigma(f_d)f - f_d\sigma(f) \in \mathbb{K}(x)[h]$. Note that $\deg(g) < d$ by construction. Moreover,

$$\tau'(g) = \tau(\sigma(f_d))\tau'(f) - \tau(f_d)\tau'(\sigma(f)).$$

Since $\tau'(f) = \mathbf{0}$ by assumption and $\tau'(\sigma(f)) = \mathcal{S}(\tau'(f)) = \mathcal{S}(\mathbf{0}) = \mathbf{0}$, it follows that $\tau'(g) = \mathbf{0}$. By the minimality of $\deg(f)$, $g = 0$, i.e., $\sigma(f_d)f - f_d\sigma(f) = 0$, or equivalently, $\frac{\sigma(f)}{f} = \frac{\sigma(f_d)}{f_d} \in \mathbb{K}(x)$. As $f \notin \mathbb{K}(x)$, this contradicts (Kar81, Theorem 4).

EXAMPLE 4.4. Take the rational difference field $(\mathbb{K}(x), \sigma)$ and the $\mathbb{K}$-monomorphism τ with defining function ev and o-function L from Example 4.2, and consider the Π-extension $(\mathbb{K}(x)(b), \sigma)$ of $(\mathbb{K}(x), \sigma)$ with $\sigma(b) = \frac{x+1}{2(2x+1)}b$. We get a $\mathbb{K}$-homomorphism $\tau' : \mathbb{K}(x)[b] \to S(\mathbb{K})$ with its defining function specified by $\text{ev}'(b, k) = \prod_{i=1}^{k} \frac{i}{2(2i-1)} = \binom{2k}{k}^{-1}$ and (4.8) where $t := b$; note that $\tau'(b)$ has no zero entries by construction. We take $L'(\sum_{i=0}^{d} f_i b^i) = \max(L(f_i)|0 \leq i \leq d))$ as o-function. By similar arguments as in Ex. 4.3 it follows that τ is injective.

More generally, we arrive at the following result; see (Sch09) for a detailed proof.

LEMMA 4.5. *Let $(\mathbb{F}(t_1) \cdots (t_e)(t), \sigma)$ be a polynomial $\Pi\Sigma^*$-extension of $(\mathbb{F}, \sigma)$ with $\mathbb{K} := \text{const}_\sigma\mathbb{F}$ and $\sigma(t) = \alpha\, t + \beta$. Let $\tau : \mathbb{F}[t_1] \cdots [t_e] \to S(\mathbb{K})$ be a $\mathbb{K}$-monomorphism with a defining function* ev *as in* (4.2)*; let L be an o-function for* ev *and let Z be a z-function for* $\text{ev}|_{\mathbb{F}}$ *(*ev *is restricted on $\mathbb{F}$). Then:*

(1) There is a $\mathbb{K}$-monomorphism $\tau' : \mathbb{F}[t_1] \cdots [t_e][t] \to S(\mathbb{K})$ with a defining function ev' *such that* $\text{ev}'|_{\mathbb{F}[t_1,\dots,t_e]} = \text{ev}$*; if $\beta = 0$,* $\text{ev}'(t, k) \neq 0$ *for all $k \geq r$ for some $r \in \mathbb{N}$. Such a τ' is uniquely determined by*

$$\text{ev}'(t, k) = \begin{cases} c\displaystyle\prod_{i=r}^{k} \text{ev}(\alpha, i-1) & \text{if } \sigma(t) = \alpha\, t \\ \displaystyle\sum_{i=r}^{k} \text{ev}(\beta, i-1) + c & \text{if } \sigma(t) = t + \beta, \end{cases} \tag{4.7}$$

up to the choice of $r \in \mathbb{N}$ and $c \in \mathbb{K}$; we require $c \neq 0$ if $\beta = 0$.

(2) Fixing (4.7) *we obtain, e.g., the following defining function for τ':*

$$\text{ev}'(\sum_{i=0}^{d} f_i t^i, k) := \sum_{i=0}^{d} \text{ev}(f_i, k)\text{ev}'(t, k)^i \quad \forall k \in \mathbb{N}. \tag{4.8}$$

(3) In particular, there is an o-function L' for ev' *with $L'|_{\mathbb{F}[t_1,\dots,t_e]} = L$; if L and Z are computable, L' can be computed. We can choose (as a constructive example)*

$$r = \begin{cases} \max(L(\alpha), Z(\alpha)) + 1 & \text{if } \sigma(t) = \alpha\, t \\ L(\beta) + 1 & \text{if } \sigma(t) = t + \beta. \end{cases} \tag{4.9}$$

REMARK 4.6. Let $(\mathbb{F}(t_1) \cdots (t_e), \sigma)$ be a polynomial $\Pi\Sigma^*$-extension of $(\mathbb{F}, \sigma)$ with $\mathbb{K} = \text{const}_\sigma\mathbb{F}$ and let $\tau : \mathbb{F}[t_1] \cdots [t_e] \to S(\mathbb{K})$ be a $\mathbb{K}$-homomorphism with a defining function ev. Then there is implicitly a z-function for $\text{ev}|_{\mathbb{F}}$; see (Sch09, Lemma 4.3).

Applying Lemma 4.5 iteratively produces the following result.

THEOREM 4.7. *Let* $(\mathbb{F}(y_1)\cdots(y_r)(t_1)\cdots(t_e),\sigma)$ *be a polynomial* $\Pi\Sigma^*$*-extension of* $(\mathbb{F},\sigma)$ *with* $\mathbb{K}:=\mathrm{const}_\sigma\mathbb{F}$*; let* $\tau:\mathbb{A}\to S(\mathbb{K})$ *with* $\mathbb{A}=\mathbb{F}[y_1,\dots,y_r]$ *be a* $\mathbb{K}$*-embedding with a defining function* (4.2) *and with an o-function* L.
Then there is a $\mathbb{K}$*-embedding* $\tau':\mathbb{A}[t_1]\cdots[t_e]\to S(\mathbb{K})$ *with a defining function* ev' *and with an o-function* L' *such that* $\mathrm{ev}'|_{\mathbb{A}}=\mathrm{ev}$ *and* $L'|_{\mathbb{A}}=L$.

This construction that leads in Theorem 4.7 to ev' is called *canonical* if it is performed iteratively as described in (4.7) and (4.8) of Lemma 4.5. Note that any defining function ev' with (4.7) evaluates as (4.8) if k is chosen big enough; in our canonical construction we assume that (4.8) holds for all $k\geq 0$.

Since τ' in Lemma 4.5 is uniquely determined by (4.7) the following holds.

THEOREM 4.8. *Let* $(\mathbb{F},\sigma)$ *be a* $\Pi\Sigma^*$*-field over* $\mathbb{K}$ *and let* $(\mathbb{F}(y_1)\cdots(y_r),\sigma)$ *be a polynomial* $\Pi\Sigma^*$*-extension of* $(\mathbb{F},\sigma)$. *Let* $\tau:\mathbb{F}[y_1,\dots,y_r]\to S(\mathbb{K})$ *be a* $\mathbb{K}$*-embedding with a defining function* ev *together with an o-function for* ev. *Take the measured sequence domain* $(\mathbb{F},\mathrm{ev},\delta_{\mathbb{K}})$. *Then for any* $f\in\mathbb{F}[y_1,\dots,y_r]$, *there is an* $F\in$ ProdSum$(\mathbb{F})$ *such that* $\tau(f)=\langle F(k)\rangle_{k\geq 0}$ *and* $\mathfrak{d}(F)=\delta_{\mathbb{K}}(F)$.

Summarizing, given such a difference ring $(\mathbb{F}[y_1,\dots,y_r],\sigma)$ and $\mathbb{K}$-monomorphism, one can rephrase the elements of $\mathbb{F}[y_1,\dots,y_r]$ as expressions from ProdSum(X) such that the depth of both domains are identical.

EXAMPLE 4.9. Take the $\Pi\Sigma^\delta$-field $(\mathbb{Q}(x),\sigma)$ with $\sigma(x)=x+1$ together with the $\mathbb{Q}$-embedding $\tau:\mathbb{Q}(x)\to S(\mathbb{Q})$ with defining function (2.1) as carried out in Ex. 4.2 ($\mathbb{K}=\mathbb{Q}$); let $(\mathrm{Sum}(\mathbb{Q}(x)),\mathrm{ev},\mathfrak{d})$ be the sum sequence domain over $\mathbb{Q}(x)$. Moreover, consider the $\Pi\Sigma^\delta$-field $(\mathbb{Q}(x)(h)(h_2)(h_3)(h_4),\sigma)$ from Ex. 3.13. Then we can construct the $\mathbb{Q}$-embedding $\tau':\mathbb{Q}(x)[h,h_2,h_3,h_4]\to S(\mathbb{Q})$ with the defining function ev' which is canonically given by $\mathrm{ev}'|_{\mathbb{Q}(x)}=\mathrm{ev}$ and by

$$\begin{aligned}\mathrm{ev}'(h,k)&=\mathrm{ev}(\mathrm{Sum}(1,\tfrac{1}{x}),k)=H_k,\\ \mathrm{ev}'(h_j,k)&=\mathrm{ev}(\mathrm{Sum}(1,\tfrac{1}{x^j}),k)=H_k^{(j)}\ \text{for } j\in\{2,3,4\}.\end{aligned}\tag{4.10}$$

Note that $\mathfrak{d}(\mathrm{Sum}(1,\frac{1}{x}))=\delta_{\mathbb{Q}}(h)$ and $\mathfrak{d}(\mathrm{Sum}(1,\frac{1}{x^j}))=\delta_{\mathbb{Q}}(h_j)$ for $j\in\{2,3,4\}$.

Recall that we want to solve problem DOS for a measured sequence domain $(X,\mathrm{ev},\mathfrak{d})$. In Section 5 we shall solve this problem for the following general setting. Loosely speaking, the terms of indefinite nested sums and products X are modeled by polynomials from $\mathbb{F}[y_1,\dots,y_r]$ and the reinterpretation of the corresponding product-sum expressions is accomplished by its $\mathbb{K}$-monomorphism from $\mathbb{F}[y_1,\dots,y_r]$ into the ring of sequences; in particular, the depth of such a product-sum expression is equal to the depth of the corresponding polynomial from $\mathbb{F}[y_1,\dots,y_r]$.

EXAMPLE 4.10. We start as in Example 4.9, but now we take the Π-extension $(\mathbb{F}(y_1)\cdots(y_r),\sigma)$ of $(\mathbb{F},\sigma)$ such that $\alpha_j=\frac{\sigma(y_j)}{y_j}\in\mathbb{F}$ for $1\leq j\leq r$. Then we can extend the $\mathbb{K}$-embedding τ to $\tau':\mathbb{F}[y_1,\dots,y_r]\to S(\mathbb{K})$ with the defining function ev' canonically given by $\mathrm{ev}'|_{\mathbb{F}}=\mathrm{ev}$ and

$$\mathrm{ev}'(y_j,k)=c_j\prod_{i=r_j}^{k}\alpha_j(i)$$

for all $1\leq j\leq r$ with $r_j\geq Z(\alpha_j)$ and $c_j\in\mathbb{K}^*$. Note: with $F_j=c_j\mathrm{Prod}(r_j,\alpha_j)$ we have $F_j(k)=\mathrm{ev}'(y_j,k)$ and $\mathfrak{d}(F_j)=\delta_{\mathbb{K}}(y_j)$. Moreover, we can model a finite set of hypergeometric terms in the sequence domain $(X,\mathrm{ev}',\delta_{\mathbb{K}})$ with $X:=\mathbb{F}[y_1,\dots,y_r]$.

Similarly, we are in the position to handle q–hypergeometric sequences or mixed hypergeometric sequences. More generally, we can handle the following case.

GENERAL CASE 4.11. **The ground field**[a]**.** Let $(\mathbb{F}, \sigma)$ be a $\Pi\Sigma^\delta$-field over $\mathbb{K}$, let $\tau_0 : \mathbb{F} \to S(\mathbb{K})$ be a $\mathbb{K}$-embedding with a defining function $\mathrm{ev}_0 : \mathbb{F} \times \mathbb{N} \to \mathbb{K}$, and let $L_0 : \mathbb{F} \to \mathbb{N}$ be an o-function and $Z : \mathbb{F} \to \mathbb{N}$ be a z-function for ev_0; moreover, consider the sequence domain $(\mathbb{F}, \mathrm{ev}_0, \delta_\mathbb{K})$.
A polynomial extension. In addition, choose a polynomial $\Pi\Sigma^\delta$-extension $(\mathbb{F}(y_1)\cdots(y_r), \sigma)$ of $(\mathbb{F}, \sigma)$ and set $X := \mathbb{F}[y_1, \dots, y_r]$. Then extend the $\mathbb{K}$-embedding τ_0 to $\tau : X \to S(\mathbb{K})$ by extending the defining function ev_0 canonically to $\mathrm{ev} : X \times \mathbb{N} \to \mathbb{K}$ and by extending the o-function L_0 to L following Lemma 4.5; if L_0 is computable, also L is computable. By construction it follows that for $1 \leq i \leq r$ there exist $F_i \in \mathrm{ProdSum}(\mathbb{F})$ such that

$$\mathrm{ev}(y_i, k) = F_i(k) \quad \forall k \geq 0 \quad \text{and } \mathfrak{d}(F_i) = \delta_\mathbb{K}(y_i). \tag{4.11}$$

In particular, for each $f \in X$, one gets explicitly an $F \in \mathrm{ProdSum}(X)$ such that $\mathrm{ev}(f, k) = F(k)$ for all $k \in \mathbb{N}$ and such that $\mathfrak{d}(F) = \delta_\mathbb{K}(f)$.
The sequence domain. We obtain the sequence domain $(X, \mathrm{ev}, \delta_\mathbb{K})$ which models the product-sum expressions (4.11) with the depth given by $\delta_\mathbb{K}$.

[a]E.g., we can take the q–mixed difference field $(\mathbb{F}, \sigma)$ with $\mathbb{F} = \mathbb{K}(x, x_1, \dots, x_m)$ from Ex. 3.2, and we can take the $\mathbb{K}$-embedding $\tau_0 : \mathbb{F} \to S(\mathbb{K})$ where $\mathrm{ev}_0 := \mathrm{ev}$ is defined as in (2.2); note that the measured sequence domain $(\mathbb{F}, \mathrm{ev}_0, \delta_\mathbb{K})$ has been presented in Ex. 2.2. From the point of view of computation we assume that $\mathbb{K}$ is of the form as stated in Remark 3.15

5. Combining the steps: Finding optimal nested sum representations

E.g., for the q–mixed sequence domain $(\mathbb{F}, \mathrm{ev}, \mathfrak{d})$ from Ex. 2.2 with $X = \mathbb{F} = \mathbb{K}(x, x_1, \dots, x_m)$ we will solve problem DOS for $A \in \mathrm{Sum}(\mathbb{F})$ as follows; here we assume that $\mathbb{K}$ is of the form as stated in Remark 3.15

Take the q–mixed difference field $(\mathbb{F}, \sigma)$ over $\mathbb{K}$ with the automorphism σ defined in Example 3.2. Moreover, take the the $\mathbb{K}$-embedding $\tau : \mathbb{F} \to S(\mathbb{K})$ with the defining function ev given in (2.2), and choose a computable o-function L and a computable z-function Z for ev. Then by Theorem 5.1 below the following construction can be carried out algorithmically.

Step I: Reduction to a $\Pi\Sigma^\delta$-field. Given $A \in \mathrm{Sum}(\mathbb{F})$, construct a polynomial Σ^δ-extension $(\mathbb{F}(s_1)\cdots(s_u), \sigma)$ of $(\mathbb{F}, \sigma)$ and extend the $\mathbb{K}$-monomorphism τ to $\tau' : \mathbb{F}[s_1, \dots, s_u] \to S(\mathbb{K})$ with a defining function ev' such that the following additional property holds: We can take explicitly an $a \in \mathbb{F}[s_1, \dots, s_u]$ and a $\lambda \in \mathbb{N}$ such that

$$\mathrm{ev}'(a, k) = A(k) \quad \forall k \geq \lambda \quad \text{and}^5\ \delta_\mathbb{K}(a) \leq \mathfrak{d}(A). \tag{5.1}$$

Note that we rely on the fact that all our sums are represented in $\Pi\Sigma^\delta$-fields; for general $\Pi\Sigma^*$-fields $\delta_\mathbb{K}(a)$ might be bigger than $\mathfrak{d}(A)$, see Example 3.6.

Step II: Reinterpretation as a product-sum expression. In particular, by the concrete construction of the $\mathbb{K}$-monomorphism based on the iterative application of Lemma 4.5, construct a $B \in \mathrm{Sum}(\mathbb{F})$ such that

$$\mathrm{ev}'(a, k) = B(k) \quad \forall k \geq 0 \quad \text{and } \delta_\mathbb{K}(a) = \mathfrak{d}(B). \tag{5.2}$$

[4]Recall that $A(k) = \mathrm{ev}(A, k)$; here ev is the evaluation function of the sequence domain $(X, \mathrm{ev}, \mathfrak{d})$ where $X = \mathbb{F}$ (or $X = \mathbb{F}[y_1, \dots, y_r]$ as defined in the general case 4.11).

Then due to the properties of the $\Pi\Sigma^\delta$-field and the fact that τ' is a $\mathbb{K}$-monomorphism (in particular, that τ' is injective), we will show in Theorem 5.5 that the depth of $B \in \text{Sum}(\mathbb{F})$ is $\text{Sum}(\mathbb{F})$-optimal, i.e., B together with λ are a solution of problem DOS.

We will solve problem DOS for the general case 4.11 by applying exactly the same mechanism as sketched above.

THEOREM 5.1. *Let $(\mathbb{F}(y_1)\cdots(y_r),\sigma)$ with $X = \mathbb{F}[y_1,\dots,y_r]$ be a $\Pi\Sigma^\delta$-field over $\mathbb{K}$, let $\tau : X \to S(\mathbb{K})$ be $\mathbb{K}$-embedding with* ev, *L and Z, and let $(X, \text{ev}, \delta_{\mathbb{K}})$ be a sequence domain as stated in the general case 4.11; in particular let $(\text{Sum}(X), \text{ev}, \mathfrak{d})$ be the sum sequence domain over X. Then for any $A \in \text{Sum}(X)$ there is a Σ^δ-extension $D := (\mathbb{F}(y_1)\cdots(y_r)(s_1)\cdots(s_u),\sigma)$ of $(\mathbb{F}(y_1)\cdots(y_r),\sigma)$, where D is a polynomial extension of $(\mathbb{F},\sigma)$, and there is a $\mathbb{K}$-embedding*

$$\tau' : X[s_1,\dots,s_u] \to S(\mathbb{K}) \tag{5.3}$$

where the defining function ev$'$ *and its o-function L' are extended from X to $X[s_1,\dots,s_u]$, with the following property: There are $\lambda \in \mathbb{N}$ and $a \in \mathbb{A}$ such that* (5.1)*; in particular, for any $h \in X[s_1,\dots,s_u]$ there is an $H \in \text{Sum}(X)$ such that*

$$\text{ev}'(h,k) = H(k) \quad \forall k \geq 0 \quad \textit{and } \delta_{\mathbb{K}}(h) = \mathfrak{d}(H). \tag{5.4}$$

This extension, the defining function ev$'$ *for τ', λ, and a can be given explicitly, if L and Z are computable and if $\mathbb{K}$ has the form as stated in Remark 3.15.*

PROOF. We show the theorem by induction on the depth. If $A \in \text{Sum}(X)$ with $\mathfrak{d}(A) = 0$, then $A \in \mathbb{K}$ and the statement clearly holds. Now suppose that we have shown the statement for expressions with depth$\leq d$ and take $A \in \text{Sum}(X)$ with $\mathfrak{d}(A) = d+1$. Let $A_1,\dots,A_l$ be exactly those subexpressions of A which do not occur inside of a sum and which cannot be split further by $\oplus$ and $\otimes$, i.e., for $1 \leq i \leq l$, either $A_i \in X$ or A_i is a sum. First we consider A_1. If $A_1 \in X$, then for $r_1 = 0$, $a_1 = A_1$, $\text{ev}' = \text{ev}$ and $i = 1$ we have

$$\text{ev}'(a_i,k) = A_i(k) \quad \forall k \geq r_i, \quad \text{and } \delta_{\mathbb{K}}(a_i) \leq \mathfrak{d}(A_i). \tag{5.5}$$

Moreover, the property (5.4) for any $h \in X$ holds by choosing $H := h$.
Otherwise, $A_1 = \text{Sum}(\lambda_1, F_1)$ for some $\lambda_1 \in \mathbb{N}$ and $F_1 \in \text{Sum}(X)$ with $\mathfrak{d}(F_1) \leq d$. If $\mathfrak{d}(A_1) \leq d$, we get by induction a Σ^δ-extension $D := (\mathbb{F}(y_1)\cdots(y_r)(s_1)\cdots(s_u),\sigma)$ of $(\mathbb{F}(y_1)\cdots(y_r),\sigma)$ where D is a polynomial extension of $(\mathbb{F},\sigma)$; moreover we can extend τ to a $\mathbb{K}$-embedding $\tau' : \mathbb{D} \to S(\mathbb{K})$ with $\mathbb{D} := X[s_1,\dots,s_u]$ where its defining function ev$'$ is extended from X to $\mathbb{D}$, and we can extend the o-function L to an o-function L' for ev$'$ such that the following holds: There are $a_1 \in \mathbb{D}$ and $r_1 \in \mathbb{N}$ such that for $i = 1$ we have (5.5); in particular, for any $h \in \mathbb{D}$, there is an $H \in \text{Sum}(X)$ such that (5.4).
If $\mathfrak{d}(A_1) = d+1$, we can take by the same reasoning such an extension D of $(\mathbb{F},\sigma)$ with $\mathbb{D} := X[s_1,\dots,s_u]$ and τ' with a defining function ev$'$ together with an o-function L' in which we can take $f_1 \in \mathbb{D}$ and $l_1 \in \mathbb{N}$ such that

$$\text{ev}'(f_1,k) = F_1(k) \quad \forall k \geq l_1, \quad \text{and } \delta_{\mathbb{K}}(f_1) \leq \mathfrak{d}(F_1).$$

By Theorem 3.14 take a Σ^δ-extension $E := (\mathbb{F}(y_1)\cdots(y_r)(s_1)\cdots(s_u)(t_1)\cdots(t_v),\sigma)$ of $(\mathbb{F}(y_1)\cdots(y_r)(s_1)\cdots(s_u),\sigma)$ such that E is a polynomial extension of $(\mathbb{F},\sigma)$ and in which we have $g \in \mathbb{E}$ with $\mathbb{E} := \mathbb{D}[t_1, \dots, t_v]$ such that

$$\sigma(g) = g + \sigma(f_1).$$

Moreover, by iterative application of Lemma 4.5 we can extend the $\mathbb{K}$-embedding τ' from $\mathbb{D}$ to a $\mathbb{K}$-embedding $\tau' : \mathbb{E} \to S(\mathbb{K})$ by extending the defining function ev′ canonically from $\mathbb{D}$ to $\mathbb{E}$, and we can extend L' to an o-function for ev′; note that this construction can be performed such that for any $h \in \mathbb{E}$ there is $H \in \mathrm{Sum}(X)$ with (5.4). Now take[6] $r_1 := \max(l_1, L'(f_1), L'(g)+1)$ and define $c := \sum_{k=l_1}^{r_1-1} F(k) - \mathrm{ev}'(g, r_1 - 1) \in \mathbb{K}$. Then for all $n \geq r_1$,

$$\begin{aligned}\mathrm{ev}'(g+c,n) &= \mathrm{ev}(\sigma^{-1}(g) + f_1, n) + c = \mathrm{ev}'(g, n-1) + \mathrm{ev}'(f_1, n) + c \\ &= \mathrm{ev}'(g, n-1) + F_1(n) + c = \cdots = \mathrm{ev}'(g, r_1 - 1) + \sum_{k=r_1}^{n} F_1(k) + c = A_1(n).\end{aligned}$$

Set $a_1 := g + c \in \mathbb{E}$. Then (5.5) for $i = 1$. Moreover, since $\delta_{\mathbb{K}}(g) \leq \delta_{\mathbb{K}}(\sigma(f_1)) + 1$ by Thm. 3.12,

$$\delta_{\mathbb{K}}(a_1) = \delta_{\mathbb{K}}(g) \leq \delta_{\mathbb{K}}(\sigma(f_1)) + 1 = \delta_{\mathbb{K}}(f_1) + 1 \leq \mathfrak{d}(F_1) + 1 \leq d + 1.$$

We continue to consider $A_2, \dots, A_l$ and finally arrive at a polynomial Σ^{δ}-extension $(\mathbb{F}(y_1)\cdots(y_r)(s_1)\cdots(s_u)(t_1)\cdots(t_{v'}), \sigma)$ of $(\mathbb{F}, \sigma)$; during this construction we can extend τ to a $\mathbb{K}$-embedding $\tau' : \mathbb{A} \to S(\mathbb{K})$ with $\mathbb{A} := X[s_1, \dots, s_u][t_1, \dots, t_{v'}]$ and with a defining function ev′ and we can extend L to an o-function L' for ev′ such that the following holds. We can take $a_1, \dots, a_l \in \mathbb{A}$ and $r_1, \dots, r_l \in \mathbb{N}$ such that for $1 \leq i \leq l$ we have (5.5); in particular, for any $h \in \mathbb{A}$, there is an $H \in \mathrm{Sum}(X)$ such that (5.4).
Finally, we construct $a \in \mathbb{A}$ by applying in the given A the substitution $A_i \to a_i$ for all $1 \leq i \leq l$ and by replacing $\otimes$ and $\oplus$ with the field operations $\cdot$ and $+$, respectively. Then it follows that (5.1) for $\lambda := \max(r_1, \dots, r_l, L'(a_1), \dots L'(a_l))$. This completes the induction step. Note that all the construction steps can be carried out by algorithms if L and Z are computable and if $\mathbb{K}$ has the form as stated in Remark 3.15. In particular, ev′, a and λ can be given explicitly. ☐

EXAMPLE 5.2. For the input sum (1.1) the presented procedure in Theorem 5.1 carries out simultaneously the constructions from Examples 3.13 and 4.9: We obtain the $\Pi\Sigma^{\delta}$-field $(\mathbb{Q}(x)(h)(h_2)(h_3)(h_4), \sigma)$ over $\mathbb{Q}$ together with the $\mathbb{Q}$-embedding $\tau' : \mathbb{Q}(x)[h, h_2, h_3, h_4] \to S(\mathbb{Q})$ where the defining function ev′ is canonically given by $\mathrm{ev}'|_{\mathbb{Q}(x)} = \mathrm{ev}$ for ev as in (2.1) and by (4.10). By construction, we can link the sums in (3.4) with $h, s', h_2, t', a' \in \mathbb{Q}(x)[h, h_2, h_3, h_4]$ from Example 3.13: for $n \in \mathbb{N}$,

$$\begin{aligned} &\mathrm{ev}'(h,n) = H_n, \quad \mathrm{ev}'(s',n) = S(n), \\ &\mathrm{ev}'(h_2,n) = H_n^{(2)}, \quad \mathrm{ev}'(t',n) = T(n), \quad \mathrm{ev}'(a',n) = A(n). \end{aligned} \tag{5.6}$$

The depths of $H_k, S, H_k^{(2)}, T, A$ are $2, 3, 2, 3, 4$, respectively. The corresponding depths $\delta_{\mathbb{Q}}(h) = \delta_{\mathbb{Q}}(s') = \delta_{\mathbb{Q}}(h_2) = \delta_{\mathbb{Q}}(t') = \delta_{\mathbb{Q}}(a') = 2$ in the $\Pi\Sigma^{\delta}$-field are the same or have been improved. Using (4.10) we can reinterpret, e.g, s' as the sum expression $F := \frac{1}{2}(\mathrm{Sum}(1, 1/x)^2 + \mathrm{Sum}(1, 1/x^2)) \in \mathrm{Sum}(\mathbb{Q}(x))$ with $F(n) = S(n)$ for all $n \geq 0$ and $\mathfrak{d}(F) = \delta_{\mathbb{Q}}(s')$; this leads to the identity $s(n) = \frac{1}{2}(H_n^2 + H_n^{(2)})$. In

[6]Remark: If f_1 is a hypergeometric term, r_1 can be also obtained by analyzing only f_1 (without knowing the telescoping solution g); for more details see (AP05).

the same way we obtain sum expressions in $\mathrm{Sum}(\mathbb{Q}(x))$ for the $\Pi\Sigma^\delta$-field elements h, h_2, t', a' in (5.6), and we arrive at the following identities. For $n \geq 0$,

$$\begin{aligned} H_n &= H_n, \quad s(n) = \tfrac{1}{2}(H_n^2 + H_n^{(2)}), \\ H_n^{(2)} &= H_n^{(2)}, \quad T(n) = \tfrac{1}{3}\left(H_n^3 + 3H_n^{(2)}H_n + 2H_n^{(3)}\right), \quad A(n) = B(n) \end{aligned}$$

where B is given as in (1.2).

We remark that this translation mechanism presented in Theorem 5.1 is implemented in the summation package Sigma. Namely, given the general case 4.11 ($\mathbb{K}$ as stated in Remark 3.15) and given[7] $A \in \mathrm{Sum}(X)$, Sigma computes the following ingredients:

- A Σ^δ-extension $(\mathbb{F}(y_1)\cdots(y_r)(s_1)\cdots(s_u),\sigma)$ of $(\mathbb{F}(y_1)\cdots(y_r),\sigma)$ and a $\mathbb{K}$-embedding (5.3) with a defining function ev' and o-function L'.
- $a \in X[s_1,\dots,s_u]$ and $\lambda \in \mathbb{N}$ such that (5.1).
- $\tau'(a)$ is given explicitly by a[8] $B \in \mathrm{Sum}(X)$ such that (5.2).

Then Sigma outputs for a given $A \in \mathrm{Sum}(X)$ the result $B \in \mathrm{Sum}(X)$ with λ.

The final goal is to prove Theorem 5.5 which guarantees that the output B is indeed a solution of problem DOS. We start with the following lemma.

LEMMA 5.3. *Let $(\mathbb{F}(y_1)\cdots(y_r),\sigma)$ be a $\Pi\Sigma^\delta$-field over $\mathbb{K}$ and let τ be a $\mathbb{K}$-monomorphism with* ev *and L as in the general case 4.11; let $(\mathbb{F}(t_1)\cdots(t_e),\sigma)$ be a polynomial Σ^*-extension of $(\mathbb{F},\sigma)$ and let $\rho : \mathbb{F}[t_1,\dots,t_e] \to S(\mathbb{K})$ be a $\mathbb{K}$-embedding with a defining function ev_ρ and an o-function L_ρ such that $\mathrm{ev}_\rho|_{\mathbb{F}} = \mathrm{ev}$.*
Then there is a Σ^*-extension $D := (\mathbb{F}(y_1)\cdots(y_r)(z_1)\cdots(z_l),\sigma)$ of $(\mathbb{F}(y_1)\cdots(y_r),\sigma)$ where D is a polynomial Σ^*-extension of $(\mathbb{F},\sigma)$ with the following properties:

(1) There is a difference monomorphism $\phi : \mathbb{F}(t_1,\dots,t_e) \to \mathbb{F}(y_1,\dots,y_r)(z_1,\dots,z_l)$ such that for all $a \in \mathbb{F}(t_1,\dots,t_e)$,

$$\delta_{\mathbb{K}}(\phi(a)) \leq \delta_{\mathbb{K}}(a) \tag{5.7}$$

and such that for all $a \in \mathbb{F}[t_1,\dots,t_e]$,

$$\phi(a) \in \mathbb{F}[y_1,\dots,y_r][z_1,\dots,z_l]. \tag{5.8}$$

(2) There is a $\mathbb{K}$-embedding $\tau' : \mathbb{F}[y_1,\dots,y_r][z_1,\dots,z_l] \to S(\mathbb{K})$ with a defining function ev' *and an o-function L' such that $\mathrm{ev}'|_{\mathbb{F}[y_1,\dots,y_r]} = \mathrm{ev}$ and such that for all $a \in \mathbb{F}[t_1,\dots,t_e]$,*

$$\tau'(\phi(a)) = \rho(a). \tag{5.9}$$

PROOF. The base case $e = 0$ holds with $\phi(a) = a$ for all $a \in \mathbb{F}$ and $\tau' := \rho$. Suppose the lemma holds for ϵ extensions $(\mathbb{H},\sigma)$ with $\mathbb{H} = \mathbb{F}(t_1)\cdots(t_e)$ and let $(\mathbb{D},\sigma)$ with $\mathbb{D} := \mathbb{F}(y_1)\dots(y_r)(z_1)\cdots(z_l)$, τ' with ev' and L', ρ with ev_ρ, and ϕ as stated above; set $\mathbb{E} = \mathbb{F}[y_1,\dots,y_r,z_1,\dots,z_l]$. Now let $(\mathbb{H}(t),\sigma)$ be a Σ^*-extension of $(\mathbb{H},\sigma)$ with $f := \sigma(t) - t \in \mathbb{F}[t_1\dots,t_e]$, and take a $\mathbb{K}$-embedding

[7] In Sigma A is inserted, e.g., in the form (1.1) without using evaluation functions like (2.1) or (2.2); this implies that the lower bounds of the involved sums and products must be chosen in such a way that no zeros occur in the denominators during any evaluation.

[8] In Sigma the lower bounds of the sums and products are computed by (4.9). Looking closer at this construction, no zeros occur in the involved denominators of B when performing the evaluation $B(n)$ for $n \geq \lambda$. Hence the output can be returned, e.g., in the form like (1.1) or (1.2) which is free of any explicit evaluation functions like (2.1) or (2.2).

$\rho' : \mathbb{F}[t_1, \dots, t_e][t] \to S(\mathbb{K})$ with a defining function ρ' and an o-function $L_{\rho'}$ such that $\mathrm{ev}_{\rho'}|_{\mathbb{F}[t_1,\dots,t_e]} = \mathrm{ev}_\rho$.

Case 1: If there is no $g \in \mathbb{D}$ such that

$$\sigma(g) - g = \phi(f), \tag{5.10}$$

we can take the Σ^*-extension $(\mathbb{D}(y), \sigma)$ of $(\mathbb{D}, \sigma)$ with $\sigma(y) = y + \phi(f)$ by Theorem 3.5.1 and we can define a difference field monomorphism $\phi' : \mathbb{H}(t) \to \mathbb{E}(y)$ such that $\phi'(a) = \phi(a)$ for all $a \in \mathbb{H}$ and such that $\phi'(t) = y$. By construction, $\delta_{\mathbb{K}}(y) = \delta_{\mathbb{K}}(\phi(f)) + 1$. Since

$$\delta_{\mathbb{K}}(\phi(f)) + 1 \leq \delta_{\mathbb{K}}(f) + 1 = \delta_{\mathbb{K}}(t), \tag{5.11}$$

$\delta_{\mathbb{K}}(\phi(a)) \leq \delta_{\mathbb{K}}(a)$ for all $a \in \mathbb{H}(t)$. Moreover, since $\phi(f) \in \mathbb{E}$, it follows that $(\mathbb{E}(y), \sigma)$ is a polynomial extension of $(\mathbb{F}, \sigma)$. Moreover, for all $a \in \mathbb{F}[t_1, \dots, t_e, t]$, $\phi(a) \in \mathbb{E}[y]$. This proves part (1).

Now we extend the $\mathbb{K}$-embedding τ' from $\mathbb{E}$ to $\tau' : \mathbb{E}[y] \to S(\mathbb{K})$ with the defining function ev', where $\mathrm{ev}'(f, k) = \mathrm{ev}(f, k)$ for all $f \in \mathbb{E}$ and $\mathrm{ev}'(y, k)$ is defined as in the right hand side of (4.7); here we choose $\beta = \phi(f)$ and $c = \mathrm{ev}_{\rho'}(t, r-1)$ for some $r \in \mathbb{N}$ properly chosen. In particular, we can extend the o-function L' for our extended ev' by Lemma 4.5 (note that there is a z function for ev' restricted to $\mathbb{F}$ by Remark 4.6). Then for all $k \geq r$ (r is chosen big enough with L'_ρ and L'),

$$\begin{aligned} \mathrm{ev}'(\phi'(t), k) = \mathrm{ev}'(y, k) &= \sum_{i=r}^{k} \mathrm{ev}'(\phi(f), i-1) + \mathrm{ev}_{\rho'}(t, r-1) \\ &= \sum_{i=r}^{k} \mathrm{ev}_\rho(f, i-1) + \mathrm{ev}_{\rho'}(t, r-1) = \sum_{i=r+1}^{k} \mathrm{ev}_\rho(f, i-1) + h(r) \end{aligned} \tag{5.12}$$

with $h(r) = \mathrm{ev}_\rho(f, r-1) + \mathrm{ev}_{\rho'}(t, r-1) = \mathrm{ev}_{\rho'}(f + t, r-1) = \mathrm{ev}_{\rho'}(\sigma(t), r-1) = \mathrm{ev}_{\rho'}(t, r)$. Applying this reduction $k - r + 1$ times shows that

$$\mathrm{ev}'(\phi'(t), k) = \sum_{i=r+1}^{k} \mathrm{ev}_\rho(f, i-1) + \mathrm{ev}_{\rho'}(t, r) = \dots = \mathrm{ev}_{\rho'}(t, k).$$

Hence $\tau'(\phi'(t)) = \rho'(t)$, and thus $\tau'(\phi'(a)) = \rho'(a)$ for all $a \in \mathbb{F}[t_1 \dots, t_e][t]$.

Case 2: Otherwise, if there is a $g \in \mathbb{D}$ s.t. (5.10), then $g \in \mathbb{E}$ by Theorem 3.7. In particular, $\delta_{\mathbb{K}}(g) \leq \delta_{\mathbb{K}}(\phi(f)) + 1$ by Theorem 3.12. With (5.11), it follows that

$$\delta_{\mathbb{K}}(g) \leq \delta_{\mathbb{K}}(t). \tag{5.13}$$

Now observe that

$$\mathcal{S}(\rho'(t)) - \rho'(t) = \rho'(\sigma(t) - t) = \rho(f) = \tau'(\phi(f)) = \tau'(\sigma(g) - g) = \mathcal{S}(\tau'(g)) - \tau'(g).$$

Hence $\mathcal{S}(\rho'(t) - \tau'(g)) = \rho'(t) - \tau'(g)$, i.e., $\rho'(t) - \tau'(g)$ is a constant in $S(\mathbb{K})$. Thus, $\rho'(t) = \tau'(g) + \langle c \rangle_{n \geq 0}$ for some $c \in \mathbb{K}$. Since $(\phi(\mathbb{H})(g), \sigma)$ is a difference field (it is a sub-difference field of $(\mathbb{D}, \sigma)$), g is transcendental over $\phi(\mathbb{H})$ by Theorem 3.5.1. In particular, we can define the difference field monomorphism $\phi' : \mathbb{H}(t) \to \mathbb{D}$ with $\phi'(a) = \phi(a)$ for all $a \in \mathbb{H}$ and $\phi'(t) = g + c$. Since $g \in \mathbb{E}$, $\phi'(t) \in \mathbb{E}$, and therefore $\phi(a) \in \mathbb{E}$ for all $a \in \mathbb{F}[t_1, \dots, t_e][t]$. With (5.13) and our induction assumption it follows that $\delta_{\mathbb{K}}(\tau'(a)) \leq \delta_{\mathbb{K}}(a)$ for all $a \in \mathbb{H}(t)$. This proves part (1). Note that $\tau'(\phi'(t)) = \tau'(g) + c = \rho'(t)$ by construction. Hence, $\tau'(\phi'(a)) = \rho(a)$ for all $a \in \mathbb{F}[t_1 \dots, t_e][t]$. This proves part (2) and completes the induction step. □

Note that the proof of Lemma 5.3 is constructive, if the underlying o- and z-functions of τ are computable and if $\mathbb{K}$ is given as in Remark 3.15. For simplicity, ingredients like r_1 (needed, e.g., for (5.12)) have not been specified explicitly.

In Theorem 5.4 we can show the following: The nested depth of an element in a $\Pi\Sigma^{\delta}$-field over $\mathbb{K}$ is smaller than or equal to the depth of an element in a $\Pi\Sigma^{*}$-field over $\mathbb{K}$ provided that both elements can be mapped to the same sequence $\boldsymbol{s} \in S(\mathbb{K})$ by appropriate $\mathbb{K}$-monomorphisms.

THEOREM 5.4. *Let $(\mathbb{F}(y_1)\cdots(y_r),\sigma)$ be a $\Pi\Sigma^{\delta}$-field over $\mathbb{K}$ and let τ be a $\mathbb{K}$-embedding with* ev *and L as in the general case 4.11. Let $(\mathbb{F}(t_1)\cdots(t_e),\sigma)$ be a polynomial Σ^{*}-extension of $(\mathbb{F},\sigma)$ and let $\rho:\mathbb{F}[t_1,\dots,t_e]\to S(\mathbb{K})$ be a $\mathbb{K}$-embedding with a defining function ev_{ρ} and an o-function for ev_{ρ}. Then for any $\boldsymbol{s} \in \tau(\mathbb{F}[y_1,\dots,y_r])\cap\rho(\mathbb{F}[t_1,\dots,t_e])$ we have*

$$\delta_{\mathbb{K}}(\tau^{-1}(\boldsymbol{s})) \le \delta_{\mathbb{K}}(\rho^{-1}(\boldsymbol{s})).$$

PROOF. Take a Σ^{*}-extension $(\mathbb{F}(y_1)\cdots(y_r)(z_1)\cdots(z_l),\sigma)$ of $(\mathbb{F}(y_1)\cdots(y_r),\sigma)$, a monomorphism $\phi:\mathbb{F}(t_1)\cdots(t_e)\to\mathbb{F}(y_1)\cdots(y_r)(z_1)\cdots(z_l)$ s.t. (5.7) and (5.8) for all $a\in\mathbb{F}[t_1,\dots,t_e]$, and take a $\mathbb{K}$-embedding $\tau':\mathbb{F}[y_1,\dots,y_r,z_1,\dots,z_l]\to S(\mathbb{K})$ with $\tau'|_{\mathbb{F}[y_1,\dots,y_r]}=\tau$ such that (5.9) for all $a\in\mathbb{F}[t_1,\dots,t_e]$; this is possible by Lemma 5.3. Now let $\boldsymbol{s}\in\tau(\mathbb{F}[y_1,\dots,y_r])\cap\rho(\mathbb{F}[t_1,\dots,t_e])$, and set $f:=(\tau')^{-1}(\boldsymbol{s})=\tau^{-1}(\boldsymbol{s})\in\mathbb{F}[y_1,\dots,y_r]$ and $g:=\rho^{-1}(\boldsymbol{s})\in\mathbb{F}[t_1\dots,t_e]$. To this end, define $g':=\phi(g)\in\mathbb{F}[y_1,\dots,y_r,z_1,\dots,z_l]$. Then by (5.7) we have $\delta_{\mathbb{K}}(g')\le\delta_{\mathbb{K}}(g)$. Since $\tau'(g')=\rho(g)=\boldsymbol{s}$ and $\tau'(f)=\boldsymbol{s}$, and since τ' is injective, $g'=f$. Thus, $\delta_{\mathbb{K}}(f)=\delta_{\mathbb{K}}(g')\le\delta_{\mathbb{K}}(g)$. □

Finally, Theorem 5.5 shows that the constructed $B\in\mathrm{Sum}(X)$ with (5.2) is indeed a solution of problem DOS.

THEOREM 5.5. *Let $(\mathbb{F}(y_1)\cdots(y_r),\sigma)$ with $X=\mathbb{F}[y_1,\dots,y_r]$ be a $\Pi\Sigma^{\delta}$-field over $\mathbb{K}$, let $\tau:X\to S(\mathbb{K})$ be $\mathbb{K}$-embedding with* ev *and L, and let $(X,\mathrm{ev},\delta_{\mathbb{K}})$ be a sequence domain as stated in the general case 4.11; let $D:=(\mathbb{F}(y_1)\cdots(y_r)(s_1)\cdots(s_u),\sigma)$ be a Σ^{δ}-extension of $(\mathbb{F}(y_1)\cdots(y_r),\sigma)$ where D is a polynomial extension of $(\mathbb{F},\sigma)$ and let (5.3) be a $\mathbb{K}$-embedding with a defining function and o-function. Moreover, let $A\in\mathrm{Sum}(X)$ and $a\in\mathbb{A}$ such that $\tau(a)=\langle A(k)\rangle_{k\ge0}$. Then $\delta_{\mathbb{K}}(a)$ is the* $\mathrm{Sum}(X)$*-optimal depth of A.*

PROOF. Take an expression of $\mathrm{Sum}(X)$ that produces $\boldsymbol{s}:=\langle A(k)\rangle_{k\ge0}$ from a certain point on and that has minimal depth, say d. By Theorem 5.1 we can take a Σ^{δ}-extension $D:=(\mathbb{F}(y_1)\cdots(y_r)(t_1)\cdots(t_e),\sigma)$ of $(\mathbb{F}(y_1)\cdots(y_r),\sigma)$ such that D is a polynomial extension of $(\mathbb{F},\sigma)$ and we can assume that there is a $\mathbb{K}$-embedding $\rho:X[t_1,\dots,t_e]\to S(\mathbb{K})$ with a defining function and an o-function with the following property. There is an $a'\in X[t_1,\dots,t_e]$ such that $\rho(a')=\boldsymbol{s}$ and $\delta_{\mathbb{K}}(a')\le d$. By Theorem 5.4 (applied twice), $\delta_{\mathbb{K}}(a)=\delta_{\mathbb{K}}(a')$. Moreover, $\rho(a')$ and $\tau(a)$ can be defined by elements from $\mathrm{Sum}(X)$ with depth $\delta_{\mathbb{K}}(a)$ by Theorem 4.8. Since d is minimal, $\delta_{\mathbb{K}}(a)=d$. □

6. Application: Simplification of d'Alembertian solutions

The d'Alembertian solutions (Nör24; AP94; Sch01), a subclass of Liouvillian solutions (HS99), of a given recurrence relation are computed by factorizing the recurrence into linear right hand factors as much as possible. Given this factorization, one can read off the d'Alembertian solutions which are of the form

$$h(n) \sum_{k_1=c_1}^{n} b_1(k_1) \sum_{k_2=c_2}^{k_2} b_2(k_2) \cdots \sum_{k_s=c_s}^{k_{s-1}} b_s(k_s) \tag{6.1}$$

for lower bounds $c_1, \dots, c_s \in \mathbb{N}$; here the $b_i(k_i)$ and $h(n)$ are given by the objects form the coefficients of the recurrence or by products over such elements. Note that such solutions can be represented in $\Pi\Sigma^\delta$-fields if the occurring products can be rephrased accordingly in Π-extensions. Then, applying our refined algorithms to such solutions (6.1), we can find sum representations with minimal nested depth. Typical examples can be found, e.g., in (DPSW06; Sch07; MS07; BBKS08; OS09; BKKS09a; BKKS09b).

In the following we present two examples with detailed computation steps that have been provided by the summation package Sigma.

6.1. An example from particle physics. In massive higher order calculations of Feynman diagrams (BBKS07) the following task of simplification arose. Find an alternative sum expression of the definite sum

$$S(n) = \sum_{i=1}^{\infty} \frac{H_{i+n}^2}{i^2} \tag{6.2}$$

such that the parameter n does not occur inside of any summation quantifier and such that the arising sums are as simple as possible. In order to accomplish this task, Sigma computes in a first step the recurrence relation

$$\begin{aligned}
&-(n+2)(n+1)^3\left(n^2+7n+16\right)S(n)\\
&\quad+(n+2)\left(5n^5+62n^4+318n^3+814n^2+1045n+540\right)S(n+1)\\
&\quad-2\left(5n^6+84n^5+603n^4+2354n^3+5270n^2+6430n+3350\right)S(n+2)\\
&\quad+2\left(5n^6+96n^5+783n^4+3478n^3+8906n^2+12530n+7610\right)S(n+3)\\
&\quad-(n+4)\left(5n^5+88n^4+630n^3+2318n^2+4453n+3642\right)S(n+4)\\
&\quad+(n+4)(n+5)^3\left(n^2+5n+10\right)S(n+5)=-\frac{4(n+7)}{(n+3)(n+4)}H_n\\
&\qquad-\frac{2\left(2n^7+35n^6+235n^5+718n^4+824n^3-283n^2-859n+10\right)}{(n+1)(n+2)(n+3)^2(n+4)^2(n+5)}
\end{aligned}$$

by a generalized version (Sch07) of Zeilberger's creative telescoping (Zei91). Given this recurrence, Sigma computes the d'Alembertian solutions

$$A_1 = 1, \quad A_2 = H_n, \quad A_3 = H_n^2,$$

$$A_4 = \sum_{i=2}^{n} \frac{\displaystyle\sum_{j=2}^{i} \frac{(2j-1)\displaystyle\sum_{k=1}^{j} \frac{1}{(2k-3)(2k-1)}}{(j-1)j}}{i},$$

$$A_5 = \sum_{i=3}^{n} \frac{\displaystyle\sum_{j=3}^{i} \frac{(2j-1)\displaystyle\sum_{k=3}^{j} \frac{2(k-2)(k-1)kH_k - (2k-1)\left(3k^2-6k+2\right)}{(k-2)(k-1)k(2k-3)(2k-1)}}{(j-1)j}}{i},$$

$$B = \sum_{i=4}^{n} \frac{\displaystyle\sum_{j=4}^{i} \frac{(2j-1)\displaystyle\sum_{k=4}^{j} \frac{\displaystyle\sum_{l=4}^{k} \frac{(2l-3)\left(l^2-3l+6\right)\tilde{B}(l)}{(l-3)(l-2)(l-1)l}}{(2k-3)(2k-1)}}{(j-1)j}}{i}$$

where

$$\tilde{B} = \sum_{r=3}^{l} -\frac{2\left(2r^6-27r^5+117r^4-254r^3+398r^2+2(r-3)(r-2)(r-1)(r+2)H_r r-446r+204\right)}{(r-2)(r-1)r(r^2-5r+10)(r^2-3r+6)}.$$

To be more precise, $A_i \in \mathrm{Sum}_n(\mathbb{Q}(x))$, $1 \le i \le 5$, are the five linearly independent solutions of the homogeneous version of the recurrence, and $B \in \mathrm{Sum}_n(\mathbb{Q}(x))$ is one particular solution of the recurrence itself; the depths of $A_1, \dots, A_5, B$ are $0, 2, 2, 4, 5, 7$, respectively. As a consequence, we obtain the general solution

$$G := B + c_1 A_1 + c_2 A_2 + c_3 A_3 + c_4 A_4 + c_5 A_5 \tag{6.3}$$

for constants c_i. Checking initial values shows that we have to choose[9]

$$c_1 = \tfrac{17}{10}\zeta_2^2,\ c_2 = \tfrac{1}{12}(48\zeta_3 - 67),\ c_3 = \tfrac{31}{12},\ c_4 = \tfrac{1}{4}(23 - 8\zeta_2),\ c_5 = -\tfrac{1}{2}$$

in order to match (6.3) with $S(n) = G(n)$ for all $n \in \mathbb{N}$.

Finally, Sigma simplifies the derived expressions further and finds sum representations with minimal nested depth (see problem DOS). Following the approach described in the previous sections, it computes the $\Pi\Sigma^\delta$-field $(\mathbb{Q}(x)(h)(h_2)(h_4)(H), \sigma)$ with $\sigma(x) = x + 1$ and

$$\sigma(h) = h + \tfrac{1}{x+1},\ \sigma(h_2) = h_2 + \tfrac{1}{(x+1)^2},\ \sigma(h_4) = h_4 + \tfrac{1}{(x+1)^4},\ \sigma(H) = H + \tfrac{\sigma(h)}{(x+1)^2};$$

in addition, it delivers a $\mathbb{Q}$-embedding $\tau : \mathbb{Q}(x)[h, h_2, h_4, H] \to S(\mathbb{Q})$ with the defining function $\mathrm{ev} : \mathbb{Q}(x)[h, h_2, h_4, H] \times \mathbb{N} \to \mathbb{Q}$ canonically given by (2.1) for all $f \in \mathbb{Q}(x)$ and by

$$\mathrm{ev}(h, n) = H_n, \quad \mathrm{ev}(h_2, n) = H_n^{(2)}, \quad \mathrm{ev}(h_4, n) = H_n^{(4)}, \quad \mathrm{ev}(H, n) = \sum_{k=1}^{n} \frac{H_k}{k^2}.$$

Moreover, it finds

$$a_1 = 1, \quad a_2 = h, \quad a_3 = h^2, \quad a_4 = \frac{1}{2}\left(h_2 - h^2\right), \quad a_5 = \frac{1}{2}\left(-h^2 + 2h_2\, h - h\right),$$
$$b = \frac{1}{24}\left(h^2 - 48hH + 128h - 12h_2^2 + (12h - 69)\, h_2 - 12h_4\right)$$

[9] ζ_k denotes the Riemann zeta function at k; e.g., $\zeta_2 = \pi^2/6$.

such that $\mathrm{ev}'(a_i, n) = A_i(n)$ for $1 \le i \le 5$ and $\mathrm{ev}'(b, n) = B(n)$. These computations lead to the following identities: For $n \ge 0$,
(6.4)

$$A_1(n) = 1, \quad A_2(n) = H_n, \quad A_3(n) = H_n^2, \quad A_4(n) = \frac{1}{2}\left(H_n^{(2)} - H_n^2\right),$$
$$A_5(n) = \frac{1}{2}\left(-H_n^2 + 2H_n^{(2)} H_n - H_n\right),$$
$$B(n) = \tfrac{1}{24} H_n^2 - 2H_n \sum_{k=1}^{n} \frac{H_k}{k^2} + \tfrac{16}{3} H_n - \tfrac{1}{2}\left(H_n^{(2)}\right)^2 + \left(\tfrac{1}{2} H_n - \tfrac{69}{24}\right) H_n^{(2)} - \tfrac{1}{2} H_n^{(4)};$$

in particular, due to Theorem 5.5, the sum expressions on the right–hand sides of (6.4) have the minimal depths 0, 2, 2, 2, 2, 3, respectively. To this end, we obtain the following identity (BBKS07, equ. 3.14): for $n \ge 0$,

$$\sum_{i=1}^{\infty} \frac{H_{i+n}^2}{i^2} = \frac{17}{10}\zeta_2^2 + 4H_n\zeta_3 + H_n^2\zeta_2 - H_n^{(2)}\zeta_2 - \frac{1}{2}\left(\left(H_n^{(2)}\right)^2 + H_n^{(4)}\right) - 2H_n \sum_{k=1}^{n} \frac{H_k}{k^2}.$$

6.2. A nontrivial harmonic sum identity. We look for an indefinite nested sum representation of the definite sum

$$S(n) = \sum_{k=0}^{n} \binom{n}{k}^2 H_k^2; \tag{6.5}$$

for similar problems see (DPSW06). First, Sigma finds with creative telescoping the recurrence relation

$$\begin{aligned}
&8(n+1)(2n+1)^3\left(64n^4 + 480n^3 + 1332n^2 + 1621n + 735\right) S(n) - 4(768n^8 + 8832n^7 \\
&+ 43056n^6 + 115708n^5 + 186452n^4 + 183201n^3 + 106442n^2 + 33460n + 4533) S(n+1) \\
&+ 2(n+2)(384n^7 + 4224n^6 + 18968n^5 + 44610n^4 + 58679n^3 + 42775n^2 + 16084n + 2616) S(n+2) \\
&- (n+2)(n+3)^3\left(64n^4 + 224n^3 + 276n^2 + 141n + 30\right) S(n+3) = \\
&- 3\left(576n^6 + 4896n^5 + 16660n^4 + 28761n^3 + 26171n^2 + 11574n + 1854\right).
\end{aligned}$$

Solving the recurrence in terms of d'Alembertian solutions and checking initial values yield the identity

$$S(n) = \binom{2n}{n}\left(\frac{1}{2} A_1(n) - \frac{19}{28} A_2(n) + B(n)\right) \quad \forall n \ge 0$$

with

$$A_1 = \sum_{i=1}^{n} \frac{4i-3}{i(2i-1)}, \quad A_2 = \sum_{i=2}^{n} \frac{(4i-3)\sum_{j=2}^{i} \frac{64j^4 - 288j^3 + 468j^2 - 323j + 84}{(j-1)j(2j-3)(4j-7)(4j-3)}}{i(2i-1)},$$

$$B = -\sum_{i=2}^{n} \frac{(4i-3)\sum_{j=2}^{i} \frac{\left(64j^4 - 288j^3 + 468j^2 - 323j + 84\right)\tilde{B}(j)}{(j-1)j(2j-3)(4j-7)(4j-3)}}{i(2i-1)}$$

where

$$\tilde{B} = \sum_{k=1}^{j} -\frac{3(2k-3)(2k-1)(4k-7)\left(576k^6 - 5472k^5 + 20980k^4 - 41559k^3 + 44882k^2 - 25113k + 5760\right)}{k(64k^4 - 544k^3 + 1716k^2 - 2379k + 1227)(64k^4 - 288k^3 + 468k^2 - 323k + 84)\binom{2k}{k}}.$$

So far, this alternative sum representation of (6.5) might not be considered as really convincing. For further simplifications, we construct the $\Pi\Sigma^{\delta}$-field $(\mathbb{Q}(x)(b), \sigma)$ with $\sigma(x) = x+1$ and $\sigma(b) = \frac{x+1}{2(2x+1)}b$, and we take the $\mathbb{Q}$-monomorphism $\tau' : \mathbb{Q}(x)[b] \to S(\mathbb{Q})$ from Example 4.4 ($\mathbb{K} = \mathbb{Q}$). Note that the sums $A_1, A_2, B \in \mathrm{Sum}_n(\mathbb{Q}(x)[b])$ have the depths 2, 3, 5, respectively. Finally, activating our machinery in this setting (we represent the sums in a $\Pi\Sigma^{\delta}$-field and reinterpret the result by an appropriate $\mathbb{Q}$-monomorphism), we arrive at the following identities: For $n \geq 0$,

$$A_1(n) = 2\left(2H_n - H_{2n}\right),$$

$$A_2(n) = 2\left(4H_n^2 + 4H_n + H_{2n}^2 + (-4H_n - 2)\,H_{2n} - H_{2n}^{(2)}\right),$$

$$B(n) = \frac{3}{14}\left(44H_n^2 + 16H_n + 11H_{2n}^2 - (44H_n + 8)\,H_{2n} - 11H_{2n}^{(2)} + 14\sum_{i=1}^{n}\frac{1}{i^2\binom{2i}{i}}\right).$$

By Theorem 5.5 we have solved problem DOS for the expressions A_1, A_2 and B, i.e., we found sum representations with optimal depths 2, 2, 3, respectively. Finally, this leads to the following identity: for $n \geq 0$,

$$\sum_{k=0}^{n}\binom{n}{k}^2 H_k^2 = \binom{2n}{n}\left(4H_n^2 - 4H_{2n}H_n + H_{2n}^2 - H_{2n}^{(2)} + 3\sum_{i=1}^{n}\frac{1}{i^2\binom{2i}{i}}\right).$$

References

[AP94] S.A. Abramov and M. Petkovšek, *D'Alembertian solutions of linear differential and difference equations*, Proc. ISSAC'94 (J. von zur Gathen, ed.), ACM Press, 1994, pp. 169–174.

[AP05] ———, *Gosper's algorithm, accurate summation and the discrete Newton–Leibniz formula*, Proc. ISSAC'05 (M. Kauers, ed.), ACM, 2005, pp. 5–12 (english).

[BBKS07] I. Bierenbaum, J. Blümlein, S. Klein, and C. Schneider, *Difference equations in massive higher order calculations*, Proc. ACAT 2007, vol. PoS(ACAT)082, 2007, [arXiv:hep-ph/0707.4659].

[BBKS08] I. Bierenbaum, J. Blümlein, S. Klein, and C. Schneider, *Two–loop massive operator matrix elements for unpolarized heavy flavor production to $o(\epsilon)$*, Nucl.Phys. B **803** (2008), no. 1-2, 1–41, [arXiv:hep-ph/0803.0273].

[BK99] J. Blümlein and S. Kurth, *Harmonic sums and Mellin transforms up to two-loop order*, Phys. Rev. **D60** (1999).

[BKKS09a] J. Blümlein, M. Kauers, S. Klein, and C. Schneider, *Determining the closed forms of the $O(a_s^3)$ anomalous dimensions and Wilson coefficients from Mellin moments by means of computer algebra.* To appear in Comput. Phys. Comm. (2009), [arXiv:hep-ph/0902.4091].

[BKKS09b] ———, *From moments to functions in quantum chromodynamics*, Proc. ACAT 2008, vol. PoS(ACAT08)106, 2009, [arXiv:hep-ph/0902.4095].

[BL82] B. Buchberger and R. Loos, *Algebraic Simplification*, Computer Algebra – Symbolic and Algebraic Computation (B. Buchberger, G. E. Collins, and R. Loos, eds.), Springer, Vienna - New York, 1982, pp. 11–43.

[BP99] A. Bauer and M. Petkovšek, *Multibasic and mixed hypergeometric Gosper-type algorithms*, J. Symbolic Comput. **28** (1999), no. 4–5, 711–736.

[DPSW06] K. Driver, H. Prodinger, C. Schneider, and J.A.C. Weideman, *Padé approximations to the logarithm III: Alternative methods and additional results*, Ramanujan J. **12** (2006), no. 3, 299–314.

[Gos78] R.W. Gosper, *Decision procedures for indefinite hypergeometric summation*, Proc. Nat. Acad. Sci. U.S.A. **75** (1978), 40–42.

[HS99] P.A. Hendriks and M.F. Singer, *Solving difference equations in finite terms*, J. Symbolic Comput. **27** (1999), no. 3, 239–259.

[Kar81] M. Karr, *Summation in finite terms*, J. ACM **28** (1981), 305–350.

[Kar85] ———, *Theory of summation in finite terms*, J. Symbolic Comput. **1** (1985), 303–315.

[KS06] M. Kauers and C. Schneider, *Indefinite summation with unspecified summands*, Discrete Math. **306** (2006), no. 17, 2021–2140.

[MS07] S. Moch and C. Schneider, *Feynman integrals and difference equations*, Proc. ACAT 2007, vol. PoS(ACAT)083, 2007, [arXiv:hep-ph/0709.1769].

[MUW02] S. Moch, P. Uwer, and S. Weinzierl, *Nested sums, expansion of transcendental functions, and multiscale multiloop integrals*, J. Math. Phys. **6** (2002), 3363–3386.

[Nör24] N.E. Nörlund, *Differenzenrechnung*, Springer, Berlin, 1924.

[NP97] I. Nemes and P. Paule, *A canonical form guide to symbolic summation*, Advances in the Design of Symbolic Computation Systems (A. Miola and M. Temperini, eds.), Texts Monogr. Symbol. Comput., Springer, Wien-New York, 1997, pp. 84–110.

[OS09] R. Osburn and C. Schneider, *Gaussian hypergeometric series and extensions of supercongruences*, Math. Comp. **78** (2009), no. 265, 275–292.

[PR97] P. Paule and A. Riese, *A Mathematica q–analogue of Zeilberger's algorithm based on an algebraically motivated aproach to q–hypergeometric telescoping*, Special Functions, q-Series and Related Topics (M. Ismail and M. Rahman, eds.), vol. 14, Fields Institute Toronto, AMS, 1997, pp. 179–210.

[PWZ96] M. Petkovšek, H. S. Wilf, and D. Zeilberger, $a = b$, A. K. Peters, Wellesley, MA, 1996.

[Sch01] C. Schneider, *Symbolic summation in difference fields*, Ph.D. thesis, RISC-Linz, J. Kepler University, Linz, May 2001.

[Sch05a] ———, *Finding telescopers with minimal depth for indefinite nested sum and product expressions*, Proc. ISSAC'05 (M. Kauers, ed.), ACM, 2005, pp. 285–292 (english).

[Sch05b] ———, *Product representations in* $\Pi\Sigma$*-fields*, Ann. Comb. **9** (2005), no. 1, 75–99.

[Sch05c] ———, *Solving parameterized linear difference equations in terms of indefinite nested sums and products*, J. Differ. Equations Appl. **11** (2005), no. 9, 799–821.

[Sch07] ———, *Symbolic summation assists combinatorics*, Sém. Lothar. Combin. **56** (2007), 1–36, Article B56b.

[Sch08] ______, *A refined difference field theory for symbolic summation*, J. Symbolic Comput. **43** (2008), no. 9, 611–644, [arXiv:0808.2543v1].

[Sch09] ______, *Parameterized telescoping proves algebraic independence of sums*, To appear in Ann. Comb. (2009), [arXiv:0808.2596].

[Ver99] J.A.M. Vermaseren, *Harmonic sums, Mellin transforms and integrals*, Int. J. Mod. Phys. **A14** (1999), 2037–2976.

[Zei91] D. Zeilberger, *The method of creative telescoping*, J. Symbolic Comput. **11** (1991), 195–204.

Research Institute for Symbolic Computation, Johannes Kepler University, A-4040 Linz, Austria

E-mail address: Carsten.Schneider@risc.uni-linz.ac.at

Clay Mathematics Proceedings
Volume 12, 2010

Logarithmic Structures and TQFT

Simon Scott

Abstract. This is an expository article on logarithmic structures on semigroups and categories. Characters of logarithmic representations of semigroups coincide with a number of fundamental topological and spectral invariants. An area of current research is the extension of this construction to TQFT, which incorporates a further idea of sewing together invariants, requiring an extension of the notion of abstract logarithms to non-linear maps between categories. The motivating ideas and the basic definition for such a structure are given here along with some examples.

Contents

2010 *Mathematics Subject Classification.* Primary 58J52; Secondary 18D10, 18G55, 19D55, 57R56, 58J22, 58J40.

Key words and phrases. logarithmic structures, trace, determinant structure, TQFT, K-theory, category, cobordism.

1. The categorization of analysis

An interesting idea that has been impulsed in mathematics, at least in part, by the study of the structures that drive quantum field theory is the need to understand certain familiar mathematical concepts in rather more abstract contexts. For example, cyclic homology frees up de Rham cohomology from the commutative algebra of smooth functions on a manifold to general noncommutative algebras. On the other hand, topological quantum field theory suggests the essential structures of geometric analysis and index theory need to be understood in the way that they relate to the construction of representations of the cobordism category.

Such categorization may be seen, perhaps, as commencing with Grothendieck's idea of motives as a proposal for a universal cohomology theory (encompassing various competing cohomology theories in algebraic geometry). Likewise, the use of n-categories (or ∞-categories) as a multi-layer approach to representations of categories is a quite old idea. But in both cases these structures have in recent times become the focus of much work in homotopy theory and geometry, in topological quantum field theory and its spin-offs into geometric analysis.

The purpose of this paper is to outline how the analytical concept of a 'determinant' and, more fundamentally, a 'logarithm' might abstract itself naturally into this discussion. Logarithmic structures appear to lie behind many basic topological invariants and it is of interest to determine what the logarithmic structures are in topological quantum field theory, corresponding in principle to certain 'higher' sewing formulae.

2. Sewing formulae

The way in which topological and spectral geometric invariants add up relative to the division of a manifold into two (or more) pieces is of some importance in geometric analysis and functorial (possibly topological) QFT. Ideally one would aim to consider decompositions into simple submanifolds, roughly expressing the manifold as a CW-complex of lower dimensional simplices, and then be able to compute on those submanifolds before using sewing formulae to reconstitute the invariant on the complicated parent manifold. Such processes are inherently cohomological. However, even in the purely topological case where analytic considerations are left to one side, such a general objective is in dimensions greater than 2 currently perhaps still at the stage of a useful, if mathematically potent, myth needed to think about representations of cobordism n-categories. Nevertheless, deep progress has been made at a fundamental category-theoretic level (see Lurie and Hopkins [**13**]) along with the discovery in dimension 2 of exotic cohomology theories (such as elliptic cohomology, Chas-Sullivan products).

2.1. Analytic sewing formulae. At the analytic level one might begin with a more modest objective. One might ask simpler questions, such as what the codimension zero splitting formula might be for the generalized zeta-function quasi-trace at zero $\zeta(A, Q, 0)$ for classical pseudodifferential operators A and Q, which, it is known, coincides with interesting spectral-geometric and topological invariants (such as the index of an elliptic operator). Here, Q is assumed to be elliptic operator of positive order $q > 0$ whose complex powers Q_θ^{-z} are defined. The operator AQ_θ^{-z} is trace class when the real part of z is large, and its trace (or supertrace given a $\mathbb{Z}_2$-grading) admits an extension $\zeta(A, Q, z) := \operatorname{Tr}(AQ_\theta^{-z})|^{\mathrm{mer}}$ to a meromorphic function defined on all of $\mathbb{C}$ which has a Laurent expansion around $z = 0$ of the form

$$\zeta(A, Q, z) = \frac{1}{q}\operatorname{res} A\,\frac{1}{z} + \zeta(A, Q, 0) + \zeta'(A, Q, 0)\, z + \cdots \qquad z \text{ near } 0. \tag{2.1.1}$$

The pole coefficient $\operatorname{res} A$ is the residue trace of A. In terms of sewing formulae relative to a splitting of the manifold rather little is known about these coefficients in general; the first few coefficients provide key invariants in geometric index theory while the higher coefficients are largely mysterious and highly non-local.

For the first term, the residue trace on manifolds with boundary with local Boutet de Monvel boundary conditions was looked into in [**10**] and [**11**], while less is known for global APS type boundary conditions. In either case pasting formulae for $\operatorname{res} A$ in terms of such boundary problems relative to a splitting of M have not yet, it appears, been studied. The next term in the expansion $\zeta(A, Q, 0)$ is in general rather complicated and non-local and no general pasting formulae are known. However, in certain cases progress has been made. Most outstandingly that is so for the eta-invariant

$$\eta(D, 0) := \zeta(D|D|^{-1}, |D|, 0)$$

of a self-adjoint Dirac type operator which arises as the correction term in the APS index theorem. In this case there is an exact pasting formula proved originally by Wojciechowski [**25, 26, 27**] and Bunke [**7**], which has been reproved and refined by other authors in numerous subsequent works, in particular in [**5**] and more from a TQFT view point in [**9**].

When $A = I$ write $\zeta(Q, z) := \zeta(I, Q, z)$. The expansion (2.1.1) in this case becomes

(2.1.2)

$$\zeta_\theta(Q, z) \;=\; \underbrace{-\left(\frac{1}{q}\operatorname{res}(\log_\theta Q) - \operatorname{tr}(\Pi_Q)\right)}_{=\zeta(Q,0)} z^0 \;+\; \underbrace{(\log\det{}_\zeta Q)}_{=-\zeta'(Q,0)}\, z \;+\; \cdots \qquad z \text{ near } 0.$$

where Π_Q is a projector onto the generalized kernel of Q. Then the first two terms of (2.1.2) are (logarithms of) determinants. The first is an exotic determinant, the 'residue determinant' equal to zero on ψdos with a well-defined classical determinant. Properties of the residue determinant and the extension of the residue trace to the log-classical ψdo $\log_\theta Q$ are detailed in [**18**] and [**20**]. The second is the log-zeta determinant, a quasi-determinant – an extension of the classical determinant but with a loss of the multiplicativity property. Pasting formulae for the first term $\zeta(Q, 0)$, relative to elliptic boundary problems for the restrictions of Q to M_1, M_2 given a partition $M = M_1 \cup_Y M_2$, have not yet been established; certainly this involves non-local terms associated to the dividing hypersurface Y, though the pasting formula does not appear to be hard to resolve using standard methods. For the next term a great deal more is known, at least for self-adjoint Dirac type operators. The original contribution in this direction for the case of APS boundary problems is the adiabatic pasting formula of Wojciechowski and Park [**16**], and has lead to a steady stream of refinements and extensions by other authors. It was suggested by Wojciechowski that a similar formula ought to hold for the curvature 2-form of the determinant line bundle for a family of Dirac operators, much as it does in the case of b-calculus [**15**], and recent work suggests that that is correct; see [**23**] for more on the curvature of the determinant line bundle in this case.

2.2. Sewing formulae in TQFT. A rather interesting way to think about sewing formulae has been thrown into the ring by mathematical topological quantum field theory (TQFT). Roughly speaking, recall that the idea of TQFT is that though the Feynman path integral (PI) formulation of QFT is currently eludes mathematical precision, one may nevertheless study the mathematical imprint that is left by its passing, the structures that would be left in its wake if the PI were rigorously defined.

2.3. Path integral formalism. Witten explained this motivation in the following concrete way. Consider a smooth compact n-dimensional manifold M with connected boundary $\partial M = Y$, then a PI has the general form

$$Z_M : \Gamma(Y) \to \mathbb{C}, \quad Z_M(f) = \int_{\Gamma_f(M)} e^{-S(\psi)} \mathcal{D}\psi, \tag{2.3.1}$$

where $\mathcal{D}\psi$ is a formal measure. $S : \Gamma(M) \to \mathbb{C}$ is the classical action functional on a space $\Gamma(M)$ of fields on M (the fields could, for example, be the space of smooth functions on M, or the space of sections of a vector bundle over M), while the subspace $\Gamma_f(M) = \{\psi \in \Gamma(M) \mid \psi_{|X} = f\}$ consists of smooth fields on M with boundary value $f \in \Gamma(Y)$. Thus we may view

$$Z_M \in Z(Y) = \text{a Hilbert space of distributions } \{u : \Gamma(Y) \to \mathbb{C}\}.$$

The precise class of distributions is to be specified, but $Z(Y)$ is supposed to be the Hilbert space of the theory so it might, for example, be a suitable Sobolev completion of $\Gamma(Y)$.

If M, on the other hand, has disconnected boundary $\partial M = \overline{Y_0} \sqcup Y_1$ then the PI defines the Schwartz kernel distribution

$$\mathbb{K}_M : \Gamma(Y_0) \times \Gamma(Y_1) \to \mathbb{C}, \qquad \mathbb{K}_M(f_0, f_1) = \int_{\Gamma_{(f_0, f_1)}(M)} e^{-S(\psi)} \mathcal{D}\psi,$$

or, what is formally the same thing, the linear operator

$$Z_M \in \operatorname{Hom}(Z(Y_0), Z(Y_1)), \qquad Z_M(u_0)(f_1) = \int_{\Gamma(Y_0)} \mathbb{K}_M(f_0, f_1) u_0(f_0)\, \mathcal{D} f_0.$$

If $Y_0 = Y_1 = Y$ this determines a bilinear form

$$\langle\ ,\ \rangle : Z(Y_0) \times Z(Y_1) \to \mathbb{C}, \qquad \langle u_0, u_1 \rangle = \int_{\Gamma(Y)} u_1(f) Z_M(u_0)(f)\, \mathcal{D} f.$$

Consider a compact boundaryless manifold $M = M_0 \cup_Y M_1$ partitioned into two halves by a connected hypersurface Y. Then the partition functions on M_0, M_1 have the form (2.3.1). Since the fields on M can be written as a fibre product

$\Gamma(M) = \Gamma(M_0) \times_{\Gamma(Y)} \Gamma(M_1)$, this suggests an equality

$$\int_{\Gamma(M)} e^{-S(\psi)}\mathcal{D}\psi = \int_{\Gamma(Y)} \left(\int_{\Gamma_f(M_0)} e^{-S(\psi_0)}\mathcal{D}\psi_0 \int_{\Gamma_f(M_1)} e^{-S(\psi_1)}\mathcal{D}\psi_1 \right) \mathcal{D}f$$

$$= \int_{\Gamma(Y)} Z_{M_0}(f) Z_{M_1}(f)\, \mathcal{D}f, \tag{2.3.2}$$

or, more schematically,

$$Z_M = \langle Z_{M_0}, Z_{M_1} \rangle. \tag{2.3.3}$$

The Hamiltonian of the theory is defined by the Euclidean time evolution operator $e^{-tH} = Z_{Y\times[0,t]} \in \mathrm{End}\,(Z(Y))$; in a 'topological theory' $H := 0$ so one then expects

$$Z_{Y\times[0,t]} = I \in \mathrm{End}\,(Z(Y)). \tag{2.3.4}$$

3. Categorical abstraction - homotopy theory

The above characterization suggests that the PI in dimension n may be viewed abstractly as a (particular) map Z which takes a boundaryless compact $(n-1)$-manifold Y to a vector space $Z(Y)$ and each n manifold M whose boundary is Y to a vector $Z_M \in Z(Y)$. If $Y = \emptyset$ is the empty $(n-1)$-manifold then from (2.3.1) Z_M is a number and $Z(\emptyset) = \mathbb{C}$. If $\partial M = \overline{X} \sqcup Y$ then Z_M defines a linear map between the vector spaces $Z(X)$ and $Z(Y)$.

Succinctly, the PI functor Z defines a representation of the cobordism category Cob_n. That is, Z is a functor $\mathsf{Cob}_n \to \mathsf{Vect}(k)$ to the category of vector spaces and linear maps over a field k.

Here, Cob_n is the category whose objects are smooth boundaryless compact $(n-1)$-manifolds and whose morphisms are smooth compact manifolds modulo orientation preserving diffeomorphisms. Thus a morphism $M \in \mathsf{mor}_n(X,Y)$ in Cob_n is a smooth compact n-manifold M equipped with a diffeomorphism $\partial M \cong \overline{X} \sqcup Y$; any two such morphisms are identified up to an orientation preserving diffeomorphism equal to the identity on the boundary. Composition of morphisms

$$\mathsf{mor}_n(X,Y) \times \mathsf{mor}_n(Y,Z) \to \mathsf{mor}_n(X,Z), \qquad (M,N) \mapsto M \cup_Y N,$$

is by pasting along the common boundary Y (up to specifying the smooth structure on $M \cup_Y N$). There is a symmetric monoidal product $\mathsf{Cob}_n \times \mathsf{Cob}_n \to \mathsf{Cob}_n$ defined by disjoint union of manifolds; with unit ('monoidal') element the empty manifold $\emptyset$. Likewise the usual tensor product defines a symmetric monoidal product on $\mathsf{Vect}(k)$ with identity the ground field k; again in $\mathsf{Vect}(k)$ identifications are made up to equivalence (isomorphism).

The formal definition, then, of a TQFT is a functor $Z : \mathsf{Cob}_n \to \mathsf{Vect}(k)$ preserving the symmetric monoidal structures.

Specifically, Z functorially maps $M \in \mathsf{mor}_n(X,Y)$ to a linear homomorphism $Z_M \in \mathrm{Hom}(Z(X), Z(Y))$, which can be equivalently written

$$Z_M \in Z(\overline{X}) \otimes Z(Y), \qquad Z(\emptyset) = k, \qquad Z(\overline{X}) \cong Z(X)^* \tag{3.0.5}$$

with

$$Z_{M\cup_Y N} = Z_M \circ Z_N. \tag{3.0.6}$$

In particular, since Cob_n is modulo diffeomorphisms then taking $M = N = X \times I$ gives that $Z_{X\times I}$ is an idempotent, which the functoriality says must, in fact, be the identity, corresponding to (2.3.4).

Since the boundary of a high-dimensional manifold M is really no topologically simpler than the manifold itself, it is important to allow for reasons of computability higher codimensional decompositions of M using manifolds with corners; in principle this may allow matters to be reduced to computing on k-simplices. Indeed, an implicit expectation is that a given TQFT ought to correspond to (a possibly exotic) generalized cohomology theory. In this respect the functorial sewing properties in a TQFT ought likewise to provide a computational tool along the lines of a generalized Mayer-Vietoris property. With this motivation, an *extended TQFT* with values in a symmetric monoidal n-category C is defined to be a symmetric monoidal functor $Z : \mathsf{Cob}_n^{\mathrm{ext}} \to \mathsf{C}$, where $\mathsf{Cob}_n^{\mathrm{ext}}$ is the extended cobordism n-category whose objects are points, whose (1-)morphisms are oriented 1-manifolds (cobordisms of points), whose 2-morphisms are cobordisms of 1-morphisms, and so on.

These definitions are due principally to Segal [**24**], Atiyah [**1**], Witten [**28**], and Hopkins and Lurie [**13**].

TQFT in dimensions 1 and 2 is relatively straightforward to characterize in a generic sort of way. A functor $Z : \mathsf{Cob}_1 \to \mathsf{Vect}(k)$ is determined by its possible evaluations on $c = [0,1]$. There are for c two possible boundary 0-manifolds p and $\overline{p}$ (with opposite orientations), to which Z assigns a finite dimensional vector space $V = Z(p)$ and its dual $V^* = Z(\overline{p})$. The meaning of $Z(c)$ changes according to whether it is regarded as a morphism in $\mathsf{mor}_n(p,p)$ in which case $Z(c) = I \in \mathrm{End}\,V$, or in $\mathsf{mor}_1(\emptyset, p \sqcup \overline{p})$ in which case $Z(c)$ is the map $\mathbb{C} \to \mathrm{End}\,(V), x \mapsto xI$ with I the identity, or in $\mathsf{mor}_1(p \sqcup \overline{p}, \emptyset)$ in which case $Z(c)$ is the trace map $\mathrm{tr} : \mathrm{End}\,(V) \to \mathbb{C}$. From this one can compute the value $Z(S^1)$ on the circle by writing $S^1 = S^1_+ \cup S^1_-$ as the union of its upper and lower semicircles, considering $S^1_+ \in \mathsf{mor}_1(\emptyset, p \sqcup \overline{p})$ and $S^1_- \in \mathsf{mor}_1(p \sqcup \overline{p}, \emptyset)$. Then according to (3.0.5), (3.0.6), $Z(S^1)$ is the composition of the maps $\mathbb{C} \to \mathrm{End}\,(V)$ and $\mathrm{End}\,(V) \to \mathbb{C}$ above, and hence $Z(S^1) = \dim V$. This is the description given in [**13**].

To give a 2-dimensional TQFT $Z : \mathsf{Cob}_2 \to \mathsf{Vect}(k)$, on the other hand, is the same thing as to specify a unital associative tracial algebra $(\mathcal{A}, \tau)$ (see below for more on traces). This is essentially because any real compact surface is a composition of copies of the disc D and copies of the 'pair of pants' surface P (a genus zero surface with two incoming and one outgoing boundary). Setting $\mathcal{A} = Z(S^1)$, then, Z_P defines according to the axioms of the TQFT an associative multiplication $\mathcal{A} \times \mathcal{A} \to \mathcal{A}$, while regarding D as an element of $\mathsf{mor}_2(\emptyset, S^1)$ defines the unit in $\mathcal{A}$ as the image of $1 \in k$, while as an element of $\mathsf{mor}_2(S^1, \emptyset)$ it defines a trace $\tau : \mathcal{A} \to k$. For more on this see [**24**], [**13**], and for a detailed account of the far richer open-closed theory with boundary conditions see Moore and Segal [**14**].

3.1. Introducing additional structure. The basic assumptions of a TQFT can be modified or extended in many interesting ways. Simple modifications are, for example, to specify additional topological structure, such as spin structures, framings, characters and so forth. Alternatively, the target category $\mathsf{Vect}(k)$ could be refined to a category of chain complexes, as considered in [**14**] and [**8**]. On the other hand, quantizing classical field theories can require the use of fibred-TQFT in which Cob_n is replaced by the category $\mathbb{C}\mathsf{ob}_n$ of smooth fibrations, with objects of fibre dimension $n-1$ and morphisms which are fibrations whose total space is a manifold with boundary with fibre diffeomorphic to a compact manifold with boundary of dimension n, all modulo fibrewise orientation preserving diffeomorphism (thus Cob_n is the subcategory of $\mathbb{C}\mathsf{ob}_n$ for the case of a fibration over a point). Composition of morphisms in $\mathbb{C}\mathsf{ob}_n$ is by fibrewise sewing. Then $\mathsf{Vect}(k)$ is replaced by the category $\mathbb{V}\mathsf{ect}(k)$ of vector bundles and bundle homomorphisms, modulo isomorphism. The symmetric monoidal structures extend to the fibred versions and a fibred-TQFT is defined to be a symmetric monoidal functor $Z : \mathbb{C}\mathsf{ob}_n \to \mathbb{V}\mathsf{ect}(k)$, with obvious extensions to their n-category counterparts.

Additional structure introduced through 'boundary conditions' is incorporated via labels attached to the morphisms in the category, defining an open-closed TQFT. The meaning here of a boundary condition is possibly more abstract, or at least more geometric, than in the usual PDE sense. Specifically, D-branes refer to the set of closed, connected, oriented, smooth submanifolds of a given manifold M and one can consider the space of maps $\gamma : [0,1] \to M$ with $\gamma(0) \in X$, $\gamma(1) \in Y$ in given D-branes X and Y. The homology of the corresponding mapping space for a given set of labels is, roughly, what is meant by 'string topology' and this is known to define an open-closed conformal field theory (CFT). See [**4**] for more on this and references.

Similarly, but moving into more uncharted areas, there is no difficulty in formally contemplating marked and degenerate morphisms. A marked morphism

$$M \in \mathsf{mor}_n^{\Sigma}(X,Y)$$

is a cobordism endowed with an embedding $\Sigma \hookrightarrow \overset{o}{M}$ of a submanifold (of arbitrary codimension) into the interior of M; thus, in connecting X to Y the morphism M is constrained to pass through Σ. The motivation for this is more easily explained when discussing logarithmic structures (below). On the other hand, there is no particular reason to suppose a more complete theory of quantum fields can be restricted to manifolds of constant dimension; a degenerate morphism is one which may be of non-constant dimension, the associated TQFT requires the use of n-categories and such morphisms may be thought of as compositions of k-morphisms for varying k.

In a different direction, one may abandon purely topological considerations in favour of introducing more analytic structure, such as metrics, linear boundary conditions, connections, and so forth, meaning in practise dropping diffeomorphism invariance. The resulting invariants may then be geometric or spectral rather than topological and the Hilbert spaces of the theory will then in general be infinite dimensional (for a TQFT $Z(X)$ is necessarily finite dimensional and so the dual has a unique meaning, but for infinite dimensional $Z(X)$ then analytic choices must be made as to which spaces of distributions are being contemplated.)

This is appropriate for understanding the role of some spectral-geometric (which in some cases turn out to be topological) invariants such as those mentioned in § 2.1. An example of this is functorial QFT $Z : \mathsf{Cob}_1^B \to \mathsf{Vect}(k)$ with Cob_1^B whose objects are the same as Cob_1 (see the characterization of 1-dimensional TQFTs in § 3), and whose non-closed morphisms $(c_{a,b}, P) \in \mathsf{mor}_1^B$ consist of (unions of) $c_{a,b} = [a,b]$, up to an orientation preserving diffeomorphism equal to the identity at a and b, and a projection P in the Grassmannian $\mathrm{Gr}(\mathbb{C}^{2m}) = \{P \in \mathrm{End}\,\mathbb{C}^{2m} \mid P^2 = P, P^* = P\}$

parametrizing linear subspaces of $\mathbb{C}^{2m}$. Consider, then, the formal fermionic PI

$$Z(c_{a,b}, P) = \int_{C^\infty(c_{a,b},\mathbb{C}^m)} e^{-\langle\psi, D_P\psi\rangle}\, d\psi$$

defined by the constrained Yang-Mills action, where $D = i\nabla_{d/dx}$ with ∇ a Hermitian covariant derivative (a Dirac operator in dimension 1) and D_P denotes its restriction to those ψ satisfying the linear elliptic boundary condition $P(\psi(a) \oplus \psi(b)) = 0$. Then, formally, $Z(c_{a,b}, P) = \det D_P$. Since $\mathrm{Gr}(\mathbb{C}^{2m})$ is a Kähler manifold with canonical Kähler form equal to the i times the curvature of the canonical line bundle $L \to \mathrm{Gr}(\mathbb{C}^{2m})$, then, as instructed by geometric quantization, we define the Hilbert space $Z(a \sqcup b)$ to be dual of the space of the holomorphic sections of $L^* \to \mathrm{Gr}(\mathbb{C}^{2m})$. Then $Z(a \sqcup b)$ is canonically isomorphic to the exterior algebra $\bigwedge \mathbb{C}^{2m}$. The point is that $\det D_P$, as P varies defines a section of the determinant line bundle, and the determinant line bundle is canonically isomorphic to L. If we restrict to the real submanifold identified with the unitary group $U(m) \hookrightarrow \mathrm{Gr}_m(\mathbb{C}^{2m})$, $g \mapsto P_g$ the projection onto graph$(g : \mathbb{C}^m \to \mathbb{C}^m)$, over which L (and the determinant line bundle) are naturally trivial, then there is an essentially canonical identification

$$Z(c_{a,b}, P) = \det D_{P_g} = \det\,(I - g^{-1}h)$$

(for example by zeta function regularization) where $h \in U(n) := U(\mathbb{C}^n)$ is the parallel transport of ∇ along $c_{a,b}$. Alternatively, considered as a functorial QFT, we may, via the Plücker embedding, regard the determinant line as a ray in the Fock space $Z(a \sqcup b)$ and the determinant as defined 'absolutely' without boundary condition as the 'vacuum vector' $\det D \in Z(a \sqcup b)$. With regard to the trivialization over $U(n)$ one finds defining $Z(a) = Z(b) = \bigwedge \mathbb{C}^m$

$$\det D \leftrightarrow \bigwedge h \in Z(a \sqcup b) \cong \mathrm{End}\,(\bigwedge \mathbb{C}^m),$$

where $\bigwedge h := \sum_{k=0}^m \bigwedge^k h$ relative to $\bigwedge \mathbb{C}^m := \sum_{k=0}^m \bigwedge^k \mathbb{C}^m$.

On the other hand, $Z(S^1) = \det\,(I - h_{S^1})$ with h_{S^1} the holonomy around the circle. Write, as in the example in § 2.1, $S^1 = S^1_+ \cup S^1_-$. Then there is a canonical pairing (2.3.3), (3.0.6)

$$Z(-1 \sqcup 1) \otimes Z(1 \sqcup -1) \to Z(\emptyset) = \mathbb{C}, \qquad A \otimes B \mapsto \mathrm{tr}\,(AB).$$

Applied to our vacuum vectors this gives

$$Z(S^1) = \langle Z(S^1_+), Z(S^1_-)\rangle,$$

which relative to the trivializations is the identity

$$\det\,(I - h_{S^1}) = \mathrm{tr}\,(\bigwedge h_+ \circ \bigwedge h_-),$$

which really is just the identity $\det\,(I - h_{S^1}) = \mathrm{tr}\,(\bigwedge h_{S^1})$, with $h_\pm$ the parallel transports along $S^1_\pm$, consequent on $h_{S^1} = h_- \circ h_+$. On the other hand, there is a quite precise identification of this with the formal PI formula (2.3.2) via the isometry $\mathrm{End}\,(\bigwedge \mathbb{C}^m) \to L^2(U(m))$ given by $T \mapsto f_T$ with $f_T(g) = \mathrm{tr}\,(T \circ \bigwedge g)$. Since, via the Peter-Weyl theorem and Schur's lemma, this is an isometry it may be applied to the elements $\bigwedge h_+$, $\bigwedge h_-$, using Haar measure dg, to give the sewing formula for the circle for this symmetric monoidal functorial QFT as

$$\underbrace{\det(I - h_{S^1})}_{\text{pairing on End}\,(\bigwedge \mathbb{C}^m)} = \underbrace{\int_{U(n)} \det(I - g^{-1}h_+)\cdot\det(I + gh_-)\,dg,}_{\text{pairing on } L^2(U(m))} \tag{3.1.1}$$

which is a rigorous version of (2.3.2) with the boundary condition f in that formula replaced here by P_g.

Thus one may say that this 1-dimensional QFT yields the fundamental representation of the Lie group $U(n)$, while the sewing formula (2.3.2) is the orthogonality relation (3.1.1). The 2-dimensional version of this theory produces the fundamental loop group representations of $LU(n)$. This is far more subtle; in particular, the determinant line bundle over $LU(n)$ is non-trivial (unlike over $U(n)$ in dimension 1) leading to a projective representation. See [**19, 24**] for more on loop group representations and determinant lines. (Indeed, there is an analogue of this theory for gauge groups in all dimensions). An alternative, possibly analytically better formulation would be to consider b-morphisms, in the sense of Melrose's b-calculus.

Note, however, that this is not a topological QFT; $Z(S^1)$ is not, here, a topological invariant of S^1. There is though an obvious topological invariant associated to the pair (S^1, ∇); namely the winding number

$$w(\nabla) := \frac{1}{2\pi}\int_{S^1} \mathrm{tr}\,(h(t)^{-1}dh(t)) \ \in \mathbb{Z}.$$

of the map $S^1 \to \mathbb{C}^*$, $t \mapsto \det h(t)$. But considered in the sense of cobordism this is not multiplicative (as would be expected in TQFT) but rather is additive. Precisely, with

$$\log c_{a,b} := \frac{1}{2\pi}\int_{c_{a,b}} h(t)^{-1}dh(t)$$

one has for $a < b < b'$

$$\log c_{a,b'} = \log c_{a,b} + \log c_{b,b'} \tag{3.1.2}$$

and in particular

$$\log S^1 = \log S^1_+ + \log S^1_-. \tag{3.1.3}$$

Taking the trace of these finite matrices gives, setting $\log\det c := \mathrm{tr}\,(\log c)$,

$$\log\det S^1 = \log\det S^1_+ + \log\det S^1_-. \tag{3.1.4}$$

One could just exponentiate and use $\det h_{a,b}$, with $h_{a,b}$ the transport along $c_{a,b}$ to get a multiplicative TQFT, and this is what is generically done when additive invariants are encountered, but doing that loses sight of interesting structure: there is an operator-valued logarithm and a trace, which when combined on a closed manifold give a topological invariant $\log\det S^1$. Indeed, the winding number is a homotopy invariant identifying the Bott isomorphism $\pi_1(U(m)) \cong \mathbb{Z}$, which is naturally additive rather than multiplicative.

4. Categorization and logarithms

This brings us, then, to the idea of logarithmic structures and, in particular, their role in constructing invariants as logarithmic functors on the cobordism category.

Such structures lie behind a number of familiar topological and analytic invariants. Index invariants are a prime example of such a thing, both in odd and even K-theory. It is, however, the case that logarithm operators taking values in an algebra $\mathcal{A}$ tend to be rather special and quite hard to find, though there may be many traces on $\mathcal{A}$ – giving rise potentially to many determinants once a logarithm is identified.

With respect to logarithmic structures on the cobordism category, the identity (3.1.4) decomposing the topological winding number as the sum of two non-topological numbers is repeated for any local invariant; for example, the index of a Dirac operator $\eth$ on a compact boundaryless spin manifold M, viewed as the $\widehat{A}$-genus, can always be broken up additively as $\operatorname{ind}\eth = \int_{M_1} \widehat{A}(x) + \int_{M_2} \widehat{A}(x)$ with respect to a codimension zero partition $M = M_1 \cup_Y M_2$. The terms $\int_{M_i} \widehat{A}(x)$, however, are not of the form $\operatorname{tr}(\log T_i)$ with $\log T_i$ trace-class logarithmic operators on M_i. Though this is consequently not a logarithmic structure, there is one of the form $\operatorname{ind}\eth = \operatorname{ind}\eth_1 + \operatorname{ind}\eth_2$ where $\eth_1, \eth_2$ are APS-type elliptic boundary problems for $\eth$ on M_1 and M_2, providing a basic instance of a log-determinant functor on Cob_n. There is more on this below.

Log-determinant structures lie behind many ψdo spectral invariants. For example, if we look again at the expansion (2.1.2) then the first term is a log-determinant structure, the so-called residue determinant. The second term, the zeta determinant, comes from an honest logarithmic operator composed with what is only a quasi-trace defined by zeta function regularization; its log-determinant properties are therefore anomalous. Likewise, spectral flow, suspended eta-invariants, the odd and even Chern characters, analytic torsion are all characters of logarithmic representations.

In the remaining sections we look first at the appearance of logarithmic structures in a number of standard invariants, before briefly outlining the categorical formulation along the lines of functorial QFT.

4.1. Logarithms. Determinants will be studied here as characters of logarithmic representations of semigroups taking values in a tracial algebra. Let $\mathcal{Z}$ be a topological semigroup and let $\mathcal{B}$ be a unital locally convex topological algebra. A *global logarithm operator* is a map

$$\log : \mathcal{Z} \to \mathcal{B}, \qquad a \mapsto \log a, \tag{4.1.1}$$

which for $a, b \in \mathcal{Z}$ satisfies

$$\log ab - \log a - \log b \ \in \ [\mathcal{B}, \mathcal{B}]. \tag{4.1.2}$$

That is, $\log ab = \log a + \log b - \sum_{j=1}^{N} [b_j, b_j']$ some $b_j, b_j' \in \mathcal{B}$. Thus, the space of global logarithms with values in $\mathcal{B}$ is

$$\mathbb{L}\mathrm{og}(\mathcal{Z}, \mathcal{B}) \ := \ \mathrm{Hom}(\mathcal{Z}, \mathcal{B}/[\mathcal{B}, \mathcal{B}]) = \mathrm{Hom}(\mathcal{Z}, HC_0(\mathcal{B})). \tag{4.1.3}$$

relative to the linear structure of $\mathcal{B}/[\mathcal{B}, \mathcal{B}]$ ($HC_0(\mathcal{B})$ is the degree zero cyclic homology group). We may likewise consider the distributional subspace of continuous log functionals.

We may write (4.1.2) as $\log ab \approx \log a + \log b$. It follows that if $\mathcal{Z}$ is unital with identity element I (so that $\mathcal{Z}$ is a monoid) then $\log I \approx 0$ and hence for $b \in \mathrm{GL}(\mathcal{Z})$ that $\log b^{-1} \approx -\log b$ and $\log(bab^{-1}) \approx \log a$. If $\mathcal{Z}$ is unital with identity element I it can be advantageous to use linearity to refine $\log \in \mathbb{L}\mathrm{og}(\mathcal{Z},\mathcal{B})$ to the relative $\mathrm{Log} \in \mathbb{L}\mathrm{og}(\mathcal{Z},\mathcal{B})$ by $\mathrm{Log}\, a := \log a - \log I$ so that $\mathrm{Log}\, I = 0$.

Counting logs, as with counting traces, is done projectively insofar as 'uniqueness' is up to a scalar multiple. Precisely, since $\mathcal{B}$ is a linear space over $\mathbb{C}$ so therefore is the space (4.1.3) of logs

$$\log_1,\ \log_2 \in \mathbb{L}\mathrm{og}(\mathcal{Z},\mathcal{B}) \quad\Rightarrow\quad \lambda\log_1 + \mu\log_2 \in \mathbb{L}\mathrm{og}(\mathcal{Z},\mathcal{B}) \tag{4.1.4}$$

any $\lambda,\mu \in \mathbb{C}$. Thus, the number of log maps $\mathcal{Z}\to\mathcal{B}$ means the dimension over $\mathbb{C}$ of $\mathbb{L}\mathrm{og}(\mathcal{Z},\mathcal{B})$.

A logarithm whose construction depends on extraneous choices may only be defined in a neighbourhood of each element of the semigroup $\mathcal{Z}$, though each such local choice, or 'branch', is required to not be visible to any consequent determinant. A *local logarithm operator* on $\mathcal{Z}$ with values in $\mathcal{B}$ is an operator which for each $a \in \mathcal{Z}$ can be defined on some open neighbourhood $\mathcal{U}$ of a, called a branch of the log, as a map

$$\log_{\mathcal{U}} : \mathcal{U} \to \mathcal{B}, \qquad c \mapsto \log_{\mathcal{U}} c,$$

such that for any $a, b \in \mathcal{Z}$ there exist respective neighbourhoods $\mathcal{U},\mathcal{V},\mathcal{W}$ of a, b, ab in $\mathcal{Z}$ for which

$$\log_{\mathcal{W}} ab - \log_{\mathcal{U}} a - \log_{\mathcal{V}} b \in [\mathcal{B},\mathcal{B}]. \tag{4.1.5}$$

The space of local logarithms on $\mathcal{Z}$ with values in $\mathcal{B}$ will be denoted $\mathbb{L}\mathrm{og}_{\mathrm{loc}}(\mathcal{Z},\mathcal{B})$. An element of $\mathbb{L}\mathrm{og}_{\mathrm{loc}}(\mathcal{Z},\mathcal{B})$ may be denoted simply by $\log$.

A trace on the algebra $\mathcal{B}$ taking values in a vector space V is a linear map $\tau : \mathcal{B} \to V$ such that $\tau([a,b]) = 0$ for $a, b \in \mathcal{B}$. Thus

$$\mathbb{T}\mathrm{races}(\mathcal{B}, V) := \mathrm{Hom}\,(\mathcal{B}/[\mathcal{B},\mathcal{B}], V)$$

and so

$$\text{scalar valued traces} = (\mathcal{B}/[\mathcal{B},\mathcal{B}])^*$$

A basic question in studying traces is whether

$$\tau(a) = 0 \quad\overset{??}{\Longrightarrow}\quad a = \sum_{j=1}^{m} [\, b_j,\, c_j\,],$$

or, equivalently, whether $\mathrm{Ker}\,(\tau) = [\mathcal{B},\mathcal{B}]$. It is readily verified that for scalar traces the following are equivalent:

- $\tau : \mathcal{B} \to \mathbb{C}$ the unique non-trivial trace on $\mathcal{B}$
- $\mathcal{B}/[\mathcal{B},\mathcal{B}]$ is 1-dimensional
- Any $a \in \mathcal{B}$ can be written w.r.t $q \in \mathcal{B}$ with $\tau(q) \neq 0$ as

$$a = \sum_{j=1}^{J} [\, b_j,\, c_j\,] + \frac{\tau(a)}{\tau(q)}\, q.$$

A global (resp. local) *log-determinant* is defined by an element $\log \in \mathbb{L}\mathrm{og}(\mathcal{Z},\mathcal{B})$ (resp. $\log \in \mathbb{L}\mathrm{og}_{\mathrm{loc}}(\mathcal{Z},\mathcal{B})$) along with a trace $\tau : \mathcal{B} \to R$ to a ring R via

$$\mathcal{Z} \xrightarrow{\ \log\ } \mathcal{B} \xrightarrow{\ \tau\ } R$$

which, in view of (4.1.2), has the log-multiplicativity property

$$\tau(\log ab) = \tau(\log a) + \tau(\log b) \qquad \text{for all } a, b \in \mathcal{Z}. \tag{4.1.6}$$

Thus the log-determinant $\tau(\log a)$ is the τ-character of the logarithmic representation log of the semigroup $\mathcal{Z}$ in $\mathcal{B}$, and the space of log-determinants $\mathcal{Z} \to R$ factored through $\mathcal{B}$ is the linear product

$$\log\det(Z, R) = \mathbb{L}\mathrm{og}(Z, \mathcal{B}) \times \mathbb{T}\mathrm{races}(\mathcal{B}, R).$$

A *global determinant structure* is a triple $(\log, \tau, e)$ where $e : R \to R'$ is a homomorphism of unital rings with the exponential property $e(x+y) = e(x) \cdot e(y)$. The multiplicative determinant functional associated to $(\log, \tau, e)$ is

$$\det_{\tau, e} := e \circ \tau \circ \log : \mathcal{Z} \to R', \qquad \det_{\tau, e}(a) := e(\tau(\log a)). \tag{4.1.7}$$

Given an exponential map, the log property (4.1.2) may be relaxed to

$$\log ab - \log a - \log b \ \in \ \mathrm{Ker}\ (e \circ \tau) \tag{4.1.8}$$

and still define a multiplicative determinant.

There is, then, the (rough and not necessarily finite) bound on the number of log-determinants $\mathcal{Z} \to V$

$$\#\ \text{log-dets } \mathcal{Z} \to V \ \leq \ \dim \mathbb{L}\mathrm{og}(\mathcal{Z}, \mathcal{B}) \ \times \ \dim \mathbb{T}\mathrm{races}(\mathcal{B}, V). \tag{4.1.9}$$

This need not be an equality in general since a trace may be identically zero on the range of some log-maps.

The linearity of the spaces $\mathbb{L}\mathrm{og}(\mathcal{Z}, \mathcal{B})$ and $\mathbb{T}\mathrm{races}(\mathcal{B}, \mathbb{C})$ define via composition with the exponential map $\exp : \mathbb{C} \to \mathbb{C}$ the canonical commutative semigroup structure on the space of determinants $\mathbb{D}\mathrm{ets}(\mathcal{Z}, \mathbb{C})$

$$\det{}_1, \det{}_2 \in \mathbb{D}\mathrm{ets}(\mathcal{Z}, \mathbb{C}) \quad \Rightarrow \quad \det{}_1 \cdot \det{}_2 \in \mathbb{D}\mathrm{ets}(\mathcal{Z}, \mathbb{C}), \tag{4.1.10}$$

$$(\det{}_1 \cdot \det{}_2)(a) := \det{}_1(a) \cdot \det{}_2(a).$$

4.2. Determinant structures on DGAs. Let $\mathcal{B} = (\Omega, d)$ be a differential graded algebra (DGA). A (graded) logarithm operator sensitive to d is a map $\log : \mathcal{Z} \to (\Omega, d)$, $a \mapsto \log a$, which satisfies

$$\log ab - \log a - \log b \ \in \ \underbrace{[\Omega, \Omega]}_{\text{graded commutator}} \ + \ d\Omega. \tag{4.2.1}$$

The space of such logarithms is

$$\mathbb{L}\mathrm{og}(\mathcal{Z}, (\Omega, d)) \ := \ \mathrm{Hom}(\, \mathcal{Z},\ \Omega/([\Omega, \Omega] + d\Omega)\,).$$

A *closed graded determinant structure* means a triple $(\log : \mathcal{Z} \to (\Omega, d), \tau, e)$ with $\tau : \Omega \to R$ a closed graded trace. The closure of τ leads again to the log-multiplicativity property (4.1.6) and, further, that for $d^t\tau(\log a) := \tau(d \log a) = 0$ for all $a \in \mathcal{Z}$. The associated graded determinant is defined as before by $\det_{\tau, e}(a) := e(\tau(\log a))$.

5. Examples of logarithmic structures

We list here some examples of log-determinant structures; details may be found in [**22**].

5.1. Fredholm index — an exotic determinant. A Fredholm operator on a Hilbert space H is a bounded operator which is far from zero and close to the Banach Lie group $\mathrm{GL}(H) := \mathsf{B}(H)^{\times}$ of invertible elements in $\mathsf{B}(H)$. As such, the multiplicative semigroup $\mathrm{Fred}\, H$ of Fredholm operators on H lies at an opposite extreme in $\mathsf{B}(H)$ to any proper ideal, and in particular to the ideal $\mathsf{F}(H)$ of finite rank operators. Precisely, $A \in \mathsf{B}(H)$ is Fredholm if there is a parametrix $P \in B(H)$, so that $L_A := PA - I \ \in \mathsf{F}(H)$ and $R_A := AP - I \ \in \mathsf{F}(H)$. $\mathrm{Fred}\, H$ is a semigroup with respect to operator composition, but is highly non-linear, having homotopy type $\mathrm{Fred}\, H \simeq \mathbb{Z} \times \mathrm{BGl}(\infty)$.

The index defines a local determinant structure in which $\mathcal{Z} = \mathrm{Fred}\, H$ and $\mathcal{B} = \mathsf{F}(H)$ endowed with the classical trace tr, while $\log \in \mathbb{L}\mathrm{og}_{\,\mathrm{loc}}(\mathrm{Fred}\, H, (\mathsf{F}(H), \mathrm{Tr}\,))$ is defined pointwise by $\log_{\,P} A := [A, P]$ for P a parametrix for A. There is not a unique choice of parametrix and for this reason the logarithm so defined is local. To see that this is logarithmic (4.1.5), let $A, B \in \mathrm{Fred}$ and let $P, Q \in B(H)$ be respective parametrices, then for any parametrix R for AB one has

$$\log_{\,R} AB = \log_{\,P} A + \log_{\,Q} B + \underbrace{[L_A\, B,\, Q\,]}_{\in [\mathsf{F}(H), \mathsf{F}(H)]} + [A\, R_B,\; P\,] + [AB, R - QP]. \tag{5.1.1}$$

The character of the logarithm is the index of A

$$\mathrm{tr}\,(\log_{\,P} A) = \mathrm{tr}\,([A, P]) = \mathrm{ind}\, A, \tag{5.1.2}$$

and its log-multiplicativity, consequent to (5.1.1), is the additivity property of the index

$$(4.1.6) \quad \longleftrightarrow \quad \mathrm{ind}\, AB = \mathrm{ind}\, A + \mathrm{ind}\, B.$$

The associated determinant functional $\det_{\mathrm{ind}} A := e^{\mathrm{Tr}(\log_P A)} = e^{\mathrm{ind} A} \in \mathbb{Z}$ is constant on the connected components of $\mathrm{Fred}\, H$. The logarithm may be refined to the composite logarithm

$$\log\; : \mathrm{Fred}\, H \to \mathsf{F}(H) \to \mathsf{F}(H)/[\mathsf{F}(H), \mathsf{F}(H)] \tag{5.1.3}$$

which is independent of the choice of P and hence a global logarithm. The classical trace is the canonical generator of the complex line $(\mathsf{F}(H)/[\mathsf{F}(H), \mathsf{F}(H)])^*$ which evaluated on (5.1.3) maps $\log A$ to the index.

This determinant structure is *exotic* insofar as the log-determinant vanishes on the subdomain of the classical Fredholm determinant.

5.2. Restricted general linear group. Let $H = H^+ \oplus H^-$ be a $\mathbb{Z}_2$-graded Hilbert space with grading defined by an indempotent $F^2 = I$ with $F\xi = \pm\xi$ for $\xi \in H^{\pm}$. For $(\mathsf{J}, \|\cdot\|_{\mathsf{J}}, \tau)$ a (proper) normed trace ideal in $(B(H), \|\cdot\|)$ one has the restricted general linear group

$$\mathrm{GL}_{\mathrm{res},\mathsf{J}}(H) = \{A \in \mathrm{GL}(H) \mid [F, A] \in \mathsf{J}\}.$$

$\mathrm{GL}_{\mathrm{res},\mathsf{J}}(H)$ is a Banach lie group and with respect to the grading $H = H^+ \oplus H^-$, its elements have the form $\begin{pmatrix} a & b \\ c & d \end{pmatrix}$ with $b, c \in \mathsf{J}$. If $A^{-1} = \begin{pmatrix} x & y \\ z & w \end{pmatrix}$ in this representation then $ax = I_+ - bz$, and $xa = I_+ - yc$ with I_+ the identity operator on H^+. The element x is thus a canonical J-parametrix for a and we have homotopy equivalences

$$\mathrm{GL}_{\mathrm{res},\mathsf{J}}(H) \stackrel{\simeq}{\to} \mathrm{Fred}_{\mathsf{J}}(H^+) \stackrel{\simeq}{\to} \mathrm{Fred}(H^+), \qquad A \mapsto a, \tag{5.2.1}$$

where $\mathrm{Fred}_{\mathsf{J}}(H^+)$ is the semigroup of bounded operators invertible modulo J. The element x provides a canonical choice of parametrix for a and so we may define the *global* logarithm

$$\log : \mathrm{GL}_{\mathrm{res},\mathsf{J}}(H) \to \mathrm{Fred}_{\mathsf{J}}(H^+), \qquad \log A := [a, x]. \tag{5.2.2}$$

Higher logarithms $\log_k \in \mathbb{L}\mathrm{og}(\mathrm{GL}_{\mathrm{res},\mathsf{J}}(H), (\Omega, d_{\mathrm{F}}))$ can be constructed with respect to the DGA differential $d_{\mathrm{F}}(A) = [F, A]$ by $\log_k A = (A^{-1}[F, A])^{2k+1}$. The associated log-determinant is the odd Chern supertrace τ_s-character

$$\mathsf{ch}^-_{k,\tau}(A) := \tau_s((A^{-1}[F, A])^{2k+1}).$$

5.3. A universal logarithm. Let $\mathcal{A}$ be an associative algebra $\mathcal{A}$ with unit 1 over a ring R. The following universal DGA $\Omega^*(\mathcal{A})$ has a role in the construction of trace character invariants for $\mathcal{A}$. It is defined by setting $\Omega^0(\mathcal{A}) = \mathcal{A}$ and

$$\Omega^1(\mathcal{A}) = \mathcal{A} \otimes_R (\mathcal{A}/R).$$

Then $\Omega^1(\mathcal{A})$ has an $\mathcal{A}$-bimodule structure defined by

$$x \cdot (a \otimes_R b) \cdot y = xa \otimes_R by - xab \otimes_R y, \qquad a, b, x, y \in \mathcal{A}, \tag{5.3.1}$$

and there is the degree one differential $d : \mathcal{A} \to \Omega^1(\mathcal{A})$ define by $da := 1 \otimes a$. By construction $d(1) = 0$. It has the universality property that if Φ is an $\mathcal{A}$-bimodule with a derivation $\delta : \mathcal{A} \to \Phi$ with $\delta(1) = 0$, then there is a bimodule homomorphism $\rho : \Omega^1(\mathcal{A}) \to \Phi$ such that $\delta = \rho \circ d$.

From (5.3.1) we obtain that d is a bimodule derivation $d(ab) = da \cdot b + a \cdot db$. Define the linear spaces

$$\Omega^n(\mathcal{A}) = \underbrace{\Omega^1(\mathcal{A}) \otimes_{\mathcal{A}} \Omega^1(\mathcal{A}) \otimes_{\mathcal{A}} \cdots \otimes_{\mathcal{A}} \Omega^1(\mathcal{A})}_{n \text{ factors}}. \tag{5.3.2}$$

Any element of $\Omega^n(\mathcal{A})$ can be written as a linear sum of elements of the form $a_0 da_1 \cdots da_n$ for some $a_0, \ldots, a_n \in \mathcal{A}$. The graded derivation $d : \Omega^n(\mathcal{A}) \to \Omega^{n+1}(\mathcal{A})$, $a_0\, da_1 \cdots da_n \mapsto da_0\, da_1 \cdots da_n$ gives $\Omega^*(\mathcal{A}) = \sum_{k \geq 0} \Omega^k(\mathcal{A})$ the structure of a DGA of degree 1 (However, $\Omega^*(\mathcal{A})$ is not an exterior algebra, it is called the algebra of non-commutative differential forms.) There is a natural algebra isomorphism

$$\Omega^n(\mathcal{A}) \cong \mathcal{A} \otimes_R (\mathcal{A}/R)^{\otimes n}, \qquad a_0 da_1 \cdots da_n \longleftrightarrow \quad a_0 \otimes_R (a_1 \otimes_R \cdots \otimes_R a_n),$$

In particular, $da_1 \cdots da_n \leftrightarrow 1 \otimes_R (a_1 \otimes_R \cdots \otimes_R a_n)$.

Set $\mathrm{GL}_\infty(\mathcal{A}) := \bigcup_n \mathrm{Gl}_n(\mathcal{A})$ with the direct limit topology, where $\mathrm{Gl}_n(\mathcal{A})$ consists of invertible elements of the set $M_\infty(\mathcal{A})$ of $n \times n$ matrices with entries in $\mathcal{A}$. Summing the diagonal elements gives the pre-trace (or Denis trace) $\mathtt{Tr}_{\mathbf{pre}} : M_\infty(\mathcal{A}) \to \mathcal{A}$. It is then not hard to show the following.

Theorem 1. *Associated to a DGA* $(\Omega^* = \sum_{k \geq 0} \Omega^k, d)$ *and a homomorphism* $\rho : \mathcal{A} \to \Omega^0$ *there is for each* $k = 1, 2, \ldots$ *a global logarithm map*

$$\log_k \in \mathbb{L}og(\mathrm{GL}(\mathcal{A}), (\Omega^*, d))$$

$$\log_k : \mathrm{GL}(\mathcal{A}) \to \Omega^{2k-1}, \qquad \log_k a := (a^{-1} da)^{2k-1}, \tag{5.3.3}$$

where we have written $(a^{-1}da)^{2k-1} := (\rho(a)^{-1} d\rho(a))^{2k-1}$. *The even degree form* $a \mapsto \gamma_k(a) = (a^{-1}da)^{2k}$ *also is logarithmic but is trivial insofar as* $\gamma_k(a) \in d\,\Omega^*$.

It follows that there is a canonical odd logarithm associated to any unital algebra $\mathcal{A}$

$$\log_k : \mathrm{GL}_\infty(\mathcal{A}) \to \Omega^{2k-1}(\mathcal{A}), \qquad \log_k a := \mathtt{Tr}_{\mathbf{pre}}\big((a^{-1}da)^{2k-1}\big). \tag{5.3.4}$$

The resulting canonical pairing with cyclic cohomology to the space of logarithmic characters on $\mathrm{GL}(\mathcal{A})$ defined by associating to $(\log, \tau)$ the log-determinant $\tau \circ \log$

$$\begin{array}{ccc} \mathbb{L}\mathrm{og}(\mathrm{GL}(\mathcal{A}), (\Omega^*(\mathcal{A}), d)) \otimes HC^*(\mathcal{A}) & \longrightarrow & \mathbb{L}\mathrm{og}(\mathrm{GL}(\mathcal{A}), \mathbb{C}) \\ & & \big\downarrow \exp \\ & & \mathrm{Hom}(\mathrm{GL}(\mathcal{A}), \mathbb{C}) \end{array} \tag{5.3.5}$$

and then by exponentiation to the space of determinants on $\mathrm{GL}(\mathcal{A})$ may be viewed as an odd Chern character pairing.

5.4. Logs and $K_1(\mathcal{A})$. If $\mathcal{Z} = \mathsf{G}$ is a topological group, then one has:

Lemma 1.

$$\mathbb{L}og(\mathsf{G}, \mathcal{B}) = \mathrm{Hom}(\mathsf{G}/\mathsf{G}^{(1)}, HC_0(\mathcal{B})) \tag{5.4.1}$$

with $\mathsf{G}^{(1)}$ *the commutator subgroup.*

In the case of graded logarithms on G *with values in* (Ω, d), (5.4.1) *becomes*

$$\mathbb{L}og(\mathsf{G}, (\Omega, d)) = \mathrm{Hom}\left(\mathsf{G}/\mathsf{G}^{(1)}, \Omega/([\Omega,\Omega] + d\,\Omega)\right). \tag{5.4.2}$$

For example, if $\mathcal{Z} = \mathsf{G} = \pi_1(X)$ with X a smooth path-connected topological space and $\mathcal{B} = \mathbb{Z}$ then $\mathsf{G}/\mathsf{G}^{(1)}$, the abelianization of π_1, is the integer coefficient singular homology group $H_1(X,\mathbb{Z})$. Hence $\mathbb{L}\mathrm{og}(\pi_1(X), \mathbb{Z}) \overset{(5.4.1)}{=} \mathrm{Hom}(H_1(X,\mathbb{Z}),\mathbb{Z}) \cong H^1(X,\mathbb{Z})$, the final equality holding since $\mathrm{Ext}(H_0(X,\mathbb{Z}),\mathbb{Z})) = 0$, by the universal coefficient theorem. This implies $H^1(X,\mathbb{Z}) = \{f : \pi_1(X) \to \mathbb{Z} \mid f([\gamma] \circ [\gamma']) = f([\gamma]) + f([\gamma'])\}$ where a homotopy class $[\gamma]$ is defined by a continuous (or smooth) loop $\gamma : S^1 \to X$.

Algebraic K-functors $K_m(\mathcal{A})$ exist for each $m \in \mathbb{N}$ and these provide a source of higher invariants. The K_1 functor from the category of algebras to the category of abelian groups has the realization $K_1(\mathcal{A}) = \mathrm{GL}_\infty(\mathcal{A})/\mathrm{GL}_\infty(\mathcal{A})^{(1)}$. Equivalently, $K_1(\mathcal{A})$ is the first de Rham homology group $H_1(\mathrm{GL}_\infty(\mathcal{A}),\mathbb{Z})$. Taking $\mathsf{G} = \mathrm{GL}_\infty(\mathcal{A})$ in (5.4.1), a logarithm map on $\mathrm{GL}_\infty(\mathcal{A})$ is the same thing as a group homomorphism from $K_1(\mathcal{A})$ to $\mathcal{B}/[\mathcal{B},\mathcal{B}]$ (with respect to the multiplicative structure on the former and the linear space structure on the latter), so that

$$\mathbb{L}\mathrm{og}(\mathrm{GL}_\infty(\mathcal{A}), \mathcal{B}) = \mathrm{Hom}(K_1(\mathcal{A}), HC_0(\mathcal{B})). \tag{5.4.3}$$

A canonical homological invariant of $K_1(\mathcal{A})$ is the odd Chern character map constructed from the logarithms

$$\mathsf{ch}_k^-([u]) = \mathtt{Tr}_{\mathbf{pre}}\left((u^{-1}du)^{2k-1}\right) \tag{5.4.4}$$

of (5.3.3). The target space for the odd Chern character ch^- is the negative cyclic homology group $HC_1^-(\mathcal{A})$, just as the target for the Chern character on $K_0(\mathcal{A})$ is $HC_0^-(\mathcal{A})$

Proposition 1. *The odd Chern character*

$$\mathit{ch}^- : K_1(\mathcal{A}) \to HC_1^-(\mathcal{A})$$

is the logarithm map defined by

$$\mathit{ch}^- = \sum_{k\geq 0} t^k\, k!\, \mathit{ch}_k^-, \tag{5.4.5}$$

that is,

$$\mathit{ch}^-([u]) = \sum_{k\geq 0} t^k\, k!\ \mathtt{Tr}_{\mathbf{pre}}\left((u^{-1}du)^{2k+1}\right). \tag{5.4.6}$$

It follows that

$$\mathsf{ch}^- \in \mathrm{Hom}\left(K_1(\mathcal{A}), \frac{\Omega^{\,\mathrm{tot}}}{([\Omega^{\,\mathrm{tot}}, \Omega^{\,\mathrm{tot}}] + d\,\Omega^{\,\mathrm{tot}})}\right),$$

where $\Omega^{\,\mathrm{tot}} := \Omega^{\,\mathrm{tot}}(\mathcal{A})^-_{[t]}$ is the graded algebra of negative cyclic chains. The odd Chern character is consequently a logarithm

$$\mathsf{ch}^- \in \mathrm{Hom}\left(K_1(\mathcal{A}), HC_1^-(\mathcal{A})\right).$$

In the case where $(\mathcal{A}, \|\cdot\|)$ is a unital Banach algebra, such as the C^*-algebra $C(M)$ of continuous functions on a compact manifold M, there is an alternative candidate for the odd K-theory of $\mathcal{A}$ defined by the topological quotient group

$$K_{-1}(\mathcal{A}) = \mathrm{GL}_\infty(\mathcal{A})/\,\mathrm{GL}_\infty(\mathcal{A})_0,$$

where $\mathrm{GL}_\infty(\mathcal{A})_0$ is the subgroup of elements of $\mathrm{GL}_\infty(\mathcal{A})$ homotopic to the identity. This is topological odd K-theory. For example, $K_{-1}(M) := K_{-1}(C(M)) = [M, Gl(\infty)]$ is the relevant odd K-theory for geometric index theory and the corresponding Chern character $\mathsf{ch}_{-1} : K_{-1}(M) \to H^{\mathrm{odd}}(M)$ equal to the trace of (5.4.5) is given by

$$\mathsf{ch}_{-1}(g) = \sum_{k=1}^{\infty} (-1)^{k-1} \frac{(k-1)!}{(2k-1)!}\ \mathrm{Tr}\,(g^{-1}dg)^{2k-1}.$$

In particular each of the Bott isomorphisms $\pi_{2k+1}(Gl(\infty)) \cong \mathbb{Z}$ arise as log determinant structures. We refer to [**22**] for more details.

5.5. Trace ideals and de Rham. If H is an infinite dimensional Hilbert space then $\mathsf{B}(H) = [\mathsf{B}(H), \mathsf{B}(H)]$, every bounded operator is a commutator, and so there are no non-trivial traces on $\mathsf{B}(H)$, and so every determinant is trivial, even though there are non-trivial logarithm operators. Likewise, the ideal of compact operators $\mathsf{C}(H) = [\mathsf{C}(H), \mathsf{C}(H)]$ has no non-trivial trace. There is, however, an intricate system of proper trace ideals $\mathsf{F}(H) \subset (\mathsf{J}, \tau) \subset \mathsf{C}(H)$; for example, $(\mathsf{C}_1(H), \mathrm{Tr}\,)$ with $\mathsf{C}_1(H)$ the first Schatten ideal and Tr the classical trace, and $(\mathsf{C}^{1,\infty}(H), \tau_\infty)$ where $\mathsf{C}^{1,\infty}(H)$ is the Macaev ideal and τ_∞ is a Dixmier trace. It would be interesting to better understand the resulting Dixmier determinants and, more generally, the log-determinant structure of the $\mathsf{B}(H)$ trace ideals.

In particular, on a sub-semigroup $\mathcal{Z}_{\mathsf{J}}$ of (J, τ) of operators having an Agmon angle θ there is a candidate logarithm for $\mathbb{L}\mathrm{og}_{\,\mathrm{loc}}(\mathrm{GL}(\mathcal{Z}), \mathcal{B})$

$$\log_{\,\theta} a = \int_{\mathcal{C}} \log_{\,\theta}\lambda\ (a-\lambda)^{-1}\ \text{đ}\lambda,$$

where $\bar{d}\lambda = (i/2\pi)d\lambda$, with associated log-determinant $a \mapsto \tau(\log_\theta a)$. The (classical) Fredholm determinant arises this way

$$\log_{\mathbb{C}}\det_{\mathsf{F}} : \mathrm{GL}(I + \mathsf{C}_1(H)) \to \mathbb{C}, \quad \log_{\mathbb{C}}\det_{\mathsf{F}}(I + A) := \mathrm{Tr}\,(\mathrm{Log}_\theta(I + A))).$$

There are, on the other hand, many Dixmier determinants,

$$\log_\theta : I + \mathsf{C}^{1,\infty}(H) \to \mathsf{C}^{1,\infty}(H), \qquad \log_{\mathbb{C}}\det_\infty(I + A) = \tau_\infty \log(I + A),$$

but these are currently largely mysterious.

The de Rham algebra $\Omega(M, \mathrm{End}\, E)$ on a compact manifold M of dimension n is a DGA with infinitely many traces. Specifically, for example, we have the trace

$$\mathrm{tr}_{Y,\sigma}(a) := \int_Y \sigma \wedge \mathrm{tr}\,(a_{[k]}) \qquad \text{for } \sigma \in \Omega(M),\ Y \subset M.$$

For $Y = \{x_0\}$ a point and $\sigma = 1$ this is the delta distribution $\mathrm{tr}_{x_0}(a) := \mathrm{tr}\,(a_{[0]}(x_0))$, or, for $Y = M, \sigma = 1$, it is the standard trace $\mathrm{tr}_M(a) = \int_M \mathrm{tr}\,(a_{[n]})$, while on a spin manifold the coupled Atiyah-Singer density is the trace $\mathrm{tr}_{M,\widehat{A}}(\exp F^2)$.

On the de Rham algebra there are relatively few logarithmic structures, though there are many determinants since there are many traces. We may, for example, consider the logarithm on admissible endomorphism-valued forms $\mathsf{P} \in \Omega(M, \mathrm{End}\, E)$ defined as above by

$$\log_\theta \mathsf{P} = \int_C \log_\theta \lambda\, (\mathsf{P} - \lambda)^{-1}\, \bar{d}\lambda \ \in \Omega(M, \mathrm{End}\, E).$$

The resolvent has a relatively simple structure; writing $\mathsf{P} = A + \mathsf{Q}$ with $A := \mathsf{P}_{[0]}$ the 0-form component

$$(\mathsf{P} - \lambda I)^{-1} = (A - \lambda)^{-1} + \sum_{k=1}^{\dim M} (-1)^k (A - \lambda)^{-1} \left(\mathsf{Q}(A - \lambda)^{-1}\right)^k.$$

Evaluating this using the trace tr_M, the resulting log-determinant structure is the Chern class on a (super)connection $\mathsf{c}(\mathbb{A}^2) = \mathrm{sdet}(I + \mathbb{A}^2) := \exp\left(\mathrm{tr_s} \log(I + \mathbb{A}^2)\right)$, giving the Chern class as a determinant structure

$$c : K^0(M) \to H^*(M).$$

On the other hand, for $g \in C^\infty(M, \mathrm{GL}(I + \mathsf{C}_1(H))$ we have a graded logarithm operator $\log_m g := (g^{-1}dg)^{2m-1}$; that is,

$$\log_m gh - \log_m g - \log_m h \ \in [\Omega^{\mathrm{odd}}, \Omega^{\mathrm{odd}}] + d\,\Omega^{\mathrm{odd}}.$$

From this one may build as in § 5.4 the odd topological Chern character logarithm

$$\mathsf{ch}^-(g) \in \mathbb{L}\mathrm{og}\left(C^\infty(M, \mathrm{GL}(I + \mathsf{C}_1(H)), (\Omega^{\mathrm{odd}}(M, \mathsf{C}_1(H)), d)\right).$$

Combined with any trace on $\Omega(M, \mathrm{End}\, E)$ defines a log-determinant. For example, on an odd-dimensional spin manifold the log-determinant

$$\mathrm{tr}_{\widehat{A}}(\mathsf{ch}^-(g)) = \int_M \widehat{A}(M)\mathsf{ch}^-(g)$$

computes spectral flow of a Dirac operator twisted by g.

5.6. Pseudodifferential log structures. A pseudodifferential operator (ψdo) of order m acting on the sections of a vector bundle $E \to M$ is a continuous operator $A : C^\infty(M,E) \to C^\infty(M,E)$ whose Schwartz kernel $k_A \in \mathcal{D}'(M \times M, E^* \boxtimes E)$ is an oscillatory integral of the form $k_A(x,y) = \int e^{i\langle x-y,\xi\rangle}\, s(x,y,\xi)\, d\xi$ with amplitude s an (x,y)-symbol of order m. Equivalently,

$$k_A(x,y) = \int e^{i\langle x-y,\xi\rangle} a(x,\xi)\, d\xi \qquad \mod \Psi^{-\infty}(M,E).$$

for a reduced order m symbol $a(x,\xi)$ (independent of y). We restrict to the algebra $\Psi^{\mathbb{Z}}(M,E)$ of integer order classical ψdos; 'classical' means that in each localization on M there is an asymptotic expansion $a(x,\xi) \sim \sum_j \mathsf{a}_{m-j}(x,\xi)$ with $\mathsf{a}_{m-j}(x,t\xi) = t^{m-j}\mathsf{a}_{m-j}(x,\xi)$ for $t, |\xi| \geq 1$. Then there is a unique (and exotic) scalar trace

$$\text{res} : \Psi^{\mathbb{Z}}(M,E) \to \mathbb{C}, \qquad \text{res}\, A = \int_M \text{res}_x(A),$$

the *residue trace*, where the residue density on M is

$$\text{res}_x(A) = \int_M \int_{|\xi|=1} \text{tr}\,(a_{-n}(x,\eta))\, d_S\eta\ |dx|.$$

(It is 'exotic' insofar as it vanishes on trace-class ψdos, such as smoothing operators.) This means that these locally defined expressions, for local coordinates x in a neighbourhood on M with $|dx|$ meaning local Lebesgue measure, patch together to define a globally defined density on M.

But there are numerous other traces on subalgebras of $\Psi^{\mathbb{Z}}(M,E)$. For example, the classical trace

$$\text{Tr} : \Psi^{-\infty}(M,E) \to \mathbb{C}, \ \text{Tr}\, A = \int_M \text{tr}\,(k_A(y,y)),$$

is the unique trace on the subideal of smoothing operators.

There are various ways of constructing logarithms on ψdos and hence defining determinant structures, and these may provide significant spectral, and possibly topological, invariants. We mention only the 'classical' logarithm here. Specifically, for $A \in \Psi^m(M,E)$ with Agmon angle we have the logarithm operator

$$\log_\theta A := -\left.\frac{d}{ds}\right|_{s=0} A_\theta^{-s} \in \Psi^{0,1}_{\log}(M,E).$$

Here, $\Psi^{k,l}(M,E)$ is the space of log-classical ψdos of order k and log-degree 1, meaning that P in local coordinates has symbol of the form

$$\mathsf{p}(x,\xi) \sim \sum_{j\geq 0} \sum_{l=0}^{k} \mathsf{a}_{k-j,\,l}(x,\xi) \log^l |\xi|,$$

while $\Psi^{0,1}_{\log}(M,E)$ is the subspace of $\Psi^{0,1}(M,E)$ of order zero log-classical ψdos of log-degree 1 with local symbol of the form $c \log|\xi| + \mathsf{a}_0(x,\xi)$ with $\mathsf{a}_0(x,\xi) \sim \sum_{j\geq 0} \mathsf{a}_{-j}(x,\xi)$ classical of order zero; so $\log_\theta A$ is almost, but not quite, classical of order 0 (but has log-degree 1). It is known that

$$\log_\theta AB - \log_\phi A - \log_\psi B \in [\Psi^*, \Psi^*]$$

is a commutator of classical ψdos and hence is a logarithmic structure on admissible ψdos. The residue trace is known to extend to a linear functional on $\Psi^{0,1}_{\log}(M,E)$ (but in general no further into $\Psi^{k,l}(M,E)$) with the tracial property that although

$\Psi^{0,1}_{\log}(M,E)$ is not an algebra it vanishes on $[\Psi^{0,1}_{\log}(M,E),\Psi^{0,1}_{\log}(M,E)]$. Hence one has a log-determinant structure

$$\log\det{}_{\rm res}A := \mathrm{res}\,(\log A)$$

for A,B classical and A,B,AB with principal angles, called the *residue determinant*. Thus

$$\det{}_{\rm res}(AB) = \det{}_{\rm res}(A)\cdot\det{}_{\rm res}(B).$$

The residue determinant has the interesting property that

$$\log\det{}_{\rm res}A = -\alpha\,(\zeta(A,0)\,+\dim\mathrm{Ker}\,(A)). \tag{5.6.1}$$

For admissible operators A,B,AB, the log-determinant property therefore implies the identity

$$(\alpha+\beta)\,\zeta(AB,0) = \alpha\,\zeta(A,0)+\beta\,\zeta(B,0)$$

with $\alpha,\beta>0$ the orders of A,B, respectively.

This log-determinant structure is a local invariant and hence reasonably computable. For example, it is not hard to compute on a closed surface Σ with Laplacian $\Delta_g = -\sum_{i,j} g^{ij}(x)\nabla_i\nabla_j + \varepsilon_x(\Delta_g)$ that

$$\log\det{}_{\rm res}(\Delta_g+tI) = \frac{\mathrm{Area}(\Sigma)\,\mathrm{rk}(E)}{2\pi}\,t - \frac{1}{2\pi}\int_\Sigma \mathrm{tr}\,(\varepsilon_x(\Delta_g))\,dx - \frac{\chi(\Sigma)\,\mathrm{rk}(E)}{3}.$$

That computability may be used to give an elementary (insofar as it uses only symbol computations), if not short, proof of the 'local Atiyah-Singer Index formula' for a coupled Dirac operator $\eth : C^\infty(M,S^+\otimes E)\to C^\infty(M,S^-\otimes E)$. Precisely, from (5.6.1) we obtain:

Theorem 2. *There is an equality of densities, or n-forms,*

$$-\frac{1}{2}\left(res_x(\log\eth^-\eth^+) - res_x(\log\eth^+\eth^-)\right) = \frac{1}{(2\pi i)^{\frac{n}{2}}}\left(\widehat{A}(M,R)\,ch(E,F)\right)_{[n]}$$

Here, F is the curvature form of E while $\widehat{A}(M,R)=\det^{1/2}\left(\frac{R/2}{\sinh(R/2)}\right)$ is the $\widehat{A}$-genus form with respect to the Riemannian curvature R of M.

A proof of this (using joint work with Don Zagier) may be found in [**22**].

The quasi log-determinant structure obtained by evaluating the zeta function extension of the classical trace on the logarithm $\log_\theta A$ has just a simple pole at $z=0$ (the second order pole vanishes for this particular log-classical operator). Precisely, one computes

$$\begin{aligned}\mathsf{TR}(\log_\theta A\cdot Q_\theta^z)|^{\rm mer} &= \frac{1}{z}\left(\frac{1}{\alpha}\mathrm{res}\,(\log A) - \frac{\alpha}{q^2}\,\mathrm{res}\log Q\right)\\ &+\left(\int_M \mathrm{Tr}_x(\log_\theta A) - \frac{1}{q}\mathrm{res}_x(\log_\theta A\log_\theta Q) + \frac{\alpha}{2q^2}\mathrm{res}_x(\log^2_\theta Q)\right)z^0 + \cdots\end{aligned}$$

So the residue of $\mathsf{TR}(\log_\theta A\cdot Q_\theta^{-z})$ is 'the' residue determinant

$$\log\det{}_{Q,\rm res}A = \frac{1}{\alpha}\mathrm{res}\,(\log A) - \frac{\alpha}{q^2}\,\mathrm{res}\log Q \tag{5.6.2}$$

or, formally, '$\det{}_{Q,\rm res}A = \det{}_{\rm res}(A^{1/\alpha})/\det{}_{\rm res}(Q^{\alpha/q^2})$'. Thus the residue determinant arises as a pole in $\zeta(\log_\theta A,Q,z)$ just as the residue trace arises as a pole in (2.1.1). Of course, $\log\det{}_{Q,\rm res}A$ is slightly different from the residue determinant defined above; in fact, it is a log-determinant structure of the form consisting of the residue trace evaluated on $\log_\theta A$ plus the residue trace evaluated on a second logarithm, but we will omit details here.

There are other interesting log-determinant structures on $\Psi^*(M,E)$ which we will not pursue here; for example, leading symbol traces on $\Psi^{0,1}(M,E)$ evaluated on $\log_\theta A$ define the 'leading symbol determinant' considered in [**12**], while other pseudodifferential logarithms combined with these traces define a raft of other invariants. Important pseudodifferential invariants which turn out to be log-determinant structures are Melrose's suspended eta invariant, spectral flow, and analytic torsion.

6. Log structures on the cobordism category

Here, we briefly outline the categorical formulation of a log-determinant structure along the lines of functorial QFT. Unlike TQFT this cannot be a linear representation of Cob_n. Logarithms are inherently non-linear, unlike (path) integrals, and so logarithms are not going to be functors in the usual sense. Nevertheless, with the notion of logarithmic representation at hand the way to proceed is essentially clear.

We will give only a brief indication of the structures here; a more detailed account will appear shortly.

6.1. Definition of a log on Cob_n. A logarithmic structure on the category Cob_n over a field k means that:

[1] For every object $X \in \mathsf{cbj}_n := \mathsf{obj}(\mathsf{Cob}_n)$ (closed compact $(n-1)$-manifold) there is an algebra $\mathcal{A}_X$. In particular, $\mathcal{A}_\emptyset = k$.

[2] For each pair $X, Y \in \mathsf{obj}(\mathsf{Cob}_n)$ there is an inclusion $i_X : \mathcal{A}_X \to \mathcal{A}_{X\sqcup Y}$ splitting a projection map $p_{X\sqcup Y,X} : \mathcal{A}_{X\sqcup Y} \to \mathcal{A}_X$ (that is, $p_{X\sqcup Y,X} \circ i_X$ is the identity map on $\mathcal{A}_X$).

[3] For each marked morphism $M \in \mathsf{mor}_n^W(X,Y)$ (thus $\partial M \cong \overline{X} \sqcup Z$ and there is an embedding $W \hookrightarrow \overset{o}{M}$) there is an element

$$\log_W M \in \mathcal{A}_{X\sqcup W\sqcup Z}.$$

[4] With respect to the composition by sewing

$$\mathsf{mor}_n^W(X,Y) \times \mathsf{mor}_n^\Sigma(Y,Z) \to \mathsf{mor}_n^{W\sqcup Y\sqcup\Sigma}(X,Z), \qquad (M,M') \mapsto M \cup_Y M',$$

one has

$$\log_{W\sqcup Y\sqcup\Sigma}(M \cup_Y M') - \log_W M - \log_\Sigma M' \in [\mathcal{A}_\sqcup, \mathcal{A}_\sqcup] \tag{6.1.1}$$

where $\mathcal{A}_\sqcup := \mathcal{A}_{X\sqcup W\sqcup Y\sqcup\Sigma\sqcup Z}$.

[5] If Cob_n is considered modulo diffeomorphisms, if [1] is replaced by the requirement that for every object $X \in \mathsf{obj}_n := \mathsf{obj}(\mathsf{Cob}_n)$ there is an *abelian group* $\mathcal{A}_X$, and if (6.1.1) is tightened to

$$\log_{W\sqcup Y\sqcup\Sigma}(M \cup_Y M') = \log_W M + \log_\Sigma M' \tag{6.1.2}$$

then the logarithmic structure is said to be *topological.*

If the $\mathcal{A}_X$ are algebras, then they are assumed to be algebras over the field k.

Marked morphisms were mentioned in § 3.1. The equalities (6.1.1), (6.1.2) are with respect to the inclusions of $\log_W M$ and $\log_\Sigma M'$ into $\mathcal{A}_\sqcup$.

In the first four axioms, for brevity, we are implicitly supposing that cobordisms may be regarded as different unless they actually coincide – that is, normally one requires everything modulo diffeomorphism (as in [5]).

A difference here, then, from TQFT, is that one expects a generalized relative-cohomology theory; that is, relative to the marking on M. It is essential to use marked morphisms; because logarithms have memory, they know where they were sewn together. This is in contrast to (3.0.6) which forgets the partition of the manifold. The idea here is that it is necessary to distinguish between a closed n-manifold M and, on the other hand, M with an embedded submanifold W. There are therefore two possibilities (given W)

$$M \in \mathsf{mor}_n(\emptyset, \emptyset) \quad \Rightarrow \quad \text{scalar valued } \log_k M \in \mathcal{A}_\emptyset = k$$

or

$$M \in \mathsf{mor}_n^W(\emptyset, \emptyset) \quad \Rightarrow \quad \text{operator valued } \log_W M \in \mathcal{A}_W.$$

In view of [2] there is a 'trace' $\tau_W : \mathcal{A}_W \to \mathcal{A}_\emptyset = k$ and we expect

$$\log_k M = \tau_W(\log_W M).$$

There are a number of variations one can contemplate including in the definition, and various consequences that must be explained, but for this short review we must be content to leave matters for a future exposition.

6.2. Example: Dirac index. The most elementary example of a scalar valued logarithmic structure is to take the relative Euler number $\chi(M, \partial M)$, which is additive with respect to sewing together manifolds. The trace-log structure is seen at the chain complex level.

On the other hand, for a general compatible Dirac operator $\eth$ acting on a bundle of Clifford modules $E \to M$ over a closed manifold M define

$$\log_k M = \operatorname{ind} \eth,$$

while to $M = M_0 \cup_Y M_1 \in \mathsf{mor}_n^Y(\emptyset, \emptyset)$ one assigns as follows. Let $\eth^i$ be the restriction of $\eth$ to M_i (which we assume has a collar neighbourhood near the boundary). Let P_{K_i} be the corresponding Calderón projector in the space $\mathsf{B}(L^2(Y, E_{|Y}))$ of boundary sections. Then set

$$\log_Y M := [P_{K_1}^\perp + P_{K_0}, L_{0,1}] : L^2(Y, E_Y) \to L^2(Y, E_Y)$$

where $L_{0,1}$ is a parametrix for the Fredholm operator $P_{K_1}^\perp + P_{K_0}$. The character of this operator is the index $\operatorname{ind} \eth$, but this particular logarithm operator depends on the choice of splitting codimension 1 manifold Y. To the restriction $\eth^i$ of $\eth$ to M^i one sets

$$\log_Y M_i := [P_{K_i}^\perp + \Pi_i, L_i] : L^2(Y, E_Y) \to L^2(Y, E_Y),$$

with Π_i is the APS projection for M_i and L_i a parametrix for $P_{K_i}^\perp + \Pi_i$. This logarithm operator has character equal to the index of the elliptic boundary problem $\eth^i_{\Pi_i}$.

It is then not hard to see using well known identifications that this defines a logarithmic structure on Cob_n with log-multiplicativity the well known sewing formula for the index

$$\operatorname{ind} \eth = \operatorname{ind}(\eth^0_{\Pi_0}) + \operatorname{ind}(\eth^1_{\Pi_1}).$$

This extends to the geometric fibre cobordism category $\mathbb{C}\mathsf{ob}_n$, defined in § 3.1 but here also endowed with a vertical metric on each fibration and spin structure, with compatibility with boundary structures. In this case an object is a geometric fibration $(N,B) := \pi : N = \bigcup_{b\in B} N_b \to B \in \mathsf{obj}(\mathsf{Cob}_n)$ of compact closed manifolds and we define

$$\log(N,B) = K_0(B)$$

the even K-theory on the base; this depends therefore only on the parameter manifold B. To a morphism $(M,B) \in \mathsf{mor}_n((N,B),(N',B))$, a geometric fibration $M \to B$ of manifolds with boundary with $\partial M \cong \overline{N} \sqcup N'$, one assigns the index bundle

$$\log(M,B) = \mathrm{Ind}\,(\eth_\Pi) \in K_0(B)$$

where $\eth_\Pi$ is now the family of Dirac operators with APS boundary conditions defined by the geometric fibration. Take $\partial M = \emptyset$ with a fibrewise partition $M = M_0 \cup_Y M_1 \to B$. Let $\eth^i$ be the restriction of $\eth$ to M_i. Then it is not hard to see that

$$\mathrm{Ind}\,(\eth^0_\Pi) + \mathrm{Ind}\,(\eth^1_\Pi) = \mathrm{Ind}\,(\eth_\Pi) \quad \text{in } K_0(B).$$

A similar logarithmic sewing formula holds when $\partial M \neq \emptyset$.

A natural trace map in this case defining the log-determinant is the Chern class $c : K_0(B) \to H^{\mathrm{even}}(B)$.

The logarithmic structure here uses the following quite general log-determinant property.An object in the category $\mathsf{C}_{\mathsf{Fred}}$ is a vector bundle $\mathcal{H} = \bigcup_{x\in X} H_x \to X$ of Hilbert spaces H_x over a compact manifold X. A morphism $\mathsf{B} \in \mathtt{mor}(\mathcal{H}^0,\mathcal{H}^1)$ is a bundle homomorphism $\mathcal{H}^0 \to \mathcal{H}^1$ defined by a continuous family of Fredholm operators, assigning to each $x \in X$ a Fredholm operator $B_x := \mathsf{B}(x) \in \mathrm{Fred}(H^0_x, H^1_x)$. A log map is defined on $\mathsf{C}_{\mathsf{Fred}}$ with values in the ring $A = K_0(X)$ by associated the index bundle Ind B. The log-property is then topological (exact)

$$\mathrm{Ind}\,(\mathsf{A}\circ\mathsf{B}) = \mathrm{Ind}\,\mathsf{A} + \mathrm{Ind}\,\mathsf{B} \qquad \text{in } K_0(X). \tag{6.2.1}$$

An 'exponential map' e in this case is to take the top exterior power giving the determinant line bundle

$$\det_{0,1} : \mathrm{mor}(\mathcal{H}^0,\mathcal{H}^1) \to \mathrm{Vect}(X), \qquad \mathsf{B} \mapsto \mathrm{Det\,Ind}\,\mathsf{B} = \Lambda^{\max}(\mathrm{Ker}\ \mathsf{B})^* \otimes \Lambda^{\max}\mathrm{Cok}\,\mathsf{B}$$

which inherits the determinant property

$$\mathrm{Det\,Ind}\,(\mathsf{A}\circ\mathsf{B}) \cong \mathrm{Det\,Ind}\,\mathsf{A} \otimes \mathrm{Det\,Ind}\,\mathsf{B}.$$

The logarithmic and trace structure may be refined to the smooth categories by defining the logarithm to be the differential form valued family of vertical ψdos

$$\log M = \log_\pi(I + \mathbb{A}^2) \in \mathcal{A}(M, \mathrm{End}\,(E))$$

and the trace to be integration over the fibre $\int_{M/B} \circ \mathrm{tr} : \mathcal{A}(M,\mathrm{End}\,(E)) \to \mathcal{A}(B)$; this is straightforward and well understood for the case of closed manifolds, and appears to hold for the non-empty boundary case without difficulty but has not been written down in detail in the literature. For more on this aspect see [**2**], [**3**], [**21, 22**].

6.3. Example: various. We mention without further detail that the *spectral flow* and *analytic torsion* define natural logarithmic representations of Cob_n.

In a different way one may construct logarithmic representations of Cob_2 using the logarithmic structures in § 5 applied to the *2D topological quantum field theory* with tracial algebra $\mathcal{A} = (Z(S^1),\tau)$ – this case is simplified by there being only one compact connected non-empty boundary manifold of dimension one.

References

[1] Atiyah, M.F.: 1989, Topological quantum field theories. Inst. Hautes Etudes Sci. Publ. Math. **68**, 175.

[2] Berline, N., E. Getzler, M. Vergne: 'Heat Kernels and Dirac Operators'. Grundlehren der Mathematischen Wissenschaften **298**, Springer-Verlag, Berlin, 1992.

[3] Bismut, J-M.: 1986, 'The Atiyah-Singer index theorem for families of Dirac operators: Two heat equation proofs', Invent. Math **83**, 91-151.

[4] Blumberg, A., Cohen, R., and Teleman, C.: 2009, 'Open-closed field theories, string topology, and Hochschild homology'.

[5] Brüning, J., Lesch, M.:1999, 'On the eta-invariant of certain non-local boundary value problems', Duke Math. J. **96**, 425-468.

[6] Booß–Bavnbek, B., Wojciechowski, K.P.: 1993, 'Elliptic Boundary Problems for Dirac Operators', Birkhäuser, Boston.

[7] Bunke, U.: 1977, 'On the gluing problem for the eta invariant': 1995, J. Diff. Geom. **41**, 397–448.

[8] Costello, K.: 2007, 'Topological conformal field theories and Calabi-Yau categories', Adv. Math. **210**, 165-214.

[9] Dai, X., Freed, D.: 1994, 'Eta invariants and determinant lines', J. Math. Phys. **35** 5155-5194.

[10] Fedosov, B., Golse, F., Leichtnam, E., Schrore, E. : 1996, 'The noncommutative residue for manifolds with boundary', J. Funct. Anal. **142**, 1-31.

[11] Grubb, G., Schrohe, E.: 2004, 'Traces and quasi-traces on the Boutet de Monvel algebra', Ann. Inst. Fourier **54**, 1641-1696.

[12] Lescure, J-M. and Paycha, S: (2007). 'Uniqueness of multiplicative determinants on elliptic pseudodifferential operators', Proc. Lond. Math. Soc., **94**, 772-812.

[13] Lurie, J.: 2009, 'On the classification of topological field theories', Current Developments in Mathematics **2008**, 129-280.

[14] Moore, G. and Segal, G.: 'D-branes and K-theory in 2D topological field theory', preprint arXiv hep-th/0609042.

[15] Piazza, P.: 1996, 'Determinant bundles, manifolds with boundary and surgery I', Comm. Math. Phys. **178**, 597-626.

[16] Park, J., Wojciechowski, K.P.: 2002, 'Scattering theory and adiabatic decomposition of the ζ-determinant of the Dirac Laplacian', Math. Res. Lett. **9**, 17-25

[17] Paycha, S., Scott, S.: 2007, 'A Laurent expansion for regularized integrals of holomorphic symbols', Geom. and Funct. Anal. **17**, 491-536.

[18] Okikiolu, K. : 1995, 'The multiplicative anomaly for determinants of elliptic operators', Duke Math. Journ. **79**, 723-750.

[19] Pressley, A. and Segal, G.B.: *Loop Groups*. OUP, Math. Monographs, 1986.

[20] Scott, S. 2005. 'The residue determinant', Commun. Part. diff. Equ. **30**, 483-507.

[21] Scott, S.: 2007, 'Zeta forms and the local family index theorem', Trans. Am. Math. Soc. **359**, 1925 -1957.

[22] Scott, S.: 2010, 'Traces and Determinants of Pseudodifferential Operators', OUP, Math. Monographs, to appear.

[23] Scott, S.: 2010, 'Eta forms and determinant line bundles', Adv. Math., to appear.

[24] Segal, G.: 2000, 'Lectures on QFT', Stanford Lectures.

[25] Wojciechowski, K.P.: 1994, 'The additivity of the η-invariant: The case of an invertible tangential operator', Houston J. Math. **20**, 603-621.

[26] Wojciechowski, K.P.: 1995, 'The additivity of the η-invariant. The case of a singular tangential operator', *Comm. Math. Phys.* **169**, 315-327.

[27] Wojciechowski, K.P.: 1999, 'The ζ-determinant and the additivity of the η-invariant on the smooth, self-adjoint Grassmannian', Comm. Math. Phys. **201**, 423-444.

[28] Witten, E: 1987, 'Physics and geometry', Proc. Internat. Congr. Math., Berkeley, 1986, Amer. Math. Soc., Providence, R. I., 267-302.

Department of Mathematics, King's College London, London WC2R 2LS, UK
E-mail address: `simon.scott@kcl.ac.uk`

Clay Mathematics Proceedings
Volume 12, 2010

Renormalization Hopf algebras for gauge theories and BRST-symmetries

Walter D. van Suijlekom

ABSTRACT. The structure of the Connes–Kreimer renormalization Hopf algebra is studied for gauge theories, with particular emphasis on the BRST-formalism. We work in the explicit example of quantum chromodynamics, the physical theory of quarks and gluons.

A coaction of the renormalization Hopf algebra is defined on the coupling constants and the fields. In this context, BRST-invariance of the action implies the existence of certain Hopf ideals in the renormalization Hopf algebra, encoding the Slavnov–Taylor identities for the coupling constants.

1. Introduction

Quantum gauge field theories are most successfully described perturbatively, expanding around the free quantum field theory. In fact, at present its non-perturbative formulation seems to be far beyond reach so it is the only thing we have. On the one hand, many rigorous results can be obtained [**2, 3**] using cohomological arguments within the context of the BRST-formalism [**4, 5, 6, 17**]. On the other hand, renormalization of perturbative quantum field theories has been carefully structured using Hopf algebras [**14, 7, 8**]. The presence of a gauge symmetry induces a rich additional structure on these Hopf algebras, as has been explored in [**15, 16, 1**] and in the author's own work [**18, 19, 20**]. All of this work is based on the algebraic transparency of BPHZ-renormalization, with the Hopf algebra reflecting the recursive nature of this procedure.

In this article we study more closely the relation between the renormalization Hopf algebras and the BRST-symmetries for gauge theories. We work in the explicit case of quantum chromodynamics (QCD), a Yang–Mills gauge theory with gauge group SU(3) that describes the strong interaction between quarks and gluons. We will shortly describe this in a little more detail, as well as the appearance of BRST-symmetries.

After describing the renormalization Hopf algebra for QCD, we study its structure in Section 3. The link between this Hopf algebra and the BRST-symmetries acting on the fields is established in Section 4.

Acknowledgements

The author would like to thank the participants of the workshop "DIAMANT meets GQT" at the Lorentz Center in Leiden.

2. Quantum chromodynamics

In order to keep the discussion in this article as explicit as possible, we will work in the setting of quantum chromodynamics (QCD). This is an example of a Yang–Mills gauge theory, as introduced originally in [**23**]. It is the physical theory that successfully describes the so-called strong interaction between quarks and gluons. Let us make more precise how these particles can be described mathematically, at least in a perturbative approach.

One of the basic principles in the dictionary between the (elementary particle) physicists' and mathematicians' terminology is that

"particles are representations of a Lie group."

In the case of quantum chromodynamics, this Lie group – generally called the gauge group – is SU(3). In fact, the *quark* is a $\mathbb{C}^3$-valued function $\psi = (\psi_i)$ on spacetime M. This 'fiber' $\mathbb{C}^3$ at each point of spacetime is the defining representation of SU(3). Thus, there is an action on ψ of an SU(3)-valued function on M; let us write this function as U, so that $U(x) \in \mathrm{SU}(3)$. In physics, the three components of ψ correspond to the so-called *color* of the quark, typically indicated by red, green and blue.

The *gluon*, on the other hand, is described by an $\mathfrak{su}(3)$-valued one-form on M, that is, a section of $\Lambda^1(\mathfrak{su}(3)) \equiv \Lambda^1 \otimes (M \times \mathfrak{su}(3))$. We have in components

$$A = A_\mu dx^\mu = A_\mu^a dx^\mu T^a$$

where the $\{T^a\}_{a=1}^8$ form a basis for $\mathfrak{su}(3)$. The structure constants $\{f_c^{ab}\}$ of $\mathfrak{g}$ are defined by $[T^a, T^b] = f_c^{ab} T^c$ and the normalization is such that $\mathrm{tr}\,(T^aT^b) = \delta^{ab}$. It is useful to think of A as a connection one-form (albeit on the trivial bundle $M \times \mathrm{SU}(3)$). The group SU(3) acts on the second component $\mathfrak{su}(3)$ in the adjoint representation. Again, this is pointwise on M, leading to an action of $U = U(x)$ on A. In both cases, that is, for quarks and gluons, the transformations

$$\psi_i \mapsto U_{ij}\psi_j, \qquad A_\mu \mapsto g^{-1}U^{-1}\partial_\mu U + U^{-1}A_\mu U \tag{1}$$

are called *gauge transformations*. The constant g is the so-called *strong coupling constant*.

As in mathematics, also in physics one is after *invariants*, in this case, one looks for functions – or, rather, functionals – of the quark and gluon fields that are invariant under a local (i.e. x-dependent) action of SU(3). We are interested in the following action functional:

$$S(A,\psi) = \frac{1}{8\pi}\int_M F_{\mu\nu}^a F_a^{\mu\nu} + \overline{\psi}_i(i\gamma^\mu\partial_\mu + \gamma^\mu A_\mu^a T_{ij}^a + m)\psi_j \tag{2}$$

with $F \equiv F(A) := dA + gA^2$ the curvature of A; it is an $\mathfrak{su}(3)$-valued 2-form on M. Before checking that this is indeed invariant under SU(3), let us explain the notation in the last term. The γ^μ are the Dirac matrices, and satisfy

$$\gamma^\mu\gamma^\nu + \gamma^\nu\gamma^\mu = -2\delta^{\mu\nu}$$

Clearly, this relation cannot be satisfied by complex numbers (which are never anti-commuting). In fact, the representation theory of the algebra with the above relation (i.e. the Clifford algebra) is quite rich. The idea is that the fields ψ are not only $\mathbb{C}^3$-valued, but that actually each of the components ψ_i is itself a 4-vector, called *spinors*. This is so as to accommodate a representation of the Clifford algebra: in 4 spacetime dimensions the Dirac matrices are 4-dimensional (although in general this dimension is $2^{[n/2]}$ for n spacetime dimensions). Besides this matrix multiplication, the partial derivative ∂_μ acts componentswise, as does the *mass* m which is really just a real number. Finally, for our purposes it is sufficient to think of $\overline{\psi}$ as the (componentswise) complex conjugate of ψ. The typical Grassmannian nature of these fermionic fields is only present in the current setup through the corresponding *Grassmann degree* of $+1$ and -1 that is assigned to both of them.

Introducing the notation $\partial\!\!\!/ = \gamma^\mu \partial_\mu$ and $A\!\!\!/ = \gamma^\mu A_\mu$, we can write

$$S(A,\psi) = -\langle F(A), F(A)\rangle + \langle \psi, (i\partial\!\!\!/ + A\!\!\!/ + m)\psi\rangle,$$

in terms of appropriate inner products. Essentially, these are combinations of spinorial and Lie algebra traces and the inner product on differential forms. For more details, refer to the lectures by Faddeev in [**9**]. The key observation is that the SU(3)-valued functions $U(x)$ act by unitaries with respect to this inner product.

2.1. Ghost fields and BRST-quantization. In a path integral quantization of the field theory defined by the above action, one faces the following problem. Since gauge transformations are supposed to act as symmetries on the theory, the gauge degrees of freedom are irrelevant to the final physical outcome. Thus, in one way or another, one has to quotient by the group of gauge transformations. However, gauge transformations are SU(3)-valued function on M, yielding an infinite dimensional group. In order to deal with this infinite redundancy, Faddeev and Popov used the following trick. They introduced so-called *ghost fields*, denoted by ω and $\overline{\omega}$. In the case of quantum chromodynamics, these are $\mathfrak{su}(3)[-1]$ and $\mathfrak{su}(3)[1]$-valued functions on M, respectively. The shift $[-1]$ and $[+1]$ is to denote that ω and $\overline{\omega}$ have *ghost degree* 1 and -1, respectively. Consequently, they have *Grassmann degree* 1 and -1, respectively. In components, we write

$$\omega = \omega^a T^a; \qquad \overline{\omega} = \overline{\omega}^a T^a.$$

Finally, an auxiliary field h – also known as the Nakanishi–Lautrup field – is introduced; it is an $\mathfrak{su}(3)$-valued function (in ghost degree 0) and we write $h = h^a T^a$.

The dynamics of the ghost fields and their interaction with the gauge field are described by the rather complicated additional term:

$$S_{\text{gh}}(A,\omega,\overline{\omega},h) = -\langle A, dh\rangle + \langle d\overline{\omega}, d\omega\rangle + \frac{1}{2}\xi\langle h,h\rangle + g\langle d\overline{\omega}, [A,\omega]\rangle,$$

where $\xi \in \mathbb{R}$ is the so-called *gauge parameter*.

The essential point about the ghost fields is that, in a path integral formulation of quantum gauge field theories, their introduction miraculously takes care of the *fixing of the gauge*, i.e. picking a point in the orbit in the space of fields under the action of the group of gauge transformations. The ghost fields are the ingredients in the BRST-formulation that was developed later by Becchi, Rouet, Stora and independently by Tyutin in [**4, 5, 6, 17**]. Let us briefly describe this formalism.

Because the gauge has been fixed by adding the term S_{gh}, the combination $S + S_{\text{gh}}$ is not invariant any longer under the gauge transformations. This is of course precisely the point. Nevertheless, $S+S_{\rm gh}$ possesses another symmetry, which is the BRST-symmetry. It acts on function(al)s in the fields as a ghost degree 1 derivation s, which is defined on the generators by

$$sA = d\omega + g[A,\omega], \qquad s\omega = \frac{1}{2}g[\omega,\omega], \qquad s\overline{\omega} = h \tag{3}$$
$$sh = 0 \qquad s\psi = g\omega\psi, \qquad s\overline{\psi} = g\overline{\psi}\omega.$$

Indeed, one can check (eg., see [**21**, Sect. 15.7] for details) that $s(S+S_{\rm gh}) = 0$.

The form degree and Grassmann degree of the fields are combined in the *total degree* and summarized in the following table:

	A	ω	$\overline{\omega}$	h	ψ	$\overline{\psi}$
Grassmann degree	0	+1	−1	0	+1	−1
form degree	+1	0	0	0	0	0
total degree	+1	+1	−1	0	+1	−1

The fields generate an algebra, the algebra of local forms $\mathrm{Loc}(\Phi)$. With respect to the above degrees, it decomposes as before into $\mathrm{Loc}^{(p,q)}(\Phi)$ with p the form degree and q the Grassmann degree. The total degree is then $p+q$ and $\mathrm{Loc}(\Phi)$ is a graded Lie algebra by setting

$$[X,Y] = XY - (-1)^{\deg(X)\deg(Y)}YX,$$

with the grading given by this total degree. Note that the present graded Lie bracket is of degree 0 with respect to the total degree, that is, $\deg([X,Y]) = \deg(X) + \deg(Y)$. It satisfies graded skew-symmetry, the graded Leibniz identity and the graded Jacobi identity:

$$[X,Y] = -(-1)^{\deg(X)\deg(Y)}[Y,X],$$
$$[XY,Z] = X[Y,Z] + (-1)^{\deg(Y)\deg(Z)}[X,Z]Y.$$
$$(-1)^{\deg(X)\deg(Z)}[[X,Y],Z] + (\text{cyclic perm.}) = 0,$$

LEMMA 1. *The BRST-differential, together with the above bracket, gives* $\mathrm{Loc}(\Phi)$ *the structure of a graded differential Lie algebra.*

Moreover, the BRST-differential s and the exterior derivative d form a double complex, that is, $d \circ s + s \circ d = 0$ and

$$\begin{array}{ccccccc}
\vdots & & \vdots & & \vdots & & \\
\uparrow s & & \uparrow s & & \uparrow s & & \\
\mathrm{Loc}^{(0,1)} & \xrightarrow[d]{} & \mathrm{Loc}^{(1,1)} & \xrightarrow[d]{} & \mathrm{Loc}^{(2,1)} & \xrightarrow[d]{} & \cdots \\
\uparrow s & & \uparrow s & & \uparrow s & & \\
\mathrm{Loc}^{(0,0)} & \xrightarrow[d]{} & \mathrm{Loc}^{(1,0)} & \xrightarrow[d]{} & \mathrm{Loc}^{(2,0)} & \xrightarrow[d]{} & \cdots \\
\uparrow s & & \uparrow s & & \uparrow s & & \\
\mathrm{Loc}^{(0,-1)} & \xrightarrow[d]{} & \mathrm{Loc}^{(1,-1)} & \xrightarrow[d]{} & \mathrm{Loc}^{(2,-1)} & \xrightarrow[d]{} & \cdots \\
\uparrow s & & \uparrow s & & \uparrow s & & \\
\vdots & & \vdots & & \vdots & &
\end{array}$$

This double complex has a quite interesting structure in itself, and was the subject of study in [**2, 3**]. This contained further applications in renormalization and the description of anomalies.

3. Renormalization Hopf algebra for QCD

As we discussed previously, quantum chromodynamics describes the interaction between quarks and gluons. In order to do this successfully at a quantum level, it was necessary to introduce ghost fields. We will now describe how the dynamics and interaction of and between these fields, naturally give rise to Feynman graphs. These constitute a Hopf algebra which encodes the procedure of renormalization in QCD. We will describe this Hopf algebra, and study its structure in terms of the so-called Green's functions.

3.1. Hopf algebra of Feynman graphs. First of all, the quark, ghost and gluon fields are supposed to *propagate*, this we will denote by a straight. dotted and curly line or *edges* as follows:

$e_1 =$  $e_2 =$ ········ $e_3 =$  .

The interactions between the fields then naturally appear as *vertices*, connecting the edges corresponding to the interacting fields. The allowed interactions in QCD are the following four:

$v_1 =$  , $v_2 =$  , $v_3 =$  , $v_4 =$  .

In addition, since the quark is supposed to have a mass, there is a *mass term*, which we depict as a vertex of valence two:

$v_5 =$  .

We can make the relation between these edges (vertices) and the propagation (interaction) more precise through the definition of a map ι that assigns to each of the above edges and vertices a (monomial) functional in the fields. In fact, the assignment $e_i \mapsto \iota(e_i)$ and $v_j \mapsto \iota(v_j)$ is

	e_1	e_2	e_3	v_1	v_2	v_3	v_4	v_5
edge/vertex		·······						
monomial ι	$i\overline{\psi}\partial\!\!\!/\psi$	$d\overline{\omega}d\omega$	$dAdA$	$\psi A\!\!\!/\psi$	$\overline{\omega}[A,\omega]$	$2dAA^2$	A^4	$m\overline{\psi}\psi$

FIGURE 1. QCD edges and vertices, and (schematically) the corresponding monomials in the fields.

REMARK 2. *We have not assigned an edge to the field h; this is because it does not interact with any of the other fields. Its only – still crucial – effect is on the propagator of the gluon, through the terms $-\langle A, dh\rangle$ and $\frac{1}{2}\xi\langle h, h\rangle$.*

A *Feynman graph* is a graph built from these vertices and edges. Naturally, we demand edges to be connected to vertices in a compatible way, respecting their straight, dotted or curly character. As opposed to the usual definition in graph theory, Feynman graphs have no external vertices. However, they do have *external lines* which come from vertices in Γ for which some of the attached lines remain vacant (i.e. no edge attached).

If a Feynman graph Γ has two external quark (straight) lines, we would like to distinguish between the propagator and mass terms. Mathematically, this is due to the presence of the vertex of valence two. In more mathematical terms, since we have vertices of valence two, we would like to indicate whether a graph with two external lines corresponds to such a vertex, or to an edge. A graph Γ with two external lines is dressed by a bullet when it corresponds to a vertex, i.e. we write $\Gamma_\bullet$. The above correspondence between Feynman graphs and vertices/edges is given by the *residue* $\text{res}(\Gamma)$. It is defined for a general graph as the vertex or edge it corresponds to after collapsing all its internal points. For example, we have:

res 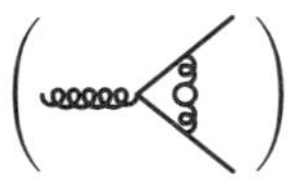 =  and res 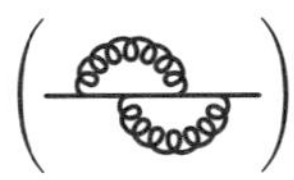 =

but

res 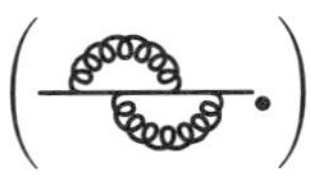 =  .

For the definition of the Hopf algebra of Feynman graphs, we restrict to *one-particle irreducible* (1PI) Feynman graphs. These are graphs that are not trees and cannot be disconnected by cutting a single internal edge.

DEFINITION 3 (Connes–Kreimer [**7**]). *The Hopf algebra H_{CK} of Feynman graphs is the free commutative algebra over $\mathbb{C}$ generated by all 1PI Feynman graphs with residue in $R = R_V \cup R_E$, with counit $\epsilon(\Gamma) = 0$ unless $\Gamma = \emptyset$, in which case $\epsilon(\emptyset) = 1$, and coproduct*

$$\Delta(\Gamma) = \Gamma \otimes 1 + 1 \otimes \Gamma + \sum_{\gamma \subsetneq \Gamma} \gamma \otimes \Gamma/\gamma,$$

where the sum is over disjoint unions of 1PI subgraphs with residue in R. The quotient Γ/γ is defined to be the graph Γ with the connected components of the subgraph contracted to the corresponding vertex/edge. If a connected component γ' of γ has two external lines, then there are possibly two contributions corresponding to the valence two vertex and the edge; the sum involves the two terms $\gamma'_\bullet \otimes \Gamma/(\gamma' \to \bullet)$ and $\gamma' \otimes \Gamma/\gamma'$. The antipode is given recursively by

$$S(\Gamma) = -\Gamma - \sum_{\gamma \subsetneq \Gamma} S(\gamma)\Gamma/\gamma. \tag{4}$$

Two examples of this coproduct are:

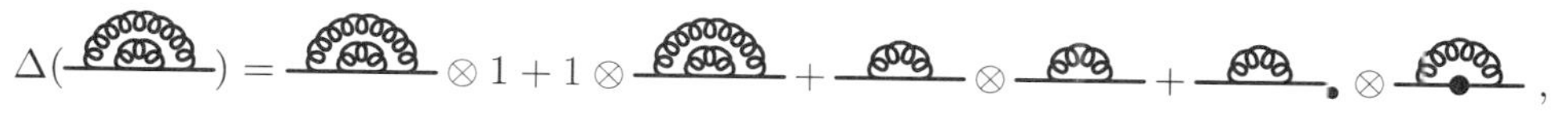

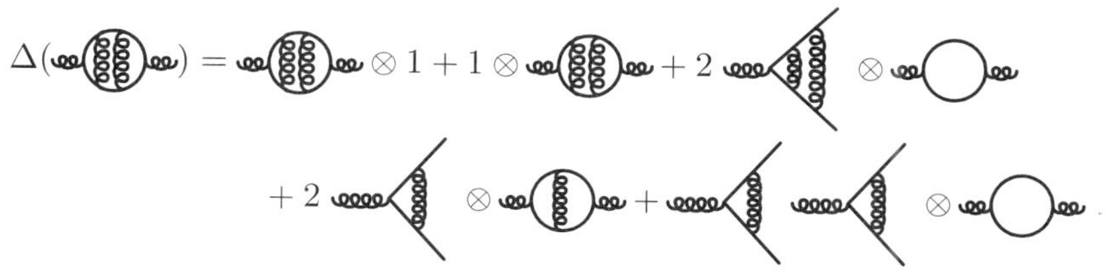

The above Hopf algebra is an example of a connected graded Hopf algebra: it is graded by the *loop number* $L(\Gamma)$ of a graph Γ. Indeed, one checks that the coproduct (and obviously also the product) satisfy the grading by loop number and H^0_{CK} consists of complex multiples of the empty graph, which is the unit in H_{CK}, so that $H^0_{\mathrm{CK}} = \mathbb{C}1$. We denote by q_l the projection of H_{CK} onto H^l_{CK}.

In addition, there is another grading on this Hopf algebra. It is given by the number of vertices and already appeared in [**7**]. However, since we consider vertices and edges of different types (straight, dotted and curly), we extend to a multigrading as follows. For each vertex v_j $(j = 1, \dots, 5)$ we define a degree d_j as

$$d_j(\Gamma) = \#\text{vertices } v_j \text{ in } \Gamma - \delta_{v_j,\mathrm{res}(\Gamma)}$$

The multidegree indexed by $j = 1, \dots, 5$ is compatible with the Hopf algebra structure, since contracting a subgraph $\Gamma \mapsto \Gamma/\gamma$ creates a new vertex. With this one easily arrives at the following relation:

$$d_j(\Gamma/\gamma) = d_j(\Gamma) - d_j(\gamma)$$

Moreover, $d_j(\Gamma\Gamma') = d_j(\Gamma) + d_j(\Gamma')$ giving a decomposition as vector spaces:

$$H_{\mathrm{CK}} = \bigoplus_{(n_1,\dots,n_5)\in\mathbb{Z}^5} H^{n_1,\dots,n_5}_{\mathrm{CK}},$$

We denote by $p_{n_1,\dots,n_5}$ the projection onto $H^{n_1,\dots,n_5}_{\mathrm{CK}}$. Note that also $H^{0,\dots,0}_{\mathrm{CK}} = \mathbb{C}1$.

LEMMA 4. *There is the following relation between the grading by loop number and the multigrading by number of vertices:*

$$\sum_{j=1}^{5} (N(v_j) - 2) d_j = 2L$$

where $N(v_j)$ is the valence of the vertex v_j.

PROOF. This can be easily proved by induction on the number of internal edges using invariance of the quantity $\sum_j (N(v_j) - 2)d_j - 2L$ under the adjoint of an edge. □

The group $\mathrm{Hom}_{\mathbb{C}}(H_{\mathrm{CK}}, \mathbb{C})$ dual to H_{CK} is called the *group of diffeographisms* (for QCD). This name was coined in general in [**8**] motivated by its relation with the group of (formal) diffeomorphisms of $\mathbb{C}$ (see Section 4 below). Stated more precisely, they constructed a map from the group of diffeographisms to the group of formal diffeomorphisms. We have established this result in general (i.e. for any quantum field theory) in [**20**]. Below, we will make a similar statement for Yang–Mills gauge theories.

3.2. Birkhoff decomposition. We now briefly recall how renormalization is an instance of a Birkhoff decomposition in the group of characters of H as established in [**7**]. Let us first recall the definition of a Birkhoff decomposition.

We let $\gamma : C \to G$ be a loop with values in an arbitrary complex Lie group G, defined on a smooth simple curve $C \subset \mathcal{P}_1(\mathbb{C})$. Let $C_\pm$ be the two complements of C in $\mathcal{P}_1(\mathbb{C})$, with $\infty \in C_-$. A *Birkhoff decomposition* of γ is a factorization of the form

$$\gamma(z) = \gamma_-(z)^{-1}\gamma_+(z); \qquad (z \in C),$$

where $\gamma_\pm$ are (boundary values of) two holomorphic maps on $C_\pm$, respectively, with values in G. This decomposition gives a natural way to extract finite values from a divergent expression. Indeed, although $\gamma(z)$ might not holomorphically extend to C_+, $\gamma_+(z)$ is clearly finite as $z \to 0$.

Now consider a Feynman graph Γ in the Hopf algebra H_{CK}. Via the so-called Feynman rules – which are dictated by the Lagrangian of the theory – one associates to Γ the Feynman amplitude $U(\Gamma)(z)$. It depends on some regularization parameter, which in the present case is a complex number z (dimensional regularization). The famous divergences of quantum field theory are now 'under control' and appear as poles in the Laurent series expansion of $U(\Gamma)(z)$.

On a curve around $0 \in \mathcal{P}^1(\mathbb{C})$ we can define a loop γ by $\gamma(z)(\Gamma) := U(\Gamma)(z)$ which takes values in the group of diffeographisms $G = \text{Hom}_{\mathbb{C}}(H_{\text{CK}}, \mathbb{C})$. Connes and Kreimer proved the following general result in [**7**].

THEOREM 5. *Let H be a graded connected commutative Hopf algebra with character group G. Then any loop $\gamma : C \to G$ admits a Birkhoff decomposition.*

In fact, an explicit decomposition can be given in terms of the group $G(K) = \text{Hom}_{\mathbb{C}}(H, K)$ of K-valued characters of H, where K is the field of convergent Laurent series in z.[1] If one applies this to the above loop associated to the Feynman rules, the decomposition gives exactly renormalization of the Feynman amplitude $U(\Gamma)$: the map γ_+ gives the renormalized Feynman amplitude and the γ_- provides the counterterm.

Although the above construction gives a very nice geometrical description of the process of renormalization, it is a bit unphysical in that it relies on individual graphs that generate the Hopf algebra. Rather, in physics the probability amplitudes are computed from the full expansion of Green's functions. Individual graphs do not correspond to physical processes and therefore a natural question to pose is how the Hopf algebra structure behaves at the level of the Green's functions. We will see in the next section that they generate Hopf subalgebras, i.e. the coproduct closes on Green's functions. Here the so-called Slavnov–Taylor identities for the couplings will play a prominent role.

3.3. Structure of the Hopf algebra. In this subsection, we study the structure of the above Hopf algebra of QCD Feynman graphs. In fact, from a dual point of view, the group of diffeographisms turns out to be related to the group of formal diffeomorphisms of $\mathbb{C}^5$. Moreover, we will establish the existence of Hopf ideals, which correspond on the group level to subgroups.

---

[1] In the language of algebraic geometry, there is an affine group scheme G represented by H in the category of commutative algebras. In other words, $G = \text{Hom}_{\mathbb{C}}(H, \cdot\,)$ and $G(K)$ are the K-points of the group scheme.

We define the *1PI Green's functions* by

$$(5) \qquad G^{e_i} = 1 - \sum_{\text{res}(\Gamma)=e_i} \frac{\Gamma}{\text{Sym}(\Gamma)}, \qquad G^{v_j} = 1 + \sum_{\text{res}(\Gamma)=v_j} \frac{\Gamma}{\text{Sym}(\Gamma)}$$

with $i = 1,2,3$ and $j = 1,\ldots,5$. The restriction of the sum to graphs Γ at loop order $L(\Gamma) = l$ is denoted by G^r_l, with $r \in \{e_i, v_j\}_{i,j}$.

REMARK 6. *Let us explain the meaning of the inverse of Green's functions in our Hopf algebra. Since any Green's function G^r starts with the identity, we can surely write its inverse formally as a geometric series. Recall that the Hopf algebra is graded by loop number. Hence, the inverse of a Green's function at a fixed loop order is in fact well-defined; it is given by restricting the above formal series expansion to this loop order. More generally, we understand any real power of a Green's function in this manner.*

We state without proof the following result of [**20**].

PROPOSITION 7. *The coproduct takes the following form on (real powers of) the Green's functions:*

$$\Delta\big((G^{e_i})^\alpha\big) = \sum_{n_1,\ldots,n_5} (G^{e_i})^\alpha Y_{v_1}^{n_1} \cdots Y_{v_5}^{n_5} \otimes p_n\big((G^{e_i})^\alpha\big),$$

$$\Delta\big((G^{v_j})^\alpha\big) = \sum_{n_1,\ldots,n_5} (G^{v_j})^\alpha Y_{v_1}^{n_1} \cdots Y_{v_5}^{n_5} \otimes p_n\big((G^{v_j})^\alpha\big),$$

with $\alpha \in \mathbb{R}$. Consequently, the algebra H generated by the Green's functions (in each vertex multidegree) G^{e_i} ($i = 1,2,3$) and G^{v_j} ($j = 1,\ldots,5$) is a Hopf subalgebra of H_{CK}.

Denote by $N_k(r)$ the number of edges e_k attached to $r \in \{e_i, v_j\}_{i,j}$; clearly, the total number of lines attached to r can be written as $N(r) = \sum_{i=1,2,3} N_i(r)$. With this notation, define for each vertex v_j an element in H by the formal expansion:

$$Y_{v_j} := \frac{G^{v_j}}{\prod_{i=1,2,3} (G^{e_i})^{N_i(v_j)/2}}.$$

We remark that alternative generators for the Hopf algebra H are G^{e_j} and Y_{v_j}, a fact that we will need later on.

COROLLARY 8. *The coproduct on the elements Y_v is given by*

$$\Delta(Y_{v_j}) = \sum_{n_1,\ldots,n_5} Y_{v_j} Y_{v_1}^{n_1} \cdots Y_{v_5}^{n_5} \otimes p_{n_1\cdots n_5}(Y_{v_j}),$$

where $p_{n_1\cdots n_5}$ is the projection onto graphs containing n_k vertices v_k ($k = 1,\ldots,5$).

PROOF. This follows directly by an application of the formulas in Proposition 7 to $\Delta(Y_{v_j}) = \Delta(G^{v_j}) \prod_{i=1,2,3} \Delta\big((G^{e_i})^{-N_i(v_j)/2}\big)$. □

Quite remarkably, this formula coincides with the coproduct in the Hopf algebra dual to the group $\overline{\text{Diff}}(\mathbb{C}^5, 0)$ of formal diffeomorphisms tangent to the identity in 5 variables, closely related to the Faà di Bruno Hopf algebra (cf. for instance the short review [**10**]). In other words, the Hopf subalgebra generated by $p_{n_1,\ldots,n_5}(Y_{v_j})$ is dual to (a subgroup of) the group $\overline{\text{Diff}}(\mathbb{C}^5, 0)$. This will be further explored in Section 4 below.

COROLLARY 9. [19] *The ideal J in H generated by* $q_l\left(Y_{v_k}^{N(v_j)-2} - Y_{v_j}^{N(v_k)-2}\right)$ *for any $l \geq 0$ and $j, k = 1, \ldots, 4$ is a Hopf ideal, i.e.*

$$\Delta(J) \subset J \otimes H + H \otimes J.$$

PROOF. Fix j and k two integers between 1 and 5. Applying the formulas in Proposition 7 to the coproduct on $Y_{v_k}^{N(v_j)-2}$ yields

$$\Delta\left(Y_{v_k}^{N(v_j)-2}\right) = \sum_{n_1,\ldots,n_5} Y_{v_k}^{N(v_j)-2} Y_{v_1}^{n_1} \cdots Y_{v_5}^{n_5} \otimes p_{n_1\cdots n_5}(Y_{v_k}^{N(v_j)-2}),$$

Now, module elements in J, we can write

$$Y_{v_2}^{n_2} = Y_{v_1}^{n_2 \frac{N(v_2)-2}{N(v_1)-2}},$$

and similarly for v_3 and v_4 so that

$$Y_{v_1}^{n_1} \cdots Y_{v_5}^{n_5} = \left(Y_{v_1}^{1/N(v_1)-2}\right)^{\sum_k n_k(N(v_k)-2)} = \left(Y_{v_1}^{1/(N(v_1)-2)}\right)^{2l}.$$

by an application of Lemma 4. Note that this is independent of the n_i but only depends on the total loop number l. For the coproduct, this yields

$$\Delta\left(Y_{v_k}^{N(v_j)-2}\right) = \sum_{l=0}^{\infty} Y_{v_k}^{N(v_j)-2}\, Y_{v_1}^{\frac{2l}{N(v_1)-2}} \otimes q_l(Y_{v_k}^{N(v_j)-2}),$$

Of course, a similar formula holds for the other term defining J, upon interchanging j and k. For their difference we then obtain

$$\Delta\left(Y_{v_k}^{N(v_j)-2} - Y_{v_j}^{N(v_k)-2}\right) = \sum_{l=0}^{\infty}\left(Y_{v_k}^{N(v_j)-2} - Y_{v_j}^{N(v_k)-2}\right) Y_{v_1}^{\frac{2l}{N(v_1)-2}} \otimes q_l(Y_{v_k}^{N(v_j)-2})$$
$$+ \sum_{l=0}^{\infty} Y_{v_k}^{N(v_j)-2}\, Y_{v_1}^{\frac{2l}{N(v_1)-2}} \otimes q_l\left(Y_{v_k}^{N(v_j)-2} - Y_{v_j}^{N(v_k)-2}\right).$$

This is an element in $J \otimes H + H \otimes J$, which completes the proof. □

REMARK 10. *An equivalent set of generators for J is given by* $Y_{v_i} - Y_{v_1}^{N(v_i)-2}$ *with $i = 2, 3, 4$.*

In this Hopf ideal, the reader might have already recognized the Slavnov–Taylor identities for the couplings. Indeed, in the quotient Hopf algebra H/J these identities hold. Moreover, since the character $U : H \to \mathbb{C}$ given by the regularized Feynman rules vanishes on J (these are exactly the Slavnov–Taylor identities) and thus factorizes over this quotient (provided we work with dimensional regularization, or another gauge symmetry preserving regularization scheme). Now, the Birkhoff decomposition for the group $\mathrm{Hom}_{\mathbb{C}}(H/J, \mathbb{C})$ gives the counterterm map C and the renormalized map R as characters on H/J. Thus, they also satisfy the Slavnov–Taylor identities and this provides a purely algebraic proof of the compatibility of the Slavnov–Taylor identities for the couplings with renormalization, an essential step in proving renormalizablity of gauge theories.

Below, we shall give a more conceptual (rather then combinatorial) explanation for the existence of these Hopf ideals, after establishing a connection between H and the fields and coupling constants.

4. Coaction and BRST-symmetries

The fact that we encountered diffeomorphism groups starting with Feynman graphs is not very surprising from a physical point of view. Indeed, Feynman graphs are closely involved in the running of the coupling constants described by the renormalization group. In the next subsection, we will clarify this point by defining a coaction of the Hopf algebra H on the coupling constants and the fields. Dually, this will lead to an action of the diffeomorphism group. It contains a subgroup that respects the BRST-invariance of the action, which will be related to the Hopf ideal of the previous section. Finally, its relation with the renormalization group is further described.

4.1. Coaction on the coupling constants and fields. In this section, we will establish a connection between the Hopf algebra of Feynman graphs defined above and the fields, coupling constants and masses that characterize the field theory. This allows for a derivation of the Hopf ideals encountered in the previous section from the so-called master equation satisfied by the Lagrangian.

Let us first introduce formal variables $\lambda_1, \lambda_2, \ldots, \lambda_5$, corresponding to the vertices describing the five possible interactions in QCD. Also, we write $\phi_1 = A$, $\phi_2 = \psi$, $\phi_3 = \omega$ and $\phi_4 = h$ for the fields, in accordance with the labelling of the edges (see Figure 3.1 above). We denote by $\mathcal{F} = \mathrm{Loc}(\phi_1, \phi_2, \phi_3, \phi_4) \otimes \mathbb{C}[[\lambda_1, \ldots, \lambda_5]]$ the algebra of local functionals in the fields ϕ_i (and their conjugates), extended linearly by formal power series in the λ_j. Recall that a local functional is an integral of a polynomial in the fields and their derivatives, and the algebra structure is given by multiplication of these integrals.

THEOREM 11. *The algebra $\mathcal{F}$ is a comodule algebra over the Hopf algebra H The coaction $\rho : \mathcal{F} \to \mathcal{F} \otimes H$ is given on the generators by*

$$\rho : \lambda_j \longmapsto \sum_{n_1, \ldots, n_5} \lambda_j \lambda_1^{n_1} \cdots \lambda_5^{n_5} \otimes p_{n_1 \cdots n_5}(Y_{v_j}), \qquad (j = 1, \ldots, 5);$$

$$\rho : \phi_i \longmapsto \sum_{n_1, \ldots, n_5} \phi \, \lambda_1^{n_1} \cdots \lambda_5^{n_5} \otimes p_{n_1 \cdots n_5}(\sqrt{G^{e_i}}), \qquad (i = 1, 2, 3),$$

while it commutes with partial derivatives on ϕ.

PROOF. Since we work with graded Hopf algebras, it suffices to establish that $(\rho \otimes 1) \circ \rho = (1 \otimes \Delta) \circ \rho$. We claim that this follows from coassociativity (i.e. $(\Delta \otimes 1) \circ \Delta = (1 \otimes \Delta) \circ \Delta$) of the coproduct Δ of H. Indeed, the first expression very much resembles the form of the coproduct on Y_j as derived in Corollary 8: replacing therein each Y_{v_k} $(k = 1, \ldots, 5)$ on the first leg of the tensor product by λ_k and one Δ by ρ gives the desired result. A similar argument applies to the second expression, using Proposition 7 above. □

COROLLARY 12. *The Green's functions $G^{v_j} \in H$ can be obtained when coacting on the interaction monomial $\int \lambda_j \iota(v)(x) d\mu(x) = \int \lambda_j \partial_{\vec{\mu}_1} \phi_{i_1}(x) \cdots \partial_{\vec{\mu}_N} \phi_{i_N}(x) d\mu(x)$ for some index set $\{i_1, \ldots, i_N\}$.*

For example,

$$\rho\Big(\lambda_2\langle d\overline{\omega},[A,\omega]\rangle\Big) = \sum_{n_1\cdots n_5} \lambda_2\lambda_1^{n_1}\cdots\lambda_5^{n_5}\langle d\overline{\omega},[A,\omega]\rangle \otimes p_{n_1\cdots n_5}\Big(Y \sqrt{G} \, G \Big)$$
$$= \sum_{n_1,\ldots,n_5} \lambda_2\lambda_1^{n_1}\cdots\lambda_5^{n_5}\langle d\overline{\omega},[A,\omega]\rangle \otimes p_{n_1\cdots n_5}\Big(G\Big)$$

Actually, the first equation in Theorem 11 can be interpreted as an action of a subgroup of formal diffeomorphisms in 5 variables on $\mathbb{C}[[\lambda_1,\ldots,\lambda_5]]$. Let us make this more precise.

Consider the group $\overline{\text{Diff}}(\mathbb{C}^5,0)$ of formal diffeomorphisms in 5 dimensions (coordinates $x_1,\ldots,x_5$) that leave the five axis-hyperplanes invariant. In other words, we consider maps

$$f(x) = \big(f_1(x),\ldots,f_5(x)\big)$$

where each f_i is a formal power series of the form $f_i(x) = x_i(\sum a^{(i)}_{n_1\cdots n_5}(f)x_1^{n_1}\cdots x_5^{n_5})$ with $a^{(i)}_{0,\ldots,0}=1$ and $x=(x_1,\ldots,x_5)$. The group multiplication is given by composition, and is conveniently written in a dual manner, in terms of the coordinates. In fact, the $a^{(i)}_{n_1\cdots n_5}$ generate a Hopf algebra with the coproduct expressed as follows. On the formal generating element $A_i(x) = x_i(\sum a^{(i)}_{n_1\cdots n_k}x_1^{n_1}\cdots x_k^{n_k})$:

$$\Delta(A_i(x)) = \sum_{n_1,\ldots,n_k} A_i(x)\,(A_1(x))^{n_1}\cdots(A_k(x))^{n_k}\otimes a^{(i)}_{n_1\cdots n_k}.$$

Thus, by mapping the $a^{(j)}_{n_1,\ldots,n_5}$ to $p_{n_1,\ldots,n_5}(Y_{v_j})$ in H we obtain a surjective map from H to the Hopf algebra dual to $\overline{\text{Diff}}(\mathbb{C}^5,0)$. In other words, $\text{Hom}(H,\mathbb{C})$ is a subgroup of $\overline{\text{Diff}}(\mathbb{C}^5,0)$ and, in fact, substituting $a^{(j)}_{n_1,\ldots,n_5}$ for $p_{n_1,\ldots,n_5}(Y_{v_j})$ in the first equation of Theorem 11 yields (dually) a group action of $\overline{\text{Diff}}(\mathbb{C}^5,0)$ on $\mathbb{C}[[\lambda_1,\ldots,\lambda_5]]$ by $f(a) := (1\otimes f)\rho(a)$ for $f\in\overline{\text{Diff}}(\mathbb{C}^5,0)$ and $a\in\mathbb{C}[[\lambda_1,\ldots,\lambda_5]]$. In fact, we have the following

PROPOSITION 13. *Let G' be the group consisting of algebra maps $f:\mathcal{F}\to\mathcal{F}$ given on the generators by*

$$f(\lambda_j) = \sum_{n_1\cdots n_5} f^{v_j}_{n_1,\ldots,n_5}\lambda_j\lambda_1^{n_1}\cdots\lambda_5^{n_5}; \qquad (j=1,\ldots,5),$$
$$f(\phi_i) = \sum_{n_1\cdots n_5} f^{e_i}_{n_1,\ldots,n_5}\phi_i\lambda_1^{n_1}\cdots\lambda_5^{n_5}; \qquad (i=1,\ldots,3),$$

where $f^{v_j}_{n_1\cdots n_5}, f^{e_i}_{n_1\cdots n_5}\in\mathbb{C}$ are such that $f^{v_j}_{0\cdots 0} = f^{e_i}_{0\cdots 0} = 1$. Then the following hold:

(1) *The character group G of the Hopf algebra H generated by $p_{n_1\cdots n_5}(Y_v)$ and $p_{n_1\cdots n_5}(\sqrt{G^e})$ with coproduct given in Proposition 7, is a subgroup of G'.*
(2) *The subgroup $N := \{f : f(\lambda_j) = \lambda_j, j=1,\ldots,5\}$ of G' is normal and isomorphic to $(\mathbb{C}[[\lambda_1,\ldots,\lambda_5]]^\times)^3$.*
(3) *$G' \simeq (\mathbb{C}[[\lambda_1,\ldots,\lambda_5]]^\times)^3 \rtimes \overline{\text{Diff}}(\mathbb{C}^5,0)$.*

PROOF. From Theorem 11, it follows that a character χ acts on $\mathcal{F}$ as in the above formula upon writing $f^{v_j}_{n_1\cdots n_5} = \chi(p_{n_1\cdots n_5}(Y_v))$ for $j=1,\ldots,5$ and $f^{e_i}_{n_1\cdots n_5} = \chi(p_{n_1\cdots n_5}(\sqrt{G^{\phi_i}}))$ for $i=1,2,3$.

For (2) one checks by explicit computation that N is indeed normal and that each series f^{e_i} defines an element in $\mathbb{C}[[\lambda_1,\dots,\lambda_5]]^\times$ of invertible formal power series.

Then (3) follows from the existence of a homomorphism from G' to $\overline{\mathrm{Diff}}(\mathbb{C}^5,0)$. It is given by restricting an element f to $\mathbb{C}[[\lambda_1,\dots,\lambda_5]]$. This is clearly the identity map on $\overline{\mathrm{Diff}}(\mathbb{C}^5,0)$ when considered as a subgroup of G and its kernel is precisely N. □

The action of (the subgroup of) $(\mathbb{C}[[\lambda_1,\dots,\lambda_k]]^\times)^3 \rtimes \overline{\mathrm{Diff}}(\mathbb{C}^5,0)$ on $\mathcal{F}$ has a natural physical interpretation: the invertible formal power series act on a field as wave function renormalization whereas the diffeomorphisms act on the coupling constants $\lambda_1,\dots,\lambda_5$.

4.2. BRST-symmetries. We will now show how the previous coaction of the Hopf algebra H on the algebra $\mathcal{F}$ gives rise to the Hopf ideal J encountered before. For this, we choose a distinguished element in $\mathcal{F}$, namely the action S. It is given by

$$\begin{aligned}(6)\quad S[\phi_i,\lambda_j] = -\langle dA,dA\rangle - 2\lambda_3\langle dA,A^2\rangle - \lambda_4\langle A^2,A^2\rangle + \langle\psi,(\partial\!\!\!/ + \lambda_1 A\!\!\!/ - \lambda_5)\psi\rangle \\ - \langle A,dh\rangle + \langle d\overline{\omega},d\omega\rangle + \frac{1}{2}\xi\langle h,h\rangle + \lambda_2\langle d\overline{\omega},[A,\omega]\rangle.\end{aligned}$$

in terms of the appropriate inner products. Note that the action has finitely many terms, that is, it is a (local) polynomial functional in the fields and coupling constants rather than a formal power series.

With the BRST-differential given in Equation (3) (involving the 'fundamental' coupling constant g), we will now impose the BRST-invariance of S, by setting

$$s(S) = 0.$$

Actually, we will define an ideal $I = \langle s(S)\rangle$ in $\mathcal{F}$ that implements the relations between the λ_j's. Strictly speaking, the fundamental coupling g is not an element in $\mathcal{F}$; we will instead set $g \equiv \lambda_1$. The remaining 'coupling' constant λ_5 is interpreted as the quark mass m.

PROPOSITION 14. *The ideal I is generated by the following elements:*

$$\lambda_1 - \lambda_2; \qquad \lambda_2 - \lambda_3; \qquad \lambda_3 - \lambda_4^2.$$

PROOF. This follows directly by applying s (involving g) to the action S. □

A convenient set of (equivalent) generators for the ideal I is $\lambda_i - g^{N(v_i)-2}$ for $i = 1,\dots 4$. Thus, the image of S in the quotient $\mathcal{F}/I$ is BRST-invariant, that is, $s(S)$ is identically zero.

Let us return to the group $G \subset (\mathbb{C}[[\lambda_1,\dots,\lambda_5]]^\times)^3 \rtimes \overline{\mathrm{Diff}}(\mathbb{C}^5,0)$, acting on $\mathcal{F}$. Consider the subgroup G^I of G consisting of elements f that leave invariant the ideal I, i.e., such that $f(I) \subseteq I$. It is clear from the above generators of I that this will involve a diffeomorphism group in 2 variables, instead of 5. More precisely, we have the following

THEOREM 15 ([**20**]). *Let J be the ideal from Theorem 9.*

(1) *The group G^I acts on the quotient algebra $\mathcal{F}/I$.*
(2) *The image of G^I in $\mathrm{Aut}(\mathcal{F}/I)$ is isomorphic to $\mathrm{Hom}_{\mathbb{C}}(H/J,\mathbb{C})$ and H/J coacts on $\mathcal{F}/I$.*

Consequently, (the image in $\mathrm{Aut}(\mathcal{F}/I)$ *of)* G^I *is a subgroup of the semidirect product* $(\mathbb{C}[[g,\lambda_5]]^\times)^3 \rtimes \overline{\mathrm{Diff}}(\mathbb{C}^2,0)$.

PROOF. The first claim is direct. For the second, note that an element $f \in G$ acts on the generators of I as

$$f\left(\lambda_i - g^{N(v_i)-2}\right) = \sum_{n_1,\ldots,n_5} \lambda_1^{n_1}\cdots\lambda_5^{n_5}\left[\lambda_i f\left(p_{n_1\cdots n_5}(Y_{v_i})\right) - g^{N(v_i)-2} f\left(p_{n_1\cdots n_5}(Y_{v_1}^{N(v_i)-2})\right)\right],$$

since $g \equiv \lambda_1$. We will reduce this expression by replacing λ_i by $g^{N(v_i)-2}$, modulo terms in I. Together with Lemma 4 this yields

$$f\left(\lambda_i - g^{N(v_i)-2}\right) = \sum_{l=0}^{\infty} g^{2l+N(v_i)-2}\, f\left(q_l\left(Y_{v_i} - Y_{v_1}^{N(v_i)-2}\right)\right) \mod I.$$

The requirement that this is an element in I is equivalent to the requirement that f vanishes on $q_l(Y_{v_i} - Y_{v_1}^{N(v_i)-2})$, i.e. on the generators of J, establishing the desired isomorphism. One then easily computes

$$\rho(I) \subset I \otimes H + \mathcal{F} \otimes J$$

so that H/J coacts on $\mathcal{F}$ by projecting onto the two quotient algebras. □

In fact, the last claim of the above Theorem can be strengthened. Focusing on the subgroup of the formal diffeomorphism group $\overline{\mathrm{Diff}}(\mathbb{C}^5,0)^I$ that leaves invariant the ideal I we have:

$$1 \to (1+I)^5 \to \overline{\mathrm{Diff}}(\mathbb{C}^5,0)^I \to \overline{\mathrm{Diff}}(\mathbb{C}^2,0) \to 1.$$

Here, an element $(1+B_i)_{i=1,\ldots,5}$ in $(1+I)^5$ acts on the generators $\lambda_1,\ldots,\lambda_5$ by right multiplication. This sequence actually splits, leading to a full description of the group $\overline{\mathrm{Diff}}(\mathbb{C}^5,0)^I$. Indeed, by the simple structure of the ideal I, a one-sided inverse of the map $\overline{\mathrm{Diff}}(\mathbb{C}^5,0)^I \to \overline{\mathrm{Diff}}(\mathbb{C}^2,0)$ can be easily constructed. A similar statement holds for the above subgroup G^I of the semidirect product $G \simeq (\mathbb{C}[[\lambda_1,\ldots,\lambda_5]]^\times)^3 \rtimes \overline{\mathrm{Diff}}(\mathbb{C}^5,0)$.

In any case, the contents of Theorem 15 have a very nice physical interpretation: the invertible formal power series act on the three fields as wave function renormalization whereas the diffeomorphisms act on one fundamental coupling constant g. We will appreciate this even more in the next section where we discuss the renormalization group flow.

4.3. Renormalization group. We will now establish a connection between the group of diffeographisms and the renormalization group à la Gell-Mann and Low [**11**]. This group describes the dependence of the renormalized amplitudes $\phi_+(z)$ on a mass scale that is implicit in the renormalization procedure. In fact, in dimensional regularization, in order to keep the loop integrals $d^{4-z}k$ dimensionless for complex z, one introduces a factor of μ^z in front of them, where μ has dimension of mass and is called the *unit of mass.* For a Feynman graph Γ, Lemma 4 shows that this factor equals $\mu^{z\sum_i (N(v_i)-2))\delta_{v_i}(\Gamma)/2}$ reflecting the fact that the coupling constants appearing in the action get replaced by

$$\lambda_i \mapsto \mu^{z\sum_k (N(v_k)-2))/2}\lambda_i$$

for every vertex v_i $(i = 1,\ldots,5)$.

As before, the Feynman rules define a loop $\gamma_\mu : C \to G \equiv G(\mathbb{C})$, which now depends on the mass scale μ. Consequently, there is a Birkhoff decomposition for each μ:

$$\gamma_\mu(z) = \gamma_{\mu,-}(z)^{-1}\gamma_{\mu,+}(z); \qquad (z \in C),$$

As was shown in [**8**], the negative part $\gamma_{\mu,-}(z)$ of this Birkhoff decomposition is independent of the mass scale, that is,

$$\frac{\partial}{\partial \mu}\gamma_{\mu,-}(z) = 0.$$

Hence, we can drop the index μ and write $\gamma_-(z) := \gamma_{\mu,-}(z)$. In terms of the generator θ_t for the one-parameter subgroup of $G(K)$ corresponding to the grading l on H, we can write

$$\gamma_{e^t\mu(z)} = \theta_{tz}\left(\gamma_\mu(z)\right), \qquad (t \in \mathbb{R}).$$

A proof of this and the following result can be found in [**8**].

PROPOSITION 16. *The limit*

$$F_t := \lim_{z\to 0} \gamma_-(z)\theta_{tz}\left(\gamma_-(z)^{-1}\right)$$

exists and defines a 1*-parameter subgroup of* G *which depends polynomially on* t *when evaluated on an element* $X \in H$.

In physics, this 1-parameter subgroup goes under the name of *renormalization group*. In fact, using the Birkhoff decomposition, we can as well write

$$\gamma_{e^t\mu,+}(0) = F_t\ \gamma_{\mu,+}(0), \qquad (t \in \mathbb{R}).$$

This can be formulated in terms of the generator $\beta := \frac{d}{dt}F_t|_{t=0}$ of this 1-parameter group as

$$\mu\frac{\partial}{\partial \mu}\gamma_{\mu,+}(0) = \beta\gamma_{\mu,+}(0). \tag{7}$$

Let us now establish that this is indeed the beta-function familiar from physics by exploring how it acts on the coupling constants λ_i. First of all, although the name might suggest otherwise, the coupling constants depend on the energy or mass scale μ. Recall the action of G on $\mathbb{C}[[\lambda_1,\ldots,\lambda_5]]$ defined in the previous section. In the case of $\gamma_{\mu,+}(0) \in G$, we define the (renormalized) *coupling constant at scale* μ to be

$$\lambda_i(\mu) = \gamma_{\mu,+}(0)(\lambda_i).$$

This function of μ (with coefficients in $\mathbb{C}[[\lambda_1,\ldots,\lambda_5]]$) satisfies the following differential equation:

$$\beta\left(\lambda_i(\mu)\right) = \mu\frac{\partial}{\partial\mu}\left(\lambda_i(\mu)\right)$$

which follows easily from Eq. (7). This is exactly the renormalization group equation expressing the flow of the coupling constants λ_i as a function of the energy scale μ. Moreover, if we extend β by linearity to the action S of Eq. (6), we obtain Wilson's continuous renormalization equation [**22**]:

$$\beta(S(\mu)) = \mu\frac{\partial}{\partial\mu}\left(S(\mu)\right)$$

This equation has been explored in the context of renormalization Hopf algebras in [**12, 13**].

Equation (7) expresses β completely in terms of $\gamma_{\mu,+}$; as we will now demonstrate, this allows us to derive that for QCD all β-functions coincide. First, recall that the maps γ_μ are the Feynman rules dictated by S in the presence of the mass scale μ, which we suppose to be BRST-invariant: $s(S) = 0$. In other words, we are in the quotient of $\mathcal{F}$ by $I = \langle s(S) \rangle$. If the regularization procedure respects gauge invariance, it is well-known that the Feynman amplitude satisfy the Slavnov–Taylor identities for the couplings. In terms of the ideal J defined in the previous section, this means that $\gamma_\mu(J) = 0$. Since J is a Hopf ideal (Theorem 9), it follows that both $\gamma_{\mu,-}$ and $\gamma_{\mu,+}$ vanish on J. Indeed, the character γ given by the Feynman rules factorizes through H/J, for which the Birkhoff decomposition gives two characters γ_+ and γ_- of H/J. In other words, if the unrenormalized Feynman amplitudes given by γ_μ satisfy the Slavnov–Taylor identities, so do the counterterms and the renormalized Feynman amplitudes.

In particular, from Eq. (7) we conclude that β vanishes on the ideal I in $\mathbb{C}[[\lambda_1, \ldots, \lambda_5]]$. This implies the following result, which is well-known in the physics literature:

PROPOSITION 17. *All (QCD) β-functions (for $i = 1, \ldots, 4$) are expressed in terms of $\beta(g)$ for the fundamental coupling constant g:*

$$\beta(\lambda_i) = \beta(g^{N(v)-2}).$$

References

[1] G. van Baalen, D. Kreimer, D. Uminsky, and K. Yeats. The QED beta-function from global solutions to Dyson- Schwinger equations. *Ann. Phys* 324 (2009) 205–219.

[2] G. Barnich, F. Brandt, and M. Henneaux. Local BRST cohomology in the antifield formalism. I. General theorems. *Commun. Math. Phys.* 174 (1995) 57–92.

[3] G. Barnich, F. Brandt, and M. Henneaux. Local BRST cohomology in the antifield formalism. II. Application to Yang-Mills theory. *Commun. Math. Phys.* 174 (1995) 93–116.

[4] C. Becchi, A. Rouet, and R. Stora. The abelian Higgs-Kibble model. Unitarity of the S operator. *Phys. Lett.* B52 (1974) 344.

[5] C. Becchi, A. Rouet, and R. Stora. Renormalization of the abelian Higgs-Kibble model. *Commun. Math. Phys.* 42 (1975) 127–162.

[6] C. Becchi, A. Rouet, and R. Stora. Renormalization of gauge theories. *Annals Phys.* 98 (1976) 287–321.

[7] A. Connes and D. Kreimer. Renormalization in quantum field theory and the Riemann-Hilbert problem. I: The Hopf algebra structure of graphs and the main theorem. *Comm. Math. Phys.* 210 (2000) 249–273.

[8] A. Connes and D. Kreimer. Renormalization in quantum field theory and the Riemann-Hilbert problem. II: The beta-function, diffeomorphisms and the renormalization group. *Commun. Math. Phys.* 216 (2001) 215–241.

[9] P. Deligne et al. *Quantum fields and strings: A course for mathematicians. Vol. 1, 2.* AMS, Providence, USA, 1999.

[10] H. Figueroa, J. M. Gracia-Bondía, and J. C. Várilly. Faà di Bruno Hopf algebras, math/0508337.

[11] M. Gell-Mann and F. E. Low. Quantum electrodynamics at small distances. *Phys. Rev.* 95 (1954) 1300–1312.

[12] F. Girelli, T. Krajewski, and P. Martinetti. An algebraic Birkhoff decomposition for the continuous renormalization group. *J. Math. Phys.* 45 (2004) 4679–4697.

[13] T. Krajewski and P. Martinetti. Wilsonian renormalization, differential equations and Hopf algebras. arXiv:0806.4309.

[14] D. Kreimer. On the Hopf algebra structure of perturbative quantum field theories. *Adv. Theor. Math. Phys.* 2 (1998) 303–334.

[15] D. Kreimer. Anatomy of a gauge theory. *Ann. Phys.* 321 (2006) 2757–2781.

[16] D. Kreimer and K. Yeats. An étude in non-linear Dyson-Schwinger equations. *Nuclear Phys. B Proc. Suppl.* 160 (2006) 116–121.
[17] I. V. Tyutin. Gauge invariance in field theory and statistical physics in operator formalism. LEBEDEV-75-39.
[18] W. D. van Suijlekom. Renormalization of gauge fields: A Hopf algebra approach. *Commun. Math. Phys.* 276 (2007) 773–798.
[19] W. D. van Suijlekom. Renormalization of gauge fields using Hopf algebras. In J. T. B. Fauser and E. Zeidler, editors, *Quantum Field Theory*. Birkhäuser Verlag, Basel, 2008. [arXiv:0801.3170].
[20] W. D. van Suijlekom. The structure of renormalization Hopf algebras for gauge theories. I: Representing Feynman graphs on BV-algebras. *Commun. Math. Phys.* 290 (2009) 291–319.
[21] S. Weinberg. *The quantum theory of fields. Vol. II.* Cambridge University Press, 1996.
[22] K. G. Wilson. Renormalization group methods. *Advances in Math.* 16 (1975) 170–186.
[23] C.-N. Yang and R. L. Mills. Conservation of isotopic spin and isotopic gauge invariance. *Phys. Rev.* 96 (1954) 191–195.

INSTITUTE FOR MATHEMATICS, ASTROPHYSICS AND PARTICLE PHYSICS, FACULTY OF SCIENCE, RADBOUD UNIVERSITY NIJMEGEN, HEYENDAALSEWEG 135, 6525 AJ NIJMEGEN, THE NETHERLANDS
E-mail address: waltervs@math.ru.nl